EDITORIAL BOARD

Editor in Chief

E. Julius Dasch, Ph.D.
RSC International
Washington, D.C.

Associate Editors

Faye Anderson
Assistant Director, School of Public Affairs
University of Maryland
College Park, Maryland

Dennis O. Nelson, Ph.D.
Groundwater Coordinator
Oregon Department of Human Services, Drinking Water Program
Springfield, Oregon

Martha R. Scott, Ph.D.
Associate Professor
Texas A & M University, Department of Oceanography
College Station, Texas

Consulting Editor

Vita Pariente, Ph.D.
College Station, Texas

EDITORIAL AND PRODUCTION STAFF

Cindy Clendenon, *Project Editor*
Kathy Edgar, Chris Lopez, Mark Mikula, Kate Millson, Brad Morgan, Jaime E. Noce, Angela Pilchak, Mark Springer, Nicole Watkins, Jennifer Wisinski, *Editorial Support*
Kristin May, Matthew Nowinski, *Editorial Interns*
Bill Atkins, Bruce Owens, *Copyeditors*
Bill Atkins, *Proofreader*
Wendy Allex, *Indexer*
Michelle DiMercurio, *Senior Art Director*
Wendy Blurton, *Senior Manufacturing Specialist*
Cindy Clendenon, *Photo Researcher and Editor*
Margaret A. Chamberlain, *Permissions Specialist*
Leitha Etheridge-Sims, *Image Cataloger, Imaging and Multimedia Content*
Kelly A. Quin, *Image Acquisition Editor, Imaging and Multimedia Content*
Lezlie Light, *Imaging Coordinator, Imaging and Multimedia Content*
Dan Newell, *Imaging Specialist, Imaging and Multimedia Content*
Randy Bassett, *Imaging Supervisor, Imaging and Multimedia Content*

Macmillan Reference USA

Frank Menchaca, *Vice President and Publisher*
Hélène Potter, *Director, New Product Development*

water
Science and Issues

Volume 3/Land–Pricing

E. Julius Dasch, Editor in Chief

**MACMILLAN
REFERENCE
USA**™

New York • Detroit • San Diego • San Francisco • Cleveland • New Haven, Conn. • Waterville, Maine • London • Munich

Water: Science and Issues
E. Julius Dasch

© 2003 by Macmillan Reference USA.
Macmillan Reference USA is an Imprint of
The Gale Group, Inc., a division of
Thomson Learning, Inc.

Macmillan Reference USA™ and Thomson
Learning™ are trademarks used herein under
license.

For more information, contact
Macmillan Reference USA
300 Park Avenue South, 9th Floor
New York, NY 10010
Or you can visit our Internet site at
http://www.gale.com

ALL RIGHTS RESERVED
No part of this work covered by the copyright hereon may be reproduced or used in any form or by any means—graphic, electronic, or mechanical, including photocopying, recording, taping, Web distribution, or information storage retrieval systems—without the written permission of the publisher.

For permission to use material from this product, submit your request via Web at http://www.gale-edit.com/permissions, or you may download our Permissions Request form and submit your request by fax or mail to:

Permissions Department
The Gale Group, Inc.
27500 Drake Rd.
Farmington Hills, MI 48331-3535
Permissions Hotline:
248-699-8006 or 800-877-4253, ext. 8006
Fax: 248-699-8074 or 800-762-4058

Cover photographs reproduced by permission of the following sources: hydropower turbine photo ©Bob Rowan, Progressive Image/Corbis; Hindu women praying in Old Brahmaputra River (Bangladesh) photo by Mufty Munir ©AFP/Corbis; Glen Canyon Dam photo ©George D. Lepp/Corbis; Yellowstone Grand Prismatic Hot Spring photo ©Raymond Gehman/Corbis; wave photo ©Digital Vision.

Library of Congress Cataloging-in-Publication Data

Dasch, E. Julius.
　　Water : science and issues / E. Julius Dasch.
　　　　p. cm.
　　Includes bibliographical references and index.
　　　ISBN 0-02-865611-3 (set hardcover : alk. paper) -- ISBN
0-02-865612-1 (v. 1 : alk. paper) -- ISBN 0-02-865613-X (v. 2 : alk.
paper) -- ISBN 0-02-865614-8 (v. 3 : alk. paper) -- ISBN 0-02-865615-6
(v. 4 : alk. paper)
　　1. Water—Encyclopedias. 2. Hydrology—Encyclopedias. 3.
Hydrogeology—Encyclopedias. I. Title.
　　　GB655.D37 2003
　　　553.7'03--dc21
　　　　　　　　　　　　　　　　　　　　　　　　　　　　　　　　　2003001309

Printed in Canada
10 9 8 7 6 5 4 3 2 1

Table of Contents

VOLUME 1

TABLE OF CONTENTS	v
PREFACE	xi
TOPICAL OUTLINE	xv
FOR YOUR REFERENCE	xxiii
CONTRIBUTORS	xxvii
Acid Mine Drainage	1
Acid Rain	5
Agassiz, Louis	11
Agriculture and Water	12
Algal Blooms, Harmful	16
Algal Blooms in Fresh Water	21
Algal Blooms in the Ocean	24
Amphibian Population Declines	28
Aquaculture	31
Aquariums	36
Aquifer Characteristics	39
Archaeology, Underwater	43
Army Corps of Engineers, U.S.	47
Artificial Recharge	49
Arts, Water in the	52
Astrobiology: Water and the Potential for Extraterrestrial Life	58
Attenuation of Pollutants	62
Balancing Diverse Interests	65
Bays, Gulfs, and Straits	67
Beaches	71
Biodiversity	77
Birds, Aquatic	80
Bivalves	82
Bottled Water	86
Bretz, J Harlen	88
Bridges, Causeways, and Underwater Tunnels	91
Brines, Natural	94
Bureau of Reclamation, U.S.	97
California, Water Management in	99
Canals	103
Carbon Dioxide in the Ocean and Atmosphere	107
Careers in Environmental Education	112
Careers in Environmental Science and Engineering	116
Careers in Fresh-Water Chemistry	117
Careers in Fresh-Water Ecology	118
Careers in Geospatial Technologies	120
Careers in Hydrology	121
Careers in International Water Resources	123
Careers in Oceanography	125
Careers in Soil Science	131
Careers in Water Resources Engineering	132
Careers in Water Resources Planning and Management	134
Carson, Rachel	136
Cavern Development	138
Cephalopods	141
Chemical Analysis of Water	144
Chemicals: Combined Effect on Public Health	147
Chemicals from Agriculture	150
Chemicals from Consumers	155
Chemicals from Pharmaceuticals and Personal Care Products	158
Chesapeake Bay	164
Clean Water Act	169
Climate and the Ocean	174
Climate Moderator, Water as a	179
Coastal Ocean	183
Coastal Waters Management	187

Table of Contents

Colorado River Basin	190
Columbia River Basin	194
Comets and Meteorites, Water in	198
Conflict and Water	201
Conservation, Water	206
Cook, Captain James	210
Corals and Coral Reefs	212
Cost–Benefit Analysis	220
Cousteau, Jacques	221
Crustaceans	224
Dams	227
Darcy, Henry	231
Data, Databases, and Decision–Support Systems	232
Davis, William Morris	236
Demand Management	237
Desalinization	239
Desert Hydrology	242
Developing Countries, Issues in	246
Douglas, Marjory Stoneman	251
Drinking Water and Society	253
Drinking-Water Treatment	257
Drought Management	260
PHOTO CREDITS	267
GLOSSARY	273
INDEX	335

VOLUME 2

TABLE OF CONTENTS	v
PREFACE	xi
TOPICAL OUTLINE	xv
FOR YOUR REFERENCE	xxiii
CONTRIBUTORS	xxvii
Earle, Sylvia	1
Earth: The Water Planet	3
Earth's Interior, Water in the	5
Ecology, Fresh-Water	7
Ecology, Marine	11
Economic Development	15
El Niño and La Niña	17
Endangered Species Act	21
Energy from the Ocean	24
Environmental Movement, Role of Water in the	27
Environmental Protection Agency, U.S.	31
Erosion and Sedimentation	33
Estuaries	37
Ethics and Professionalism	42
Everglades	44
Fish	48
Fish and Wildlife Issues	50
Fish and Wildlife Service, U.S.	54
Fisheries, Fresh-Water	56
Fisheries, Marine	61
Fisheries, Marine: Management and Policy	65
Fishes, Cartilaginous	69
Floodplain Management	71
Florida, Water Management in	76
Food from the Sea	79
Food Security	82
Forest Hydrology	85
Fresh Water, Natural Composition of	89
Fresh Water, Natural Contaminants in	94
Fresh Water, Physics and Chemistry of	100
Garrels, Robert	103
Geological Survey, U.S.	105
Geospatial Technologies	106
Geothermal Energy	111
Glaciers and Ice Sheets	115
Glaciers, Ice Sheets, and Climate Change	118
Global Warming and Glaciers	123
Global Warming and the Hydrologic Cycle	126
Global Warming and the Ocean	129
Global Warming: Policy-Making	134
Globalization and Water	137
Great Lakes	141
Groundwater	149
Groundwater, Age of	157
Groundwater Supplies, Exploration for	158
Hem, John D.	162
Hoover Dam	163
Hot Springs and Geysers	165
Hot Springs on the Ocean Floor	169
Hubbert, Marion King	173
Human Health and the Ocean	174
Human Health and Water	180
Hutton, James	186
Hydroelectric Power	187
Hydrogeologic Mapping	191
Hydrologic Cycle	194
Hydropolitics	198
Hydrosolidarity	200

Ice Ages	202
Ice at Sea	206
Ice Cores and Ancient Climatic Conditions	210
Infrastructure, Water-Supply	213
Instream Water Issues	217
Integrated Water Resources Management	220
International Cooperation	223
Irrigation Management	227
Irrigation Systems, Ancient	232
Islands, Capes, and Peninsulas	235
Isotopes: Applications in Natural Waters	239
Karst Hydrology	243
Lake Formation	247
Lake Health, Assessing	251
Lake Management Issues	254
Lakes: Biological Processes	259
Lakes: Chemical Processes	262
Lakes: Physical Processes	267
PHOTO CREDITS	271
GLOSSARY	277
INDEX	339

VOLUME 3

TABLE OF CONTENTS	v
PREFACE	xi
TOPICAL OUTLINE	xv
FOR YOUR REFERENCE	xxiii
CONTRIBUTORS	xxvii
Land Use and Water Quality	1
Land-Use Planning	7
Landfills: Impact on Groundwater	11
Landslides	14
Law, International Water	18
Law of the Sea	24
Law, Water	26
Legislation, Federal Water	28
Legislation, State and Local Water	31
Leonardo da Vinci	34
Leopold, Luna	37
Lewis, Meriwether and William Clark	39
Life in Extreme Water Environments	43
Life in Water	48
Light Transmission in the Ocean	52
Marginal Seas	54
Mariculture	58
Marine Mammals	60
Markets, Water	66
Mars, Water on	68
Microbes in Groundwater	72
Microbes in Lakes and Streams	75
Microbes in the Ocean	78
Mid-Ocean Ridges	83
Mineral Resources from Fresh Water	85
Mineral Resources from the Ocean	88
Mineral Waters and Spas	92
Minorities in Water Sciences	93
Mississippi River Basin	98
Modeling Groundwater Flow and Transport	102
Modeling Streamflow	104
Moorings and Platforms	106
Nansen, Fridtjof	110
National Environmental Policy Act	112
National Oceanic and Atmospheric Administration	114
National Park Service	117
Navigation at Sea, History of	119
Nutrients in Lakes and Streams	123
Ocean Basins	126
Ocean Biogeochemistry	130
Ocean Chemical Processes	134
Ocean Currents	138
Ocean Health, Assessing	145
Ocean Mixing	146
Ocean-Floor Bathymetry	148
Ocean-Floor Sediments	152
Oceanography, Biological	157
Oceanography, Chemical	159
Oceanography from Space	161
Oceanography, Geological	163
Oceanography, Physical	165
Oceans, Polar	167
Oceans, Tropical	172
Ogallala Aquifer	173
Oil Spills: Impact on the Ocean	176
Petroleum from the Ocean	181
Plankton	186
Planning and Management, History of Water Resources	190
Planning and Management, Water Resources	194
Plate Tectonics	201

Table of Contents

Policy-Making Process	206
Pollution by Invasive Species	209
Pollution of Groundwater	217
Pollution of Groundwater: Vulnerability	223
Pollution of Lakes and Streams	225
Pollution of Streams by Garbage and Trash	229
Pollution of the Ocean by Plastic and Trash	233
Pollution of the Ocean by Sewage, Nutrients, and Chemicals	236
Pollution Sources: Point and Nonpoint	242
Population and Water Resources	246
Ports and Harbors	249
Powell, John Wesley	255
Precipitation and Clouds, Formation of	256
Precipitation, Global Distribution of	259
Pricing, Water	262
PHOTO CREDITS	265
GLOSSARY	271
INDEX	333

VOLUME 4

TABLE OF CONTENTS	v
PREFACE	xi
TOPICAL OUTLINE	xv
FOR YOUR REFERENCE	xxiii
CONTRIBUTORS	xxvii
Prior Appropriation	1
Privatization of Water Management	3
Public Participation	5
Pumps, Modern	8
Pumps, Traditional	11
Radioactive Chemicals	17
Radionuclides in the Ocean	21
Rainwater Harvesting	26
Reclamation and Reuse	29
Recreation	35
Reisner, Marc	39
Religions, Water in	41
Reptiles	43
Reservoirs, Multipurpose	47
Rights, Public Water	49
Rights, Riparian	52
River Basin Planning	54
Rivers, Major World	58
Runoff, Factors Affecting	62
Safe Drinking Water Act	66
Salmon Decline and Recovery	69
Sea Level	72
Sea Water, Freezing of	76
Sea Water, Gases in	77
Sea Water, Physics and Chemistry of	79
Security and Water	84
Senses, Fresh Water and the	89
Septic System Impacts	94
Solar System, Water in the	98
Sound Transmission in the Ocean	101
Space Travel	104
Sports	106
Springs	107
Stream Channel Development	111
Stream Ecology: Temperature Impacts on	114
Stream Erosion and Landscape Development	117
Stream Health, Assessing	121
Stream Hydrology	125
Stream, Hyporheic Zone of a	129
Streamflow Variability	132
Stumm, Werner	137
Submarines and Submersibles	137
Supplies, Protecting Public Drinking-Water	144
Supplies, Public and Domestic Water	147
Supply Development	150
Survival Needs	153
Sustainable Development	156
Sverdrup, Harald	160
Tennessee Valley Authority	161
Theis, Charles Vernon	164
Tides	165
Tourism	172
Tracers in Fresh Water	175
Tracers of Ocean-Water Masses	177
Transboundary Water Treaties	180
Transportation	182
Tsunamis	188
Twain, Mark	191
Uses of Water	191
Utility Management	197
Volcanoes and Water	200
Volcanoes, Submarine	203

War and Water 206	Weathering of Rocks 234
Wastewater Treatment and Management 209	Wells and Well Drilling 236
Waterfalls 213	Wetlands 241
Watershed, Restoration of a 216	White, Gilbert 247
Watershed, Water Quality in a 219	Women in Water Sciences 249
Waterworks, Ancient 221	PHOTO CREDITS 255
Waves 224	GLOSSARY 261
Weather and the Ocean 231	INDEX 323

Preface

Reflecting for this Preface, I realized my experiences with water in all its forms undoubtedly parallel those of most earth scientists, and most humans in general, for that matter. I became keenly interested in geology as a Boy Scout, and carried this interest through to my doctorate degree at Yale University. So my training has always been shaped by an appreciation of scenery and the mighty influence of liquid water and ice.

What about my personal adventures with water? Once my geology field partner and I lost a Jeep® in a flash flood in West Texas: a bright blue sky was overhead, but torrential rains upstream had quickly filled the streambed we were trying to cross. Then there was a voyage from Iceland to eastern Greenland on an icebreaker, crunching its way through the sea ice to reach the remote Skaergaard igneous rocks. And a flyover of the then-underwater (currently emerged) Kovachi volcano in the South Pacific's Solomon Islands.

The most spectacular experience with water? That would have to be 5 weeks on the ice of Antarctica, searching for meteorites. In my tent during the sunlit "night," I wondered at the occasional cracking noises of the vast but slowly moving continental glacier on which I slept.

Why Water?

My adventures with water have given me a keen appreciation for this simple molecule. After all, it creates much of the impressive scenery on planet Earth—from clouds, oceans, streams, and glaciers, to erosional and depositional landforms such as steep cliffs and river plains. It is Earth's most ubiquitous and most effective dissolving agent, whether in the cells of plants, animals, and humans; in a stream; or in the deep plumbing system of a hydrothermal vent. Water quenches thirst and enables the growth of food and fiber for Earth's 6.1 billion human inhabitants. Put simply, water offers the medium for the origin, development, and maintenance of life as we know it.

But why should water have an entire encyclopedia devoted to it? Why should students, educators, decisionmakers, scientists, and general readers want to learn more about this critical and multifaceted topic? And why now?

It is precisely the necessity—indeed, the urgency—of water resources that makes this encyclopedia a timely contribution. Daily news reports tell the story: droughts, floods, damaged ecosystems, invasive species, chemical pollution, human health threats, and water shortages, to name a few. In 2002 alone, headlines included the severe drought in Canada; the massive floods in Europe; the "dead zones" of Lake Erie and the Gulf of Mexico; the highly invasive snakehead fish in the United States; natural arsenic contamination of groundwater wells in Bangladesh; the West Nile virus in North America; and inadequate drinking-water supplies in many developing countries.

But headlines can only hint at the importance of this vast topic. Water's key role in human civilization is without dispute. Consider the following:

- The history of civilization cannot be discussed apart from water. Water is interwoven with humanity's physical, social, economic, and cultural spheres. It runs like a thread through each person's life.
- The Earth is undergoing rapid and unprecedented change. Humans are truly changing the face of the planet: degrading fresh-water and marine ecosystems; depleting natural water-supply sources; and influencing global climate.
- Human consumption of water rose by a factor of six in the last century—twice the rate of global population growth. Humans now use more than half of the readily available fresh water, which already is in short supply: less than 1 percent of Earth's water is readily usable for human or agricultural needs. (The rest is in the salty oceans or locked up as ice.)
- Worldwide, more than 1 billion people do not have safe water to drink, and 2 to 3 billion lack access to basic sanitation (sewerage) services. Between 3 and 5 million people, mostly children, die each year from water-related diseases. By the year 2025, one-third of the world's population in approximately 50 countries likely will face severe water scarcity. In fact, water scarcity is the greatest threat to global food production, and has been deemed by some experts as the global security issue of the twenty-first century.

In a nutshell, human societies are challenged with assuring the quantity and quality of our most precious water resource while maintaining or improving its environmental integrity. But we cannot meet the challenge in a vacuum. We need a broad understanding of water in its varied forms, distribution, occurrence, and quality—and all within a human context. The encyclopedia *Water: Science and Issues* offers a vehicle to enhance such understanding.

The World of Water (in Four Volumes)

Because the interdisciplinary topic of water covers a wide range of subjects, our development of encyclopedia material was a challenge. The editors chose a three-way organization: fresh waters (groundwater, lakes, streams, and ice); marine waters; and policy and management. Although the entries appear alphabetically, they reflect this threefold categorization. The Topical Outline following this Preface clusters the entries by major themes.

The complexities of water are made understandable in just over 300 essays written by water scientists, professors, educators, and professional communicators. Entries addressing key concepts, current issues, traditional and emerging research, and major legislation are integrated with historical overviews, biographical sketches, and career information.

Preface

The table of contents reflects a breadth of topics not found in any other work at this level: namely, a scientific reference work tailored for nonspecialist readers, yet suitable for people already knowledgeable about water topics. Entries ranging from 500 words to 2,500 words cover hydrology, geology, chemistry, ecology, environmental science, waterways and waterbodies, engineering, earth science, oceanography, economics, policy, planning, management, law, rights, and more.

The table of contents also reveals aspects of water never before addressed in a comprehensive water-related encyclopedia. Topics such as security, globalization, sustainability, global warming, pollution, and water scarcity are not new, but have been thrust to the forefront as the twenty-first century opened. *Water: Science and Issues* addresses subtopics as diverse as pharmaceuticals and personal care products in water supplies; caffeine as a tracer; the search for water on Mars; hydrosolidarity; the ocean's role in human health (good and bad); protecting the water-supply infrastructure; issues in developing countries; survival needs; the search for drinking water; and water's role in war.

Our goal is to tell the interdisciplinary story of water in a format accessible to a wide readership. *Water: Science and Issues* is geared toward high school students and a general audience, but also forays into discussions appropriate for undergraduates and water resource professionals seeking concise overviews of complex subjects. Hence, the audiences include students, educators, communicators, decisionmakers, scientists, and the interested public.

More than 575 color photographs and illustrations help tell this interdisciplinary story. Selected glossary definitions, sidebars, cross-references, and a short bibliography accompany each entry. Reference aids in the frontmatter, a comprehensive glossary in the backmatter, and a high-quality cumulative index provide additional tools.

Acknowledgements

First I thank my wife Pat for her many contributions. And special thanks go to my associate academic editors, who provided tremendous expertise in their respective areas of specialty. The editors and I collectively acknowledge the thoughtful and professional contributions made by members of Macmillan Reference USA and the Gale Group. Hélène Potter and former publisher Elly Dickason (now retired) were instrumental in launching and nurturing the project. Cindy Clendenon has been especially helpful in her editing and managing of the components associated with the 304 articles. Her training, knowledge, and keen interest in the field have resulted in a markedly better product.

E. Julius Dasch

Topical Outline

AGRICULTURE

Agriculture and Water
Aquaculture
Chemicals from Agriculture
Food Security
Irrigation Management
Irrigation Systems, Ancient
Mariculture
Pollution of the Ocean by Sewage, Nutrients, and Chemicals

AQUATIC ANIMALS

Aquariums
Birds, Aquatic
Bivalves
Cephalopods
Crustaceans
Fish
Fishes, Cartilaginous
Marine Mammals
Reptiles
Salmon Decline and Recovery

BIOGRAPHIES

Agassiz, Louis
Bretz, J Harlen
Carson, Rachel
Cook, Captain James
Cousteau, Jacques
Darcy, Henry
Davis, William Morris
Douglas, Marjory Stoneman
Earle, Sylvia
Garrels, Robert
Leopold, Luna
Lewis, Meriwether and William Clark
Hem, John D.
Hubbert, Marion King
Nansen, Fridtjof
Powell, John Wesley
Reisner, Marc
Stumm, Werner
Sverdrup, Harald
Theis, Charles Vernon
White, Gilbert

CAREERS

Careers in Environmental Education
Careers in Environmental Science and Engineering
Careers in Fresh-Water Chemistry
Careers in Fresh-Water Ecology
Careers in Geospatial Technologies
Careers in Hydrology
Careers in International Water Resources
Careers in Oceanography
Careers in Soil Science
Careers in Water Resources Engineering
Careers in Water Resources Planning and Management
Ethics and Professionalism
Minorities in Water Sciences
Women in Water Sciences

CHEMICAL AND PHYSICAL PROPERTIES, PROCESSES, AND APPLICATIONS

Acid Mine Drainage
Acid Rain
Attenuation of Pollutants
Beaches
Brines, Natural
Carbon Dioxide in the Ocean and Atmosphere

Chemical Analysis of Water
Chemicals: Combined Effect on Public Health
Chemicals from Agriculture
Chemicals from Consumers
Chemicals from Pharmaceuticals and Personal Care Products
Climate Moderator, Water as a
Coastal Ocean
Erosion and Sedimentation
Fresh Water, Physics and Chemistry of
Hot Springs and Geysers
Hot Springs on the Ocean Floor
Hydrologic Cycle
Isotopes: Applications in Natural Waters
Lakes: Chemical Processes
Lakes: Physical Processes
Land Use and Water Quality
Light Transmission in the Ocean
Ocean Biogeochemistry
Ocean Chemical Processes
Ocean Currents
Ocean Mixing
Oceanography, Chemical
Oceanography, Physical
Oceans, Polar
Precipitation and Clouds, Formation of
Radioactive Chemicals
Radionuclides in the Ocean
Sea Level
Sea Water, Freezing of
Sea Water, Gases in
Sea Water, Physics and Chemistry of
Sound Transmission in the Ocean
Stream Channel Development
Stream Erosion and Landscape Development
Tides
Tracers in Fresh Water
Tracers of Ocean-Water Masses
Waves
Weathering of Rocks

CLIMATE CHANGE

Amphibian Population Declines
Carbon Dioxide in the Ocean and Atmosphere
Climate and the Ocean
El Niño and La Niña
Glaciers, Ice Sheets, and Climate Change
Global Warming and Glaciers
Global Warming and the Hydrologic Cycle
Global Warming and the Ocean
Global Warming: Policy-Making
Ice Ages
Ice at Sea
Ice Cores and Ancient Climatic Conditions

DRINKING WATER

Bottled Water
Desalinization
Drinking Water and Society
Drinking-Water Treatment
Groundwater
Groundwater Supplies, Exploration for
Human Health and Water
Ice at Sea (Iceberg Harvesting)
Infrastructure, Water-Supply
Safe Drinking Water Act
Security and Water
Supplies, Protecting Public Drinking-Water
Supplies, Public and Domestic Water
Survival Needs
Utility Management

EARTH AND BEYOND

Astrobiology: Water and the Potential for Extraterrestrial Life
Climate Moderator, Water as a
Comets and Meteorites, Water in
Earth: the Water Planet
Earth's Interior, Water in the
Fresh Water, Natural Composition of
Hot Springs and Geysers
Hot Springs on the Ocean Floor
Hydrologic Cycle
Life in Extreme Water Environments
Mars, Water on
Ocean Currents
Oceanography from Space
Plate Tectonics
Precipitation, Global Distribution of
Space Travel
Solar System, Water in the
Volcanoes and Water
Volcanoes, Submarine
Weather and the Ocean

ECOLOGY AND ECOSYSTEMS

Algal Blooms in Fresh Water
Algal Blooms in the Ocean

Biodiversity
Corals and Coral Reefs
Desert Hydrology
Ecology, Fresh-Water
Ecology, Marine
Estuaries
Fish and Wildlife Issues
Forest Hydrology
Lakes: Biological Processes
Life in Water
Life in Extreme Water Environments
Microbes in Groundwater
Microbes in Lakes and Streams
Microbes in the Ocean
Plankton
Pollution by Invasive Species
Salmon Decline and Recovery
Stream Ecology: Temperature Impacts on
Stream, Hyporheic Zone of a
Wetlands

ECONOMICS AND COMMERCE

Bottled Water
Canals
Cost–Benefit Analysis
Demand Management
Economic Development
Energy from the Ocean
Fisheries, Fresh-Water
Fisheries, Marine
Geothermal Energy
Globalization and Water
Hydroelectric Power
Markets, Water
Mineral Resources from Fresh Water
Mineral Resources from the Ocean
Petroleum from the Ocean
Ports and Harbors
Pricing, Water
Privatization of Water Management
Transportation

FEDERAL AGENCIES (U.S.)

Army Corps of Engineers, U.S.
Bureau of Reclamation, U.S.
Environmental Protection Agency, U.S.
National Oceanic and Atmospheric
    Administration
National Park Service
Tennessee Valley Authority

GEOLOGIC PROPERTIES, PROCESSES, AND APPLICATIONS

Acid Mine Drainage
Aquifer Characteristics
Bays, Gulfs, and Straits
Beaches
Brines, Natural
Cavern Development
Earth's Interior, Water in the
Erosion and Sedimentation
Fresh Water, Natural Composition of
Geothermal Energy
Glaciers and Ice Sheets
Glaciers, Ice Sheets, and Climate Change
Groundwater
Groundwater Supplies, Exploration for
Hot Springs and Geysers
Hot Springs on the Ocean Floor
Hydrogeologic Mapping
Islands, Capes, and Peninsulas
Karst Hydrology
Lake Formation
Landslides
Mid-Ocean Ridges
Ocean Basins
Ocean-Floor Bathymetry
Ocean-Floor Sediments
Oceanography, Geological
Plate Tectonics
Runoff, Factors Affecting
Springs
Stream Channel Development
Stream Erosion and Landscape Development
Tsunamis
Volcanoes and Water
Volcanoes, Submarine
Waterfalls
Weathering of Rocks

GROUNDWATER

Aquifer Characteristics
Artificial Recharge
Fresh Water, Natural Composition of
Fresh Water, Natural Contaminants in
Groundwater
Groundwater, Age of
Groundwater Supplies, Exploration for
Hydrogeologic Mapping
Karst Hydrology
Landfills: Impact on Groundwater

Microbes in Groundwater
Modeling Groundwater Flow and Transport
Ogallala Aquifer
Pollution of Groundwater
Pollution of Groundwater: Vulnerability
Springs
Supplies, Protecting Public Drinking-Water
Wells and Well Drilling
Wetlands

HUMAN EXPERIENCE

Acid Mine Drainage
Acid Rain
Agriculture and Water
Algal Blooms, Harmful
Aquariums
Archaeology, Underwater
Arts, Water in the
Balancing Diverse Interests
Bottled Water
Conflict and Water
Developing Countries, Issues in
Drinking Water and Society
Drought Management
Economic Development
Environmental Movement, Role of Water in the
Ethics and Professionalism
Fisheries, Marine: Management and Policy
Floodplain Management
Glaciers, Ice Sheets and Climate Change
Globalization and Water
Global Warming: Policy-Making
Human Health and the Ocean
Human Health and Water
Hydrologic Cycle
Hydropolitics
Hydrosolidarity
International Cooperation
Irrigation, Ancient
Land-Use Planning
Law, International Water
Law, Water
Mineral Waters and Spas
Minorities in Water Sciences
Nansen, Fridtjof
Navigation at Sea, History of
Ogallala Aquifer
Planning and Management, History of Water Resources
Planning and Management, Water Resources
Policy-Making Process
Population and Water Resources
Prior Appropriation
Public Participation
Religions, Water in
Rights, Riparian
Rights, Public Water
River Basin Planning
Security and Water
Senses, Fresh Water and the
Space Travel
Sports
Submarines and Submersibles
Supplies, Public and Domestic Water
Survival Needs
Sustainable Development
Transboundary Water Treaties
Uses of Water
War and Water
Waterworks, Ancient
Women in Water Sciences

HYDROLOGY AND HYDROGEOLOGY

Acid Mine Drainage
Aquifer Characteristics
Desert Hydrology
Estuaries
Forest Hydrology
Groundwater
Groundwater Supplies, Exploration for
Hydrogeologic Mapping
Hydrologic Cycle
Karst Hydrology
Landfills: Impact on Groundwater
Modeling Groundwater Flow and Transport
Modeling Streamflow
Rivers, Major World
Springs
Stream Channel Development
Stream Erosion and Landscape Development
Stream Hydrology
Stream, Hyporheic Zone of a
Streamflow Variability
Tracers in Fresh Water
Weathering of Rocks

Topical Outline

ICE (ON EARTH)

Glaciers and Ice Sheets
Glaciers, Ice Sheets, and Climate Change
Ice Ages
Ice at Sea
Ice Cores and Ancient Climatic Conditions
Oceans, Polar
Sea Water, Freezing of

INTERESTING WATER FEATURES

Astrobiology: Water and the Potential for Extraterrestrial Life
Bays, Gulfs, and Straits
Corals and Coral Reefs
Estuaries
Hot Springs and Geysers
Hot Springs on the Ocean Floor
Ice at Sea
Islands, Capes, and Peninsulas
Karst Hydrology
Life in Extreme Water Environments
Mineral Waters and Spas
Pumps, Traditional
Springs
Stream, Hyporheic Zone of a
Tsunamis
Volcanoes and Water
Volcanoes, Submarine
Waterfalls
Wetlands

ISSUES: NATIONAL AND INTERNATIONAL

Acid Rain
Algal Blooms, Harmful
Amphibian Population Declines
Chemicals: Combined Effect on Public Health
Chemicals from Agriculture
Chemicals from Consumers
Chemicals from Pharmaceuticals and Personal Care Products
Conflict and Water
Dams
Developing Countries, Issues in
Drinking Water and Society
Drought Management
Fish and Wildlife Issues
Fisheries, Marine: Management and Policy
Floodplain Management
Food Security
Globalization and Water
Global Warming: Policy-Making
Great Lakes
Human Health and the Ocean
Human Health and Water
Hydropolitics
Hydrosolidarity
Instream Water Issues
International Cooperation
Law, International Water
Law of the Sea
Pollution by Invasive Species
Pollution of Groundwater
Pollution of Lakes and Streams
Pollution of the Ocean by Sewage, Nutrients, and Chemicals
Pollution Sources: Point and Nonpoint
Population and Water Resources
Rainwater Harvesting
Rights, Riparian
Rights, Public Water
Salmon Decline and Recovery
Security and Water
Supplies, Protecting Public Drinking-Water
Supplies, Public and Domestic Water
Survival Needs
Sustainable Development
Transboundary Water Treaties
War and Water

LAKES AND STREAMS

Fresh Water, Natural Composition of
Fresh Water, Natural Contaminants in
Great Lakes
Lake Formation
Lake Health, Assessing
Lake Management Issues
Lakes: Biological Processes
Lakes: Chemical Processes
Lakes: Physical Processes
Microbes in Lakes and Streams
Modeling Streamflow
Nutrients in Lakes and Streams
Rivers, Major World
Runoff, Factors Affecting
Stream Channel Development
Stream Ecology: Temperature Impacts on
Stream Erosion and Landscape Development
Stream Health, Assessing
Stream Hydrology

Stream, Hyporheic Zone of a
Streamflow Variability
Waterfalls
Wetlands

LEGISLATION, POLICY, AND LAW

Clean Water Act
Endangered Species Act
Hydropolitics
International Cooperation
Instream Water Issues
Law, International Water
Law of the Sea
Law, Water
Legislation, Federal Water
Legislation, State and Local Water
National Environmental Policy Act
Planning and Management, History of Water Resources
Policy-Making Process
Prior Appropriation
Rights, Public Water
Rights, Riparian
Safe Drinking Water Act
Transboundary Water Treaties

MICROBES: ECOSYSTEMS AND HUMAN IMPACTS

Algal Blooms, Harmful
Algal Blooms in Fresh Water
Algal Blooms in the Ocean
Human Health and the Ocean
Human Health and Water
Microbes in Groundwater
Microbes in Lakes and Streams
Microbes in the Ocean
Plankton

OCEAN SCIENCE

Algal Blooms, Harmful
Algal Blooms in the Ocean
Carbon Dioxide in the Ocean and Atmosphere
Climate and the Ocean
Ecology, Marine
Estuaries
Ice at Sea
Ocean Basins
Ocean Biogeochemistry
Ocean Chemical Processes
Ocean Currents
Ocean Mixing
Ocean-Floor Bathymetry
Ocean-Floor Sediments
Oceanography, Biological
Oceanography, Chemical
Oceanography from Space
Oceanography, Geological
Oceanography, Physical
Oceans, Polar
Oceans, Tropical
Plankton
Radionuclides in the Ocean
Sea Level
Sea Water, Freezing of
Sea Water, Gases in
Tides
Tracers of Ocean-Water Masses
Waves
Weather and the Ocean

PLANNING AND MANAGEMENT

Balancing Diverse Interests
California, Water Management in
Chesapeake Bay
Coastal Waters Management
Colorado River Basin
Columbia River Basin
Conflict and Water
Conservation, Water
Cost-Benefit Analysis
Data, Databases, and Decision-Support Systems
Demand Management
Drought Management
Everglades
Floodplain Management
Florida, Water Management in
Great Lakes
Instream Water Issues
Integrated Water Resources Management
Lake Management Issues
Land Use and Water Quality
Land-Use Planning
Mississippi River Basin
Planning and Management, History of Water Resources
Planning and Management, Water Resources
Prior Appropriation
Public Participation

Reclamation and Reuse
Recreation
Reservoirs, Multipurpose
Rights, Public Water
Rights, Riparian
River Basin Planning
Supplies, Protecting Public Drinking-Water
Supply Development
Tourism

POLLUTION AND ENVIRONMENTAL QUALITY

Acid Mine Drainage
Acid Rain
Agriculture and Water
Algal Blooms, Harmful
Amphibian Population Declines
Attenuation of Pollutants
Chemicals: Combined Effect on Public Health
Chemicals from Agriculture
Chemicals from Consumers
Chemicals from Pharmaceuticals and Personal Care Products
Clean Water Act
Erosion and Sedimentation
Lake Health, Assessing
Land Use and Water Quality
Land-Use Planning
Landfills: Impact on Groundwater
Microbes in Groundwater
Microbes in Lakes and Streams
Microbes in the Ocean
National Environmental Policy Act
Nutrients in Lakes and Streams
Ocean Health, Assessing
Oil Spills: Impact on the Ocean
Pollution by Invasive Species
Pollution of Groundwater
Pollution of Groundwater: Vulnerability
Pollution of Lakes and Streams
Pollution of Streams by Garbage and Trash
Pollution of the Ocean by Plastic and Trash
Pollution of the Ocean by Sewage, Nutrients, and Chemicals
Pollution Sources: Point and Nonpoint
Runoff, Factors Affecting
Safe Drinking Water Act
Septic System Impacts
Stream Health, Assessing
Supplies, Protecting Public Drinking-Water
Watershed, Restoration of a
Watershed, Water Quality in a

RESOURCES: LIVING AND NONLIVING

Aquaculture
Energy from the Ocean
Fisheries, Fresh-Water
Fisheries, Marine
Food from the Sea
Geothermal Energy
Hydroelectric Power
Mariculture
Mineral Resources from Fresh Water
Mineral Resources from the Ocean
Petroleum from the Ocean

STRUCTURES AND FACILITIES

Bridges, Causeways, and Underwater Tunnels
Canals
Dams
Hoover Dam
Infrastructure, Water-Supply
Irrigation Systems, Ancient
Land-Use Planning
Moorings and Platforms
Ports and Harbors
Pumps, Modern
Pumps, Traditional
Supplies, Public and Domestic Water
Utility Management
Wastewater Treatment and Management
Waterworks, Ancient
Wells and Well Drilling

TECHNOLOGY

Archaeology, Underwater
Artificial Recharge
Dams
Data, Databases, and Decision-Support Systems
Energy from the Ocean
Geospatial Technologies
Hydroelectric Power
Modeling Groundwater Flow and Transport
Modeling Streamflow
Navigation at Sea, History of
Oceanography from Space
Pumps, Modern

Topical Outline

Pumps, Traditional
Sound Transmission in the Ocean
Submarines and Submersibles

USING WATER

Agriculture and Water
Artificial Recharge
Balancing Diverse Interests
Bottled Water
Conflict and Water
Conservation, Water
Demand Management
Desalinization
Drinking Water and Society
Energy from the Ocean
Fish and Wildlife Issues
Geothermal Energy
Ice at Sea
Infrastructure, Water-Supply
Instream Water Issues
Integrated Water Resources Management
Irrigation Management
Irrigation Systems, Ancient
Land Use and Water Quality
Land-Use Planning
Navigation at Sea, History of
Ogallala Aquifer
Pollution Sources: Point and Nonpoint
Pumps, Modern
Pumps, Traditional
Rainwater Harvesting
Reclamation and Reuse
Recreation
Reservoirs, Multipurpose
Sports
Streamflow Variability
Sustainable Development
Tourism
Transportation
Uses of Water
Utility Management
Wastewater Treatment and Management

Waterworks, Ancient
Wells and Well Drilling

WEATHER AND CLIMATE

Carbon Dioxide in the Ocean and Atmosphere
Climate and the Ocean
Climate Moderator, Water as a
El Niño and La Niña
Global Warming and Glaciers
Global Warming and the Hydrologic Cycle
Global Warming and the Ocean
Global Warming: Policy-Making
Ice Ages
Ice at Sea
Ice Cores and Ancient Climatic Conditions
Precipitation and Clouds, Formation of
Precipitation, Global Distribution
Rainwater Harvesting
Weather and the Ocean

SELECTED SIDEBARS

Agates, Geodes, and Petrified Wood
Corals as Historic Recordkeepers
Drug Benefits from the Sea
Exclusive Economic Zone
Lake Versus Ocean Beaches
Life in a Martian Meteorite?
Manganese Nodules
Marine and Fresh-Water Shipwrecks
Pathogenic Viruses in Groundwater
Pesticides, Land Use, and Groundwater
Properly Disposing of Unused and Outdated Drugs
The Myth of Underground Streams
Tragedy of the Commons
Water and the Middle East Peace Process
Water Witching
What Do Floods Cost?
Who Owns the Panama Canal?

For Your Reference

TABLE 1. SELECTED METRIC CONVERSIONS

WHEN YOU KNOW	MULTIPLY BY	TO FIND
Temperature		
Celsius (°C)	1.8 (°C) +32	Fahrenheit (°F)
Celsius (°C)	°C +273.15	Kelvin (K)
degree change (Celsius)	1.8	degree change (Fahrenheit)
Fahrenheit (°F)	[(°F) −32] / 1.8	Celsius (°C)
Fahrenheit (°F)	[(°F −32) / 1.8] +273.15	Kelvin (K)
Kelvin (K)	K −273.15	Celsius (°C)
Kelvin (K)	1.8(K −273.15) +32	Fahrenheit (°F)

WHEN YOU KNOW	MULTIPLY BY	TO FIND
Distance/Length		
centimeters	0.3937	inches
kilometers	0.6214	miles
meters	3.281	feet
meters	39.37	inches
meters	0.0006214	miles
microns	0.000001	meters
millimeters	0.03937	inches

WHEN YOU KNOW	MULTIPLY BY	TO FIND
Capacity/Volume		
cubic kilometers	0.2399	cubic miles
cubic meters	35.31	cubic feet
cubic meters	1.308	cubic yards
cubic meters	8.107×10^{-4}	acre-feet
liters	0.2642	gallons
liters	33.81	fluid ounces

WHEN YOU KNOW	MULTIPLY BY	TO FIND
Area		
hectares (10,000 square meters)	2.471	acres
hectares (10,000 square meters)	107,600	square feet
square meters	10.76	square feet
square kilometers	247.1	acres
square kilometers	0.3861	square miles

WHEN YOU KNOW	MULTIPLY BY	TO FIND
Weight/Mass		
kilograms	2.205	pounds
metric tons	2205	pounds
micrograms (µg)	10^{-6}	grams
milligrams (mg)	10^{-3}	grams
nanograms (ng)	10^{-9}	grams

For Your Reference

TABLE 2. SELECTED SYMBOLS, ABBREVIATIONS, AND ACRONYMS

Ancillary capitalization is used throughout to illustrate how the abbreviations are derived. In appropriate usage, however, most chemical names do not contain mixtures of uppercase and lowercase (e.g., PerChlorEthylene is perchlorethylene).

MEASUREMENTS

μ	"micro" (10^{-6})
m	"milli" (10^{-3})
°C	degrees Celsius
°F	degrees Fahrenheit
μg/L	micrograms per liter
psi	Pounds-force per Square Inch
ppb	Parts Per Billion (or micrograms per liter)
ppbv	Parts Per Billion by Volume
ppm	Parts Per Million (or milligrams per liter)
ppmv	Parts Per Million by Volume
ppt	Parts Per Thousand (or grams per liter)

CHEMISTRY AND POLLUTION

AMD	Acid Mine Drainage
ARD	Acid Rock Drainage
ANS	Aquatic Nuisance Species
ATP	Adenosine TriPhosphate
BOD	Biochemical Oxygen Demand
BTEX	Benzene, Toluene, Ethylbenzene, and Xylene
CFC	ChloroFluoroCarbon
COD	Chemical Oxygen Demand
DDE	DichloroDiphenyldichloroEthylene
DDT	DichloroDiphenylTrichloroethane
DIC	Dissolved Inorganic Carbon
DNA	DeoxyriboNucleic Acid
DO	Dissolved Oxygen
DOC	Dissolved Organic Carbon
MCL	Maximum Contaminant Level
MNA	Monitored Natural Attenuation
MTBE	Methyl *Tert*-Butyl Ether
NPDES	National Pollutant Discharge Elimination System
PAH	Polycyclic Aromatic Hydrocarbon
PCB	PolyChlorinated Biphenyl
PCE	PerChlorEthlyene
PPCP	Pharmaceutical and Personal Care Product
PVC	PolyVinyl Chloride
QA	Quality Assurance
QAPP	Quality Assurance Project Plan
RNA	RiboNucleic Acid
RO	Reverse Osmosis
STW	Sewage Treatment Works
TCE	TriChlorEthylene
TDS	Total Dissolved Solids
TMDL	Total Maximum Daily Load
VOC	Volatile Organic Compound

OCEAN SCIENCE

DSV	Diving Support Vehicle
ENSO	El Niño Southern Oscillation
HAB	Harmful Algal Bloom
OTEC	Ocean Thermal Energy Conversion
ROV	Remotely Operated Vehicle
SeaWiFS	Sea-viewing Wide Field-of-view Sensor
scuba (or SCUBA)	Self-Contained Underwater Breathing Apparatus
SONAR	SOund Navigation and Ranging
TAO	Tropical Atmosphere Ocean

RESOURCE MANAGEMENT

ADR	Alternative Dispute Resolution
AR	Artificial Recharge
ASR	Aquifer Storage and Recovery
CERP	Comprehensive Everglades Restoration Plan
EAA	Everglades Agricultural Area
EEZ	Exclusive Economic Zone
EIS	Environmental Impact Statement
IQ	Individual Quota
ITQ	Individual Transferable Quota
LOS	Law of the Sea
MSY	Maximum Sustainable Yield
MPA	Marine Protected Area
RDU	Rural Development Unit
TAC	Total Allowable Catch
WUA	Water User Association
WIN	Water Infrastructure Network
WMD	Water Management District

TECHNOLOGY

DSS	Decision-Support System
GIS	Geographic Information Systems
GPS	Global Positioning System
GLIMS	Global Land Ice Measurement from Space
MODIS	MODerate-resolution Imaging Spectroradiometer

[continued]

TABLE 2 (continued). SELECTED SYMBOLS, ABBREVIATIONS, AND ACRONYMS

Ancillary capitalization is used throughout to illustrate how the abbreviations are derived. In appropriate usage, however, most chemical names do not contain mixtures of uppercase and lowercase (e.g., PerChlorEthylene is perchlorethylene).

MISCELLANEOUS

3D (or 3-D)	Three Dimensional
AC	Alternating Current (also stands for Asbestos-Cement)
B.C.E.	Before the Common Era
AIDS	AutoImmune Deficiency Syndrome
B.P.	Before the Present
B.S.	Bachelor of Science
c.	Circa
C.E.	Common Era
D.C.	District of Columbia
GNP	Gross National Product
M.S.	Master of Science
Ph.D.	Doctor of Philosophy
ULV	Ultra-Low-Volume
UV	UltraViolet

LEGISLATION

CWA	Clean Water Act
CZMA	Coastal Zone Management Act
ESA	Endangered Species Act
NEPA	National Environmental Policy Act
RCRA	Resource Conservation and Recovery Act
SDWA	Safe Drinking Water Act

ORGANIZATIONS

ACE	Army Corps of Engineers
AWRA	American Water Resources Association
CIA	Central Intelligence Agency
EPA	Environmental Protection Agency
FDA	Food and Drug Administration
FEMA	Federal Emergency Management Agency
FWS	Fish and Wildlife Service
IPCC	Intergovernmental Panel on Climate Change
IJC	International Joint Commission
IWA	International Water Association
IWRA	International Water Resources Association
NASA	National Aeronautics and Space Administration
NATO	North Atlantic Treaty Organization
NMFS	National Marine Fisheries Service
NPS	National Park Service
NOAA	National Oceanic and Atmospheric Administration
NRCS	Natural Resource Conservation Service
NSF	National Science Foundation
TVA	Tennessee Valley Authority
UN	United Nations
UNESCO	United Nations Educational, Scientific and Cultural Organization
UNICEF	United Nations International Children's Fund
U.S.	United States
U.S.A.	United States of America
USBR	United States Bureau of Reclamation (also BOR or USBR)
USDA	United States Department of Agriculture
USGS	United States Geological Survey

Geologic Timescale

Era	Period		Epoch	started (millions of years ago)
Cenozoic 66.4 millions of years ago–present time	Quaternary		Holocene	0.01
			Pleistocene	1.6
	Tertiary	Neogene	Pliocene	5.3
			Miocene	23.7
		Paleogene	Oligocene	36.6
			Eocene	57.8
			Paleocene	66.4
Mesozoic 245–66.4 millions of years ago	Cretaceous		Late	97.5
			Early	144
	Jurassic		Late	163
			Middle	187
			Early	208
	Triassic		Late	230
			Middle	240
			Early	245
Paleozoic 570–245 millions of years ago	Permian		Late	258
			Early	286
	Carboniferous	Pennsylvanian		320
		Mississippian		360
	Devonian		Late	374
			Middle	387
			Early	408
	Silurian		Late	421
			Early	438
	Ordovician		Late	458
			Middle	478
			Early	505
	Cambrian		Late	523
			Middle	540
			Early	570
Precambrian time 4560–570 millions of years ago				4560

xxvi Timescale

Contributors

Barbara Johnston Adams
Amagansett, New York

Cain Allen
University of British Columbia
Vancouver, British Columbia, Canada

Anthony F. Amos
The University of Texas
Port Aransas, Texas

Faye Anderson
University of Maryland
College Park, Maryland

Gail Glick Andrews
Oregon State University
Corvallis, Oregon

William Arthur Atkins
Atkins Research and Consulting
Normal, Illinois

Jeffery A. Ballweber
Mississippi Water Resources Research
Mississippi State University
Starkville, Mississippi

Nadine G. Barlow
Northern Arizona University
Flagstaff, Arizona

Janice A. Beecher
Michigan State University
East Lansing, Michigan

Amy G. Beier
Picton, New Zealand

Paul S. Berger
U.S. Environmental Protection Agency
Washington, D.C.

Christina E. Bernal
Beaumont, Texas

Andrew R. Blaustein
Oregon State University
Corvallis, Oregon

Arthur L. Bloom
Cornell University
Ithaca, New York

Andrew J. Boulton
University of New England
Armidale, Australia

Patrick V. Brady
Sandia National Laboratories
Albuquerque, New Mexico

Amy J. Bratcher
Texas A & M University
College Station, Texas

Arthur S. Brooks
University of Wisconsin-Milwaukee
Milwaukee, Wisconsin

Scott F. Burns
Portland State University
Portland, Oregon

Piers Chapman
Louisiana State University
Baton Rouge, Louisiana

Randall Charbeneau
The University of Texas at Austin
Austin, Texas

Ralph Christensen
EGR & Associates, Inc.
Eugene, Oregon

Timothy A. Chuey
KVAL-13 Television
Eugene, Oregon

Neil Clark
The Writing Company
Watertown, MA

Jeanne Nienaber Clarke
University of Arizona
Tucson, Arizona

Flaxen D. L. Conway
Oregon State University
Corvallis, Oregon

James R. Craig
Emerald Isle, North Carolina

Ron Crouse
Interpretive Solutions
Lincoln City, Oregon

Benjamin Cuker
Hampton University
Hampton, Virginia

Michael Cummings
Portland State University
Portland, Oregon

Mark Cunnane
Western Groundwater Services, LLC
Bozeman, Montana

Scott G. Curry
Oregon Department of Human Services
Medford, Oregon

E. Julius Dasch
RSC International
Washington, D.C.

Pat Dasch
RSC International
Washington, D.C.

Christian G. Daughton
U.S. Environmental Protection Agency
Las Vegas, Nevada

Laura O. Dávalos-Lind
Baylor University
Waco, Texas

Thomas E. Davenport
U.S. Environmental Protection Agency
Chicago, Illinois

Joseph W. Dellapenna
Villanova University
Villanova, Pennsylvania

Andrew P. Diller
University of Florida
Cantonment, Florida

Terry C. Dodge
Florida Center for Environmental Studies
Palm Beach Gardens, Florida

Jane Dougan
Nova Southeastern University
Dania Beach, Florida

Lisa A. Drake
Old Dominion University
Norfolk, Virginia

Contributors

Laurie Duncan
The University of Texas at Austin
Austin, Texas

Cheryl Lyn Dybas
National Science Foundation
Arlington, Virginia

David A. V. Eckhardt
U.S. Geological Survey
Ithaca, New York

Hillary S. Egna
Oregon State University
Corvallis, Oregon

John F. Elder
U.S. Geological Survey
Middleton, Wisconsin

Audrey Eldridge
Oregon Department of Environmental Quality
Medford, Oregon

Carolyn Embach
University of Tulsa
Tulsa, Oklahoma

Richard A. Engberg
American Water Resources Association
Middleburg, Virginia

Jack D. Farmer
Arizona State University
Tempe, Arizona

Rana A. Fine
University of Miami
Miami, Florida

Denise D. Fort
University of New Mexico
Albuquerque, New Mexico

Doretha B. Foushee
North Carolina Agricultural & Technical State University
Greensboro, North Carolina

Jeffrey Frederick
Oregon Department of Human Services
Springfield, Oregon

Richard Gates
Lake Oswego, Oregon

Bart Geerts
University of Wyoming
Laramie, Wyoming

Deidre M. Gibson
Hampton University
Hampton, Virginia

Larry Gilman
Sharon, Vermont

Meredith A. Giordano
International Water Management Institute
Colombo, Sri Lanka

Michael N. Gooseff
Utah State University
Logan, Utah

Pamela J. W. Gore
Georgia Perimeter College
Clarkston, Georgia

Rick G. Graff
Graff Associates
Portland, Oregon

Neil S. Grigg
Colorado State University
Fort Collins, Colorado

M. Grant Gross
Washington College
Chestertown, Maryland

James R. Groves
Virginia Military Institute
Lexington, Virginia

Steven C. Hackett
Humboldt State University
Arcata, California

Richard Haeuber
Washington, D.C.

Pixie A. Hamilton
U.S. Geological Survey
Richmond, Virginia

Julie K. Harvey
Oregon Department of Environmental Quality
Portland, Oregon

Richard J. Heggen
University of New Mexico
Albuquerque, New Mexico

Stephen R. Hinkle
U.S. Geological Survey
Portland, Oregon

Brian D. Hoyle
Nova Scotia, Canada

Christina Hulbe
Portland State University
Portland, Oregon

Patricia S. Irle
Washington State Department of Ecology
Olympia, Washington

Richard H. Ives
U.S. Bureau of Reclamation
Washington, D.C.

Walter C. Jaap
Lithophyte Research
St. Petersburg, Florida

Jeffrey W. Jacobs
National Research Council
Washington, D.C.

Cindy Johnson
Suwannee River Water Management District
Gainesville, Florida

William W. Jones
Indiana University
Bloomington, Indiana

Jeffrey L. Jordan
University of Georgia at Griffin
Griffin, Georgia

Karen E. Kelley
Oregon Department of Human Services
Springfield, Oregon

Dana R. Kester
University of Rhode Island
Narragansett, Rhode Island

Phillip Z. Kirpich
World Bank (former staff member)
Miami Beach, Florida

Philip Koth
Atkins Research and Consulting
Normal, Illinois

F. Michael Krautkramer
Robinson & Noble, Inc.
Tacoma, Washington

David E. Kromm
Kansas State University
Manhattan, Kansas

Christopher Lant
Southern Illinois University–Carbondale
Carbondale, Illinois

Kelli L. Larson
Oregon State University
Corvallis, Oregon

Brenda Wilmoth Lerner
Lerner & Lerner, LLC
London, U.K.

K. Lee Lerner
Science Research and Policy Institute
London, U.K. and Washington, D.C.

Judith Li
Oregon State University
Corvallis, Oregon

Roberta J. Lindberg
Oregon Department of Environmental Quality
Eugene, Oregon

Kenneth E. Lite Jr.
Oregon Water Resources Department
Salem, Oregon

Steven E. Lohrenz
University of Southern Mississippi
Stennis Space Center, Mississippi

Daniel P. Loucks
Cornell University
Ithaca, New York

Robert W. Malmsheimer
State University of New York College of Environmental Science and Forestry
Syracuse, New York

Contributors

Michael Manga
University of California, Berkeley
Berkeley, California

William R. Mason
*Oregon Department of
Environmental Quality*
Eugene, Oregon

Olen Paul Matthews
University of New Mexico
Albuquerque, New Mexico

Michael J. Mattick
Oregon Water Resources Department
Springfield, Oregon

Larry W. Mays
Arizona State University
Tempe, Arizona

Sue McClurg
Water Education Foundation
Sacramento, California

Richard H. McCuen
University of Maryland
College Park, Maryland

John D. McEachran
Texas A & M University
College Station, Texas

Vincent G. McGowan
*Southern Illinois
University–Carbondale*
Carbondale, Illinois

Minerva Mercado-Feliciano
Bloomington, Indiana

Grant A. Meyer
University of New Mexico
Albuquerque, New Mexico

Donn Miller
*Oregon Water Resources
Department*
Salem, Oregon

Timothy L. Miller
U.S. Geological Survey
Reston, Virginia

Bruce Mitchell
University of Waterloo
Waterloo, Ontario, Canada

James E. T. Moncur
University of Hawaii at Manoa
Honolulu, Hawaii

Karl A. Morgenstern
Eugene Water & Electric Board
Eugene, Oregon

Earl Finbar Murphy
The Ohio State University
Columbus, Ohio

Richard W. Murray
Boston University
Boston, Massachusetts

Clifford M. Nelson
U.S. Geological Survey
Reston, Virginia

Dennis O. Nelson
*Oregon Department of Human
Services*
Springfield, Oregon

Gary Nelson
Bend, Oregon

John W. Nicklow
*Southern Illinois
University–Carbondale*
Carbondale, Illinois

Vita Pariente
College Station, Texas

Amy B. Parmenter
*Oregon Department of
Environmental Quality*
Eugene, Oregon

Richard J. Pedersen
*Oregon Department of
Environmental Quality*
Portland, Oregon

Howard A. Perlman
U.S. Geological Survey
Atlanta, Georgia

Catherine M. Petroff
University of Washington
Seattle, Washington

Laurel E. Phoenix
University of Wisconsin–Green Bay
Green Bay, Wisconsin

James L. Pinckney
Texas A & M University
College Station, Texas

Ashanti Johnson Pyrtle
University of South Florida
St. Petersburg, Florida

Timothy Randhir
University of Massachusetts
Amherst, Massachusetts

Elliot Richmond
Education Consultants
Austin, Texas

Richard Robinson
Tucson, Arizona

David M. Rohr
Sul Ross State University
Alpine, Texas

Joel S. Rubin
New England Aquarium
Boston, Massachusetts

Christopher L. Sabine
*National Oceanic and
Atmospheric Administration*
Seattle, Washington

Dorothy Sack
Ohio University
Athens, Ohio

Kari Salis
*Oregon Department of Human
Services*
Portland, Oregon

Marie Scheessele
St. Mary's Catholic School
Alexandria, Virginia

Steffen W. Schmidt
Iowa State University
Ames, Iowa

Alison Cridland Schutt
Chevy Chase, Maryland

Martha R. Scott
Texas A & M University
College Station, Texas

Ralph L. Seiler
U.S. Geological Survey
Carson City, Nevada

George H. Shaw
Union College
Schenectady, New York

N. Earl Spangenberg
*University of Wisconsin–
Stevens Point*
Stevens Point, Wisconsin

Sheree L. Stewart
*Oregon Department of
Environmental Quality*
Portland, Oregon

Robert R. Stickney
Texas A & M University
College Station, Texas

Margaret M. Streepey
University of Michigan
Ann Arbor, Michigan

Kimberly J. Swanson
*Oregon Department of Human
Services*
Springfield, Oregon

George H. Taylor
Oregon State University
Corvallis, Oregon

Robert J. Taylor
Texas A & M University
College Station, Texas

Terri A. Thomas
Carbondale, Illinois

David B. Thompson
Texas Tech University
Lubbock, Texas

Eileen Tramontana
*St. Johns River Water
Management District*
Palatka, Florida

Tas D. van Ommen
Australian Antarctic Division
Kingston, Tasmania, Australia

Steve Vandas
U.S. Geological Survey
Denver, Colorado

Joan Vernikos
Thirdage, LLC
Alexandria, Virginia

Contributors

Warren Viessman Jr.
University of Florida
Gainesville, Florida

Edward F. Vitzthum
University of Nebraska
Lincoln, Nebraska

Noam Weisbrod
*Ben-Gurion University
of the Negev*
Sde-Boqer, Israel

Gilbert F. White
University of Colorado at Boulder
Boulder, Colorado

Donald A. Wilhite
University of Nebraska
Lincoln, Nebraska

Thomas C. Winter
U.S. Geological Survey
Denver, Colorado

Ellen Wohl
Colorado State University
Fort Collins, Colorado

Aaron T. Wolf
Oregon State University
Corvallis, Oregon

Christopher J. Woltemade
Shippensburg University
Shippensburg, Pennsylvania

Jennifer Yeh
*University of California,
San Francisco*
San Francisco, California

Land Use and Water Quality

Streams, rivers, and lakes are an important part of the landscape, as they provide water supply, recreation, and transportation for humans, and a place to live for a variety of plants and animals. **Groundwater** also is an important water resource that serves as a source of drinking water for more than 140 million people in the United States.

In some areas, **contamination** from natural and human sources has affected the use of these waters. For example, naturally occurring minerals within bedrock can impair the taste of groundwater and in some cases limit its use. The spilling, leaking, improper disposal, or intentional application of chemicals at the land surface can result in runoff that contaminates nearby streams and lakes, or infiltration that contaminates underlying **aquifers** (see the illustration on page 2).

The type and severity of water contamination often is directly related to human activity, which can be quantified in terms of the intensity and type of land use in the source areas of water to streams and aquifers. The analysis of patterns of land use and population provides a tool in the investigation of sites with known contamination, and in the prediction and prevention of future contamination of downstream waters. Studies of contamination sources and transport pathways that affect surface water and groundwater draw upon several disciplines, including **hydrology**, geology, biology, soil science, agriculture, physics, chemistry, and engineering.

Groundwater Contamination and Land Use

A relatively simple way to study the effects of land use on groundwater quality is to compare the predominant land uses within a given area to the concentrations of selected contaminants in water drawn from shallow aquifers within that area. Analysis of the relation between land use and the magnitude of contamination in a specific area primarily is based on the following two assumptions.

First, it is assumed that contaminated groundwater at a well originated as uncontaminated **recharge** (precipitation) that passed through a contaminated area before reaching the well. The area from which a well derives its water (and associated contaminants) is known as the well's groundwater "contributing area." A well's contributing area can be delineated on a map

groundwater: generally, all subsurface (underground) water, as distinct from surface water, that supplies natural springs, contributes to permanent streams, and can be tapped by wells; specifically, the water that is in the saturated zone of a defined aquifer

contamination: impairment of the quality of water by natural or human-made substances to a degree that is considered undesirable for certain uses; this term usually implies a human or environmental health threat, but some types of contamination are merely non-aesthetic rather than harmful

aquifer: a water-saturated, permeable, underground rock formation that can transmit significant quantities of water under ordinary hydraulic gradients to wells and springs

hydrology: the science that deals with the occurrence, distribution, movement, and physical and chemical properties of water on Earth

recharge: the process by which precipitation infiltrates below the surface and replenishes an aquifer

Land Use and Water Quality

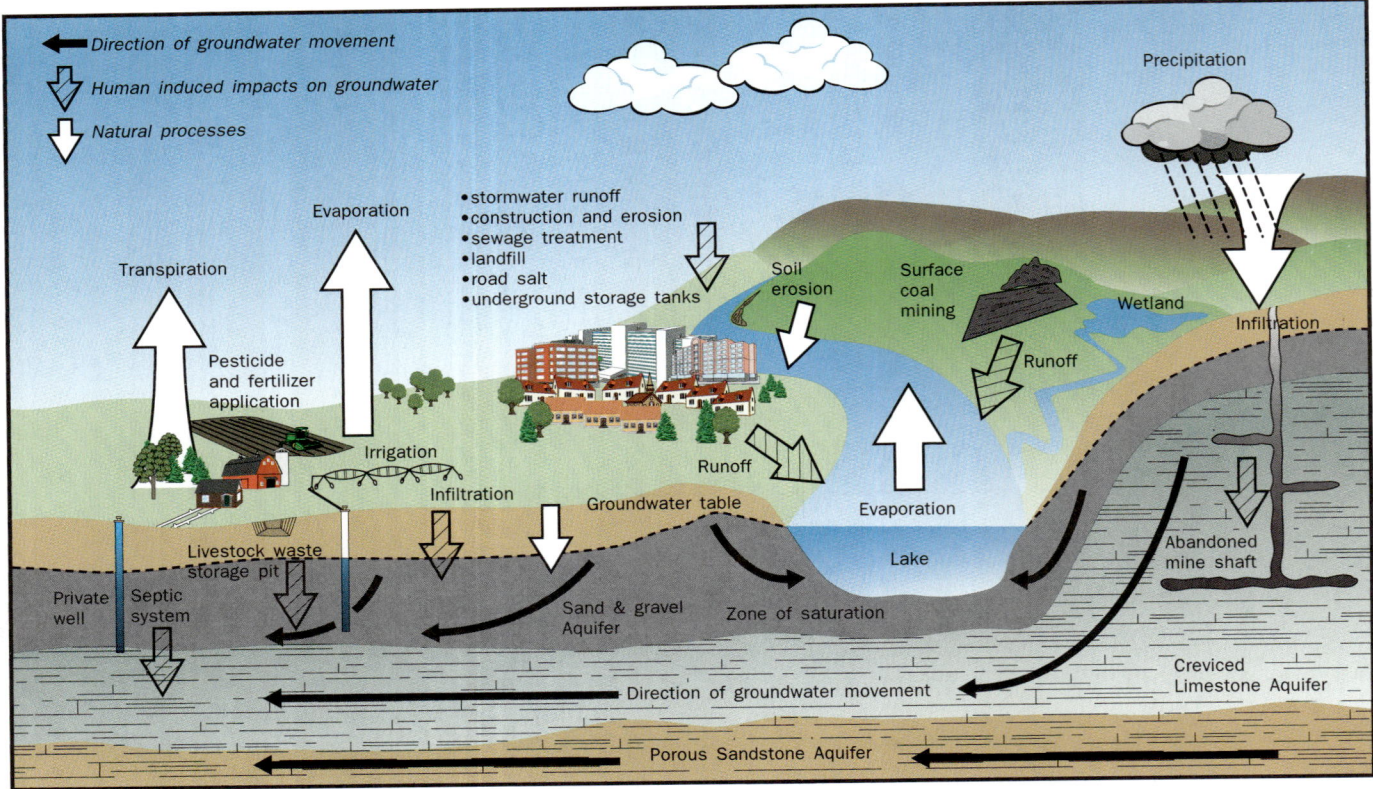

Land use and land cover largely determine the type and amount of contaminants entering streams, lakes, and underground pathways, including aquifers. Some contaminants occur and move naturally (white arrows), whereas others are produced by human activities (hatched arrows), and their movement often is accelerated as a result of rainfall that accentuates runoff and infiltration.

pesticides: a broad group of chemicals that kills or controls plants (herbicides), fungus (fungicides), insects and arachnids (insecticides), rodents (rodenticides), bacteria (bactericides), or other creatures that are considered pests

volatile organic compounds: organic compounds that can be isolated from the water phase of a sample by purging the water sample with inert gas, such as helium, and subsequently analyzed by gas chromatography

methemoglobinemia: a disease, primarily in infants, caused by the conversion of nitrates to nitrites in the intestines, and which limits the body's ability to receive oxygen; often referred to as "blue baby syndrome"

through an analysis of aquifer characteristics and the direction and velocity of groundwater flow near the well.

Second, it is assumed that the contaminants detected in groundwater were present within the well's contributing area and were transported by groundwater flow to the well. The source(s) of contaminants within a contributing area, such as buried septic systems and leaking underground fuel tanks, can be difficult to identify and locate. In many instances, these sources can be inferred from the type and intensity of land use within the contributing area.

Factors that can affect the movement of contaminants from source areas to wells are:

- The chemical nature of the contaminant;
- The physical properties of the soil and aquifer material;
- The amount and timing of recharge; and
- The direction and velocity of groundwater movement.

Despite the complexity of contaminant hydrology, the effects of certain land uses on groundwater quality have been scientifically documented in many areas. For example, relationships have been found between land use and five common groundwater contaminants: nitrogen, bacteria, road salt, **pesticides**, and **volatile organic compounds**.

Nitrate. Nitrate (a form of nitrogen) is essential for plant growth, but an overabundance can contaminate streams and groundwater. In high concentrations it causes **methemoglobinemia** in infants.

Nitrate can originate from domestic sewage and lawn fertilizers in residential areas, and from crop fertilizers and manure in agricultural areas. Land-use data on population, housing density, and agricultural practices can provide reliable indications of the likelihood of nitrate contamination of underlying aquifers.

Nitrogen-bearing lawn and crop fertilizers can readily leach through the soil and contaminate the groundwater after heavy rains or **irrigation**. As a result, nitrate concentrations in well water in residential and agricultural areas can be correlated with application rates of water and fertilizer. In agricultural areas, the nitrate concentration in groundwater also can be directly correlated with animal density. In unsewered residential areas, the rate of sewage discharge from domestic septic tanks can be estimated from population and housing density.

Bacteria. Human sewage and manure from cattle, hogs, chickens, geese, and other animals contain bacteria and other **pathogens** that can cause human illness. Many outbreaks have resulted from wells contaminated by fecal waste. Land-use data on densities of septic tanks and animals therefore are useful indicators of the presence of bacteria in wells.

Road Salt. Storm runoff and snowmelt in areas with salt-treated roads can carry sodium and calcium chlorides into the groundwater. Data on road density, salt application rate, and locations of salt storage piles can provide useful indicators for detection of elevated chloride concentrations in wells.

Pesticides. Pesticides are used to kill unwanted pests, such as termites, ants, and rodents around homes; nematodes in soil; and fungi and insects in crops. Similarly, herbicides kill undesirable weeds and grasses in lawns, along roads, and in agricultural areas.

The types and amount of pesticides used can be related to land-use factors such as population, housing density, number of roads, and the type of cropland. In recent studies, the concentrations of most pesticides in well water have rarely exceeded state or federal standards for drinking water. However, the effects of chronic, low-level exposures to pesticides on ecological and human health have not yet been fully assessed.

Volatile Organic Compounds. Volatile organic compounds (VOCs) have affected groundwater locally throughout the United States. Many VOCs are **carcinogenic**; thus, their presence in groundwater creates a serious problem for water suppliers.

VOCs commonly are detected in groundwater in industrial and commercial areas where petroleum fuels and **organic** solvents are used. A major source is leaking fuel tanks, which contaminate the underlying aquifers with compounds such as benzene, toluene, ethylbenzene, and xylenes (BTEX) and methyl *tert*-butyl ether (MTBE), an additive used in gasoline to reduce smog-producing vehicular emissions.

The presence of VOCs in groundwater is directly related to urban and suburban development. A 1995 national survey of wells in near-surface aquifers found that the presence of MTBE was directly related to the population density near the wells. Assessments of the density of industrial and commercial development also can be used to estimate the number of potential sources of VOCs such as chlorinated solvents like trichloroethene (TCE) in underlying aquifers.

---

irrigation: the controlled application of water for agricultural or other purposes through human-made systems; generally refers to water application to soil when rainfall is insufficient to maintain desirable soil moisture for plant growth

pathogen: a disease-producing agent, usually a living organism, and commonly a microbe (microorganism)

carcinogenic: describes a cancer-causing substance or agent

organic: pertaining to, or the product of, biological reactions or functions

---

PESTICIDES, LAND USE, AND GROUNDWATER

Soluble pesticides and their degradation products can readily enter the groundwater beneath the areas in which they are used. One recent study showed that the concentration of metolachlor (an herbicide widely used to control weeds and grass in croplands) in community water-supply wells in near-surface aquifers in New York state was directly related to the amount of agricultural land within a half-mile radius of the wells.

Concentrations of certain pesticides in some areas have caused restrictions on groundwater use. The degradation products of aldicarb (an insecticide and nematicide used on potato and bean crops) had affected more than 4,000 homeowner wells in agricultural areas of eastern Long Island, New York in 1979. Homeowners in the affected areas were thus required to find alternative sources of water or install costly water-treatment systems.

Land Use and Water Quality

Runoff of urban stormwater carries litter and other debris in addition to sediment and chemicals. A storm drain carries these substances to its outlet, which could be a nearby river, wetland, recharge basin, estuary, or (as shown here) an ocean beach.

WHAT ARE CAFOS?

Concentrated Animal Feeding Operations (CAFOs) are agricultural operations where large numbers of livestock or poultry are kept and raised in buildings or confined feedlots that occupy a relatively small land area. Manure and wastewater from CAFOs can contribute pollutants such as nitrogen and phosphorus, organic matter, sediment, pathogens, heavy metals, hormones, antibiotics, and ammonia to the environment.

Under the federal Clean Water Act, CAFOs are required to implement comprehensive nutrient management plans that meet the requirements of the National Pollutant Discharge Elimination System (NPDES) permit program. The nutrient management plans are designed to ensure that CAFOs are sited, constructed, and managed in an environmentally sound manner that prevents the discharge of manure and wastewater to nearby waterbodies due to accidents or excessive rain.

Surface-Water Contamination and Land Use

A variety of natural and human factors can affect the quality and use of streams, lakes, and rivers, known as surface water. One of the most important factors that can affect the quality of a surface-water body is the land use within its watershed. A number of studies have shown that the density of population and housing can affect the concentration of chloride, nitrate, and a variety of pesticides in streams that drain urban and suburban settings. For example, studies of the water supplies that serve New York City have shown that the lowest chloride and nitrate concentrations occur in water from areas dominated by forested land, whereas elevated chloride and nutrient concentrations occur in areas with high densities of housing with septic systems. Chloride concentrations in streams also were linked to the wintertime application of road salt in the populated areas.

Pollution Sources and Contaminant Pathways. Pollution sources that affect surface water may be separated into two categories: point and nonpoint. Point sources include sewage treatment plants, industrial discharges, or any other type of discharge from a specific location (commonly a pipe) into a stream. By contrast, nonpoint sources—which include runoff from lawns, roads, or fields—are diffuse sources of contaminants that are not as easily identified or measured as point sources. Typically, the contaminant concentration from nonpoint sources will increase as flow increases during storm runoff; conversely, concentrations from point sources generally decrease through dilution during storm runoff. The type and severity of these pollution sources often are directly related to human activity, which can be quantified in terms of the intensity and type of land use and the associated densities of humans and livestock in source-water areas.

Contaminants can travel from a variety of sources through multiple pathways into nearby stream channels or lakes (see previous illustration). Thus, scientific assessments of the origin of a nonpoint-source contaminant can be difficult because its source(s) usually is dispersed throughout a landscape. For example, chloride (a component of salt) can have multiple sources

Soil erosion and water runoff from cropland into nearby streams can be a major source of sediment, nutrients, and pesticides in watersheds dominated by agricultural land. This photograph shows poor cropland management in which the tilled field extends to the edge of an unvegetated (and eroding) streambank. Implementation of soil and water conservation measures, such as buffer strips of undisturbed land between cropland and adjacent streams, can provide an effective control that reduces contaminant entry into aquatic systems.

in a watershed, including road salt and sewage. Chloride can be flushed into streams when snow melts from roads where it has been applied, or it can move underground from septic systems, through the groundwater system, and into nearby streams; it also can be discharged from wastewater treatment plants directly into streams. In a similar way, nitrogen that is applied as lawn fertilizer and also is present in sewage can travel from lawns, septic systems, and sewage pipes to streams.

Sediment. Sediment is eroded and transported mostly during heavy rainfall events and the associated high streamflows, particularly floods. Sediment can become a problem because its deposition in streams and lakes can ruin the **habitat** for aquatic plants and animals; it also can fill stream channels, lakes, and harbors, which then require costly dredging. Studies have shown that the amount of **suspended** sediment in surface-water bodies can be related to natural factors such as soil type and geology. In general, however, the most important factor for sediment transport is the amount of land cleared of vegetation. Sediment sources typically are lacking in developed areas, but during tillage or construction, when little vegetative cover or pavement exists, the exposed soil can be easily eroded during storms and deposited in downstream waterways.

Sediment in rivers and lakes also is a concern because many contaminants can attach (adsorb) and move with the sediment particles. Different types of contaminants can be transported with sediment, such as phosphorus (a nutrient which can cause excessive plant growth in rivers and lakes) and persistent organochlorine compounds such as **PCBs** and **DDT**. A national study of DDT in stream sediments recently showed that nearly 30 years after DDT was banned, this persistent insecticide is still present in nearly 40 percent of the streams surveyed. DDT most often was associated with sediment in streams that drain urban areas, although it was also found in streams in agricultural areas.

Wastewater Discharges. Discharge of wastewater from municipal sewage treatment plants, industrial and commercial sources, and confined animal

habitat: the environment in which a plant or animal grows or lives; the surroundings include physical factors such as temperature, moisture, and light, together with biological factors such as the presence of food and predators

suspended: describes a particulate remaining in a fluid for a long period of time because of its slow settling velocity in water or air; for example, a fine-grained sediment remaining suspended in water, or a fine-grained volcanic ash remaining suspended in the upper atmosphere

PCBs: abbreviation for polychlorinated biphenyls, chemicals once commonly used as insulator fluid for electric condensers and as an additive for high-pressure lubricants

DDT: abbreviation for dichlorodiphenyltrichloroethane, a colorless, odorless, water-insoluble, crystalline pesticide that acts as a nerve poison and is effective at killing insects; it tends to accumulate in ecosystems, and has toxic effects on many vertebrates; use as a pesticide is now prohibited in the United States

feedlots can contain a variety of contaminants that may impair the quality of the receiving waters. Municipal sewage, for example, contains high concentrations of organic compounds that may seriously deplete the dissolved oxygen content of water downstream from the discharge. Depleted oxygen levels, combined with elevated concentrations of ammonium that are typically found in the treated wastewater, can be toxic to benthic (bottom-dwelling) fauna and fish. Municipal wastewaters also contain significant amounts of phosphorus and nitrogen, which can cause **eutrophication** of lakes and **estuaries**. The volume (and the environmental consequence) of these wastewater discharges often is directly related to urbanization within the contributing watershed.

eutrophication: the process by which lakes and streams become enriched, to varying degrees, by concentrations of nutrients such as nitrogen and phosphorus; enrichment results in increased plant growth (principally algae) and decay, the latter of which reduces the dissolved oxygen content; highly eutrophic conditions may be considered undesirable depending on the human use of the waterbody

estuary: a tidally influenced coastal area in which fresh water from a river mixes with sea water, generally at the river mouth; the resulting water is brackish, which results in a unique ecosystem

Emerging Contaminants

The environmental occurrence of recently studied "emerging contaminants" includes human and veterinary pharmaceuticals, industrial and household wastewater products (such as caffeine, detergent byproducts, and insect repellants), and reproductive and steroidal hormones. Reconnaissance studies on a national scale have shown that many organic wastewater contaminants can persist in waterbodies far downgradient of their discharge points, which commonly are found in cities and livestock production areas. Studies since the 1990s have shown that concentrations of these organic contaminants were typically low, often at trace-levels, but they frequently were found in rivers and aquifers that supply drinking water.

The toxicological significance is unknown for many of these contaminants, particularly for the effects of long-term exposures at low levels. Continued monitoring of these and other emerging contaminants will provide additional knowledge about their presence and movement in water. Further analysis of land use and the associated human population and livestock densities in watersheds will provide an additional tool for scientists as they work to define and control the diverse sources of water contaminants. SEE ALSO AQUIFER CHARACTERISTICS; CHEMICALS: COMBINED EFFECT ON PUBLIC HEALTH; CHEMICALS FROM AGRICULTURE; CHEMICALS FROM CONSUMERS; CHEMICALS FROM PHARMACEUTICALS AND PERSONAL CARE PRODUCTS; EROSION AND SEDIMENTATION; GROUNDWATER; LAKE MANAGEMENT ISSUES; LANDFILLS: IMPACT ON GROUNDWATER; LAND-USE PLANNING; POLLUTION OF GROUNDWATER; POLLUTION OF LAKES AND STREAMS; POLLUTION OF STREAMS BY GARBAGE AND TRASH; POLLUTION SOURCES: POINT AND NONPOINT; SEPTIC SYSTEM IMPACTS; SUPPLIES, PROTECTING DRINKING-WATER; WASTEWATER TREATMENT AND MANAGEMENT.

David A. V. Eckhardt

Bibliography

Barbash, J. E., and E. A. Resek. *Pesticides in Groundwater—Distribution, Trends, and Governing Factors.* Chelsea, MI: Ann Arbor Press, Inc., 1996.

Eckhardt, D. A., and P. E. Stackelberg. "Relation of Ground-Water Quality to Land Use on Long Island, New York." *Ground Water* 33, no. 6 (1995):1019–1033.

Kolpin, D. W. et al. "Occurrence of Pesticides in Shallow Ground Water of the United States: Initial Results of the National Water Quality Assessment Program." *Environmental Science & Technology* 32, no. 6 (1998):558–566.

Kolpin, D. W. et al. 2002. "Pharmaceuticals, Hormones, and Other Organic Wastewater Contaminants in U.S. Streams, 1999–2000: A National Reconnaissance." *Environmental Science & Technology* 36, no. 6 (2002):1202–1211.

Nolan, B. T., and J. D. Stoner. "Nutrients in Ground Water of the Conterminous United States, 1992–1995." *Environmental Science & Technology* 34, no. 7 (2000): 1156–1165.

Squillace, P. J. et al. "Volatile Organic Compounds in Untreated Ambient Ground Water of the United States, 1985–1995." *Environmental Science & Technology* 33, no. 23 (1995):4176–4187.

U.S. Geological Survey. "The Quality of Our Nation's Waters: Nutrients and Pesticides." Reston, VA: U.S. Geological Survey, Circular 1225 (1999).

Internet Resources

Daughton, Christian G., ed. *Pharmaceuticals and Personal Care Products (PPCPs) as Environmental Pollutants: Pollution from Personal Actions, Activities, and Behaviors.* U.S. Environmental Protection Agency. <http://www.epa.gov/nerlesd1/chemistry/pharma/>.

Emerging Water-Quality Issues Investigations. U.S. Geological Survey, Toxic Substances Hydrology Program. <http://toxics.usgs.gov/regional/emc.html>.

National Water Quality Assessment (NAWQA) Program. U.S. Geological Survey. <http://water.usgs.gov/nawqa/>.

Land-Use Planning

Land and water resources are essential for farming, grazing, forestry, wildlife, tourism, urban development, transport infrastructure, and other environmental functions. The increasing demand for land, coupled with a limitation in its supplies, is a major cause for more conflicts over land use throughout the world.

The Watershed Perspective

Each type of land use has a varying effect on the **hydrologic cycle**, thereby affecting the people and the natural resources on a landscape. A watershed perspective can be used to scientifically study the effect of land uses on water and downstream **ecosystems**. A watershed is defined as a topographically delineated area drained by a stream system; that is, the total land area above some point on a stream or river that drains past that point.

A watershed acts as a receiver, collector, and conveyer of precipitation on a landscape. Land uses affect these pathways by altering surface runoff and **groundwater** infiltration, thereby changing the quantity and quality of water resources.

Impacts and Benefits of Land Uses

Natural vegetation, such as forest cover, is usually the most benign of land uses, with higher infiltration and reduced runoff rates. The opposites of forest cover are urbanized areas, where large surface areas are impermeable, and pipes and sewer networks augment the natural channels. The impervious surfaces in urban areas reduce infiltration and can reduce the **recharge** of groundwater. In addition, urban runoff contributes to poor water quality.

Agricultural activities are major forms of land use, including row crops, rangelands, animal farms, **aquaculture**, and other agribusiness activities. Cropping activities involve soil and water manipulation through tillage and **irrigation**, thereby affecting runoff water and groundwater resources. If improperly used, fertilizer and plant protection chemicals in agricultural operations can affect water resources and ecosystems.

hydrologic cycle: the solar-driven circulation of water on and in the Earth, characterized by the ongoing transfer of water among the oceans, atmosphere, surface waters (lakes, streams, and wetlands), and groundwaters

ecosystem: the community of plants and animals within a water or terrestrial habitat interacting together and with their physical and chemical environment

groundwater: generally, all subsurface (underground) water, as distinct from surface water, that supplies natural springs, contributes to permanent streams, and can be tapped by wells; specifically, the water that is in the saturated zone of a defined aquifer

recharge: the process by which precipitation infiltrates below the surface and replenishes an aquifer

aquaculture: the science, art, and business of cultivating marine or fresh-water animals or plants under controlled conditions

irrigation: the controlled application of water for agricultural or other purposes through human-made systems; generally refers to water application to soil when rainfall is insufficient to maintain desirable soil moisture for plant growth

Land-Use Planning

Farmland continues to be lost to nonagricultural uses as population and development pressures increase. This cropland in Pennsylvania is being encroached by housing developments.

Urban and agricultural land uses contribute to what is termed nonpoint-source pollution in watersheds. Nonpoint-source pollution is defined as diffuse (spread-out) sources of contamination from a wide area of a landscape, often difficult to be attributed to a single location. Transportation infrastructure (e.g., roads and airports) is another type of land use that affects water resources through road runoff and alterations to components of the hydrologic cycle.

Benefits. Despite land-use impacts, land is required to support human and ecosystem needs. Urban areas promote economic growth and satisfy housing, industrial, and commercial needs of growing human populations. Agricultural land is critical to provide food and fiber to growing populations, and is an important source of employment in many countries. Forest areas provide raw materials for housing and the lumber industry, and are important habitats for wildlife. **Wetlands** and waterbodies cover land and are important in sustaining aquatic habitat and water supplies. Coastal fisheries, which are influenced by land-based activities, provide commercial and recreational opportunities. Thus, the basic needs of food, water, fuel, clothing, and shelter are met from the land, which increasingly is becoming limited in supply.

wetland: an area that is periodically or permanently saturated or covered by surface water or groundwater, that displays hydric soils, and that typically supports or is capable of supporting hydrophytic vegetation

What Is Land-Use Planning?

As population and human aspirations increase, land becomes an increasingly scarce resource, calling for land-use planning. Land-use planning is important to mitigate the negative effects of land use and to enhance the efficient use of resources with minimal impact on future generations.

Land-use planning is defined as a systematic assessment of land and water potential, alternatives for land use, and the economic and social condi-

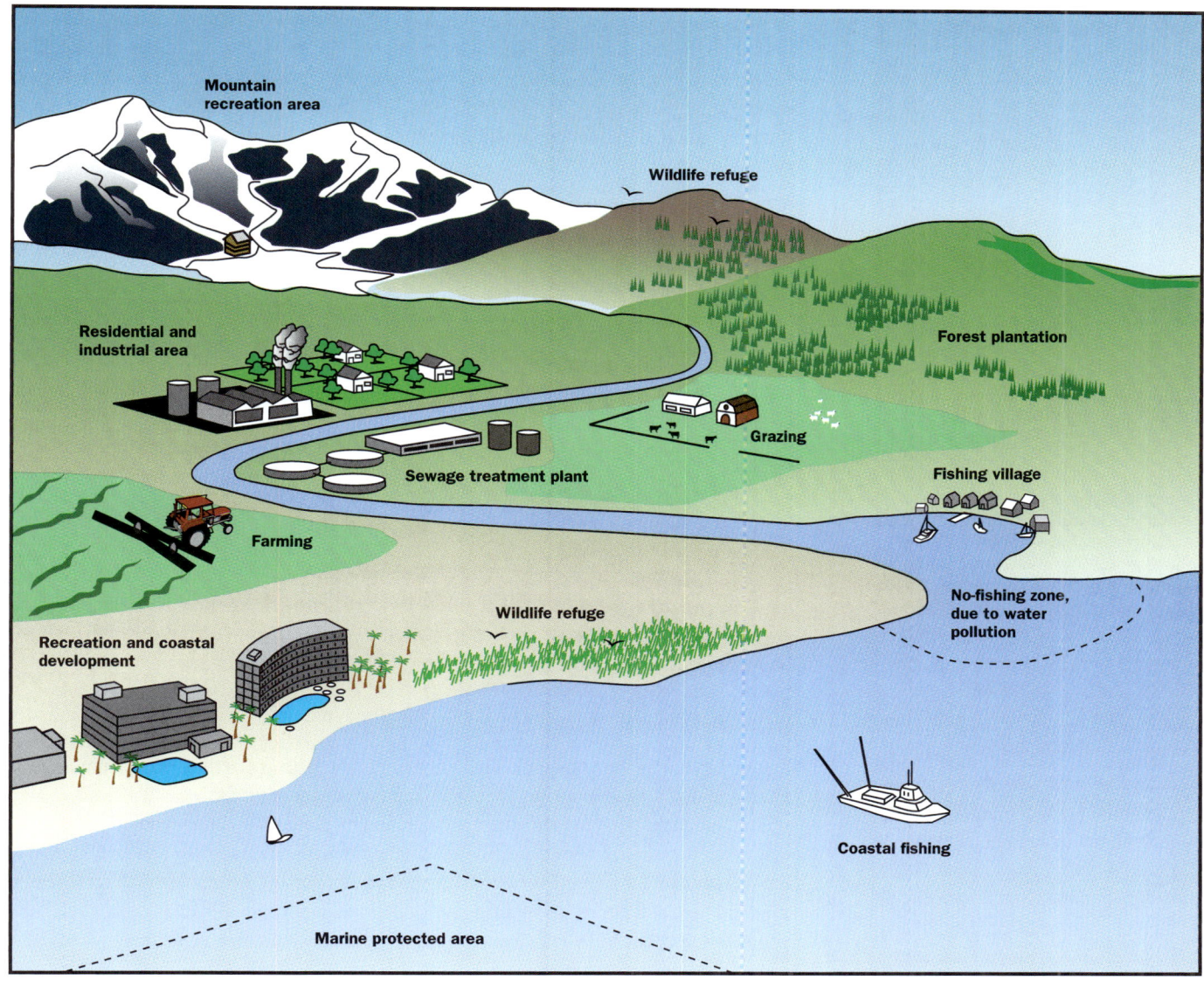

Land-use planning can help coordinate various land uses in a watershed, thereby minimizing conflicts and sustaining water quantity and quality for future generations. This schematic represents typical land uses and conservation strategies that might be found in a coastal mountain watershed.

tions required to select and adopt the best land-use options. The main objective of this planning process is to **allocate** land uses to meet the needs of people while safeguarding future resources.

The planning process is iterative (cyclically repetitive) and continuous, and three goals are used to develop a plan: efficiency, equity, and sustainability. Efficiency in land use is achieved by matching different land use with areas that will yield the greatest benefit at the least cost. Equity in land use focuses on reducing inequalities in income, food security, and housing. Sustainable land use meets the needs of the present while conserving resources for future generations.

Land-use planning aims at achieving a balance among these goals through the use of information on trade-offs, appropriate technology, and consensus-based decision-making. Effective land-use planning often involves local communities, scientific information on land resources, appropriate technologies, and integrated evaluation of resource use.

allocate: to distribute resources for a particular purpose

stakeholder: an individual or group impacted by a potential decision or action; term is usually associated with a limited number of individuals representing the interests of other like-minded individuals or groups

zoning: usually a legislative process by which a county or city is divided into separate zones or districts, each with its own unique requirements; this process can serve many purposes, including preservation of open spaces and prioritization of land uses

GIS AND NATURAL RESOURCE MANAGEMENT

Geographic information systems (GIS) technology is an information system designed to work with data referenced by spatial or geographic coordinates. GIS technology is capable of assembling, storing, manipulating, and displaying spatially and geographically referenced information (e.g., data identified according to its location). Users also regard the total GIS as including operating personnel and the data that go into the system, as well as a set of operations for analyzing such data.

GIS is an important aspect for helping government and commercial organizations manage natural resources more efficiently. GIS helps the water resource professional evaluate arrays or "layers" of information, and hence provides a tool for conducting tasks ranging from watershed planning to assessments of global climate change.

Levels and Process. Planning can be at various levels: local, town, district, state, regional, national, and international. A two-way link between these levels is important for successful planning. A "bottom-up" type of planning starts at the local level and links to the next higher level with active local participation. Local acceptability of the plan is a critical element of a successful plan.

A typical planning process involves the following steps:

- Establishing goals and a baseline;
- Inventorying and organizing resources;
- Analyzing problems;
- Establishing priorities and alternatives;
- Checking for land suitability;
- Evaluating alternatives and choosing the best option;
- Developing a land-use plan;
- Consulting and implementing the plan; and
- Revising the plan.

It is important that local people and **stakeholders** be involved in all steps of the planning process to make it a successful plan. This will also ensure local acceptability and effective use of local information.

The Future of Land-Use Planning

New ways of effective land-use planning include information management through GIS (geographic information systems), computer simulation, and spatial–temporal data modeling on present land use, alternative scenarios, and assessment of consequences. While **zoning** and regulation are the primary methods adopted by land-use planners, public education often is a neglected area that is increasingly being recognized. Other methods that planners use include economic incentives, institutional reform, and investment through multiagency cooperative projects.

Land-use planning is becoming complex and multidisciplinary as planners face multiple problems that need to be addressed within a single planning framework. Such problems include nonpoint-source pollution, water allocation, urbanization, ecosystem deterioration, global warming, poverty and unemployment, deforestation, desertification, farmland deterioration, and low economic growth. Watershed-scale planning is gaining popularity among communities and agencies so that biological, physical, and socioeconomic components of the landscape system can be integrated into the planning framework. SEE ALSO CHEMICALS FROM AGRICULTURE; FLOODPLAIN MANAGEMENT; FOOD SECURITY; GEOSPATIAL TECHNOLOGIES; LAND USE AND WATER QUALITY; PLANNING AND MANAGEMENT, WATER RESOURCES; POLLUTION SOURCES: POINT AND NONPOINT; PUBLIC PARTICIPATION; WASTEWATER TREATMENT AND MANAGEMENT; WETLANDS.

Timothy Randhir

Bibliography

Brooks, Kenneth et al. *Hydrology and Management of Watersheds.* Ames: Iowa State University Press, 1997.

Food and Agriculture Organization. *Guidelines for Land-Use Planning*. FAO Development Series 1. Rome, Italy: Food and Agriculture Organization, 1993.

Loganathan, D., D. Kibler, and T. Grizzard. "Urban Stormwater Management." In *Water Resources Handbook*, ed. Larry W. Mays. New York: McGraw-Hill, 1996.

Makepeace, D. K., D. W. Smith, and S. J. Stanley. "Urban Stormwater Quality: Summary of Contaminant Data." *Critical Reviews in Environmental Science and Technology* 25 (1995):93–129.

Landfills: Impact on Groundwater

Solid waste landfills are a necessity in modern-day society, because the collection and disposal of waste materials into centralized locations helps minimize risks to public health and safety. Solid waste landfills, which are regulated differently than **hazardous waste** landfills, may accept a variety of solid, semi-solid, and small quantities of liquid wastes. Landfills generally remain open for decades before undergoing closure and postclosure phases, during which steps are taken to minimize the risk of environmental **contamination**.

Municipal solid waste (MSW) landfills accept nonhazardous wastes from a variety of sources, such as households, businesses, restaurants, medical facilities, and schools. Many MSW landfills also can accept contaminated soil from gasoline spills, conditionally exempted hazardous waste from businesses, small quantities of hazardous waste from households, and other **toxic** wastes. Industrial facilities may utilize their own captive landfill (i.e., a solid waste landfill for their exclusive use) to dispose of nonhazardous waste from their processes, such as sludge from paper mills and wood waste from wood-processing facilities.

The Concern Over Landfill Impacts

Although landfills are an indispensable part of everyday living, they may present long-term threats to **groundwater** and also **surface waters** that are hydrologically connected. In the United States, federal standards to protect groundwater quality were implemented in 1991 and required some landfills to use plastic liners and collect and treat **leachate**. However, many disposal sites were either exempted from these rules or grandfathered (excused from the rules owing to previous usage).

Although the federal rules marked a significant improvement in the management of solid waste, some think that these rules do not go far enough. There is an increasing belief among solid waste experts that unless further steps are taken to detoxify landfilled materials, today's society will be placing a burden on upcoming generations to address future landfill impacts. Much of the concern revolves around leachate, the watery solution that results after water passes through a landfill.

Leachate Generation and Composition.
The precipitation that falls into a landfill, coupled with any disposed liquid waste, results in the extraction of the water-soluble compounds and particulate matter of the waste, and the subsequent formation of leachate. The creation of leachate, sometimes deemed "garbage soup," presents a major threat to the current and future quality of groundwater. (Other major threats include underground

hazardous waste: any solid, liquid, or gas that, when disposed, exhibits the characteristics of ignitability, corrosivity, reactivity, or toxicity, as well as any industrial waste that has been specifically listed in the federal regulations as having hazardous properties

contamination: impairment of the quality of water or the environment by natural or human-made substances to a degree that is considered undesirable for certain uses; this term usually implies a human or environmental health threat, but some types of contamination are merely nonaesthetic rather than harmful

toxic: describes chemical substances that are or may become harmful to plants, animals, or humans when the toxicant is present in sufficient concentrations

groundwater: generally, all subsurface (underground) water, as distinct from surface water, that supplies natural springs, contributes to permanent streams, and can be tapped by wells; specifically, the water that is in the saturated zone of a defined aquifer

surface water: water found above ground and open to the atmosphere, such as the oceans, lakes, ponds, wetlands, rivers, and streams

leachate: liquid that has moved through a substance, removing solids from the substance, generally by dissolution

TYPICAL LEACHATE QUALITY OF MUNICIPAL WASTE

Excludes volatile and semi-volatile organic compounds

Parameter	Typical Range (milligrams per liter, unless otherwise noted)	Upper Limit (milligrams per liter, unless otherwise noted)
Total Alkalinity (as $CaCO_3$)	730–15,050	20,850
Calcium	240–2,330	4,080
Chloride	47–2,400	11,375
Magnesium	4–780	1,400
Sodium	85–3,800	7,700
Sulfate	20–730	1,826
Specific Conductance	2,000–8,000 µmhos/cm	9,000 µmhos/cm
Total Dissolved Solids	1,000–20,000	55,000
Chemical Oxygen Demand	100–51,000	99,000
Biological Oxygen Demand	1,000–30,300	195,000
Iron	0.1–1,700	5,500
Total Nitrogen	2.6–945	1,416
Potassium	28–1,700	3,770
Chromium	0.5–1.0	5.6
Manganese	Not detected – 400	1,400
Copper	0.1–9.0	9.9
Lead	Not detected – 1.0	14.2
Nickel	0.1–1.0	7.5

SOURCE: Based on Canter et al. (1988), McGinley and Kmet (1984), and Lee and Jones (1991)

---

anion: an ion that has a negative charge

cation: an ion that has a positive charge

chemical oxygen demand: abbreviated as COD, the amount of oxygen required to degrade the organic compounds of wastewater; the larger the COD value of wastewater, the more oxygen the waste discharge demands from a waterbody

pH: a measure of the acidity of water; a pH of 7 indicates neutral water, with values between 0 and 7 indicating acidic water (0 is very acidic), and values between 7 and 14 indicating alkaline (basic) water (14 is very alkaline)

volatile organic compounds: organic compounds that can be isolated from the water phase of a sample by purging the water sample with inert gas, such as helium, and subsequently analyzed by gas chromatography

heavy metals: a group of metals that have high density and are considered toxic at specified concentrations; such metals include copper, iron, manganese, molybdenum, cobalt, zinc, cadmium, mercury, nickel, and lead

aquifer: a water-saturated, permeable, underground rock formation that can transmit significant quantities of water under ordinary hydraulic gradients to wells and springs

plume: a concentrated area or mass of a substance that is emitted from a natural or human-made point source and that spreads in the environment; a plume can be thermal, chemical, or biological in nature

unsaturated zone: the zone between the ground surface and the water table where pore spaces contain both air and water

saturated zone: an area where pore spaces within the soil are entirely filled with water

hydraulic gradient: the change in hydraulic head between two points (e.g., the difference in water level between two points divided by the distance between the two points)

---

storage tanks, abandoned hazardous waste sites, agricultural activities, and septic tanks.)

Leachate composition varies relative to the amount of precipitation and the quantity and type of wastes disposed. In addition to numerous hazardous constituents, leachate generally contains nonhazardous parameters that are also found in most groundwater systems (see above table). These constituents include dissolved metals (e.g., iron and manganese), salts (e.g., sodium and chloride), and an abundance of common **anions** and **cations** (e.g., bicarbonate and sulfate). However, these constituents in leachate typically are found at concentrations that may be an order of magnitude (or more) greater than concentrations present in natural groundwater systems.

Leachate from MSW landfills typically has high values for total dissolved solids and **chemical oxygen demand**, and a slightly low to moderately low **pH**. MSW leachate contains hazardous constituents, such as **volatile organic compounds** and **heavy metals**. Wood-waste leachates typically are high in iron, manganese, and tannins and lignins. Leachate from ash landfills is likely to have elevated pH and to contain more salts and metals than other leachates.

Leachate Release and Migration. A release of leachate to the groundwater may present several risks to human health and the environment. The release of hazardous and nonhazardous components of leachate may render an **aquifer** unusable for drinking-water purposes and other uses. Leachate impacts to groundwater may also present a danger to the environment and to aquatic species if the leachate-contaminated groundwater **plume** discharges to wetlands or streams.

Once leachate is formed and is released to the groundwater environment, it will migrate downward through the **unsaturated zone** until it eventually reaches the **saturated zone**. Leachate then will follow the **hydraulic gradient** of the groundwater system.

Monitoring wells at landfills allow scientists to determine whether contaminants in leachate are escaping into the local groundwater system. The wells are placed downgradient of the landfill at appropriate depths and at various intervals to intercept any contaminants and monitor their movement.

A number of forces may act on or react with the migrating leachate, resulting in changes of chemistry and a general reduction of strength from the original release. These forces are physical (filtration, sorption, advection, and dispersion), chemical (oxidation–reduction, precipitation–dissolution, adsorption–desorption, hydrolysis, and ion exchange), and biological (microbial degradation). The extent of these reactions depends on the materials underlying the landfill, the hydraulics of the groundwater system, and the chemistry of the leachate.

Although many of these reactions have the capability to reduce the potential impact to groundwater, some (such as microbial degradation) can actually increase the toxicity by producing by-products that are more hazardous than the original contaminant. This can be seen, for example, in the creation of vinyl chloride from the degradation of trichloroethene.

Old and New Viewpoints. Today's landfills are constructed with liners that contain leachate, and leachate collection systems that collect it. But historically, many landfills were constructed without liners or leachate collection systems.

Two philosophies previously existed regarding the placement of unlined landfills. One viewpoint was to place most of these facilities in moderately permeable materials and as close as possible to rivers or streams (i.e., surface water). This type of siting would allow the landfills to slowly (and deliberately) leak leachate into the groundwater, minimizing the length and size of the leachate plume, which would ultimately discharge into surface water.

The other theory was the exact opposite: to place landfills as far away from surface water as possible. This may have spared impacts to the surface water, but locating unlined facilities in this fashion usually resulted in creating significant leachate-contaminated groundwater plumes that followed the direction of groundwater flow.

Are Design Standards Enough? The U.S. Environmental Protection Agency issued standards for MSW landfill design and operation in 1991. These new rules were adopted to protect groundwater from the release of leachate by MSW landfills. Landfill owners who could not meet these new design standards requiring plastic liners and leachate collection were required to close and to conduct groundwater monitoring for 30 years. Groundwater protection standards were developed for all MSW landfills, setting a national precedent for solid waste management that may eventually become the standard for all types of landfills.

As the twenty-first century opened, there was a controversy whether the new rules for MSW landfills offer adequate environmental protection, especially as these facilities age. Landfill designs that utilize plastic liners below the waste and then are covered with plastic when the landfill stops placing waste into the active cell (area of waste input) are referred to as "dry tombs." These engineered systems are designed to minimize leachate generation by restricting the introduction of moisture, primarily precipitation.

It is widely recognized that even the best-installed plastic liner will succumb to deterioration and eventually will allow leachate to be created and released. However, this may not happen within the required 30 years of postclosure groundwater monitoring. Moreover, it may not be detected during the time the landfill operators are actively involved and financially obligated.

Opponents of dry-tomb landfills advocate for recycling the collected leachate through the waste, which will enhance the rate of chemical reactions inside the landfill and eventually stabilize the waste material prior to covering the landfill. This could reduce the toxic nature of the waste materials in the closed landfill and minimize the future threat posed to groundwater from these facilities. SEE ALSO GROUNDWATER; POLLUTION SOURCES: POINT AND NONPOINT.

Audrey Eldridge

Bibliography

Canter, L. W., R. C. Knox, and D. M. Fairchild. *Groundwater Quality Protection.* Chelsea, MI: Lewis Publishers, 1988.

Lee, G. Fred, and A. R. Jones. "Groundwater Pollution by Municipal Landfills: Leachate Composition, Detection and Its Water Quality Significance." Proceedings of the National Water Well Association Fifth National Outdoor Action Conference, Las Vegas, NV, 1991.

Lee, G. Fred, and A. R. Jones. "Landfills and Groundwater Quality." *Groundwater* 29 (1991):482–486.

McGinley, P. M., and P. Kmet. *Formation, Characteristics, Treatment and Disposal of Leachate from Municipal Solid Waste Landfills.* Wisconsin Department of Natural Resources Special Report, 1984.

Internet Resources

"MSW Disposal." U.S. Environmental Protection Agency, Office of Solid Waste. <http://www.epa.gov/epaoswer/non-hw/muncpl/disposal.htm>.

Landslides

Landslides are natural hazards that cause millions of dollars of damage each year and also cause many deaths. They are defined as downslope movements of soil and rock under the influence of gravity. They are triggered primarily by water, but sometimes earthquakes can lead to some spectacular land-

Landslides

As a 7.6-magnitude earthquake rocked El Salvador in January 2001, this hillside above a suburban neighborhood gave way. The landslide buried hundreds of homes and accounted for over half of the nearly 700 earthquake victims.

slides. The water comes mainly from high-rainfall storms, but also can come from rapid snowmelt.

Factors

The stability of a slope can be described as two forces working against one another. Driving forces work to cause slope materials to move downslope, whereas resisting forces act to keep the materials on the slope. When the ratio of resisting forces to driving forces (called the factor of safety) is greater than 1, the slope is stable. When it is less than 1, the slope usually fails.

Water and Vegetation. Water can increase the driving forces and reduce the resisting forces. Saturation from rainfall can increase the slope mass, thereby increasing the driving forces. Filling the slope soil pores with water also reduces soil cohesion by allowing particles to pass by one another, thereby reducing the internal resistance of the soil to sliding. To reduce landslide danger on a slope, the first thing done is to remove the water.

Vegetation also is important to slope stability because it increases resisting forces through its roots, especially tree roots, that bind the soil. Trees also act as natural pumps that remove water from the soils through **evapotranspiration**, thereby increasing slope stability.

evapotranspiration: water discharged to the atmosphere as a result of evaporation from the soil and surface-water bodies and by plant transpiration

Slope and Materials. As the slope angle increases, the driving forces also increase. Few landslides occur on slopes less than 15 degrees. Cutting a road into a slope will create an oversteepened slope prone to landslides if a wall is not built. Certain slope materials also have weak strengths and low resisting forces to landslides. For example, clay, shale, serpentine, and uncompacted fill are prone to failure.

If the slope bedrock is inclined and is somewhat parallel to the slope, it is called a dip slope. Landslides are prone along failure planes (clay beds and old soils) on these dip slopes. Examples of dip-slope failure are Italy's Vaiont Dam disaster in 1963 that killed 2,600 people, and the Gros Ventre landslide that dammed Wyoming's Gros Ventre River in 1925.

To stabilize a slope or prevent landslides, one needs to lower the slope angle, drain the slope of water, build retaining walls, plant vegetation, and avoid building on old landslides. If the slope has moved once, it has a high chance of moving again.

Classification

Names are given to different landslides depending upon the process that brings the soil and rock down the slope. Falls are the free fall of detached materials, usually rocks, which descend down a steep slope. Translational slides occur along a failure surface in the bedrock and move parallel to the surface. If the sliding mass occurs along a curved plane, it usually is called a slump.

Flows. Flows occur when material moves downslope as a viscous (thick) fluid. Most of these flows are saturated with water. Fast-moving ones move as a slurry that can be as much as 70 percent water and 30 percent sediment, and these are called debris flows (see figure). If the flow contains mostly fine-grained particles like sand, silt and clay, it is called a mudflow. They frequently follow stream canyons and pose significant hazard to life and property. Velocities can reach 55 kilometers (34 miles) per hour. Earthflows are slower flows that generally originate on hillslopes as large tongues that break away from scarps (arc-shaped steep slopes cut into a hill).

Debris Avalanches. The fastest and largest landslides are called debris avalanches, because they travel at speeds up to 300 kilometers (186 miles) per hour, travel long distances sometimes in excess of 30 kilometers (18.6 miles), and are very large. The world's largest historical landslide occurred in Washington state when Mount St. Helens erupted in 1980. ✴ In Yungay, Peru, about 22,000 people were killed in 1970 when a debris avalanche descended from the volcano Nevado Huascaran, traveling a distance of 14 kilometers (8.7 miles) in only a few minutes.

✴ See "Volcanoes and Water" for a photograph of the Mount St. Helens eruption.

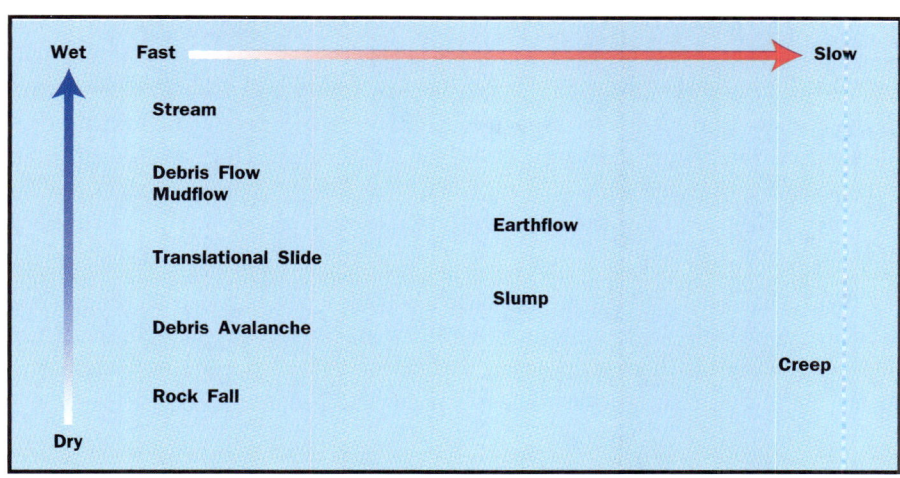

Water is critical in determining the rate and nature of landslide movement. This figure expresses the differences relating landslide classification to water content in the slide mass.

Creep. Some movement on slopes can be imperceptible, called creep. Creep is caused by gravity, and is assisted by freeze–thaw or shrink–swell processes operating on the soil sediments. Rates range from 0.1 millimeter (0.004 inch) per year to 10 centimeters (about 4 inches) per year. Evidence of movement is found in cracked walls, leaning telephone poles, and leaning trees that adjust (straighten) their direction of growth, creating so-called "trees with knees," also called pistol-butt trees.

Recognizing Landslide Terrains

Scientists "read" the landscape for signs that tell of past processes. Several diagnostic landslide features tell of past movement on the terrain. First, scarps are arcuate, steep slopes cut into the hill where the landslide has torn away. Second, hummocky topography is "bumpy" ground that has been produced by the landslide mass-weathering over time. A hummocky slope with a steep, half-moon slope at the top usually means an old landslice occurred there.

Younger landslides generally will have vegetation on the slide mass that is different than vegetation on the surrounding slopes. In coniferous forests, landslides typically are covered with deciduous trees for the first 100 years after the failure.

If one can see the actual sediments making up the landform, landslide debris and **colluvium** are unsorted, with all particles mixed together in a random fashion. Colluvium can be confused with glacial till, because both are unsorted. SEE ALSO FOREST HYDROLOGY; GROUNDWATER; VOLCANOES AND WATER.

colluvium: a general term for loose, unconsolidated material moved downslope under the influence of gravity

Scott F. Burns

Bibliography

Ritter, D. F., Kochel, R. C., and Miller, J. R. *Process Geomorphology*, 3rd ed. Dubuque, IA: Wm. C. Brown Publishers, 1995.

Turner, A. K., and R. L. Schuster, eds. *Landslides: Investigation and Mitigation*. Washington, D.C.: Transportation Research Board, National Research Council, Special Report 247 (1996).

Law, International Water

Of all the resources that people depend on, only air is more directly vital to sustaining human life than water. Deprive a person of air, and he dies in minutes. Deprive him of water, and he dies in days. Deprive him of food, and he can go on for weeks or months, depending on his physical condition at the beginning of the fast—and on whether he has adequate supplies of water.

Water is essential not only to individual humans, but also to human societies. Yet potable water is an increasingly scarce resource, particularly when viewed against the backdrop of an expanding global population, changing water technologies, and the prospect of global climate change.

Water and the Potential for Conflict

Humans have long been organized into social groups with more or less clearly defined boundaries within which they conduct their affairs. Water, however, is an **ambient** resource that neither knows nor respects human boundaries.

The world's 261 international river basins, covering 45 percent of Earth's land surface (excluding Antarctica), are shared by more than one nation.✶ Even the most cordial and cooperative of neighboring nations have found it difficult to achieve mutually acceptable arrangements to govern their transboundary **surface waters**, even in relatively humid regions where fresh water usually is found in sufficient abundance to satisfy most or all needs. When nations are located in **arid** regions, conflicts become endemic and intense despite otherwise friendly relations or even membership in a federal union. Little wonder the English language derives the word "rival" from the Latin word "*rivalis*," meaning persons who live on opposite banks of a river used for **irrigation**.

Water, it turns out, is too vital a resource for nations to fight over. Nations can, and often do, arrange to share the waters of a transboundary basin, thereby minimizing the risks of conflict. Nations have gone to war repeatedly in the twentieth century while carefully protecting each other's water facilities and respecting the promised share of a water allocation. Even in the absence of such arrangements, customary international law provides some simple rules to enable nations to coexist peacefully within a single, shared river basin.

The Customary International Law of International Waters

Customary international law arises from the practices of nations when those practices are undertaken out of a sense of legal obligation (called the *opinio juris*). The resulting law is highly decentralized and institutionally undeveloped, similar to the customary law found among subsistence farmers or nomadic tribesmen.

In order to make the evolution of customary international law clear, consider this analogy. Suppose there is a field between two villages, with no road across the field. People initially will tend to wander at will from one village to the next. Gradually, most people will follow a particular line, perhaps the shortest route, or perhaps the easiest route, or perhaps the route

---

ambient: describes an encompassing environment that is natural or nearly natural and largely unaffected by human influence (e.g., ambient air temperature)

✶ See "Hydropolitics" for a map of major international river basins.

surface water: water found above ground and open to the atmosphere, such as the oceans, lakes, ponds, wetlands, rivers, and streams

arid: describes a climate or region where precipitation is exceeded by evaporation; in these regions, agricultural crop production is impractical or impossible without irrigation

irrigation: the controlled application of water for agricultural or other purposes through human-made systems; generally refers to water application to soil when rainfall is insufficient to maintain desirable soil moisture for plant growth

most convenient to the heaviest walkers—walkers whose tread wears a path more decisively into the landscape. For whatever reason, a definite path will emerge, and gradually it will become a road. Eventually, all will agree that this road is the only right way to travel from village to village, even though no one can say precisely when this notion took hold. At that point, people will object to others as trespassers if the others choose a different path—by which time a legal norm exists and not merely a factual description of a path.

Customary international law, developing through a process of claim and counterclaim between nations, empowers international actors by legitimating their claims while also limiting the claims they are able to make. Finding the appropriate international practice is fairly easy; relevant practices can be found in a consistent set of international agreements, in votes at international assemblies, in decisions by international courts or international arbitrators, or in unilateral acts of nations. Proving that these practices arise from a sense of legal obligation rather than from mere expedience can be more difficult. As a result, international decisionmakers frequently turn to the writings of well-respected scholars of international law for evidence of such intent.

Because of the general absence of a neutral enforcement mechanism, international law often has no better method for sanctioning violations than the law of the vendetta. As a result, customary international law by itself has proven unable to solve the problem of managing transboundary water resources. Still, the pattern of national claim and counterclaim, and of national behavior intended to make such claims effective, is consistent and the ultimate outcome, except in cases of strong power imbalances among the nations sharing an international watercourse is, in general terms, entirely predictable.

Riparian Nations. All nations agree that only riparian nations—nations across which, or along which, a river flows—have any legal right, apart from an agreement, to use the water of a river. Beyond that, however, the patterns of international claim and counterclaim initially diverge sharply according to the riparian status of the nation making the claim.

The upper-riparian nations initially base their claims on *absolute territorial sovereignty*, typically claiming the right to do whatever they choose with the water regardless of its effect on other riparian nations. Downstream nations, on the other hand, generally begin with a claim to the *absolute integrity of the river*, claiming that upper-riparian nations can do nothing that affects the quantity or quality of water that flows in the watercourse. The utter incompatibility of such claims guarantees that neither claim will prevail in the end, although the process of negotiating or otherwise arriving at a solution might require decades.

The usual solution to the contradictory claims of upper and lower riparians is found in a concept known as the rule of equitable utilization. The rule of equitable utilization, based on the concept that an international drainage basin is a coherent legal and managerial unit, embodies a theory of restricted **sovereignty** under which each nation recognizes the right of all riparian nations to use water from a common source and the obligation to manage their uses so as not to interfere unreasonably with like uses in other riparian nations. Nations often allocate water under this

sovereign: possessing independent authority or power; (of a group) fully independent and able to determine its own affairs; (of affairs) subject to a specified control but without outside interference

Law, International Water

International water law helps enable nations to peacefully share a river basin and the waters it contains. The *UN Convention* (an international agreement) embraces several principles that will likely become the guiding force in managing international watercourses and resolving water conflicts.

arable: used or suitable for growing crops

theory according to some selected historic pattern of use, although occasionally some other more or less objective measure of need is advanced (e.g., population, area, **arable** land). At the extreme, the theory of equitable utilization is no more developed than the vague notion that each nation is entitled to a "reasonable share" of the water.

The UN (United Nations) Convention

That the rule of equitable utilization is required by customary international law is shown by the many treaties based on the concept, by international judicial and arbitral awards, and by the near-unanimous opinions of the most highly qualified legal scholars. The United Nations General Assembly recently codified the rule of equitable utilization in Article 5 of its United Nations Convention on the Non-Navigational Uses of International Watercourses (UN Convention). The Assembly approved the UN Convention on May 21, 1997, by a vote of 104–3. Article 5 requires watercourse nations to utilize an international watercourse in an equitable and reasonable manner with a view to attaining optimal and sustainable utilization and benefits consistent with adequate protection in the watercourse. Article 5 also provides that watercourse nations shall participate in the use, development, and protection of an international watercourse in an equitable and reasonable manner. The right to participate includes both the right to utilize the watercourse and the duty to cooperate in its protection and development.

No-Harm Rule. The UN Convention also originally embraced a second principle, termed the "no-harm rule," in its Article 7. Article 7 was very controversial during the drafting process because it seemed to contradict the rule of equitable utilization in Article 5. The final version of Article 7 makes clear that the "no-harm rule" is subordinate to the rule of equitable utilization. Article 7 requires watercourse nations, in utilizing an international watercourse, to take all "appropriate measures" to prevent the causing of significant harm to other watercourse nations. If significant harm nevertheless is caused to another watercourse nation, the nation whose use causes such harm must, in the absence of agreement for the use, take all appropriate measures, having due regard for the provisions of Articles 5 and 6 (equitable utilization) in consultation with the affected nation, to eliminate or mitigate the harm and, where appropriate, to discuss the question of compensation.

When the UN Convention comes into force (after 35 ratifications), it will unquestionably be the law governing internationally shared fresh waters at least among those nations ratifying it. It appears to be an accurate statement of the international customary law as well. This reading was confirmed by the International Court of Justice in its ruling on the *Danube River Case (Hungary v. Slovakia)* in 1997.* The Court's opinion referred twice to the rule of equitable utilization and did not mention the "no-harm rule." The Court's failure even to mention the "no-harm" rule despite Hungary's heavy reliance on the principle in its pleadings confirms that the rule of equitable utilization is primary, and that avoidance of harm is to be considered only in analyzing whether a particular use or pattern of use is equitable.

* See "Transboundary Water Treaties" for a photograph related to the Danube River case involving Hungary and Slovakia.

Equitable Sharing of Waters. What amounts to an "equitable" share of the waters of an international water basin often is not clear. Some have argued that "equitable" sharing must mean equal sharing. The merest perusal of the standards for equitable utilization demonstrates that while equal access is guaranteed, equal shares are not. The standards are found in Article 6 of the UN Convention, which contains a long list of relevant factors:

- The geographic, hydrographic, hydrologic, climatic, ecological, and other factors of a natural character;
- The social and economic needs of the watercourse nations concerned;
- The effects of the use or uses of the watercourse in one watercourse nation on other watercourse nations;
- The existing and potential uses of the watercourse;
- The conservation, protection, development and economy of use of the water resources of the watercourse and the costs of measures taken to that effect; and
- The availability of alternatives, or corresponding value, to a particular planned or existing use.

Non-lawyers, particularly engineers and hydrologists, sometimes see in this list of factors a poorly stated equation. By this view, if one simply fills in numerical values for each factor, one could somehow calculate each watercourse nation's share of the water without reference to political or other non-quantitative variables. However, the UN Convention is a legal document that ultimately is addressed to judges. Judges make judgments, and in English at least, the word "judgment" carries a strong connotation that the result is not dictated in any immediate sense by the factual and other inputs that the judge relies upon in exercising judgment. In short, any attempt to treat the list of relevant factors as an equation simply misses the point entirely.

Thus, even when each interested nation always agrees to the rule of equitable utilization, nations would still dispute what should be the common standard for sharing and the proper application of the agreed standard. The rule of equitable utilization is simply too general and too vague

A CASE STUDY OF THE NILE RIVER

The Nile Valley nations perfectly epitomize the scenario set out in the text regarding the role of customary international law in resolving international disputes over water. Egypt is not a wealthy nation; its per capita gross domestic product is only US$630/year, making it one of the poorer nations in the Middle East. Yet Egypt is wealthier than Sudan (US$540/year), and Egypt and Sudan are far wealthier than Ethiopia (US$120/year). As is commonly the case throughout the world, the lower basin nations are wealthier and more highly developed than the upper basin states. Yet without a common border, Egypt cannot easily pose a military threat to Ethiopia or otherwise set about to impose its will directly. Ethiopia, on the other hand, gets the water first. One might think, therefore, that Ethiopia is in a position simply to do as it chooses, regardless of the effect on downstream states. Ethiopia, however, is too poor and too poorly organized to construct the dams and related infrastructure necessary to exploit the Blue Nile without outside financial assistance.

Egypt has succeeded in exploiting its greater political importance to block international financing of Ethiopian dams and related works. As part of this diplomatic program, Egypt has freely deployed legal arguments, particularly the so-called no-harm rule. Egypt most recently did not object to a loan application by the Ethiopians for a small-scale irrigation project, suggesting that there might be some truth to rumors of a secret agreement between the two nations regarding development of the Nile.

Priority of Use. Ultimately, Egypt claims an absolute right to the integrity of the river because of the priority of their use. Priority of use, while undoubtedly relevant to an equitable allocation of water among national communities, has never been treated as absolutely controlling in international law. Any other approach would negate the concept

to be applied without the interested nations filling in the details in what remains merely an obligation of fairness. Yet without a legal resolution of disputes over water, the disputes can only lead back to the law of the vendetta.

Serious conflict in one form or another cannot be avoided under the rule of equitable utilization without a clear definition of the precise standards for managing the shared waters and a peaceful mechanism for the orderly investigation and resolution of the disputes characteristic of the rule. Most disputes over international river systems thus have eventually led to a treaty based on equitable utilization, and several hundred such treaties now have entered into force regarding internationally shared waters. The treaties reflect not only logic, but also need and power.

Future Challenges

Groundwater has emerged as a critical transnational resource. Groundwater comprises about 97 percent of the world's fresh water (apart from the polar ice caps and **glaciers**). But despite its transnational nature, groundwater and its sharing has received little international attention relative to the sharing of surface waters. The UN Convention limits its application to underground water either tributary to, or sharing a common terminus with, surface waters covered by the document's Articles.

Yet nations need to develop treaties to ensure the equitable utilization and management of internationally shared groundwater basins. Because hydrologic, economic, and engineering variables are essentially the same as those

groundwater: generally, all subsurface (underground) water, as distinct from surface water, that supplies natural springs, contributes to permanent streams, and can be tapped by wells; specifically, the water that is in the saturated zone of a defined aquifer

glacier: a huge mass of ice, formed on land by the compaction and recrystallization of snow, which moves slowly downslope or outward owing to its own weight

> **A CASE STUDY OF THE NILE RIVER** (CONTINUED)
>
> of "equitable utilization" that is the rule of customary international law. Furthermore, for priority in time to override all other values, or even to dominate other values, would hardly be conducive to achieving the developmental equity proclaimed under various banners at the United Nations. To accord such priority to existing uses in the Nile Basin would condemn Ethiopia to remain impoverished and dependent on international food aid to stave off mass starvation, for the benefit of the relatively richer Egyptians and Sudanese. In the Jordan Valley, this approach would condemn the Palestinians to remaining a colonial society utterly dependent on Israeli largesse, and would leave the Jordanians only marginally better off.
>
> The tension between protecting "historic rights" and providing for developmental equity can be managed only if the water is cooperatively managed by the several national communities in such a way as to assure equitable participation in the benefits derived from the water by all communities sharing the basin. Customary international law, in its somewhat primitive state of development, cannot by itself resolve the management problems of a region.
>
> While uncertainty of legal right can induce cooperation among those sharing a resource, it can also promote severe conflict. Nor can a partitioning of the waters be an adequate resolution when there simply is too little water to divide. To create the sort of regime necessary to allay conflict and optimize the use and preservation of the resource of the Nile will require a new treaty, one that includes all basin communities, creates appropriate representative basin-wide institutions, and has the clout to enforce its mandates. International practice provides numerous examples as models for institution design. As of early 2003, the UN Development Programme was promoting the negotiation of such a treaty for the Nile.

of surface waters, the principles regarding shared rivers likely apply to shared **aquifers**. The United Nations system has only begun to address the international legal issues relating to groundwater, as evidenced by the resolution of the International Law Association in 1994 calling for more study of the law governing the use of groundwater.

International water law is a complex topic, which grows increasingly important in a water-scarce world. Nations will continue to struggle to manage their transboundary resources and international agencies will continue to develop an integrated system of water law to guide them in this endeavor. As management challenges evolve, so too will international water law. SEE ALSO HYDROPOLITICS; INTERNATIONAL COOPERATION; LAW, WATER; TRANSBOUNDARY WATER TREATIES.

Joseph W. Dellapenna

aquifer: a water-saturated, permeable, underground rock formation that can transmit significant quantities of water under ordinary hydraulic gradients to wells and springs

Bibliography

Beach, Heather et al. *Transboundary Freshwater Dispute Resolution: Theory, Practice, and Annotated References.* New York: United Nations Publications, 2000.

Dellapenna, Joseph. *Middle East Water: The Potential and Limits of Law.* The Hague, Netherlands: Kluwer Press, 2002.

Wouters, Patricia, ed. *Codification and Progressive Development of International Water Law: The Work of the International Law Commission of the United Nations.* London, U.K.: Kluwer Law International, 2002.

Internet Resources

Convention on the Law of the Non-Navigational Uses of International Watercourses. 1997. <http://www.un.org/law/ilc/texts/nnavfra.htm>.

International Water Law Project Website. <http://www.internationalwaterlaw.org/>.

Law of the Sea

The oceans have long been viewed by societies as a wide-open free space—a vast frontier associated with adventure and mystery. In the seventeenth century, nations formalized this viewpoint into the Freedom of the Seas doctrine. This doctrine limited any nation's rights to the ocean to a narrow belt, traditionally 4.8 kilometers (3 miles), surrounding its coastline and declared the rest of the seas to be free to all nations and belonging to no one. This doctrine formalized views that the seas were such a vast resource that all nations could use them as they wished.

Dramatic growth in use of the oceans directly challenged this doctrine by the twentieth century. The ocean's resources were increasingly used for economic uses, and nations desired to extend their claims over offshore resources. Fishing fleets, transport ships, oil drilling, and military navies all relied on the seas for their success. Concern grew over the impact of these uses and many conflicts emerged between nations over rights to the resources.

Growing Conflict over Ocean Uses

In the United States, president Harry Truman directly challenged the Freedom of the Seas doctrine in 1945 by unilaterally extending the nation's jurisdiction to its coastal waters. He extended the United States rights to a wider band to include all of the resources on the **continental shelf**, including oil, gas, and minerals. Partly, he was acting in response to pressures from the U.S. oil industry that eyed profitable reserves offshore.

continental shelf: the relatively flat, submerged natural platform, about 1-degree slope, that extends seaward from the beach for about 70 kilometers (45 miles), with water depth up to 130 meters (425 feet) maximum, and ending where the slope and water depth increase

Many nations followed Truman's lead and extended their sovereign national rights to the seas as well. In 1950, Ecuador claimed rights to a 322-kilometer (200-mile) zone. Indonesia and the Philippines asserted rights over all waters separating their islands. The vast resources derived from the oceans countered the viewpoint that it was a free space. As new technologies increased human abilities to exploit those resources, conflicting claims multiplied and nations further desired to expand their territorial rights. The human relationship with the seas had dramatically changed.

United Nations Conference on the Law of the Sea. In 1967, Arvid Pardo, Malta's ambassador to the United Nations, called on the nations of the world to recognize their potential devastation of the oceans and the importance of the oceans to world peace. He pleaded for "an effective international regime over the seabed and the ocean floor beyond a clearly defined national jurisdiction." This began a 15-year process to create a management mechanism for the world's seas.

The United Nations Seabed Committee was formed and, after much preparation, met in 1973 in New York to draft an international treaty for the oceans. Nine years of negotiations between more than 160 nations over national rights and obligations ensued. In 1982, the United Nations Convention on the Law of the Sea was adopted at the Third UN Conference on the Law of the Sea. The Convention needed 60 states to sign on to ratify it, and it came into force in 1994.

While the Law of the Sea lays down a comprehensive regime to govern the world's oceans and seas, and establishes rules regarding all uses of the oceans and their resources, disputes often arise between nations. These disputes, most often over conflicting uses of ocean resources, may be taken to the International Tribunal for the Law of the Sea (ITLOS), to the International Court of Justice, or to arbitration. In 1999, for example, North and South Korea disagreed over which nation has rights to a crab fishing area in the Yellow Sea, and brought in their navy ships to patrol the disputed area.

Law of the Sea

The Law of the Sea (LOS) is a comprehensive treaty covering territorial sea limits, navigational rights, the legal status of the ocean's resources, economic jurisdictions, protection of the marine environment, marine research, and other facets of ocean management. It attempted to address the existing conflicts over the oceans. After its adoption, some called it "possibly the most significant legal instrument of this (the twentieth) century."

The treaty established legal principles governing ocean space, its uses and resources. The Law of the Sea treaty also set up a binding procedure for settling disputes between nations and established the International Tribunal for the Law of the Sea. Nations could now take other nations to court over perceived violations of international convention. The treaty also recognized the right to conduct marine scientific research. It addresses the main sources of ocean pollution: land and coastal activities, continental shelf drilling, seabed mining, ocean dumping, and vessel-source pollution.

The Law of the Sea also established the International Seabed Authority, which regulates activities in the deep seabed beyond national jurisdictions. One of the most contentious aspects of the Law of the Sea was the language dealing with the mining of minerals in the deep ocean floor, the part of the international seabed area beyond the national jurisdictions (Part XI). In 1998, an agreement was passed, formally known as the Agreement Related to the Implementation of Part XI of the Convention. This Agreement in jointly implemented with the LOS.

> **EXCLUSIVE ECONOMIC ZONE**
>
> An Exclusive Economic Zone (EEZ) is one of the tools defined by the Convention on the Law of the Sea. In the eighteenth century, the cannon-shot rule governed a nation's claims to territorial seas, based on the distance that projectiles could be fired from a cannon onshore—at that time, about 4.8 kilometers (3 miles). The Law of the Sea built on this idea and expanded the zone to 322 kilometers (200 miles).
>
> Today, nations have the right to develop, manage, and conserve all resources in waters, on the ocean floor, and in the subsoil in the area extending 322 kilometers (200 miles) from its shore. These EEZs bring many benefits to countries, because lucrative fishing, oil, and other reserves often lie within these zones.

The Ocean's Future. The joining of the world's countries to protect the oceans through the Law of the Sea Convention signals society's growing recognition of the importance of the oceans to life on Earth. The Law of the Sea is a good example of intergovernmental cooperation to protect an important resource from global pressures. The ocean's future depends on the abilities of nations to implement effective governance. SEE ALSO CONFLICT AND WATER; LEGISLATION, FEDERAL WATER; MINERAL RESOURCES FROM THE OCEAN; PETROLEUM FROM THE OCEAN; SUSTAINABLE DEVELOPMENT.

Faye Anderson

Bibliography

Wang, James C. F. 1991. *Ocean Politics and Law: An Annotated Bibliography*. Westport, CT: Greenwood Publishing Group, 1991.

Internet Resources

Atlas of the Oceans. United Nations. <http://www.oceansatlas.org/index.jsp>.

International Tribunal for the Law of the Sea. <http://www.itlos.org/>.

NOAA Ocean Page. National Oceanic and Atmospheric Administration. <http://www.noaa.gov/ocean.html>.

Ocean Affairs and the Law of the Sea. United Nations. <http://www.un.org/Depts/los/index.htm>.

Law, Water

Water law is a system of enforceable rules that controls the human use of water resources. In the United States, these rules are created by statutes, court decisions, and administrative regulations. Much of U.S. water law is rooted in the **common law** system inherited from England. Under this system, courts resolved disputes by setting legal precedents that were followed by subsequent courts. Today, U.S. water law is a complex mix of federal and state regulations superimposed on a system of public and private water rights.

Resolving Water-Use Conflicts

Because water is a mobile (moving) resource, many management problems are created that require legal resolution. As water moves through the **hydrologic cycle**, many people in succession can use it. For example, a hydropower plant can nonconsumptively use river water to generate electricity. Far downstream from the powerplant, and at a later date, a golf-course owner may use the same water to irrigate the fairways.

Consumptive uses alter the hydrologic cycle and may modify the environment. Because the same drop (or even the same molecule) of water can be reused many times by humans and also is needed to maintain environmental integrity, conflicts between different uses are frequent. Water law attempts to resolve these conflicts by encouraging desirable uses and discouraging undesirable ones. Uses are encouraged or discouraged through the complex interlinkings of water rights systems and the exercise of government power.

From Private Rights to Public Values. The exercise of government power and creation of water rights is based on societal **values** that have changed over time and vary from state to state. Although the U.S. common law system recognized **public rights** to navigate and to fish, most water rights his-

---

common law: a body of rules and principles based on court decisions, traditional usage, and precedent, rather than legislative enactments comprising codified written laws; contrast with statutory law

hydrologic cycle: the solar-driven circulation of water on and in the Earth, characterized by the ongoing transfer of water among the oceans, atmosphere, surface waters (lakes, streams, and wetlands), and groundwaters

consumptive use: a use which lessens the amount of water available for another use; for example, water that is used for development and growth of plant tissue or consumed by humans or animals

values: abstract concepts of what is right and wrong, and what is desirable and undesirable

public right: a right given to the public's common need, such as public rights to water (e.g., using surface waters for navigation); contrast with private (property) rights

torically were defined in terms of **private rights**. These private rights allocated water to individuals for their use. When conflicts occurred, they were resolved in court.

This water rights system oriented toward private rights often ignored public values such as recreation and environmental protection. But as societal values changed, the exercise of government power at times restricted traditional water rights.

In past decades, for example, many cities felt they had a right to use rivers for the routine disposal of raw sewage. But today, federal statutes prohibit such disposal and require sewage first be treated before the processed wastewater is discharged to waterways. In the American West, the right to remove all the water from a stream for use in irrigation was traditionally accepted. But today, federal statutes may require a minimum **instream flow** to protect **endangered** species or to maintain other designated instream uses. Public values have greater recognition today than in past decades, and are widely recognized by courts, legislation, and administrative regulations.

Relationship Between Rights.

Either state or federal laws can create public and private water rights. These water rights are relational and take on meaning when the exercise of one right conflicts with the exercise of other rights.

In addition, water rights are rarely exclusive. This need to share water results in conflicts between individuals who have private rights, between individuals with private rights and people with public rights, and between those who have federal rights and those who have state rights. Water law is used to resolve conflicts between different claimants by determining the rights and obligations of each party in a dispute.

Water Rights and the Hydrologic Cycle

Private water rights evolved as a pragmatic system to **allocate** water use in different parts of the hydrologic cycle. In a simplified hydrologic cycle, water can be classified as **groundwater**, surface water, atmospheric water, or soil moisture.✱

Each part of the hydrologic cycle is treated by the legal system as if it were disconnected. For example, the law establishing rights to surface water is often different from the law for groundwater. In some parts of the hydrologic cycle, private water rights are difficult to establish. For example, soil moisture cannot be easily extracted from the soil in which it is found. This type of water is treated as land, and no water rights are created. Although this practice ignores the reality of the hydrologic cycle, it does reduce potential conflicts with those who own land.

Until recently, capturing atmospheric waters was difficult. Property rights generally did not exist until the water reached the land surface. Although increasing precipitation through cloud-seeding is possible, assigning a water right to the "newly created water" is still problematic. Proving the amount of precipitation increase and determining where this amount actually fell makes the basis for a claim difficult to establish.

Groundwater and Surface-Water Rights.

Groundwater science developed after early policymakers had already established a system of groundwater rights. Although these common-law approaches still exist, statutory regulations have been superimposed, modifying groundwater rights substantially.

Water law plays a major role in properly allocating limited water resources. Sometimes state and federal courts must resolve disputes between claimants who are exercising what they consider to be their water rights.

private right: in terms of property rights, a right held by the owner of a resource (e.g., water) that entitles them to access, withdraw, manage, exclude others (from using the resource), and sell their ownership to someone else

✱ See "Hydrologic Cycle" for a schematic of the water cycle.

instream flow: the amount of water remaining in a stream, without diversions, that is required to satisfy a particular aquatic environment or water use, such as the water required for fish and wildlife or for navigation

endangered: describes a plant or animal species threatened with extinction by human-made or natural changes throughout all or a significant area of its range; designated in accordance with the 1973 Endangered Species Act

allocate: to distribute resources for a particular purpose

groundwater: generally, all subsurface (underground) water, as distinct from surface water, that supplies natural springs, contributes to permanent streams, and can be tapped by wells; specifically, the water that is in the saturated zone of a defined aquifer

One initial approach was to treat groundwater like part of the land, and give the owner of the land surface absolute ownership of the water below it. Another pragmatic approach, the so-called "rule of capture," was used on a limited basis. This right gave ownership to the person who pumped the water from the ground. Although these two approaches ignored the mobile nature of groundwater, the rights were easy to determine.

Today, the mobility of groundwater is accounted for in the dominant systems of water rights. In the eastern states, the doctrine of reasonable use allows water to be shared between surface owners. In the western states, the doctrine of **prior appropriation** establishes priorities between users.

prior appropriation: a concept or doctrine in water law under which the first person to take a quantity of water and put it to beneficial use has a higher priority of right than a subsequent user; that is, "first in time is first in line;" contrast with riparian water rights

natural flow: a doctrine developed by some riparian rights states that would require all water to be left in a watercourse

Surface-water rights have the most developed set of water allocation laws. In the eastern states, the ownership of land adjacent to a river gives rise to a riparian water right. Initially, the right was attached to the riparian land and only that land. But over time, two kinds of riparian water rights evolved: **natural flow** and reasonable use.

Today, statutory permit systems often are imposed on private riparian rights. In the West, the prior appropriation doctrine evolved under different conditions. The need to move water long distances led to a system of temporal (time-based) preferences designed to protect investments. The right to water did not automatically come with ownership of land. Moreover, western states claimed they either owned all the water within the state, or they held it in trust for the people of the state. Thus, a private water right could be established only by following the requirements of state law.

This complex system of water laws controls almost all aspects of water use, including environmental protection. The system has never been static and will continue to evolve. SEE ALSO CONFLICT AND WATER; LAW, INTERNATIONAL WATER; PRIOR APPROPRIATION; REISNER, MARC; RIGHTS, PUBLIC WATER; RIGHTS, RIPARIAN.

Olen Paul Matthews

Bibliography

Balleau, W. P. "Water Appropriation and Transfer in a General Hydrogeologic System." *Natural Resources Journal* 28 (1988):269–291.

Beck, Robert E., ed. *Water and Water Rights.* Charlottesville, VA: The Michie Company, 1991.

Matthews, Olen Paul. *Water Resources, Geography and Law.* Washington, D.C.: Association of American Geographers, 1984.

Reisner, Marc. *Cadillac Desert.* New York: Penguin Books, 1987.

Trelease, Frank J. "Government Ownership and Trusteeship of Water." *California Law Review* 45 (1957):638.

Worster, Donald. *Rivers of Empire.* New York: Pantheon Books, 1985.

Legislation, Federal Water

Federal involvement in water resource legislation initially addressed issues of water use, such as managing the commons (e.g., regulating fisheries) and regulating **navigable** waterways to support navigation and commerce. As the country grew westward, water legislation was used to fund massive water development projects to increase water supplies for irrigation, hydroelectricity, flood control, and municipal and industrial water supply. Finally, as water quality degraded across the country, federal legislation moved into

navigable: in general usage, describes a waterbody deep and wide enough to afford passage to small and large vessels; also can be used in the context of a specific statutory or regulatory designation

areas previously controlled by state and local governments by mandating nationwide water quality standards for waterways and standards for public drinking water. The following overview of selected statutes demonstrates this trend of legislation.

Navigation and Water Supply

The River and Harbors Act of 1899 banned dumping of nonliquid waste in navigable rivers without permission from the U.S. Army Corps of Engineers. Although focused on keeping solid trash out of waterways to prevent navigational obstructions and to reduce pier fires caused by burning trash, this act often is cited as the first act related to water quality.

The Reclamation Act of 1902 was enacted to build irrigation projects in the dry western states and territories. The act also created the Reclamation Service (now called the U.S. Bureau of Reclamation) to build dams and delivery systems that would supply the irrigation projects. An assured water supply in this **arid** region was needed to ensure successful settlement of the West so the government could maintain control over these sparsely populated lands. The Bureau has built hundreds of dams in the West, with communities and farmlands still enjoying affordable, federally subsidized water from these water projects.

The Federal Water Power Act of 1920 created the Federal Power Commission, which would control **hydroelectric** dam construction on all navigable rivers. This allowed the federal government to regulate private electric utilities by requiring operating licenses. The government also built hydroelectric dams in order to provide affordable electricity and encourage private electric companies to offer comparable prices.

Water Quality

The slow degradation of American water quality was first addressed by the Federal Water Pollution Control Act of 1948, which addressed water pollution from solid and liquid wastes. Prior to this act, water quality had been considered only a local concern. Because the U.S. Congress never appropriated money to fund the grants for planning studies and sewage treatment plants, this act was ineffective.

The 1956 amendments gave federal authority over water quality in interstate streams (i.e., streams flowing across multiple states). The 1961 amendments added that some water stored in federal **reservoirs** could be used to dilute **pollution** on some streams, but that sewage treatment plants still needed to be built. The 1965 amendments, also known as the Water Quality Act, created the first water standards and mandated a water quality assessment program of the nation's waters. However, these standards were not effective and were not enforced.

Clean Water Act.
Increasing water-quality problems eventually resulted in the 1972 amendments to the Federal Water Pollution Control Act, commonly known as the Clean Water Act. The objective of this act was to restore and maintain the physical, chemical, and biological integrity of the nation's waters. Waters were to become fishable and swimmable by 1983, and receive zero discharge of pollutants by 1985. **Toxic** chemicals in the nation's waters were recognized as a major concern.

Federal water-related legislation runs the gamut from waterborne commerce to watershed restoration. The twentieth century witnessed an evolution from a few narrowly defined interests to broader, more integrated approaches.

arid: describes a climate or region where precipitation is exceeded by evaporation; in these regions, agricultural crop production is impractical or impossible without irrigation

hydroelectric: often used synonymously with "hydropower," describes electricity generated by utilizing the power of falling water, as with water flowing through and turning turbines at a dam

reservoir: a pond, lake, basin, or tank for the storage, regulation, and control of water; more commonly refers to artificial impoundments rather than natural ones

pollution: any alteration in the character or quality of the environment, including water in waterbodies or geologic formations, which renders the environmental resource unfit or less suited for certain uses

toxic: describes chemical substances that are or may become harmful to plants, animals, or humans when the toxicant is present in sufficient concentrations

Legislation, Federal Water

The Clean Water Act created specific water quality standards, and started the National Pollutant Discharge Elimination System, wherein discharges from a point source into any waterway were required to have a permit. Even with these mandated water quality standards, more than 40 percent of assessed waters today still do not meet the standards set for them. The act requires states to develop lists of impaired waters, to develop priority rankings for action, and to develop total maximum daily loads (TMDLs) for these waters. A TMDL sets a specific amount of a pollutant that a waterbody can receive from both **point sources** and **nonpoint sources**. The U.S. Environmental Protection Agency must approve these lists.

Billions of dollars were spent in the 1970s funding sewage treatment plant construction and improvement. The 1987 amendments to the Clean Water Act added more money for sewage treatment plants; created **stormwater** regulations for treating nonpoint-source pollution from urban areas; and focused more attention toward nonpoint-source pollution prevention programs.

Drinking-Water Standards. The Safe Drinking Water Act of 1974, amended in 1986 and 1996, was the first federal law mandating drinking-water standards for all public water systems, ranging from big cities down to roadside campgrounds. Under the original legislation, public water systems were required to follow water quality standards for particular **contaminants**. Water systems must be tested for these contaminants and, if necessary, the water is treated to reduce contaminants to the maximum contaminant levels (MCLs) set for each contaminant. The amendments added new contaminants and new programs to bolster the protection of public health. The U.S. Environmental Protection Agency has primary enforcement responsibility.

Water Resources Planning

Although national water resources planning had been tried in the early twentieth century, albeit in several weak formats, a new attempt was made in the 1960s after the U.S. Bureau of Reclamation proposed building two more dams on the Colorado River. The Water Resources Planning Act of 1965 created federal–state river basin commissions to promote better water resources planning and to support economic development. It also created the Water Resources Council, a federal integrated water resource planning entity that would coordinate the seven river basin commissions. The Council would also provide biennial water resources assessments, review Army Corps of Engineers plans for navigable waters, promote research, and provide public education. This council was comprised of the secretaries of Agriculture, Interior, Army, Health, and Education, as well as the chairman of the Federal Power Commission.

The Water Resources Council produced the Principles and Guidelines, which were decision-making guidelines for evaluating federally funded water projects and **allocating** water not just by considering its economic highest use, but also by incorporating social values for how water and land were used. The Principles and Guidelines replaced the so-called "Green Book" of 1950 (whose formal name was *Proposed Practices for Economic Analysis of River Basin Projects*), the standard manual used for cost–benefit analysis of water projects that did not include social values. The council ceased to exist in 1981 when federal funding was eliminated.

---

point source: a pollutant release or discharge originating from one specific location (e.g., an outfall pipe from a factory) rather than over a wide land area (e.g., water runoff from a farm field)

nonpoint source: a pollutant release or discharge originating from a land use active over a wide land area (e.g., agriculture) rather than from one specific location (e.g., an outfall pipe from a factory)

stormwater: runoff from precipitation events in which precipitation rate exceeds infiltration rate or falls directly on an impermeable surface; stormwater often is discharged directly to streams and may carry pollutants such as bacteria, petroleum products, and metals

contaminant: as defined by the U.S. Environmental Protection Agency, any physical, chemical, biological, or radiological substance in water, including constituents that may not be harmful to the environment or human health

allocate: to distribute resources for a particular purpose

WHO IS RESPONSIBLE FOR WATER RESOURCES?

The vastness of the United States, the differences in its regional climates, and the complexity of water resources in the face of ever-changing demands make water management too complex to be left to one governmental agency. Hence, numerous agencies at the international, federal, state, and local levels must all play a part in water resources management. Moreover, every American citizen is responsible to care for the nation's finite resource. The daily actions of individuals can either enhance or hinder water conservation and quality protection; hence, the ultimate responsibility is a collective one.

Development and Restoration

The Water Resources Development Act of 1986 was different from previous omnibus (multi-component) water bills in that it emphasized a "beneficiary pays" principle wherein increasing amounts of money were to be contributed by the beneficiary in the planning and building of a project. Newly authorized projects under this act would have more local burden, because they would now have to share the project's cost. This had the effect of scaling back or killing some water projects. The act also created the National Dam Safety Program. There have been many amendments to the Water Resources Development Act, which serve to authorize new water projects built primarily by the U.S. Army Corps of Engineers.

Some recent federal legislation is trying to lessen the impacts of past environmental damage from large water projects. For example, the Water Resources Development Act of 2000 authorized the South Florida Ecosystem Restoration and the Comprehensive Everglades Restoration Plan to replenish the Everglades with some of the water it lost through previous Army Corps of Engineers development projects. Those projects had diverted water from Lake Okeechobee to Miami and other coastal cities, depriving the Everglades of its yearly flows. SEE ALSO ARMY CORPS OF ENGINEERS, U.S.; BUREAU OF RECLAMATION, U.S.; CLEAN WATER ACT; EVERGLADES; FLORIDA, WATER MANAGEMENT IN; HYDROELECTRIC POWER; INTEGRATED WATER RESOURCES MANAGEMENT; LEGISLATION, FEDERAL WATER; LEGISLATION, STATE AND LOCAL WATER; PLANNING AND MANAGEMENT, HISTORY OF WATER RESOURCES; PLANNING AND MANAGEMENT, WATER RESOURCES; POLICY-MAKING PROCESS; POLLUTION SOURCES: POINT AND NONPOINT; RIVER BASIN PLANNING; SAFE DRINKING WATER ACT; TRANSPORTATION; WASTEWATER TREATMENT AND MANAGEMENT.

Laurel E. Phoenix

Bibliography

Rogers, Peter. *America's Water: Federal Roles and Responsibilities.* Cambridge, MA: MIT Press, 1993.

Thompson, Stephen A. *Water Use, Management, and Planning in the United States.* San Diego, CA: Academic Press, 1999.

Legislation, State and Local Water

Surface water and **groundwater** are transboundary resources that often cross political boundaries. Generally, under the United States' federal system of government, no single governmental level has absolute **sovereign** authority over water. Effective implementation of water policies requires coordination among all levels of government, various administrative commissions, and regional independent agencies.

A complex legal and administrative framework controls how federal, tribal, state, and local governments share legal authority over water quality and quantity, as well as over broader water development and management issues. This system is based on **common law**, constitutional and statutory law, local custom, judicial decisions, and international treaties. Accordingly, different governmental levels have primary authority over certain water issues. The federal government has undisputed sovereignty to develop and

surface water: water found above ground and open to the atmosphere, such as the oceans, lakes, ponds, wetlands, rivers, and streams

groundwater: generally, all subsurface (underground) water, as distinct from surface water, that supplies natural springs, contributes to permanent streams, and can be tapped by wells; specifically, the water that is in the saturated zone of a defined aquifer

sovereign: possessing independent authority or power; (of a group) fully independent and able to determine its own affairs; (of affairs) subject to a specified control but without outside interference

common law: a body of rules and principles based on court decisions, traditional usage, and precedent, rather than legislative enactments comprising codified written laws; contrast with statutory law

Legislation, State and Local Water

interstate: existing or carried on between states

intrastate: existing or carried on within a state

instream flow: the amount of water remaining in a stream, without diversions, that is required to satisfy a particular aquatic environment or water use, such as the water required for fish and wildlife or for navigation

> **WHAT IS SOVEREIGNTY?**
>
> Sovereignty is the independence of a national, tribal, state, or local government combined with their right to regulate internal affairs without external approval. Under the United States' federalist system, sovereignty over water is scattered among governmental levels.
>
> Theories of federalism explain relationships between different governmental levels as their degree of sovereignty has and continues to shift to address priority water issues. A flexible view of sovereignty is necessary to understand the willingness of federal and state governments to delegate some of their sovereign authority to various regional institutions, such as federal–interstate compacts and interstate compacts formed to allocate or manage water.

land-use planning: a generic term for a wide range of legislative and regulatory activities intended to limit or direct land development for the purpose of making its usage sustainable; large-scale land-use plans often are implemented by local zoning and land-use ordinances

zoning: usually a legislative process by which a county or city is divided into separate zones or districts, each with its own unique requirements; this process can serve many purposes, including preservation of open spaces and prioritization of land uses

reclamation: in historical use, the process of converting land to a more desired use, such as draining a marsh for human development

manage navigation on **interstate** or international bodies of water used for commerce. Conversely, primarily state or local governments govern **intrastate** water quantity and quality issues. Between these extremes, each level of government vigorously guards its authority as the balance of power fluctuates to meet changing water priorities.

Riparian Doctrine versus Prior Appropriation

Across the United States, states have different water issues and concerns. States have considerable authority to establish and implement water laws, policies, and programs suited to their priority water concerns. State authority is especially paramount in allocating water rights.

Water-rich eastern states adopted the common-law riparian doctrine. This doctrine allows riparian landowners (landowners adjacent to a river or stream) to use surface water for any locally recognized beneficial purpose. Conversely, arid western states enacted the prior appropriation doctrine to allow nonriparian landowners to establish rights to divert surface water for use outside riparian lands.

Faced with growing demands for limited water, other states' legislatures modified the riparian doctrine to allow the state to obtain water rights to protect **instream flows** and other public purposes. Finally, in certain situations, the U.S. Supreme Court recognizes a federal reserved water right associated with selected types of federal public land regardless of state-granted water rights.

Regional and Local Water Authority

State legislatures have considerable authority to develop and manage water. Legislatures realize that local governments and agencies need legal authority to tailor solutions to meet local water development priorities. Hence, over the years, states implemented numerous administrative approaches to water development and management.

A historical perspective is helpful to identify general water development and management themes, as well as mechanisms for regional and local governments and agencies to cooperate with federal agencies. During the westward expansion of the United States, Congress adopted laws, policies, and programs for water development and management to encourage settlement and agriculture. State legislatures responded by delegating some of their water authority to regional or local governments and agencies.

In many instances, regional county or local city governments would have significant roles in water activities because of their authority over **land-use planning** and **zoning**. For unincorporated rural areas, regional or local agencies were necessary to undertake **reclamation** and development projects. Some common agencies were: drainage districts to drain swamps and wetlands; levee boards or flood-control districts to provide flood protection; and irrigation districts to provide water for crops.

Over time, these regional and local agencies frequently gained additional authority to pursue broader water management responsibilities. Frequently legislatures authorized additional new types of agencies such as water supply, treatment, or management districts or authorities, and soil and water conservation districts to meet emerging water issues. Often a formal statewide water policy would help guide and coordinate these agencies' activities.

Legislatures commonly used two different approaches to create these new agencies. One method was to pass general legislation outlining procedures communities had to follow to create and set the boundaries for a specific type of agency, the agencies' duties and powers, and funding mechanisms. Legislatures could create special-purpose, multi-county or regional water agencies. Regardless of how they were created, many of these agencies have overlapping jurisdictions: some possess broad powers, whereas others are limited to a single purpose. Unlike most statewide agencies, many of these water agencies' boundaries correspond roughly with natural hydrologic boundaries (e.g., drainage basin boundaries).

Cooperation of Various Water Authorities

Despite the individual states' water authority, a comprehensive approach to water development, management, and regulation requires interagency, interstate, and federal–state coordination and cooperation. The federal government has numerous programs that offer scientific, technical, and financial incentives for cooperative federal, state, and local water activities. For instance, the National Flood Insurance Program is available to state and local governments that are willing to use their land-use planning and zoning authority to manage floodplain development. Similarly, many federal flood-control laws and soil and water conservation laws require regional or local agency partners or project sponsors. The state legislation authorizing regional and local water authority usually expressly lists the federal agencies and programs with whom these agencies can cooperate.

The governance of state waters sometimes can be a source of debate. Here a protestor opposes the proposed sale of Oklahoma water to Texas via a possible state–tribal compact.

The programs described above provide mechanisms for coordinated intrastate **watershed** management. However, the federal government has a much larger role in coordinated interstate river-basin-scale management. Assorted institutional mechanisms are available to blend national, state, and local authority to address interstate water issues. The Tennessee Valley Authority (TVA) is a unique federal approach to coordinated basin management. Congress created the TVA by statute in 1933 as an independent public management corporation with broad authority to construct and operate dams and reservoirs for electric power generation, flood control, and to improve navigation on the Tennessee River.

watershed: the land area drained by a river and its tributaries; also called river basin, drainage basin, catchment, and drainage area

The use of regional **cooperative agreements** such as the Environmental Protection Agency's Chesapeake Bay Program is another approach. The region's states voluntarily agreed to formally participate in this cooperative program, which is expressly authorized by the federal Clean Water Act to develop and implement coordinated plans to improve and protect the bay's water quality and living resources.

cooperative agreement: an agreement, typically voluntary, entered into by parties to achieve common goals

compact: a formal contract or agreement between two or more parties, often governmental units

The use of federal–interstate **compacts** such as the Delaware River Basin Commission is another alternative. Such compacts typically require participating governments to transfer some of their sovereign authority to a **commission**. Congress and state legislatures must ratify agreements authorizing such a transfer. Compacts often allocate water between states, but can also be formed to meet many water management needs. Many other formal and informal **intergovernmental** institutional arrangements are in place or are

commission: as in river basin commission, an independent regional body established to manage and coordinate federal, state, and local water management policies in a river basin, particularly regional basins and basins that cross state or international boundaries; river basin commissions almost always are created by treaty or legislation that outline the commission's mission, duties, and authority to carry out those duties

intergovernmental: existing or carried on between governmental bodies

being explored to meet emerging water issues in different regions of the nation. SEE ALSO CALIFORNIA, WATER MANAGEMENT IN; CHESAPEAKE BAY; CONFLICT AND WATER; DROUGHT MANAGEMENT; FLOODPLAIN MANAGEMENT; FLORIDA, WATER MANAGEMENT IN; LAND-USE PLANNING; LAW, WATER; LEGISLATION, FEDERAL WATER; PRIOR APPROPRIATION; RIGHTS, PUBLIC WATER; RIGHTS, RIPARIAN; RIVER BASIN PLANNING; TENNESSEE VALLEY AUTHORITY.

Jeffery A. Ballweber

Bibliography

Adams, David A. *Renewable Resource Policy The Legal–Institutional Foundations*. Washington, D.C.: Island Press, 1993.

Adler, Robert W. "Addressing Barriers to Watershed Protection." *Environmental Law* 25 (1995):973–1106.

Dworsky, Leonard B. et al. "Water Resources Planning and Management in the United States Federal System: Long Term Assessment and Intergovernmental Issues." *Natural Resources Journal* 31 (1991):475–547.

Houck, Oliver A. *The Clean Water Act TMDL Program: Law, Policy and Implementation*. Washington, D.C.: Environmental Law Institute, 2000.

Reuss, Martin, ed. *Water Resources Administration in the United States Policy, Practice and Emerging Issues*. East Lansing, MI: American Water Resources Association and Michigan State University Press, 1993.

Rosen, Howard and Martin Reuss, eds. *The Flood Control Challenge: Past, Present and Future*. Chicago, IL: Public Works Historical Society, 1988.

Leonardo da Vinci

Italian Artist and Scientist
1452–1519

Leonardo da Vinci is history's foremost Renaissance man, a master of both art and science. Da Vinci is best known as the artist who created such masterpieces as the *Mona Lisa, Madonna of the Rocks,* and *The Last Supper.* Yet he was also a brilliant scientist, architect, engineer, and inventor. In fact, he was one of the best scientific minds of the Renaissance period, carrying out sophisticated research in fields ranging from architecture and civil engineering to astronomy and anatomy. The dynamics of water and the study of **hydraulics** were prominent among his many interests.

A Brief Chronology

Leonardo da Vinci was born in 1452 in the hills of Tuscany. As a child, he was a gifted artist and became an apprentice in one of the best art studios in Italy. By 1478, Leonardo became an independent master. He was hired into the court of Ludovico Sforza (Duke of Milan) when he was 30 years old and served as a painter, sculptor, musician, architect, and engineer. He also served as the principal engineer in the Duke's numerous military endeavors. Leonardo returned to Florence in 1499 when the Duke's family was driven from Milan by French forces.

In 1502, da Vinci entered the service of Duke Ceasare Borgia as his chief architect and engineer. There he completed many of his art and architectural works. The French governor summoned him back to Milan in 1506 where he became the court painter to King Louis XII of France, who had taken Milan from the Sforzas. He continued his many engineering projects during this time.

hydraulics: the scientific study of water in motion; modern hydraulics emphasizes the mechanical properties of water that describe the specific pattern and rate of movement in the natural environment or in artificial systems (for example, pipe systems)

From 1514 to 1516, Leonardo lived in Rome under the patronage of Pope Leo X, and concentrated on his scientific studies while in the Vatican. Then in 1516, da Vinci traveled to France to enter the service of King Francis I. He died in France on May 2, 1519.

Notebooks

During the Renaissance period, there was a shift from purely philosophical endeavors toward observational or **empirical** science. Da Vinci's observational skills and data recording efforts were exceptional. His scientific legacy is found in his *Notebooks*. These handwritten manuscripts (approximately 8,000 pages survive) were apparently meant to be a great encyclopedia of knowledge, but, like many of his projects, it was never finished.

empirical: based on experience or observations, as opposed to reason or conjecture

All of da Vinci's notes are written backwards, reportedly so that only someone intelligent enough to realize this fact could read them. It makes the manuscripts difficult to read, as does his use of peculiar spellings and abbreviations, and the lack of logical ordering and arrangement of the entries. For these reasons, the magnitude of the impact of his scientific work was not fully understood until later in the nineteenth century.

The Codex Leicester, written between 1506 and 1510, is the only notebook manuscript by da Vinci that is still privately owned, and the only one kept in America. (Bill Gates, Microsoft Corporation's chairman and chief software architect, paid $30.8 million in 1994 for the Codex Leicester manuscript.) This codex (unbound manuscript) was found in 1690 in an old chest in storage in Rome. Seventy-two pages in all, the Codex Leicester is a record of Leonardo's thoughts on a wide variety of topics, from astronomy to hydrodynamics, and includes his observations and theories related to the nature and properties of water. As in the rest of his notebooks, its pages feature his signature mirror writing.

Hydraulics. Leonardo da Vinci's *Notebooks* reveal that the subject of hydraulics was his most frequently studied and recorded topic. Da Vinci made the first empirical studies of streams and their velocity distribution. He used a weighted rod held afloat by an inflated animal bladder. Da Vinci traced the velocity distribution across the stream's channel by releasing the rod at different places in the stream's cross-section. His inventiveness in devising scientific experiments was well ahead of his time.

Leonardo had plenty of time to observe nature during his years of service to the Duke of Milan (1482–1499). It is reported that he was an expert on the rocks and **fossils** found in northern Italy. He was fascinated by the idea of moving mountains or piercing them with tunnels. His *Notebooks* are full of observations he made on mountains and rivers, and they reveal that he understood the principle of **sedimentation**. He explained how rocks could be formed by the deposition of sediments by water, while at the same time rivers erode rocks and carry their sediments to the sea in a grand continuous cycle.

fossil: a preserved plant or animal imprint or remains

sedimentation: in geology and geomorphology, a process in which sediment is transported and deposited in a new location

Hypotheses. During the Renaissance, there were several **hypotheses** on why shells and fossilized forms of living creatures were found in rocks on the tops of mountains. Some believed the shells to have been carried there by the Biblical Flood, while others thought that these shells had grown in the rocks. Leonardo disliked both of these explanations and refuted them

hypothesis: a statement made about the condition or behavior of a variable or event that lends itself to rigorous testing for validity

Leonardo da Vinci

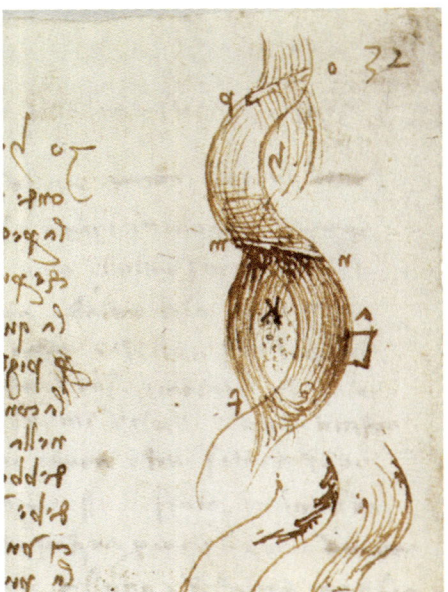 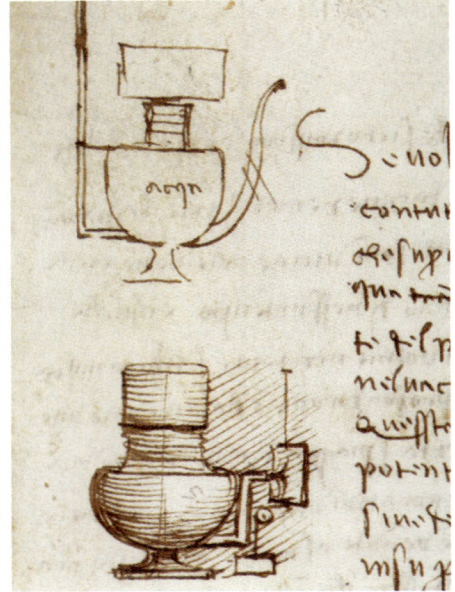

These two sheets from the Codex Leicester show Leonardo da Vinci's firm grasp of hydraulics. On left, he provides a geometric analysis of river flow and riverbank erosion. On right, he illustrates a water pressure device amidst a broader discussion of the weight and power of water. The handwriting is deliberately backwards, making it difficult to decipher.

based upon his careful observations. Leonardo doubted the existence of a worldwide flood, noting that there would have been no place for the water to go when the floodwaters receded. His observations recorded that rain falling on mountains rushed downhill, not uphill, and this suggested that any Great Flood would have carried fossils away from the land, not towards it. As for the second hypothesis of his time, he disputed it by noting the evidence that these shells had once been living organisms and therefore could not have grown without access to food, which, as shells, they would not have had if anchored in the rocks.

Leonardo's answer to how shells came to be found on the mountaintops was very close to our modern understanding. Fossils were once-living organisms that were been buried at a time before the mountains were raised. He wrote, "It must be presumed that in those places there were sea coasts, where all the shells were thrown up, broken, and divided . . . ." In other words, where there is now land, there once was ocean. Much of his knowledge and observations on flooding dynamics came from the Arno River in northern Italy.

Arno Canals

In Arno, daVinci worked with Niccolo Machiavelli (1469–1527) on his lifelong dream of building a system of canals that would make the Arno River navigable from Florence to the sea. Machiavelli was a well-known political thinker of the Renaissance and author of *The Prince*. The treatise stood apart from all other political writings of that period in that it focused on the practical problems a ruler faced in retaining power, rather than the more speculative issues that explained the foundation of political authority and the pursuit of ideals.

In addition to being a great engineering feat, the canal project had economic and military purposes. Da Vinci envisioned irrigating the Arno valley and selling water to farmers to make money for the government. If they succeeded, da Vinci and Machiavelli would have transformed Florence into a major world power of the time. But in 1504, their plan failed after a flood destroyed much of their work. Some say Leonardo's obsession with this project is the motivation for the view of this valley in the background of the *Mona Lisa*, and it also drove his lifelong quest to understand the dynamics of water.

da Vinci's Legacy

Leonardo da Vinci's contribution to hydraulics and the understanding of water resources is not often the first thing historians associate with his brilliant life. Yet his wide-ranging interests and efforts to gather data to understand the world around him led to many significant advancements in knowledge, and remain an example to today's scientists, thinkers, and visionaries. SEE ALSO CANALS; INFRASTRUCTURE, WATER-SUPPLY; WATER WORKS, ANCIENT.

Faye Anderson

Bibliography

Masters, Roger D. *Fortune is a River: Leonardo da Vinci and Niccolo Machiavelli's Magnificent Dream to Change the Course of Florentine History.* New York: The Free Press, 1998.

Mays, Larry W. *Water Resources Handbook.* New York: McGraw-Hill, 1996.

Internet Resources

Leonardo. National Museum of Science and Technology, Milan. <http://www.museoscienza.org/english/leonardo/Default.htm>.

Leonardo da Vinci. Leonardo da Vinci Museum. <http://museum.brandx.net/main.html>.

Welcome to the Louvre Museum. The Louvre Palace and Museum, Paris. <http://www.louvre.fr/>.

Leopold, Luna

American Engineer, Environmentalist, Geologist, and Hydrologist
1915–

Luna Bergere Leopold, the son of the famous wildlife ecologist Aldo Leopold (who wrote the environmental classic, *A Sand County Almanac*) has been described by the American Geological Institute as one of the most distinguished earth scientists of the last half of the twentieth century. He was a leader in the development of a better and more **quantitative** understanding of streams and the landscapes they form.

quantitative: describes the measurement of the quantity of something rather than its quality

Education and Career

Luna Leopold's educational background reflects broad multidisciplinary interests. He received a bachelor's degree in civil engineering, a master's degree in physics and meteorology, and a doctor of philosophy (Ph.D.) degree in geology from Harvard University. This background paved the way for his creative and innovative approach to the science of **hydrology**. Throughout his career, water resources remained of vital concern to him.

Leopold is perhaps best known for his 22-year career in the Water Resources Division of the U.S. Geological Survey, where he converted his

hydrology: the science that deals with the occurrence, distribution, movement, and physical and chemical properties of water on Earth

Luna Leopold is best known for his integration of engineering, science, and policy, but his legacy extends to environmental protection, land-use management, and water-control policies. He is shown here in 1979 near Pinedale, Wyoming.

geomorphology: the scientific study of the physical characteristics of the land surface and landforms that are the result of specific geologic processes

landform: a discernible natural landscape that exists as a result of wind, water, ice, or other geological activity, such as a plateau, plain, basin, or mountain

division from a data storage agency to one of the most respected research organizations in the world. His 1953 paper "The Hydraulic Geometry of Stream Channels" is credited with introducing quantitative analysis into the study of streams and **geomorphology**.

In particular, Leopold recognized a strong cause–effect relationship between stream characteristics and the development of stream channels. His work provided a framework for objective data collection on rivers everywhere and significantly increased the understanding of how stream systems operate. In 1964, he coauthored the book *Fluvial Processes in Geomorphology* that brought quantitative analysis to the study of **landforms**.

Leopold retired from his position of chief hydrologist of the U.S. Geological Survey in 1972 and joined the faculties of the Department of Geology and Geophysics and the Department of Landscape Architecture at the University of California at Berkeley, where as of 2002 he was professor emeritus.

Accomplishments

Luna Leopold has published six books and 150 scientific articles, and is a member of the National Academy of Sciences, the American Philosophical Society, and the American Academy of Arts and Sciences. For his contributions to the field of science, Leopold received the National Medal of Science. In addition, he has received numerous awards and honors, including the Warren Prize from the National Academy of Sciences; the Robert E. Horton Medal from the American Geophysical Union; the Penrose Medal from the Geological Society of America; and the Ian Campbell Award of the American Geological Institute. Interestingly, Luna's sister Estella and his brother Starker also became members of the National Academy of Sciences. Membership is one of the highest honors a U.S. scientist can receive.

Luna Leopold also is a well-known conservationist, having served on the Sierra Club's Board of Directors and the Board of Directors of the Environmental Law Institute. As of 2002, he was serving on the Board of Directors of the Aldo Leopold Foundation, Inc.

As an environmentalist, he is credited with contributing significantly to saving the Florida Everglades with his report "Environmental Impact of the Big Cypress Jetport." This report, which detailed the degradation of the environment, became the model for **environmental impact statements** throughout the country. SEE ALSO DOUGLAS, MARJORY STONEMAN; EVERGLADES; GEOLOGICAL SURVEY, U.S.; NATIONAL ENVIRONMENTAL POLICY ACT; STREAM CHANNEL DEVELOPMENT; STREAM EROSION AND LANDSCAPE DEVELOPMENT.

Dennis O. Nelson

environmental impact statement: a detailed document that outlines potential impacts of projects being considered by federal agencies and that potentially have significant environmental implications; an EIS is required by the National Environmental Policy Act

Bibliography

Leopold, Luna B. *Water, Rivers and Creeks.* Sausalito, CA: University Science Books, 1997.

Leopold, Luna B., M. Gordon Wolman, and John P. Miller. *Fluvial Processes in Geomorphology.* New York: W. H. Freeman & Co., 1964.

Lewis, Meriwether and William Clark

American Explorers
1774–1809 and 1770–1838, respectively

Meriwether Lewis and William Clark are best known for their exploration from 1804 to 1806 of the Louisiana Purchase and the Northwest Territory in what is now part of the United States. Lewis, the personal secretary to president Thomas Jefferson (1743–1826), was appointed by the president to lead the exploration of the 2 million square kilometers (about 800,000 square miles) of land purchased in 1803 by the United States from France, the Louisiana Purchase. In turn, Lewis asked William Clark, a former U.S. Army friend, to be a co-leader of the expedition.

The Explorers

Meriwether Lewis, born in 1774 in Virginia's Albemarle County, lived on a plantation near Thomas Jefferson. At an early age, Lewis moved to Georgia

Meriwether Lewis and William Clark relied on Sacagawea of the Shoshone Tribe to serve as a guide and interpreter during portions of their expedition to find the Northwest Passage. Sacagawea, with her infant son Jean-Baptiste Charbonneau, was the only woman to accompany the thirty-three members of the permanent party to the Pacific Ocean and back.

> **BICENTENNIAL COMMEMORATION**
>
> In early 2003, Monticello—the home of president Thomas Jefferson, located in Charlottesville, Virginia—hosted the launch of a nationwide observance of the 200th anniversary of Lewis and Clark's historic trek to the Pacific: it was on January 18, 1803 that Jefferson had requested the expedition. Twenty-three federal agencies signed a memorandum of understanding in support of the Lewis and Clark Bicentennial Commemoration.

and later moved back to Virginia. He liked exploring his surroundings and would roam the wildernesses of Georgia and Virginia. These early explorations would prove useful in his later life.

Lewis's military experience began with the Whiskey Rebellion in 1794. After the rebellion was halted, Lewis joined the U.S. Army. He eventually became a paymaster, traveling up and down the Ohio River, and through this position he met Captain William Clark.

William Clark, born in Virginia in 1770, moved with his family to a Kentucky plantation as a teenager. In addition to learning about the wilderness, Clark learned early about Native Americans, as they, too, lived in the area. He eventually joined the military as several of his brothers, including George Rogers, had done earlier and, at one point, served with Lewis for a short time.

Corps of Discovery

Lewis and Clark's mission, called by Jefferson the Corps of Discovery, was to find the Northwest Passage, a waterway sought by earlier explorers that would connect the Atlantic Ocean with the Pacific Ocean. Waterways were like today's highways. Jefferson thought that finding such a passage would increase trading with other countries. A system of connected waterways would allow people and goods to travel from one place to another more easily.

Jefferson wanted Lewis and Clark to keep detailed notes on everything they saw and experienced. They were to map their expedition and describe the unexplored territory, including the climate, game, and vegetation. Clark mapped the expedition while Lewis collected soil and plant samples. In addition, Jefferson told them to make peaceful relations with the Native Americans that they met.

St. Louis. After assembling a crew, Clark, using a keelboat, set off from St. Louis on May 14, 1804, and traveled upstream on the Missouri River. Lewis joined the crew on May 20. From the beginning, the crew noted that the river was rapid and muddy. The banks of the Missouri would cave in, and the deep mud would make it necessary for the men to sometimes use long poles instead of oars to push the boat forward. Clark usually stayed on the boat, since he was more knowledgeable about the river than Lewis.

Shoshone Country. The crew continued traveling until late October 1804 when they set up camp for the winter. They encountered Mandan and Hidatsa Native Americans. At this time, Sacagawea, a Shoshone, and her husband Toussaint Charbonneau, a French trader, joined the expedition to act as guides and interpreters through Shoshone country.

North Dakota. In April 1805, Lewis and Clark began their river expedition again, and by April 26 they reached the intersection of the Yellowstone and Missouri Rivers in present-day North Dakota. Lewis reported an abundance of timber, buffalo, elk, antelope, deer, bear, and wolves. He noted that wolves could catch antelope easily because antelope were not good swimmers. As the explorers encountered different terrain and climates, such as the Great Plains, the Rocky Mountains, and semi-desert conditions, the amount of game and timber varied.

Montana. On June 2, the expedition reached the junction of the Maria's and Missouri Rivers in present-day Montana. The crew noticed that the bluffs were getting lower, signaling the plains, and at this junction did not know whether the Missouri River was the north or south branch. Lewis and Clark split up in order to investigate both branches and finally decided to travel along the south branch because the water was clearer and swifter, which is indicative of being near the source. They wanted to find the source of the Missouri.

Lewis traveled ahead of Clark, and on June 13 found the waterfall of which the Native Americans had warned him. He sent word to Clark that he had found the waterfall, and, therefore, they had chosen the correct river to follow. In all, Lewis found five waterfalls.

Continental Divide. The expedition arrived at the junction of the Three Forks, the beginning of the Missouri River, on July 25. After several days they reached the Continental Divide. Expecting the Columbia River to be on the other side, they were disappointed to find more mountain ranges and no northwest passage. The need for horses became imperative if they were to continue.

Clearwater and Snake Rivers. Lewis and Clark acquired horses from the Shoshones and continued their journey through the Rocky Mountains where they encountered the Nez Perce. The Nez Perce showed the explorers how to make dugouts, or canoes, to help navigate the Clearwater and Snake Rivers. On October 16, the crew reached the Columbia River.

> **PBS SPECIAL**
>
> *Lewis and Clark: The Journey of the Corps of Discovery* (1997) is a 4-hour documentary directed by Ken Burns and originally shown on the Public Broadcasting Service (PBS). Using the journal entries of several expedition members, the film weaves the story of the members within the Corps of Discovery on their search for the Northwest Passage.
>
> Beginning with an explanation of why Lewis and Clark were appointed commanders of the expedition, viewers learn, from the explorers' own words, about the game, climate, and vegetation encountered on the trip into the Northwest Territory. Burns invites viewers to re-trace the trail of Lewis and Clark along the Missouri and Columbia Rivers to the Pacific Ocean, including an account of the crew's encounters with various Native American groups.

Lewis and Clark's outbound route is shown in red, and return route in green. The expedition occurred between May 1804 and September 1806.

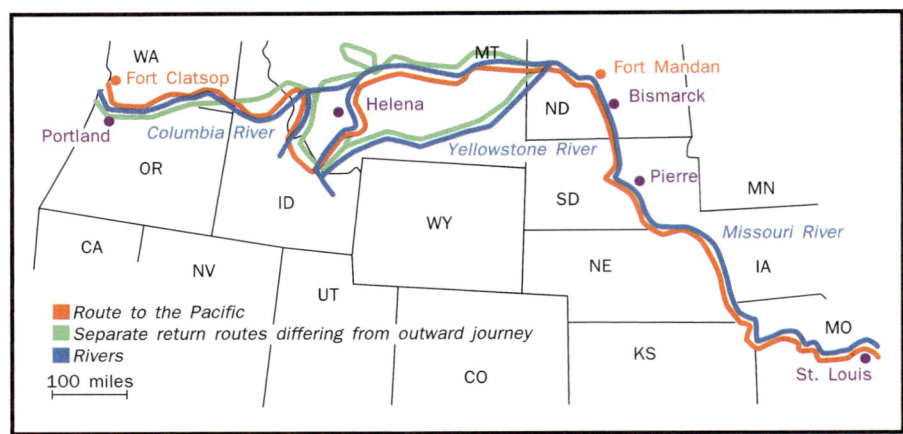

Columbia River. Along the Columbia River, the explorers encountered a rainforest climate. Fish, especially salmon, were plentiful in the river. Traveling became easier as the current flowed with the explorers now, and by November 7, the crew knew they were getting close to the Pacific Ocean because the river water became saltier, and waves rocked the boats. Finally, on November 18, 1805, the Corps of Discovery reached the Pacific Ocean.

Homeward Bound

After spending the winter in the nearby region, Lewis and Clark started their return trip on March 23, 1806. The crew split up in early July with Clark exploring the Yellowstone River while Lewis explored a shortcut to the Great Falls on the Missouri River and also explored the Maria's River. On their westbound trip, Lewis and Clark went out of their way by following the Missouri River to its source. They reunited on August 12 at the junction of the Yellowstone and Missouri Rivers. From here, the crew could travel up to 129 kilometers (80 miles) a day because they were headed downstream, and the current was with them. They arrived back in Saint Louis on September 23, 1806.

Following the expedition, Lewis was named the governor of the Louisiana Territory while Clark was given the position of Superintendent of Indian Affairs. Lewis's life began falling apart shortly after his appointment, and he died in 1809. In 1813, Clark was appointed governor of the Missouri Territory, formerly known as the Louisiana Territory. He died in 1838. SEE ALSO COLUMBIA RIVER BASIN; TRANSPORTATION.

Marie Scheessele

Bibliography

DeVoto, Bernard, ed. *The Journals of Lewis and Clark.* Boston, MA: Houghton Mifflin Company, 1981.

Herbert, Janis. *Lewis and Clark for Kids: Their Journey of Discovery with 21 Activities.* Chicago, IL: Chicago Review Press, Inc., 2000.

Lewis and Clark: The Journey of the Corps of Discovery. (videocassette) Ken Burns, dir. Washington, D.C.: Florentine Films and WETA, 1997.

Roosevelt, Theodore. *The Winning of the West: Louisiana and the Northwest, 1791–1807.* Bison Book, Vol. 4. Lincoln, NE: University of Nebraska Press, 1995.

Internet Resources

Biography of Lewis and Clark. LewisAndClarkTrail.com <http://www.lewisandclarktrail.com/biography.htm>.

Circa 1803. Lewis and Clark: Inside the Corps. PBS Online. <http://www.pbs.org/lewisandclark/inside/circa.html>.

The Ethnography of Lewis and Clark: Native American Objects and the American Quest for Commerce and Science. Peabody Museum of Archaeology and Ethnology, Harvard University. <http://www.peabody.harvard.edu/Lewis_and_Clark/>.

Life in Extreme Water Environments

Beginning in the early 1990s, scientific knowledge of the environmental limits of microbial life on Earth expanded dramatically as microbiologists applied new methods of molecular biology over a broad range of environmental extremes. Microbial species are now known to occupy a vast range of environments that previously were unimagined. New discoveries have revolutionized scientific understanding of Earth's biosphere, opened up new views of the history of terrestrial (land-based) life, and increased the possibilities that life could develop elsewhere in the cosmos.

The name applied to this new research area of biology is extremophiles research. Extremophiles (literally "extreme-loving") are defined as organisms that occupy environments judged by human standards as harsh. These encompass both physical and chemical extremes. Examples of water environments characterized as extreme are summarized in the accompanying table.

Different classes of extremophiles have been defined based on the nature of the environments where they are found. For example, extremophiles that have adapted to high temperatures are called thermophiles. Those that require cold temperatures for growth and reproduction are called psychrophiles (as opposed to other organisms that can tolerate occasional cold temperatures and are not considered extremophiles). Those that love acidic environments (i.e., with low **pH**) are called acidophiles, whereas those found in highly alkaline conditions (high pH) are alkaliphiles. Organisms that live under high pressure are called piezophiles, and those found in high-radiation environments are as yet unnamed. Some organisms occupy more than one environmental extreme simultaneously, and are known as polyextremophiles. An example is the archaebacterial species, *Sulfolobus acidocalderius*, which thrives in boiling **mudpots** at temperatures exceeding 80°C (176°F) and at acidities less than pH 3. Although mostly microbial, extremophiles include a few species of multicellular organisms such as worms, amphibians, mollusks, and crustaceans.

Physical Extremes

Temperature. Microorganisms are now known to thrive over a broad range of physical extremes in temperature. For high temperatures, this environment includes **geysers** and hot springs, boiling mudpots, and hydrothermal vents on the deep seafloor. In the latter case, where vent temperature can reach 400°C (752°F), the high **hydrostatic** pressure prevents vent water from boiling, and thermophilic species exhibit growth up to a temperature of about 114°C (237°F). (Under atmospheric pressure, water boils at 100°C, or 212°F.)

At the lower end of the temperature environment is sea ice, ground ice, permafrost (icy soils), and subglacial lake environments like Lake Vostok in Antarctica, a deep subglacial lake located more than 4 kilometers (2.5 miles)

pH: a measure of the acidity of water; a pH of 7 indicates neutral water, with values between 0 and 7 indicating acidic water (0 is very acidic), and values between 7 and 14 indicating alkaline (basic) water (14 is very alkaline)

mudpot: a hot spring that reaches the surface through water-saturated fine-grained sediments; the hot water mixes with the sediments, producing a thick, pasty mud through which hot gases periodically escape, showering the immediate area with mud globules

geyser: a periodic thermal spring that results from the expansive force of superheated steam; also, a special type of thermal spring which intermittently ejects a column of water and steam into the air with considerable force

hydrostatic: referring to the pressure exerted by water at a point, related to the weight of the water above the point

Life in Extreme Water Environments

SOME WATER ENVIRONMENTS THAT HARBOR EXTREMOPHILES

Extreme Condition	Water Environment	Common Extremophiles
Snow, freezing water, slush, ice	Continental ice sheets, sea ice, polar lakes, permafrost, glaciers, snowfields, lakes found at high elevations	ice worms, crustaceans, insects, fish, frogs, aquatic mosses, algae, bacteria (including cyanobacteria), Archaea
Hot water and rock or mud substrates	Geothermal systems of continental volcanic areas: geysers, hot springs, fumaroles, mudpots, volcanic lakes. Geothermal systems of the deep sea: hydrothermal vents associated with divergent plate margins and associated submarine volcanoes	vent crabs, mussels, tubeworms, bacteria, Archaea
Acidic waters	Acidic hot springs and mudpots, acid mine drainages, peat bogs	Archaea, bacteria, unicellular algae, fungi, insects, vascular and nonvascular plants
Alkaline waters	Soda lakes	Archaea, bacteria, unicellular algae, protists insects, crustacea
Salty waters	Hypersaline lagoons and lakes, playas, evaporation ponds at saltworks, deep subsurface brines	brine shrimp (crustacea), insects, Archaea, Bacteria (cyanobacteria), unicellular algae, protists
Anoxic waters (no oxygen)	Swamps and other hypereutrophic wetlands, anoxic lakes and streams, anoxic groundwater	Archaea, bacteria, protists, fungi
Dark waters (no solar energy)	Deep seafloor, subglacial lakes, caves, subsurface formations	Archaea, bacteria, protists, cave (troglodyte) faunas—including insects and other arthropods, fish, amphibians
High radiation	Nuclear reactors, naturally radioactive formations	bacteria, green algae
High pressure	Deep crust and seafloor	bacteria, Archaea, protists, fish, invertebrates
Low pressure	Upper atmosphere	spores of bacteria, fungi
Extreme dryness	Interior subtropical and orographic deserts, polar deserts, salt flats, playa (dry) lake basins	endolithic bacteria, Archaea, fungi, unicellular algae, yeast, tardigrades, brine shrimp, insects

brine film: a thin film of salty water surrounding grains of ice or soil that forms when salts are excluded during freezing of permafrost, ground ice, or sea ice; dissolved salts depress the freezing point of water; depending on the water's original composition, interstitial brines formed by freezing can remain liquid far below the freezing point of fresh water

lithosphere: the rigid outer layer of Earth made up of the crust and the uppermost mantle

obligate: describes organisms that require the specified condition (e.g., high pressure) for growth

beneath the Antarctic ice sheet. In salt-water environments, water can remain liquid below 0°C (32°F) because dissolved salts lower its freezing point. Some psychrophilic species, such as those found in **brine films**, are known to be active down to −15°C (5°F). Some complex multicellular organisms, like the wood frog, can tolerate the freezing of up to 65 percent of their body water during winter hibernation.

Pressure. Pressure, which is measured relative to atmospheric pressure at sea level (where 1 bar roughly equals 14.5 pounds per square inch), increases with depth in the oceans. In the ocean, this hydrostatic pressure goes up at the rate of approximately 1 bar per 100 meters. Measured within the crust, **lithospheric** pressure increases at a rate almost twice hydrostatic. Live microorganisms obtained from the Mariana Trench, the deepest place in the oceans (10.9 kilometers, or 6.8 miles), have been successfully grown under surface conditions, whereas others have been shown to be **obligate** piezophiles that grow only at high pressure.

Life in Extreme Water Environments

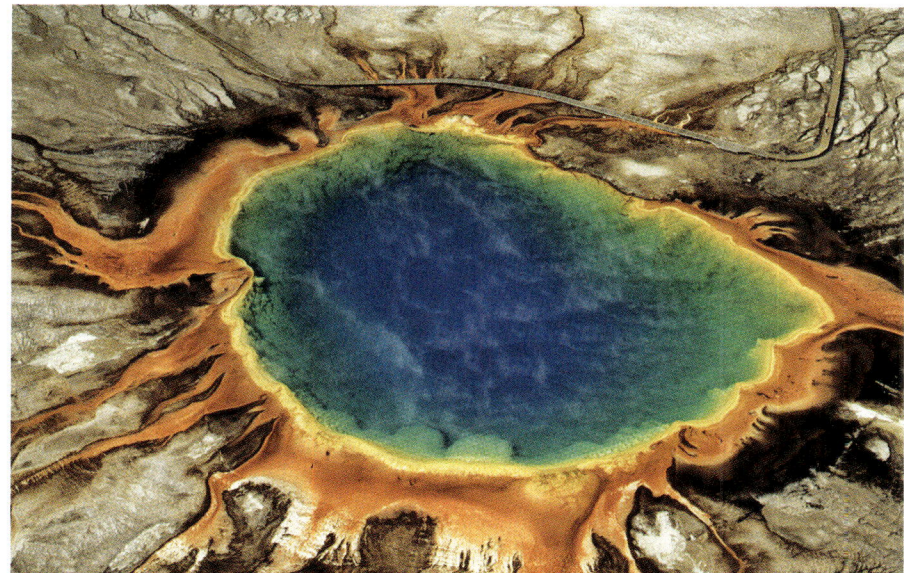

The gradations of color at Yellowstone National Park's Grand Prismatic Spring are caused by different types of algae and other microbes. As the hot spring water leaves the ground, it cools as it moves outward (downslope), creating different temperature zones, with the cooler zones located farther from the spring outlet. Each microbial species (or strain) is adapted to a very specific temperature zone, as is evident by the color patterns caused by their presence.

Pressure decreases with increasing altitude, such that at 10 kilometers (6.2 miles) above the Earth's surface, the pressure is only about one-fourth that at sea level. Organisms have been discovered growing on the top of Mount Everest, the highest point on the Earth's surface (more than 8.8 kilometers [5.4 miles]). Viable spores of bacteria and fungae have even been collected from the lower stratosphere.

Radiation. Radiation is energy that travels as either particles (e.g., high-energy neutrons, protons, electrons, or ions) or waves (e.g., X-rays, gamma rays, or ultraviolet rays). The bacterium *Deinococcus radiodurens*, which has been found growing on the fuel rods of nuclear reactors, is a famous example of an extremophile that can tolerate high levels of radiation.

Chemical Extremes

pH. Chemical extremes in the environment include pH, which ranges from values of less than 0 (extremely acidic) to more than 14 (extremely alkaline or basic). In nature, microorganisms have been shown to occupy nearly the entire range of pH. Some species of bacteria have been found living in a acid mine drainage at a pH of approximately 0.5. Others live in soda lakes, such as those found in the western United States and Egypt, where the highly alkaline waters can reach a pH of 11.

Salinity. Life also occupies an equally broad range of salinity. Salt-loving halophiles live in salt plains, evaporation ponds at saltworks, and natural salt lakes (e.g., the Dead Sea, Israel and the Great Salt Lake, Utah). Halophiles also live within hypersaline brines that exist around deep-sea vents and in deep subsurface rock formations. In nature, salinities can range from fresh water, with very low salinity, to super-saturated brines. At very high concentrations, salt **precipitates**, often entrapping microorganisms.

Desiccation. The ability to survive desiccation (extreme drying) has been demonstrated for both vegetative cells and reproductive spores of many microbial species. In the driest deserts on Earth, microbial species (so-called

precipitate: (verb) in a solution, to separate into a relatively clear liquid and a solid substance by a chemical or physical change; (noun) the solid substance resulting from this process

WATERMELON SNOW

Snow algae were noted first by Aristotle (384–322 B.C.E.). Although simple microbial communities can be associated with snowfields in either alpine or polar regions, the most common species in snow generally are single-celled green algae (chlorophytes). The most obvious sign of these algae is the pink watermelon-colored snow associated with the psychrophilic *Chlamydamonas nivialis,* a green alga that can acquire a strong reddish color produced by high levels of carotenoids (yellow, orange, or red pigments). These pigments are produced to protect the algal cells from ultraviolet irradiance, which is extremely high at the snow surface, largely because of snow's reflective properties.

Life in Extreme Water Environments

> **PHOTOGRAPHS OF EXTREME ENVIRONMENTS**
>
> This encyclopedia contains photographs of several types of extreme water environments. Entries with notable images include:
>
> | Entry | Photograph |
> |---|---|
> | "California, Water Management in" | Mono Lake (a hypersaline lake) |
> | "Desert Hydrology" | a desert |
> | "Glaciers and Ice Sheets" | snow and ice |
> | "Hot Springs and Geysers" | Old Faithful (a geyser) |
> | "Hot Springs on the Ocean Floor" | a black smoker and tubeworms (at a hydrothermal vent) |
> | "Ice at Sea" | snow and ice |
> | "Lakes: Chemical Processes" | a salt lake |
> | "Senses, Fresh Water and the" | a volcanic lake |
> | "Springs" | a hot-water coastal spring |
> | "Springs" | a salt spring and salt plain |
> | "Springs" | a mudpot |
> | "Volcanoes and Water" | volcanoes |
> | "Volcanoes, Submarine" | pillow lava |

"endoliths") often survive by living inside porous rocks where they are protected from ultraviolet radiation. The coldest desert environments on Earth are found in the dry valleys of Antarctica. These polar deserts harbor many types of endolithic communities dominated by cyanobacteria, algae, and fungi. Antarctic endoliths live just a few millimeters beneath rock surfaces in limestones or in translucent, quartz-rich sandstones. Decades may pass with no rain, but when it comes, these organisms spring to life, using the available light, water, and **nutrients** to quickly grow and reproduce before drying out and again becoming dormant (inactive).

Aphotic and Anoxic Environments

Even though **photosynthesis** accounts for more than 99 percent of the energy that powers the biosphere, thermal and chemical energy sources within the Earth can provide forms of energy capable of supporting complex **ecosystems**. Consequently, extremophiles can also be found in aphotic (non-light) environments, such as deep in the ocean or in the Earth's subsurface.

In hydrothermal vent environments on the ocean floor, complex ecosystems have been found in which the organisms (including large, multicellular animals) derive their energy entirely from chemical sources provided by the hot fluids issuing from the vent. Single-celled forms of life also survive and grow in the deep subsurface of Earth, within the tiny pore spaces and fractures of endurated rock. These are aphotic environments where sunlight does not penetrate; consequently, organisms living there must use chemical energy sources for their metabolism. Some subsurface microbes do depend

---

nutrients: a group of chemical elements or compounds needed for all plant and animal life

photosynthesis: the process by which plants manufacture food from sunlight; specifically, the conversion of water and carbon dioxide to complex sugars in plant tissues by the action of chlorophyll driven by solar energy

ecosystem: the community of plants and animals within a water or terrestrial habitat interacting together and with their physical and chemical environment

Life in Extreme Water Environments

Crabs, mussels, and tubeworms are multicellular organisms found in hydrothermal vent environments on the seafloor. These organisms and a host of microbes are specially adapted for this unusual ecosystem, which is based on chemosynthesis rather than photosynthesis.

on photosynthetically-derived **organic** matter that washes down from the surface, but many so-called lithoautotrophic species (which literally means "self-feeding from rocks") use simple byproducts of chemical **weathering** of rocks to extract energy from the environment. For example, oxidation reactions associated with the weathering of basalts in an oxygen-free environment may lead to the release of hydrogen. The hydrogen released is used by methanogens to produce methane and energy.

Although some scientists have questioned the evidence for hydrogen-based microbial ecosystems in deep basalt formations, the possibility of an active microbial community at great depths could have implications for subsurface storage of highly radioactive materials and other wastes. Microbial interactions could act to weaken containers, leading to leakage and the undesirable spreading of waste materials.

Methanogens are microbes that live in anoxic (non-oxygen) environments, which can include some swamps, rice paddies, or certain highly enriched lakes, ponds, or streams, and their sediments. Methanogens combine carbon dioxide (CO_2) and hydrogen (H_2) to produce organic matter, while releasing methane gas as a byproduct. Wetlands and rice paddies (agricultural wetlands) account for half the total methane produced globally. Methane is a **greenhouse gas**, and its rate of increase in the atmosphere is exceeding that of CO_2. Human activity has played a major role in this methane increase.

organic: pertaining to, or the product of, biological reactions or functions

weathering: the decay or breakdown of rocks and minerals through a complex interaction of physical, chemical, and biological processes; water is the most important agent of weathering; soil is formed through weathering processes

greenhouse gas: a gas in the atmosphere that traps heat and reflects it back to the planetary body

Implications and Benefits

The ability of some extremophiles to survive harsh conditions similar to those found on other planets has raised the possibility that life might exist beyond Earth. As an example of this survival ability, halophiles have been cultivated from inclusions of brine contained in salt crystals deposited hundreds of millions of years ago. Microbes also have been isolated from Siberian permafrost, where they have remained in deep freeze for more than 3 million years. Equally impressive are bacteria germinated from spores preserved in Dominican amber dated at more than 30 million years old. Given the propensity for prolonged survival in these types of environments, could an extraterrestrial biota someday be discovered within brines, salts, or ices on another planet, like Mars or a moon like Europa?

Cellular enzymes extracted from extremophiles have spawned a multibillion dollar biotechnology industry. The enzymes are used in industrial and medical applications, ranging from the production of stone-washed jeans, to creating artificial sweeteners, to genetic fingerprinting. One thermophile that lives in hot springs is the source of the heat-stable deoxyribonucleic acid (DNA) polymerase enzyme used in polymerase chain reaction (PCR). PCR forms part of the foundation of much of the biotechnology industry. Proteins produced by psychrophilic organisms may one day prove useful in cold-food preparation and in detergents for washing in cold water. SEE ALSO ACID MINE DRAINAGE; ASTROBIOLOGY: WATER AND THE POTENTIAL FOR EXTRATERRESTRIAL LIFE; BRINES, NATURAL; DESERT HYDROLOGY; EARTH'S INTERIOR, WATER IN THE; ECOLOGY, FRESH-WATER; ECOLOGY; MARINE; FRESH WATER, PHYSICS AND CHEMISTRY OF; GEOTHERMAL ENERGY; HOT SPRINGS AND GEYSERS; HOT SPRINGS ON THE OCEAN FLOOR; ICE AT SEA; LIFE IN WATER; MARS, WATER ON; MICROBES IN GROUNDWATER; MICROBES IN LAKES AND STREAMS; MICROBES IN THE OCEAN; MINERAL RESOURCES FROM FRESH WATER; MINERAL RESOURCES FROM THE OCEAN; PLANKTON; RADIOACTIVE CHEMICALS; SEA WATER, PHYSICS AND CHEMISTRY OF; SOLAR SYSTEM, WATER IN THE; SPRINGS; VOLCANOES AND WATER; VOLCANOES, SUBMARINE.

Jack D. Farmer

Bibliography

Anderson, R. T., and Chapelle, F. H. "Evidence against Hydrogen-Based Microbial Ecosystems in Basalt Aquifers." *Science* 281 (1998):976–977.

Chyba, C., and K. Hand. "Life without Photosynthesis." *Science* 292 (2001): 2026–2027.

Frederickson, J. K., and T. C. Onstott. "Microbes Deep Inside the Earth." *Scientific American* 275, no. 4 (1996):42–47.

Horikoshi, K., and W. D. Grant. *Extremophiles: Microbial Life in Extreme Environments.* New York: Wiley-Liss, 1998.

Lemonick, M. D. *Other Worlds: the Search for Life in the Universe.* New York: Simon and Schuster, 1998.

Madigan, M. T., and B. L. Marrs. "Extremophiles." *Scientific American* 276, no. 4 (1997):82–87.

Pace, N. R. "A Molecular View of Microbial Diversity and the Biosphere." *Science* 276 (1997):734–740.

Rothschild, Lynn J., and Rocco L. Mancinelli. "Life in Extreme Environments." *Nature* 409 (2001):1092–1101.

Stevens, T. O., and J. P. McKinley. "Lithoautotrophic Microbial Ecosystems in Deep Basalt Aquifers." *Science* 270 (1996):450–454.

Life in Water

Life is thought to have originated in an aquatic environment—the oceans. Living organisms have since adapted to numerous aquatic **habitats**, both

habitat: the environment in which a plant or animal grows or lives; the surroundings include physical factors such as temperature, moisture, and light, together with biological factors such as the presence of food and predators

marine and fresh-water. They occupy environments as diverse as lakes, rivers, and oceans.

Marine Environment

About 17 percent of known biological species live in oceans. Marine species are described as either pelagic or benthic. Pelagic organisms live in the water column itself. Benthic species live on the ocean bottom.

Pelagic. Pelagic organisms include plankton, which float along with currents, and nekton, which are active swimmers. Plankton are divided into phytoplankton, which include photosynthesizing species such as **algae**, and zooplankton, which are consumer species. Zooplankton consist largely of copepods (tiny crustaceans).

Although plankton generally drift with ocean currents, some plankton have limited mobility. For example, certain zooplankton species move towards the water surface at night to feed, when there is less danger of predation, and return to deeper waters during the day.

Organisms that are planktonic throughout their life cycles are known as holoplankton. Organisms that are only planktonic during the early parts of their life cycles are called meroplankton. Meroplankton include the larval or juvenile forms of many species of fish and **mollusks**. These species use the planktonic stage to disperse to new areas. Although most planktonic species are small, some are large, such as kelp and jellyfish.

Nekton are active swimmers that use diverse means to propel themselves through the water. Some species swim using fins, tails, or flippers. Other species, such as mussels, move by shooting out jets of water, known as jet propulsion. Nektonic species include fish, octopus, sea turtles, whales, seals, penguins, and many others. Many nektonic species eat high in the **food chain**, although there are plankton-eating species (e.g., some fish) and herbivorous species (e.g., sea turtles) in addition to carnivorous ones (e.g., seals and killer whales).

Pelagic marine species may also be categorized according to the depths at which they occur. Different water depths are characterized by differences in temperature, amount of sunlight received, and availability of **nutrients**. The epipelagic zone describes oceanic waters closest to the surface, and is the zone richest in marine life. In the epipelagic zone, there is enough sunlight for **photosynthesis**. For that reason, the epipelagic zone is also called the photic (light) zone. All photosynthetic species, including the phytoplankton, live in this zone, as do many of the species that feed on phytoplankton.

Below the photic zone is the aphotic zone, which is characterized by very limited light (or no light) and limited food. Species in the aphotic zone often depend on food drifting down from above. Consequently, there are many detritivores (species that feed on dead or decaying organic matter) in these habitats.

Certain deep-sea habitats can be highly diverse. In the deep-sea vents, for example, chemosynthetic bacteria (rather than photosynthetic species) form the basis of the food chain. These bacteria obtain energy from chemical sources such as hydrogen sulfide instead of from sunlight.

Sea nettles and jellyfish are pelagic and planktonic, meaning that they live in the water column and primarily float or drift as opposed to swimming.

algae: (singular, alga) simple photosynthetic organisms, usually aquatic, containing chlorophyll, and lacking roots, stems, and leaves

mollusk: an invertebrate animal with a soft, unsegmented body and usually a shell and a muscular foot; examples are clams, oysters, mussels, and octopuses

food chain: the levels of nutrition in an ecosystem, beginning at the bottom with primary producers, which are principally plants, to a series of consumers—herbivores, carnivores, and decomposers

nutrients: a group of chemical elements or compounds needed for all plant and animal life

photosynthesis: the process by which plants manufacture food from sunlight; specifically, the conversion of water and carbon dioxide to complex sugars in plant tissues by the action of chlorophyll driven by solar energy

The seahorse is pelagic and nektonic, meaning that it lives in the water column and swims rather than floating or drifting.

biomass: the total mass of living organic matter in a defined location; generally expressed as grams per unit volume or per unit area

Benthic. Benthic species live on the ocean bottoms, and represent the greatest proportion of marine species; in fact, 98 percent of marine species are benthic. Benthic species include epifauna, which live on the surface, and infauna, which burrow into seafloor sediment. Benthic epifauna include species such as oysters, scallops, sea stars, crabs, and lobsters. Examples of infauna include clams and many species of worms.

Some benthic species are sessile (non-moving), and live attached to the ocean bottom. Benthic plants generally are found only in shallow waters where there is enough sunlight for photosynthesis. However, benthic animals are found at a wide variety of depths, including in the deepest parts of the ocean. Some species, such as flounder, are capable of both benthic and nektonic existence.

Distance from Shore. The distance of a zone from shore can categorize marine environments. The neritic zone describes coastal marine regions. The neritic zone is particularly rich with life because the relatively shallow water allows for plentiful photosynthesis, and because a steady flow of nutrients is washed into the water from land.

Farther from land, areas of open ocean are described as the oceanic zone. The oceanic zone has significantly less total **biomass** than the neritic zone. The intertidal zone is the area of shore that alternates between being submerged and dry, depending on the level of the tide. Numerous species are specialized for living in the intertidal zone, including the familiar barnacles. The intertidal zone has the greatest density of living organisms among marine environments.

Fresh-Water Environment

Fresh-water habitats are extremely diverse, and include both still-water environments like lakes and ponds, and flowing-water environments like rivers and streams.

Still-Water Habitats. Like oceans, lakes have pelagic and benthic zones. The temperature of lake water varies depending on depth, and can also change dramatically over seasons. The epilimnion is the topmost layer of lake water. It is significantly warmer than deeper areas due to heating by sunlight. The hypolimnion layer describes deeper, colder lake water. Many of the nutrients in lakes collect at lake bottoms.

Turnover occurs when all the water in a lake is nearly thermally uniform and mixed, distributing nutrients throughout the water. Turnover occurs twice a year in many temperate lakes, but may occur only once in subtropical environments, or not at all in permanently stratified lakes.

Lakes also can be described as either oligotrophic or eutrophic (or in between these two extremes). Oligotrophic lakes have low levels of nutrients and low productivity. They generally contain cold, highly oxygenated water and support species adapted to these conditions. Eutrophic lakes, on the other hand, have plentiful nutrients and are highly productive. Species that inhabit eutrophic lakes must be tolerant of low oxygen levels and warm temperatures. In general, oxygen levels in lakes depend on the amount of water circulation, the surface area that is exposed to air, and levels of oxygen consumption by living organisms.

The sea anemone anchors to the bottom substrates: in this case, the volcanic rocks of Hawaii. Anemones are classified as a benthic epifauna: that is, living on the surface of the ocean bottom rather than being burrowed.

Flowing-Water Habitats. River habitats are characterized by flowing water. River species generally have special adaptations for living in water currents. Some species are sessile and live anchored to the river bottom. Other species have evolved adaptations such as suckers or hooks to keep themselves from being washed away. Still other species are strong swimmers. Many of these have flattened bodies that help them resist the pressure of the current.

Compared to lakes, rivers tend to be well-oxygenated because of the constant motion of the water. Temperatures can change quickly in rivers, but do not span as great a range as in lakes or other still water. Because there is less penetration of light in flowing water, plant diversity is generally lower in rivers than in lakes. As in other aquatic ecosystems, algae frequently occupy the base of the food chain.

Challenges of Aquatic Life

Flotation. Flotation or placement in the water column is a challenge faced by all aquatic organisms. For example, it is crucial to phytoplankton to stay in the photic zone, where there is access to sunlight. The small size of most phytoplankton, plus a special oily substance in the cytoplasm of cells, helps keep these organisms afloat.

Zooplankton use a variety of techniques to stay close to the water surface. These include the secretion of oily or waxy substances, possession of air-filled sacs similar to the swim bladders of fish, and special appendages that assist in floating. Some zooplankton even tread water.

Fish have special swim bladders, which they fill with gas to lower their body density. By keeping their body at the same density as water, a state called neutral buoyancy, fish are able to move freely up and down.

Salinity. Aquatic species also have to deal with salinity, the level of salt in the water. Some marine species, including sharks and most marine invertebrates, simply maintain the same salinity level in their tissues as is in the surrounding water. Some marine vertebrates, however, have lower salinity in their tissues than is in sea water. These species have a tendency to lose water to the environment. They make up for this by drinking sea water and excreting excess salt through their gills.

Fresh-water aquatic species have the opposite problem—a tendency to absorb too much water. These species must constantly expel water, which they do by excreting a dilute urine.

Species that occupy both fresh water and marine habitats at different stages of their life cycle must transition between two modes of maintaining water balance. Salmon hatch in fresh water, mature in the ocean, and return to fresh-water habitats to spawn. Eels, on the other hand, hatch in salt water, migrate to fresh-water environments where they mature, and return to the ocean to spawn.

Salinity is particularly variable in coastal waters, because oceans receive variable amounts of fresh water from rivers and other sources. Species in coastal habitats must be tolerant of salinity changes and are described as euryhaline. In the open ocean, salinity levels are generally constant, and species that live there cannot tolerate salinity changes. These organisms are described as stenohaline.

Aquatic snakes include amphibious species that live both on land and in the water and species that occupy marine habitats only (such as this yellow-bellied sea snake). Some have paddle-like tails that help propel them through the water. All aquatic snakes breathe using lungs, however, and must return to the water surface to obtain oxygen.

zooplankton: microscopic animals that live suspended in bodies of water and that drift about because they cannot move by themselves or because they are too small or too weak to swim effectively against a current; composed primarily of protozoans, microcrustacea (copepods, cladocera, rotifers) and larval stages of certain invertebrates

Temperature. Eurythermal species are those that can survive in a variety of temperatures. Eurythermality generally characterizes species that live near the water surface, where temperatures change depending on the seasons or the time of day. Species that occupy deeper waters generally experience more constant temperatures, are intolerant of temperature changes, and are described as stenothermal. SEE ALSO ALGAL BLOOMS IN FRESH WATER; ALGAL BLOOMS IN THE OCEAN; CORALS AND CORAL REEFS; ECOLOGY, FRESHWATER; ECOLOGY, MARINE; FISH; HOT SPRINGS ON THE OCEAN FLOOR; LAKES: BIOLOGICAL PROCESSES; LAKES: CHEMICAL PROCESSES; LIFE IN EXTREME WATER ENVIRONMENTS; OCEAN BIOGEOCHEMISTRY; OCEANS, POLAR; OCEANS, TROPICAL; PLANKTON.

Jennifer Yeh

Bibliography

Barnes, R. S. K., and R. N. Hughes. *An Introduction to Marine Ecology.* Boston, MA: Blackwell Scientific Publications, 1982.

Byatt, Andrew, Alastair Fothergill, and Martha Holmes. *The Blue Planet: A Natural History of the Oceans.* New York: DK Publishers, 2001.

Gould, James L., and William T. Keeton, with Carol Grant Gould. *Biological Science,* 6th ed. New York: W. W. Norton & Co., 1996.

Lalli, Carol M., and Timothy R. Parsons. *Biological Oceanography: An Introduction.* New York: Pergamon Press, 1993.

Nybakken, James Willard. *Marine Biology: An Ecological Approach.* New York: Harper and Row, 1982.

Light Transmission in the Ocean

Visible radiation, or light, from the Sun is important to the world's ocean systems for several reasons. It provides the energy necessary for ocean currents and wind-driven waves. Conversion of some of that energy into heat helps form the thin layer of warm water near the ocean's surface that supports the majority of marine life. Most significantly, the transmission of light in sea water is essential to the productivity of the oceans.

Visible wavelengths of light are captured by chlorophyll-bearing marine plants, which then make their own food through the process of **photosynthesis**. The organic molecules created by this process are an important energy source for many small organisms that are the base of the entire marine **food chain**. All life in the oceans is ultimately dependent upon the light and the process of photosynthesis that it initiates. Similarly, light transmission is a key factor in the **ecology** of lakes and streams, which are discussed elsewhere in this encyclopedia.

Reflection, Refraction, and Color

The uppermost, sunlit layer of the ocean where 70 percent of the entire amount of photosynthesis in the world takes place is called the euphotic zone. It generally extends to a depth of 100 meters (330 feet). Below this is the disphotic zone, between 100 and 1,000 meters (330 and 3,300 feet) deep, which is dimly lit. Some animals are able to survive here, but no plants. Although the amount of light is measurable at this range of depths, there is not enough available for photosynthesis to take place. The layer of the ocean where no light at all penetrates—over 90 percent of the entire ocean area

---

photosynthesis: the process by which plants manufacture food from sunlight; specifically, the conversion of water and carbon dioxide to complex sugars in plant tissues by the action of chlorophyll driven by solar energy

food chain: the levels of nutrition in an ecosystem, beginning at the bottom with primary producers, which are principally plants, to a series of consumers—herbivores, carnivores, and decomposers

ecology: the scientific study of the interrelationships of living things to one another and to the environment; also refers to the ecology of a given region

MEASURING LIGHT TRANSMISSION

Scientists have developed several different methods and instruments to measure light transmission in water. The simplest measurement method involves the use of the Secchi disk, a white plate about 30 centimeters (12 inches) in diameter. It is fastened horizontally to a rope marked in meters. The disk is then lowered into the sea, lake, or other waterbody. The depth at which the disk is lost to sight is noted using the rope markings. This provides a rough estimate of the depth of light penetration.

A more sophisticated device for measuring light transmission is the nephelometer, which measures the scattering of incident (incoming) light by particles in the water. The optical backscatter meter and light scattering meter work in a similar fashion by projecting a light beam into the water. A detector on the instrument measures the amount of light that is scattered back into it.

The transmissometer measures light attenuation, or the sum of scattering and absorption of light in sea water. It projects a beam of light of a known wavelength over a known distance, and the data may be used to calculate the percentage of light that is transmitted at a certain depth.

The a–c meter has separate sensors to detect absorption of light by particles and total light attenuation. It functions in a manner like that of a transmissometer.

---

on Earth—is called the aphotic zone, where depths are more than 1,000 meters (3,300 feet).

Light Penetration. A certain amount of incoming light is reflected away when it reaches the ocean surface, depending upon the state of the water itself. If it is calm and smooth, less light will be reflected.

If it is turbulent, with many waves, more light will be reflected. The light that penetrates the surface is **refracted** due to the fact that light travels faster in air than in water. Once it is within the water, light may be scattered or absorbed by solid particles. Most of the visible light spectrum is absorbed within 10 meters (33 feet) of the water's surface, and almost none penetrates below 150 meters (490 feet) of water depth, even when the water is very clear.

refraction: the change in direction of propagation that occurs when a wave passes from one medium to another

Greater abundances of solid particles in the water will decrease the depth of light penetration. Therefore, water near the seashore that is more turbid (cloudy) due to particles will show a decrease in light transmission, even in shallow water. This is due to large numbers of particles brought in by river systems, and biological production by **microorganisms**, as well as waves, tides, and other water movement picking up debris on the ocean floor.

microorganism: a microscopic organism

Light Spectrum. Water selectively scatters and absorbs certain wavelengths of visible light. The long wavelengths of the light spectrum—red, yellow, and orange—can penetrate to approximately 15, 30, and 50 meters (49, 98, and 164 feet), respectively, while the short wavelengths of the light spectrum—violet, blue and green—can penetrate further, to the lower limits of the euphotic zone. Blue penetrates the deepest, which is why deep, clear ocean water and some tropical water appear to be blue most of the time. Moreover, clearer waters have fewer particles to affect the transmission of light, and scattering by the water itself controls color. Water in shallow coastal areas tends to contain a greater amount of particles that scatter or absorb light wavelengths differently, which is why sea water close to shore may appear more green or brown in color. SEE ALSO ALGAL BLOOMS IN THE

Ocean; Carbon Dioxide in the Ocean and Atmosphere; Lakes: Biological Processes; Lakes: Chemical Processes; Lakes: Physical Processes; Plankton; Sea Water, Physics and Chemistry of.

Christina E. Bernal

Bibliography

Davis, Richard A. *Oceanography: An Introduction to the Marine Environment*, 2nd ed. Dubuque, IA: Wm. C. Brown Publishers, 1991.

Gross, M. Grant. *Oceanography: A View of the Earth*, 3rd ed. Englewood Cliffs, NJ: Prentice Hall, 1982.

Richardson, Mary Jo, and Wilford D. Gardner. "Tools of the Trade." *Quarterdeck* 5, no. 1 (1997):10–15.

Thurman, Harold V., and Elizabeth A. Burton. *Introductory Oceanography*, 9th ed. Upper Saddle River, NJ: Prentice Hall, 2001.

Marginal Seas

Marginal seas, which separate coastal zones from open oceans, often exist as large indentations into continental landmasses. Some of the major marginal seas include the Arabian Sea, Baltic Sea, Bay of Bengal, Bering Sea, Beaufort Sea, Black Sea, Gulf of California, Gulf of Mexico, Mediterranean Sea, Red Sea, Ross Sea, Weddell Sea, and all four of the Siberian Seas (Barents, Kara, Laptev, and East Siberian).

Marginal seas are similar to open oceans with respect to being created by large-scale geological processes, exhibiting **biodiversity**, and possessing layered water circulation patterns. The degree of water circulation between marginal seas and open oceans varies with respect to location. The primary differences between marginal seas and open oceans are associated with depth and proximity to landmasses. Marginal seas, which are generally shallower than open oceans, are more influenced by human activities, river runoff, climate, and water circulation.

Human Impacts

Humans utilize nearshore environments, including coastal waters and marginal seas, for food and fuel resources, as well as for various purposes related to scientific, industrial, and recreational activities. The greatest human impact on marginal seas is related to the fisheries industry. Ninety percent of the world's fisheries exist within coastal waters that are located less than 200 kilometers (124 miles) from the shoreline. Other human activities that have adversely affected marginal seas include industrial enterprises such as sewage disposal, dredge disposal, offshore oil drilling, and accidental releases of pollutants, including petroleum products, radioactive waste, detergents, and plastics.

Coastal waters and marginal seas are more susceptible to **pollution** than open ocean regions because of the high concentration of human activities near coastlines and rivers. Pollutants from the nearby landmasses are introduced into marginal seas in concentrations that are thousands of times greater than in open oceans. Pollutants enter marginal seas by way of water discharge and airborne deposition.

Biological growth of marine organisms in marginal seas can be either stimulated or inhibited depending on the nature and concentration of the partic-

biodiversity: a measure of the variety of the Earth's species, of the genetic differences within species, and of the ecosystems that support those species

pollution: any alteration in the character or quality of the environment, including water in waterbodies or geologic formations, which renders the environmental resource unfit or less suited for certain uses

The Mediterranean Sea and the Black Sea are marginal seas found in proximity to one another. The color difference shown here is due to a phytoplankton bloom occurring in the Black Sea.

ular pollutant—for example, whether it is a nutrient or a toxin. For example, the discharge of domestic sewage leads to elevated nutrient concentrations, which could result in increased **primary productivity** for phytoplankton, possibly including those species responsible for harmful algal blooms.

Biomass Production and Primary Productivity

Biomass is comprised of plants, herbivores (plant-eaters), carnivores (animal-eaters), and omnivores. Marine biomass production originates with primary productivity, which in turn is affected by the availability of sunlight, carbon dioxide, nutrients such as nitrates and phosphates, and trace elements.

primary productivity: the rate at which biomass is produced by photosynthetic and chemosynthetic organisms in the form of organic substances

Marginal Seas

upwelling: in marine environments, the movement of nutrient-rich water from great depths to the ocean surface; in hydrogeology, the upward movement of groundwater in areas of discharge (i.e., streams and springs); upward movement of water in a spring-fed pond or pool

turbidity: a measure of the cloudiness (reduced transparency) of water, determined by the amount of light reflected by particulate matter in the water

dissolved: describes the chemical breakdown of a solid in a solution into individual atoms or molecules and their dispersement in the fluid medium

suspended: describes a particulate remaining in a fluid for a long period of time because of its slow settling velocity in water or air

bathymetry: the science of measuring the depths and underwater topography of seas, oceans, lakes, and reservoirs

sill: the shallow area that separates coastal bays or marginal seas from the adjacent oceans or that separates two basins from one another

anaerobic: describes organisms able to live and grow only where there is no air or free oxygen, and conditions that exist only in the absence of air or free oxygen

Marginal seas generally exhibit intermediate levels of primary production, with the highest rates found in coastal **upwelling** regions and the lowest primary production occurring in open ocean regions. Hence, the highest biomass production rates occur in coastal upwelling zones, the lowest in open oceans regions, and intermediate rates in marginal seas.

Regional variations in primary productivity primarily occur as a result of water column chemistry (e.g., nutrient and trace elements concentrations) and dominant physical processes. For nearshore regions such as coastal waters and marginal seas, the dominant physical processes influencing primary productivity are river runoff, water column mixing, and **turbidity**. River runoff and water column mixing introduce **dissolved** nutrients, trace elements, and **suspended** particles into the photic (light) zones of nearshore regions. Although the addition of dissolved nutrients and trace elements to coastal waters and marginal seas serves to increase primary production, the addition of suspended particles increases water turbidity, which results in reduced sunlight penetration and decreased primary productivity.

Water Circulation

Water circulation patterns in marginal seas depend largely on **bathymetry**, fresh-water input (e.g., river runoff and precipitation) and evaporation. If river runoff and precipitation exceed evaporation, as is the case in the Black and Baltic Seas, the excess fresh water will tend to flow seaward near the sea surface, diluting the marginal sea. If evaporation exceeds river runoff and precipitation, as in the Mediterranean Sea, the marginal sea water becomes saltier, then sinks and flows towards the less salty open-ocean region.

Within marginal seas, these general water circulation patterns are often modified by the local bathymetry, which in some instances serves to restrict water flow. The water circulation patterns of four major marginal seas are described below.

Black and Baltic Seas. The Black Sea and Baltic Sea basins, which exhibit maximum depths of approximately 400 meters (1,312 feet) and 2,200 meters (7,216 feet), respectively, both possess **sills** that restrict subsurface-water circulation. While the surface waters of the Black and Baltic Seas are able to flow over the sills and introduce lower salinity water into the open ocean, the flow of the saltier subsurface waters is blocked by these bathymetric features. This type of subsurface-water restriction often leads to stagnation, which may eventually result in local oxygen depletion. In the Black Sea, this oxygen depletion process has led to subsurface **anaerobic** conditions and significant decrease in biomass with depth. Due to the shallowness of the Baltic Sea, anaerobic conditions only exist in the deepest areas, where stagnation occurs.

Mediterranean Sea. The Mediterranean Sea, which is divided by a 400-meter (1300-foot) sill into two subbasins, is connected to the Atlantic Ocean via the Straits of Gibraltar, to the Black Sea via the Bosporus Strait, and to the Red Sea via the manmade Suez Canal. Atlantic Ocean water enters this marginal sea through the Straits of Gibraltar as a surface flow. This ocean water replaces a fraction of the water that evaporates in the eastern Mediterranean Sea.

Upon arrival to the northern coast of the eastern basin, a portion of the Atlantic Ocean water is cooled (13°C or 56°F) and made saltier by evaporation, which results in its sinking in the Adriatic Sea. During the winter as the remaining Atlantic Ocean water continues to flow towards Cyprus, it sinks to a depth ranging from 200 to 600 meters (656 to 1,968 feet), where it forms the Mediterranean Intermediate Water. This water mass then flows along the North African coast into the North Atlantic Ocean via the Straits of Gibraltar. Because of its density, upon introduction into the North Atlantic, the Mediterranean Intermediate Water sinks to approximately 1,000 meters (3,280 feet), where it mixes with Atlantic Ocean water. This mixing process results in the formation of the Mediterranean water mass.

The circulation between the Mediterranean Sea and the Atlantic Ocean is typical of closed, restricted basins in which evaporation exceeds precipitation. As such this type of circulation pattern is often referred to as Mediterranean circulation, which is opposite of estuarine circulation.

Gulf of Mexico. Compared to the Black, Baltic and Mediterranean Seas, the Gulf of Mexico is a much less complex marginal sea. The Gulf of Mexico possesses an extensive, broad **continental shelf** and a maximum water depth of approximately 3,600 meters (11,808 feet). The Gulf of Mexico is connected to the Atlantic Ocean via the Straits of Florida and the Caribbean Sea via the Yucatán Strait.

continental shelf: the relatively flat, submerged natural platform, about 1-degree slope, that extends seaward from the beach for about 70 kilometers (45 miles), with water depth up to 130 meters (425 feet) maximum, and ends where the slope and water depth increase

Surface water in the Gulf of Mexico is as shallow as 90 meters (295 feet) in the winter and 125 meters (410 feet) during the summer. Surface salinities values for this marginal sea generally range between 36.0 and 36.3. In the northern Gulf of Mexico region, Mississippi River runoff influences surface waters as far as 150 meters away from the shore, resulting in salinities as low as 25. Gulf of Mexico surface-water temperatures range from 18°C to 21°C (64°F to 70°F) in the north and 24°C to 27°C (75°F to 81°F) in the south.

A unique feature of the Gulf of Mexico's surface circulation pattern is the Loop Current, which results from the Caribbean Current entering the Gulf of Mexico through the Yucatán Strait and upon arrival, turning in a clockwise direction and "looping" around a warm "dome" of Gulf of Mexico surface water. SEE ALSO ALGAL BLOOMS, HARMFUL; ALGAL BLOOMS IN THE OCEAN; BAYS, GULFS, AND STRAITS; BEACHES; COASTAL OCEAN; COASTAL WATERS MANAGEMENT; FISHERIES, MARINE; HUMAN HEALTH AND THE OCEAN; LIGHT TRANSMISSION IN THE OCEAN; OCEAN BASINS; OCEAN CURRENTS; OCEAN MIXING; POLLUTION OF THE OCEAN BY INDUSTRIAL WASTES; POLLUTION OF THE OCEAN BY SEWAGE.

Ashanti Johnson Pyrtle

Bibliography

Pickard, George L. and William J. Emery. *Descriptive Physical Oceanography: An Introduction*, 5th ed. New York: Pergamon Press, 1990.

Pinet, Paul. *Invitation to Oceanography*, 2nd edition. Boston, MA: Jones and Bartlett Publishers, 2000.

Thurman, Harold. *Introductory Oceanography*, 7th edition. New York: Macmillan, 1994.

Internet Resource

Environmental Sanitation Network. Sanitation Connection. <http://www.sanicon.net/index.php3.>

Mariculture

Mariculture is the farming of aquatic plants and animals in salt water. Thus, mariculture represents a subset of the larger field of aquaculture, which involves the farming of both fresh-water and marine organisms. The major categories of mariculture species are seaweeds, mollusks, crustaceans, and finfish.

Recent information indicates that the total amount of seafood (including fresh-water species and aquatic plants) is about 140 million metric tons annually. Over 20 percent of the total comes from aquatic plants (mostly seaweeds). Marine fish account for only 2 percent of the total.

Mollusks (clams, oysters, abalone, scallops, and mussels) represent the most important species cultured in marine waters. Seaweeds (brown, red, and green) are a close second. While most people do not think that they eat much (or any) seaweed, extracts from seaweeds can be found in everything from toothpaste and ice cream to automobile tires. Seaweeds themselves are dried and used directly as human food in many parts of the world.

Crustaceans include shrimp, crabs, lobsters, and crayfish. While shrimp culture has become a major industry in Asia and Latin American since the early 1980s, global production is far less than that of mollusks and seaweeds. Marine fish production is even smaller. Top finfish groups include Atlantic salmon, milkfish, sea bream, sea bass, red drum, yellowtail, striped bass, and hybrid striped bass.

The top mariculture-producing countries include the following.

Country	Species Produced
China	mollusks, shrimp
Japan	algae, mollusks, yellowtail, sea bream
Taiwan	mollusks, shrimp, eels
Philippines	algae, shrimp, milkfish
United States	mollusks, shrimp, Atlantic salmon, red drum
Norway	salmon
Ecuador	shrimp
Republic of Korea	algae, mollusks
Indonesia	algae, shrimp, milkfish

Types of Operations

Various levels of technology are involved in mariculture, the lowest giving nature the major role in producing the crop. The culturist may help prepare the growing area but does little else. For example, oyster culturists may place old shells on the bottom to provide places for a new generation of oysters to attach. The oysters feed on wild **phytoplankton** and are harvested when they reach the proper size. The next level would be to spawn oysters in a hatchery and allow the larval oysters (called spat) to settle on oyster shell, after which the shell is placed on the oyster bed in bays or suspended on ropes from a raft. Mussels and scallops also can be grown on ropes below rafts.

phytoplankton: microscopic floating plants, mainly algae, that live suspended in bodies of water and that drift about because they cannot move by themselves or because they are too small or too weak to swim effectively against a current

The culture of blue mussels on long ropes is common in the bays and inlets of Nova Scotia, Canada. This mollusk is economically important to local growers, even though it represents only a small fraction of the province's mollusk production.

Ponds. Shrimp and various species of marine fishes are often grown in ponds. The young shrimp and fish are usually produced in hatcheries, though collection of young animals from nature has been used in the past and is still used in some cases. The ponds may be filled with sea water by pumping water, or through tidal flow (the farmer opens the floodgate when the tide is rising and closes it when the pond is full). Depending on the particular species being produced and the size at stocking, the time required for the animals to reach market size can range from a few months to nearly 2 years.

Pens and Cages. In addition to ponds, marine fish also are being reared in floating pens or cages in protected bays.✳ Most cultured salmon are produced in these types of facilities, primarily in Norway, Canada, the United States, Scotland, and Chile. Various other fish species also are being produced in pens and cages in Japan, Europe, and the Middle East. In recent years, there has been interest and a limited amount of activity associated with cage culture in offshore waters.

✳ See "Aquaculture" for a photograph of a floating net pen.

Indoor Facilities. The highest level of technology is associated with indoor facilities in which the animals are grown in raceways or tanks (circular raceways) that receive pumped seawater that may be taken directly from the ocean. The water may be flowed through the tanks and discarded, or it may be recirculated, that is, reused by passing it through an elaborate water treatment system. Marine species can be reared to market size in such facilities, but they are most commonly used as hatcheries and to hold broodstock (adults used for reproduction).

Considerations

While a number of species are being reared successfully by mariculturists, several desirable ones have not yet been produced economically. This lack of commercial production is because their life cycles either are difficult to

Low-technology forms of mariculture take advantage of the nutrients and other physical and chemical characteristics of the ocean. This employee works with cultured seaweed on an East Africa plantation. The cycle of planting and harvesting occurs every 3 to 4 weeks year-round, and daily hours are spent preparing, planting, and collecting the seaweed during low tide.

control under culture conditions or are very complex. In addition, a number of popular food animals are highly cannibalistic. Various species of crabs and lobsters, for example, are difficult-to-rear species that also are cannibalistic.

Opposition to mariculture has developed in several countries since the 1980s. Many people do not want to see pens and cages in their bays, and they are concerned about possible environmental impacts associated with mariculture. Scientists are attempting to address these and a variety of other issues that have been raised. The goal is to produce high-quality seafood in an environmentally responsible manner.

Although world fish production from capture fisheries leveled off during the 1990s, demand for seafood continues to increase. This is because of the growth of the human population and also the view that seafood is healthy food. Scientists believe that natural production from the ocean will not increase; consequently, if the demand for seafood by humans is to be met in the future, both mariculture and fresh-water aquaculture production will have to increase significantly. SEE ALSO AQUACULTURE; BIVALVES; CRUSTACEANS; FISH; FISHERIES, MARINE; FOOD FROM THE SEA.

Robert R. Stickney

Bibliography

Avault, James W., Jr. *Fundamentals of Aquaculture*. Baton Rouge, LA: AVA Publishing, 1996.

Stickney, Robert R. *Principles of Aquaculture*. New York: John Wiley & Sons, 1994.

Internet Resources

The State of the World Fisheries and Aquaculture, 2000. Food and Agriculture Organization of the United Nations. <http://www.fao.org/sof/sofia/index_en.htm>.

Marine Mammals

What comes to mind when the subject of marine mammals is introduced? Most people probably only think of a few species of dolphins, whales, or

Keiko, the killer whale made famous by a Hollywood movie and by efforts to return him to the wild, is shown in 1997 at the Oregon Coast Aquarium in Newport, Oregon. As of early 2003, Keiko was being reintegrated into Iceland's wild orca population.

seals. But there are over one hundred species grouped into three orders: cetaceans, sirenians, and carnivores.

Marine mammals commonly are defined as mammals that require the ocean for most or all of their needs. Yet some scientists disagree on where to draw the line between terrestrial and marine mammals. Some regard a few species of bats and even the Arctic fox as marine mammals because they depend on food from the sea. Regardless of these more inclusive definitions, all true marine mammals have adapted to life in the water in wonderful ways.

Cetaceans

Cetaceans—whales, dolphins, and porpoises—spend their entire lives in water. Cetaceans are divided into two types: mysticetes and odontocetes.

Mysticetes. Mysticetes, such as humpback, right, minke, and gray whales, have baleen instead of teeth that are used for filtering small fish and invertebrates from sea water. Blue whales, the largest of all animals on Earth, at up to 27 meters (90 feet) long, are included in this group. Most baleen whales make yearly migrations, feeding during the summer in colder water and traveling up to thousands of miles in warmer and shallower areas to mate and give birth 1 year after mating. Baleen whales also have a thick layer of blubber, which is used both for insulation in cold water and energy storage during migration and winter fasting.

Odontocetes. The rest of the whales, as well as all dolphins and porpoises, are grouped together as odontocetes because they have teeth. Sperm whales, the largest odontocetes at up to 18 meters (60 feet) long, dive to as deep as 1.6 kilometers (1 mile), where they feed on deep-water fishes and squid. Most females of both sperm and pilot whales stay with their mothers their entire lives; but the males more often leave and form all-male groups. In sperm whales, they even live alone as "bachelors," only meeting up with other whales for breeding.

> **FREE KEIKO**
>
> Keiko, the killer whale featured in the 1993 Hollywood movie *Free Willy*, was captured in the Atlantic Ocean near Iceland in 1979. In 1996, largely in response to increased public awareness about his captive living conditions, Keiko was moved to a custom-built facility at the Oregon Coast Aquarium, then to an open-ocean pen in an Icelandic bay in 1998.
>
> With the help of his caretakers, Keiko underwent reintegration into Iceland's wild orca population through monitored interactions and ocean "walks." Keiko was allowed to wander free in January 2003, once other orcas had migrated to the same waters. Caretakers continue to support his reintegration, monitoring his health and progress.

Marine Mammals

The manatee (left), a member of the herbivorous order Sirenia, is found in warm coastal waters, including the southern United States. The elephant seal (right) belongs to the pinniped suborder of the order Carnivora, and is found as separate species in the northern Pacific, and in the southern Atlantic and Antarctic islands.

Although called a whale because of its size, the killer whale (*Orcinus orca*) is actually the largest member of the dolphin (Dephinidae) family. Killer whales, also often called orcas, have been studied most extensively in the Puget Sound off Washington state and British Columbia, where they are divided into two types: residents and transients.

Resident orcas eat fish and spend their lives in specific regions. Male and female offspring stay with their mothers their entire lives. Transient orcas eat primarily marine mammals, but even birds, turtles, or sharks are eaten on occasion. Transient orcas move between areas much more frequently; consequently, their family units are not as stable as the residents' units.

The smaller dolphins feed on a wide variety of organisms. Some, such as the pan-tropical spotted dolphin or the common dolphin, feed in the deep waters of the open ocean and are not seen by humans as often as the many species that come close to shore. Spinner dolphins have different forms: some live only in the deep ocean like those mentioned above, whereas others feed in deep water at night but rest in shallow bays during the daytime. Several dolphins, called river dolphins, live only in fresh water. Dusky dolphins of the southern hemisphere cooperate to herd schooling fish, to feed on them more successfully.

Bottlenose dolphins are very common in the wild and can be found close to shore almost anywhere except in Arctic and Antarctic waters. These dolphins feed on a wide variety of prey, mostly fishes, and many have adapted to living in areas close to humans. Flipper was a trained bottlenose dolphin that starred in a movie and a television show in the 1960s.

Porpoises are similar to dolphins, but tend to occur more often in colder waters. They also have different teeth, dorsal fin shape, and skull structure from those of true dolphins. The vaquita, or Gulf of California harbor porpoise, is endangered due to many individuals being killed in **gill nets**.

Sirenians

Manatees and dugongs are included in the order Sirenia. They are the only **herbivores** among marine mammals, feeding on sea grasses in tropical waters. They are slow-moving coastal animals, which is unfortunate for many manatees that often are struck by boats, being killed or severely injured. Be-

gill net: a net set upright in the water so that fish are caught in it when their gills become entangled in its meshes

herbivore: an animal that feeds mostly on plants

cause of their slow movements and inshore habitats, manatees often are kept successfully in large commercial aquaria.

Carnivores

The order Carnivora, of which cats and dogs are members, includes the pinnipeds (seals, sea lions, and walruses); sea and marine otters; and polar bears. All pinnipeds spend part of their time on land or ice but feed in the sea. Most mate on land or ice, and all need to be out of the water to give birth. The newborn pups are nursed anywhere from 3 days to 3 years, depending on species, and are mostly independent afterwards.

True seals, such as harbor seals, harp seals, or elephant seals, have thick blubber layers for insulation and have a strange way of getting around on land. They have to wiggle forward like an inchworm because their front flippers do not reach the ground. Despite their clumsiness on land, they are excellent swimmers. True seals have shorter infant care than the other pinnipeds.

Fur seals and sea lions have thinner blubber layers, and rely more on hair for insulation. They can walk on all four flippers, but generally do not dive as deep as the true seals. California sea lions are commonly trained to clap and play with balls in aquaria. Walrus, known for their long ivory tusks, occur only in the Arctic Ocean. They are also common entertainers in aquaria.

Sea otters and marine otters, both of the Pacific coast of the Americas, have the thickest fur of all marine mammals, and spend much of their time keeping it clean. Sea otters use their paws to gather shellfish to eat, and may even use rocks to crack open the shells. Marine otters cannot use their paws the way that sea otters do, and instead just grab prey with their mouths.

Although they do not spend as much time in the water as other marine mammals, polar bears are well adapted to life in the water. They swim with large paddle-shaped fore and hind feet, and feed on fishes in the water and seals on land or ice. Occasionally, they will even eat white whales and have been known to attack humans when threatened or approached too closely.

Marine mammals still serve as a primary food source for many aboriginal societies living near coasts worldwide. This seal meat is drying outside a residence in Savoonga, a small community on St. Lawrence Island in the Bering Sea, Alaska.

Marine Mammals

Marine mammals are prominent residents and entertainers at aquaria and entertainment complexes throughout the world. This bottlenose dolphin "smiles" for the camera with its celebrity visitors, track and field athletes Calvin and Alvin Harrison.

The Human Connection to Marine Mammals

Humans have had a long history of interaction with marine mammals, primarily with humans as hunters and marine mammals as prey. By the middle of the twentieth century, large-scale, unregulated whaling led to severe depletion of some populations. Within the last half century, recognition of this fact, combined with insights into the intelligence of marine mammals, has led to the emergence of protection and appreciation as the primary interactions between humans and marine mammals.

Aboriginal hunters of many coastal groups have long exploited whales, seals, sea lions, and other marine mammals for subsistence (survival) and, in the case of seals, fur. The effect of Native hunters on population levels was almost always small, unlike for some terrestrial mammals, because hunting large mammals on the open ocean is hard and dangerous, and an enormous amount of meat can be harvested from a single kill. Shore-dwelling seals are more susceptible to local extirpation, but even that was rare in the absence of commercial (rather than subsistence) hunting. Unregulated commercial hunting, however, can and has led to widespread decline in marine mammal populations. For example, fur seals were driven close to extinction in the 1800s by overhunting.

In colonial times, oil rendered from whale fat (blubber) was the main product that drove commercial whaling. Before the development of petroleum-based fuels, whale oil was widely used in lamps. Commercial whaling in the American colonies and elsewhere depleted coastal whale populations early on. Larger whaling ships and longer voyages further out to sea were the pattern in the 1800s, until exploitation of **fossil fuels** largely ended commercial whaling for oil.

Whaling for meat continued to be a commercially profitable enterprise, however, and the development of sea-based "factory ships" for processing

fossil fuel: substance such as coal, oil, or natural gas, found underground in deposits formed from the remains of organisms that lived millions of years ago

whales led to continued decline in the numbers of whales. Recognizing the serious depletion of whale populations, the International Whaling Commission (IWC) was founded by international agreement in 1946 to regulate the whaling industry. In 1983, the IWC decreed a moratorium on all commercial whaling, pending fuller understanding of the population dynamics and degree of endangerment of each commercial species.

Even with this moratorium, several aboriginal groups were allowed to maintain their traditional whale hunts. These include hunting of bowheads and greys by Northwest Coast and Eastern Russian Native groups; minke and fin whales by Greenlanders; and humpbacks by Caribbean Natives. Two countries—Japan and Norway—have attracted attention for continued whaling in the face of the moratorium, although each offers reasons why its whale catch falls within the few types of exceptions allowed by the IWC.✱ One such exception is for scientific research, and in part the controversy surrounds whether the research permits given out by these countries are simply used to skirt the regulations banning commercial harvesting. Despite these controversies, the ban on whaling has been an enormous success, and the populations of all types of whales have grown since the moratorium was instituted.

✱ See "Sustainable Development" for a photograph related to Japan's whale harvesting.

Other marine mammals have also been the subject of concern, most notably dolphins. While not a target of significant commercial harvesting themselves, they do get caught in large trawling nets designed to catch tuna. Like all mammals, dolphins must breathe air, and once caught in the nets, they drown. "Dolphin-friendly" tuna, which is not harvested via this fishing practice, is now marketed; international environmental monitoring groups work to ensure that companies that advertise dolphin-friendly tuna actually are using safe practices.

At the same time marine mammals were being increasingly protected from hunting, scientists increasingly came to appreciate the intelligence of marine mammals. It is no coincidence that the performing animals at Sea World and elsewhere are marine mammals: learning the tricks they display requires significant intelligence, which is not found in fish. While captivity continues to be the major environment in which humans interact with marine mammals, whale watching has become a significant tourist industry for some coastal towns, and it is even possible to "swim with the dolphins" in some warm-water bays. (In the United States, however, swimming with marine mammals is not legal.)

It is difficult to accurately compare the intelligence of different species, because the way humans measure intelligence often relies on skills possessed especially by themselves, such as language-based thinking and manipulation of objects with the hands. Despite these inherent limitations, it is clear that whales and dolphins are especially intelligent creatures, capable of solving puzzles and jumping through hoops for food rewards, and also having complex social systems and a type of "language." Researchers are attempting to understand these languages in hopes of learning more about these creatures and the societies they form. SEE ALSO ARTS, WATER IN THE; ECOLOGY, MARINE; FOOD FROM THE SEA; LIFE IN WATER; POLLUTION OF THE OCEAN BY SEWAGE, NUTRIENTS, AND CHEMICALS.

Amy G. Beier (taxonomic orders)
Richard Robinson (human connection)

Bibliography

Berta, Annalisa, and James L. Sumich. *Marine Mammals: Evolutionary Biology.* San Diego, CA: Academic Press, 1999.

Cahill, Tim. *Dolphins.* Washington D.C.: National Geographic Society, 2000.

Reynolds III, John E., and Daniel K. Odell. *Manatees and Dugongs.* New York: Facts on File, Inc., 1991.

Reynolds III, John E., Randall S. Wells, and Samantha D. Eide. *The Bottlenose Dolphin: Biology and Conservation.* Gainesville, FL: The University Press of Florida, 2000.

Riedman, M. *The Pinnipeds: Seals, Sea Lions, and Walruses.* Los Angeles: University of California Press, 1990.

Sterling, Ian. *Polar Bears.* Ann Arbor, MI: University of Michigan Press, 1988.

Internet Resources

International Whaling Commission. <http://www.iwcoffice.org>.

Markets, Water

Some people think of water as a gift of nature, with profound cultural, spiritual, and aesthetic significance, concluding that such a natural resource should not be bought and sold like fast-food hamburgers, t-shirts, or diamond rings. But water has many uses and immense total value, and if people want to use it to maximum benefit, they must have some mechanism to decide who gets it, when, for what purposes, and at what levels of quality.

At one extreme, a government might **allocate** water. Government officials would decide who gets how much, for what uses, at what times, and how to invest in the search for additional sources. At the other extreme, individuals who want to use water might be allowed to make these decisions, diverting water from streams or pumping it from wells, using it entirely as they wish and buying and selling it at will. Problems exist with both approaches, and probably no complex society has ever used either allocation procedure in its pure form.

allocate: to distribute resources for a particular purpose

Balancing Cost and Benefit

How much water should be used? In the figure below, each point on the marginal benefits (MB) curve indicates the amount consumers are willing to pay for the last unit of water purchased. Generally, the more one has, the less intensely one wants additional amounts. Hence, this curve slopes downward from left to right.

The marginal costs (MC) line shows the value of resources necessary to produce an additional unit of water. Generally, higher rates of flow can be had only at higher cost, so this line slopes upward.

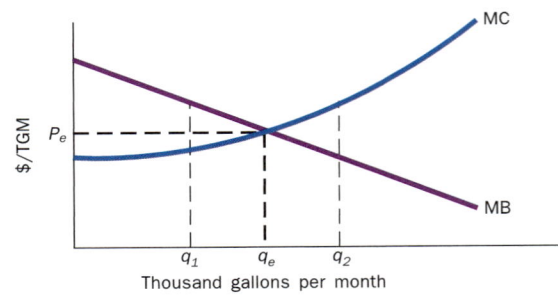

Consider the flow rate q_1. Here, MB > MC, indicating that consumers value an additional unit of water more than the resources necessary to produce it. This calls for increasing the quantity produced and consumed. At q_2, the opposite holds true. Only at q_e does MB = MC. No incentive exists to either increase or decrease the amount produced; buyers and sellers have achieved the optimum quantity. This conclusion rests on the assumption that the MB and MC curves include *all* costs and benefits, whether they accrue to the buyer or the seller or to some third party.

Consider a problem sometimes created when a farmer sells irrigation water to a distant city. Much of the water applied to fields would have seeped through the ground and eventually back into the stream, thus becoming available for reuse by people living downstream. If the city diverts the water into a canal or **aqueduct**, however, the irrigation return flows are cut off, and the downstream users, who have not been a party to the transaction, are deprived of a resource without compensation. The *private* marginal benefit and cost calculations that led to the optimum q_e may ignore "external" or third-party effects.

These effects have led many to conclude that governments must regulate water use tightly despite the high costs, delays, and controversy engendered by public decision processes. Economist Terry L. Anderson and others argue that properly defined water use rights would prevent much of the problem, in both quantity and quality dimensions—just as property rights in real estate, for example, prevent neighbors from encroaching or dumping trash into one another's yards. Indeed, the lack of well-defined water rights is the basis of much inefficiency and public controversy.

Market Institutions and Efficiency

Historically, the eastern United States generally enjoyed abundant water supplies, so the efficiency offered by **market institutions** was unnecessary (although this rosy picture darkened somewhat in the late twentieth century). Eastern states have mostly adopted variations of the riparian doctrine of water rights: owners of riparian land abutting streams have the right to withdraw water.

Market institutions have a longer history in the **arid** western states, where early settlers established efficient market systems on a type of first-come, first-serve basis known as the prior appropriation doctrine. In New Mexico, Colorado, and Utah, for example, it is possible to buy and sell rights to the use of stream flows or groundwater and divert such purchases away from any riparian (streamside) land. Transactions crossing state lines pose further difficulties, even for these states, and have been handled by rather rigid interstate **compacts**.

Market institutions require a clear definition of what it means to own the right to use water as well as the right to sell or buy such rights. Even the most efficient market entails some transactions costs, of course, to cover fees of brokers, agents, lawyers, and so on. But market transfers generally involve fewer people and clearer rules and so minimize such costs. In sum, market institutions offer an underused path to increased efficiency in water use. SEE ALSO DEMAND MANAGEMENT; LAW, WATER; PRICING, WATER; PRIOR APPROPRIATION; RIGHTS, PUBLIC WATER; RIGHTS, RIPARIAN.

James E. T. Moncur

---

aqueduct: long, canal-like or pipe-like structure, either above or below ground, for transporting water some distance

market institution: an arrangement that allows individuals to decide voluntarily how much of a good or service to produce, sell, buy, or consume based mainly on prices set by demand and supply conditions; in the alternative, the government decides who gets how much of a good or service, and at what price

arid: describes a climate or region where precipitation is exceeded by evaporation; in these regions, agricultural crop production is impractical or impossible without irrigation

compact: a formal contract or agreement between two or more parties, often governmental units

---

WATER MARKETS: THE PERVERSE CASE OF HAWAII

Relying mainly on water rights granted by the nineteenth-century monarchs, Hawaiian sugar planters developed extensive irrigation systems, according to their estimations of benefits and costs. They frequently bought or sold water rights, usually small amounts, as their needs changed, in transactions that were not much more involved or controversial than, say, buying a car. Titles were recorded officially by the government.

In 1972, however, the Supreme Court of Hawaii invalidated private water rights. Today, water transfers require elaborate and costly state approvals. No water sales have been recorded since 1972 despite vast changes in the state's water-using economy. Water allocation has been a continual major subject of public controversy, with costs for one single court case reaching millions of dollars.

Bibliography

Anderson, Terry L. *Water Rights: Scarce Resource Allocation, Bureaucracy, and the Environment.* San Francisco, CA: Pacific Institute for Public Policy Research, 1983.

Howe, Charles W., Dennis R. Schurmeier, and W. Douglas Shaw Jr. "Innovative Approaches to Water Allocation: The Potential for Water Markets." *Water Resources Research* 22 (1986):439–445.

Roumasset, James, Rodney Smith, and James Moncur. "Optimal Allocation of Ground and Surface Water on Oahu: Water Wars in Paradise." In *Conflict and Cooperation on Trans-Boundary Water Resources*, eds. Richard E. Just and Sinaia Netanyahu. Boston, MA: Kluwer Academic Publishers, 1998.

Saliba, Bonnie Colby, and David B. Bush. *Water Markets in Theory and Practice: Market Transfers, Water Values and Public Policy.* Boulder, CO: Westview Press, 1987.

Mars, Water on

Water and other easily vaporized molecules become more common as one moves outward from the Sun. Since Earth has a substantial amount of water, one can assume that Mars should as well.

Spectroscopic studies from Earth-based telescopes and Mars-orbiting spacecraft have detected water vapor in the Martian atmosphere, particularly in white clouds and fog. Temperature measurements made by the Viking Orbiter missions in the 1970s revealed that the north pole residual cap, which remains year-round, even at the height of Martian summer, was

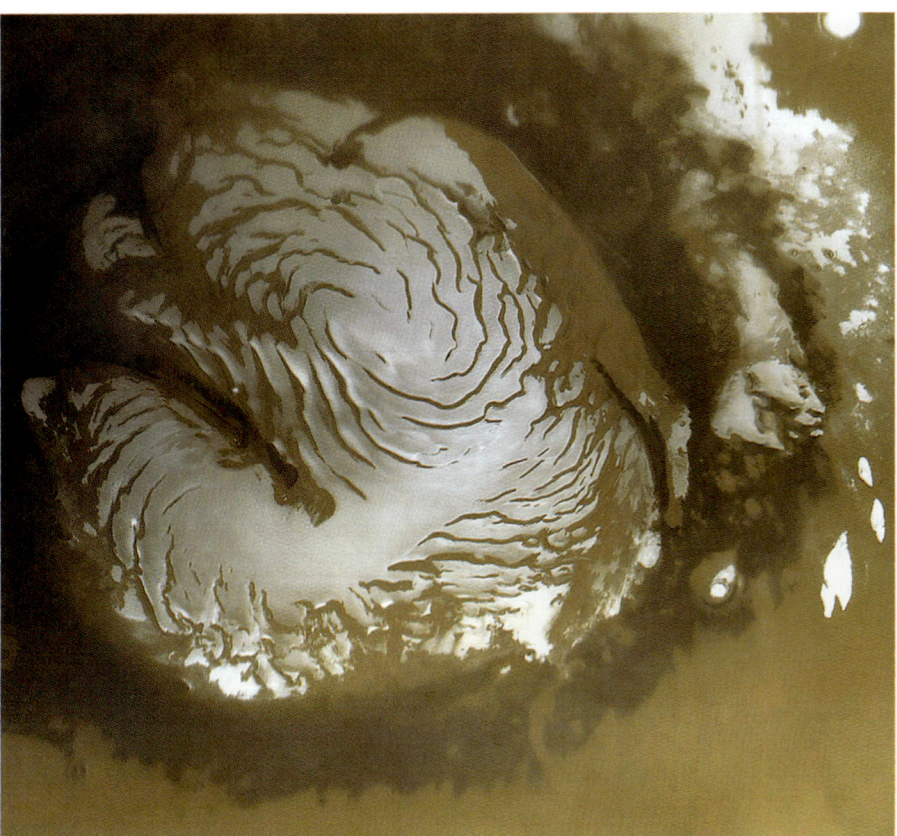

The residual north polar cap of Mars is roughly 1,100 kilometers (680 miles) across. In this photograph taken in March 1999 by the Mars Global Surveyor, the light-toned surfaces are residual water ice that remains through the Martian northern summer. The nearly circular band of dark material surrounding the cap consists mainly of sand dunes formed and shaped by the wind.

likely water ice, rather than carbon dioxide ice, which comprises the seasonal polar caps.

Although the Martian atmosphere and polar caps contain only a small amount of the water that scientists believe exists on Mars, geologic evidence from channels and craters suggests that much water (mainly ice) is buried below the surface of the planet, probably within the upper few kilometers of the surface. Recent results from Mars Odyssey seem to confirm that large amounts of water ice exist in the upper layers of Martian soil. Upcoming robotic Mars missions will directly measure the amounts and distribution of subsurface water by using ground-penetrating radar (Mars Express), seismic studies (Netlander), penetrators, and drilling.

Early Evidence: Cold and Dry

Early Mariner missions to Mars in the 1960s determined that its atmosphere is composed primarily of carbon dioxide and is extremely thin. At sea level on Earth, the atmosphere exerts a pressure of 1 **atmosphere**. At the surface of Mars, the atmosphere exerts a pressure of only 0.006 atmosphere. This very thin atmosphere has two important implications:

1. The temperature stays below freezing because the atmosphere is unable to retain heat; and
2. Liquid water cannot exist on the surface.

Consequently, the view of Mars in the 1960s was that it was a cold, dry world and that it had probably been that way throughout its history.

atmosphere: the gaseous layer surrounding the Earth, consisting of 78 percent nitrogen, 21 percent oxygen, and approximately 1 percent argon; atmospheric pressure at sea level is about 14.7 pounds per square inch, termed "one atmosphere"

Later Evidence: Geologic and Fluvial Activity

The *Mariner 9* spacecraft entered orbit around Mars in 1971 and revealed that 60 percent of the planet is very old, whereas the remaining 40 percent (located primarily in the northern hemisphere) has experienced more recent geologic activity. Among the features seen in the *Mariner 9* images were huge volcanoes, a gigantic canyon system, and channels that had apparently been formed by a flowing liquid. After considering all other options, scientists concluded that the channels had been formed by flowing water.

Two main types of channels are seen on Mars: the small valley networks, and the large outflow channels.

The valley networks are similar in appearance to channels on Earth that form from rainfall runoff. Valley networks tend to be found on ancient terrain, which led scientists to speculate that Mars had a thicker atmosphere and warmer surface conditions during its early history than it does today. Alternately, the valley networks could form by groundwater sapping, a process whereby the surface collapses as underground water is removed. Sapping could occur under present climatic conditions.

The outflow channels appear to have formed by huge floods. Outflow channels typically begin in areas of collapse, called **chaotic terrain,** where water bursts out from the subsurface, quickly carving channel pathways. Calculations show that such channels could form in only a few days and under present atmospheric conditions. By the end of the *Mariner 9* mission, scientists were beginning to see Mars as a place where limited amounts of water had affected its geology.

CANALS ON MARS?

Although a few astronomers using Earth-based telescopes in the early twentieth century reported what they believed to be canals on Mars, spacecraft imagery in the 1960s and 1970s showed that the canals were optical illusions.

chaotic terrain: Martian surface having the appearance of jumbled and broken angular slabs or blocks; may be related to the melting of subsurface ice followed by collapse of the surface

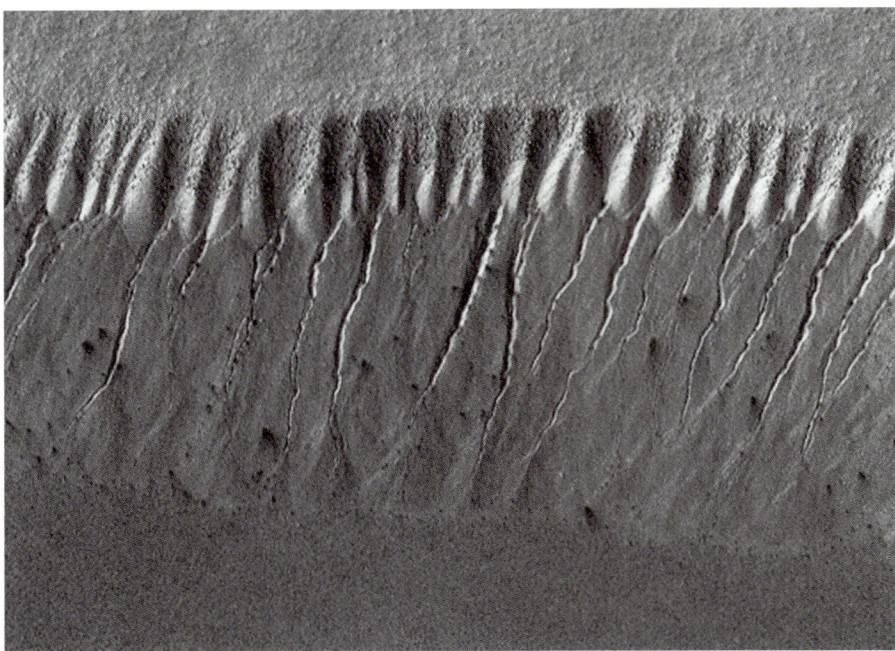

These gully landforms near the south polar region of Mars show evidence of geologically recent seepage and runoff of liquid water. Today, however, water on Mars appears to exist mainly as subsurface ice.

Other Geologic Evidence

Information obtained by the Viking Orbiters in the 1970s and 1980s led scientists to begin to look more closely at a number of mysterious geologic features and consider how water might be involved in their formation. **Ejecta blankets** surrounding fresh Martian impact craters display a fluidized appearance, quite different from the radial pattern seen around craters on dry bodies such as the Moon. Most scientists now believe that these ejecta patterns form by subsurface water and ice being vaporized during crater formation.

In the northern plains of Mars, features such as **polygonal terrain** (which look like large mud cracks) suggest that the area was once saturated with water and has since dried out. The rims of many craters in the ancient terrain are **dissected** by channels, and the floors of these craters appear to be covered by smooth sedimentary deposits.

A Martian Ocean?

Scientists proposed a radical idea in the mid-1980s: namely, that the low-lying northern plains of Mars may have contained an ocean within the past 2 billion years or so. They based this theory on a number of observations:

- Linear features along the edges of the plains that resemble shorelines;
- The concentration of polygonal terrain and fluidized ejecta **morphologies** in this region; and
- The large number of outflow channels that debouch (empty) into the northern plains.

Recent evidence from the Mars Global Surveyor (MGS) mission, in orbit around Mars since 1997, seems to support this view. MGS instruments have revealed that the northern plains are extremely flat, which is a characteristic of the deep ocean basins on Earth.

ejecta blanket: material ejected (thrown out) during impact crater formation and deposited around the crater, covering the surrounding terrain

polygonal terrain: a major morphologic component of the Utopia Planitia region of the northern plains on the planet of Mars; thought to be similar to outflow channels and a possible ancient ocean; its formation process is a subject of debate

dissect: in landscape evolution, to cut by erosion into hills and valleys or into flat upland areas separated by valleys

morphology: the external shape, structure, form, and arrangement of landforms and waterbodies

Evidence in Minerals and Sedimentary Features

The ancient terrain of Mars shows the effects of water. MGS has revealed large deposits of minerals, such as hematite, which on Earth commonly form by interaction with water. Gullies are found along crater and canyon walls and show evidence that they are recent features. Layered deposits have been identified in topographic depressions and are similar in appearance to sedimentary deposits on Earth.

Minerals found in Martian meteorites also show evidence of interaction with water. These rocks were formed and altered by geologic processes on Mars prior to being ejected off the planet by impact events (and which subsequently have fallen to Earth). Analysis of these meteorites reveals the environmental conditions under which they existed on Mars. Carbonates and other minerals formed by interactions with water, such as clays, are common in these meteorites.

Reconciling the Evidence

Scientists' perception of Mars has changed dramatically in 40 years, from that of a dry desert world to one where flowing water has played a major role in shaping its surface. How can this picture be reconciled with the current atmospheric conditions that prohibit liquid water on the surface?

Scientists now believe that Mars goes through episodes of climate change, caused by variations in the tilt of its rotational axis and orbital factors due to the gravitational effects of other planets. Right now, conditions on Mars are such that most water is tied up as ice in the polar caps or in the subsurface. However, when the Martian poles tip more than their current 25-degree tilt, increased solar **insolation** can cause the polar ices to evaporate, thickening the atmosphere and causing a greenhouse warming of the surface. Subsurface ice can then melt and appear on the surface in the form of rivers, lakes, and perhaps even oceans.

insolation: the amount of solar radiation that reaches a given area

The episodic appearance of liquid water has important implications for the question of life (either present or past) on Mars and is of great interest as a resource for future explorers who may colonize Earth's neighboring world. SEE ALSO ASTROBIOLOGY: WATER AND THE POTENTIAL FOR EXTRATERRESTRIAL LIFE; COMETS AND METEORITES, WATER IN; LIFE IN EXTREME WATER ENVIRONMENTS; SOLAR SYSTEM, WATER IN THE; STREAM EROSION AND LANDSCAPE DEVELOPMENT.

Nadine G. Barlow

Bibliography

Carr, Michael H. "Mars." In *The New Solar System*, 4th ed. Beatty, J. Kelly, Carolyn Collins Petersen, and Andrew L. Chaikin, eds. New York: Cambridge University Press, 1998.

Carr, Michael H. *Water on Mars*. New York: Oxford University Press, 1995.

Microbes and Human Health *See Algal Blooms, Harmful; Drinking Water and Society; Human Health and Water; Microbes in Groundwater; Microbes in Lakes and Streams; Microbes in the Ocean; Safe Drinking Water Act; Supplies, Protecting Public Drinking-Water.*

Microbes in Groundwater

More than 95 percent of the world's available fresh water (excluding ice caps and glaciers) is underground. This **groundwater** is valuable as a source of drinking water for most communities in the world, especially small ones. In the United States, about 15 million private wells serve fewer than 25 people each. In addition, about 92 percent of the 168,000 water systems serving 25 people or more get most or all of their drinking water from groundwater. This article focuses on the common microbes in groundwater (bacteria, protozoa, and viruses), primarily from a public health point of view.

Commonly Occurring Microbes

Groundwater near the land surface usually teems with microbial life. Bacteria, which are microscopic single-celled organisms that lack a true nucleus and normally have a cell wall, are far more numerous than any other organism in the soil and groundwater. Near the surface where plant roots are abundant, there may be 100 million to 1 billion bacteria per gram of dry soil. These values decrease dramatically with soil depth below the root zone, with densities depending on the amount of nutrients (food) and water available. Although concentrations of microbes below the root zone are lower than in the root zone itself, as many as 10 to 100 million bacteria per gram of **aquifer** material may be present. Bacteria have been found in core samples from a depth of 2.8 kilometers (1.7 miles) below the Earth's surface, and at a depth of 3.2 kilometers in South African gold mines.

Protozoa, single-celled organisms that have a nucleus but lack a cell wall, also are common in groundwater. Protozoa typically are much larger than bacteria, and many types in groundwater feed on bacteria. Molds and other fungi also are common in groundwater near the land surface, where plenty of oxygen is present. These microbes are larger than bacteria and, unlike the protozoa, have a cell wall and often grow in long filaments. Most fungi feed on dead or decaying material. **Algae** may even be present despite the absence of sunlight. Similar to the bacteria, the numbers of protozoa, fungi, and algae decrease with depth.

A large variety of viruses probably are also common in groundwater. Viruses are much smaller than other microbes and cannot be seen with ordinary microscopes. They do not have a nucleus or cell wall, and can multiply only within the cells of larger organisms. The natural viruses in groundwater are able to reproduce only by infecting and usually killing the bacteria and other larger microbes present.

Microbial Habitats. Microbes below Earth's surface must compete vigorously with each other for limited food and space. Bacteria that are not adapted to this **habitat** cannot survive long. Normally, nutrients are scarce in this environment, and thus microbes in groundwater grow slowly. Because the microbes in groundwater and other environmental habitats are well adapted to life in these habitats, few of them have the ability or the need to infect and cause illness in humans and other animals. However, a few environmental bacteria and protozoa may cause disease in humans by chance under certain circumstances, especially in individuals with a weakened immune system or major breaks in the skin that allow microbes to enter.

Microorganisms found in groundwater can be seen only through a microscope (shown here). Viewing extremely small organisms such as viruses requires a powerful microscope, such as a scanning electron microscope.

groundwater: generally, all subsurface (underground) water, as distinct from surface water, that supplies natural springs, contributes to permanent streams, and can be tapped by wells; specifically, the water that is in the saturated zone of a defined aquifer

aquifer: a water-saturated, permeable, underground rock formation that can transmit significant quantities of water under ordinary hydraulic gradients to wells and springs

algae: (singular, alga) simple photosynthetic organisms, usually aquatic, containing chlorophyll, and lacking roots, stems, and leaves

habitat: the environment in which a plant or animal grows or lives; the surroundings include physical factors such as temperature, moisture, and light, together with biological factors such as the presence of food and predators

Many naturally-occurring bacteria below the surface may serve a beneficial role by destroying human-made toxic chemicals that migrate from the surface to groundwater. Some **toxic** chemicals, such as gasoline, are easy for subsurface bacteria to break down, but others, such as dry-cleaning fluid and some pesticides, are destroyed only slowly, if at all.

Disease-Causing Pathways

The groundwater in a drinking-water well may contain a wide variety of microbes without presenting a public health risk. However, groundwater in some areas becomes contaminated by the fecal material of humans and other animals. This is a cause for concern because fecal material may contain pathogenic (disease-causing) microbes that can infect the intestinal tract of humans. Fecal pathogens may be bacterial, viral, or protozoan.

Water containing fecal material may seep into the groundwater from the land surface or from underground sources of contamination. Major surface sources include:

- Wastewater and biosolids from sewage treatment facilities that have been applied to land as a soil conditioner;
- Seepage from shallow artificial ponds (lagoons) used for processing sewage;
- Seepage from contaminated lakes and other surface-water bodies;
- Urban runoff;
- Feces from cattle and other livestock operations; and
- Improperly constructed sanitary landfills where trash and garbage are disposed.

Fecal contamination also can reach the groundwater from underground sources, such as improperly functioning septic tank systems, underground reservoirs for liquid household sewage (cesspools), or leaking underground sewer lines. About 25 million septic tanks exist in the United States.

Pathogen Movement and Persistence. Most pathogens from fecal matter remain either near the surface, or near the point of origin in the case of an underground source. However, where conditions are favorable, some pathogens travel along with the water flow though pores (tiny openings) in the surrounding soil and rock, and may enter groundwater. Whether a pathogen reaches the groundwater, especially in the area of a well intake, depends on how strongly it is retained by the soil and how long it survives. The extent to which a pathogen is retained by the soil depends on:

- Soil characteristics, such as the number and size of pores and their interconnections, and how strongly the pathogen adsorbs (binds) to the soil particles;
- Environmental factors, especially the amount of rainfall; and
- The characteristics of an organism, including its size and its tendency to attach to other organisms and organic debris.

Rapid pathogen transport may occur in **aquifers** that have large pores (such as in a gravel-dominated aquifer), as well as in fractured rock, cave systems, and **sinkholes**.

toxic: describes chemical substances that are or may become harmful to plants, animals, or humans when the toxicant is present in sufficient concentrations

aquifer: a water-saturated, permeable, underground rock formation that can transmit significant quantities of water under ordinary hydraulic gradients to wells and springs

sinkhole: a depression in the Earth's surface caused by the collapse of underlying limestone, dolomite, salt, or gypsum

PATHOGENIC VIRUSES IN GROUNDWATER

Common sources by which disease-causing viruses enter groundwater are land disposal of sewage, overflow from septic systems, and livestock waste. Leachate from solid waste landfills also can contain viruses.

Contamination of drinking water with hepatitis A virus and other viruses has been documented throughout the world. Knowing how long these viruses can remain infective is important in estimating the probability that groundwater flow will move the viruses to drinking-water wells while they are still active.

Infective viruses have been demonstrated to travel more than 50 meters in depth from septic tanks into drinking-water wells. Viruses can travel great distances horizontally in aquifers with rapid water velocity, such as karst systems or alluvium with coarse cobbles.

Hydrologists are only beginning to understand the community dynamics of groundwater microbes, particularly viruses. More knowledge of the hydrology of soils and sediments also is necessary to better assess the human health problems related to viral contamination of groundwaters.

Microbes in Groundwater

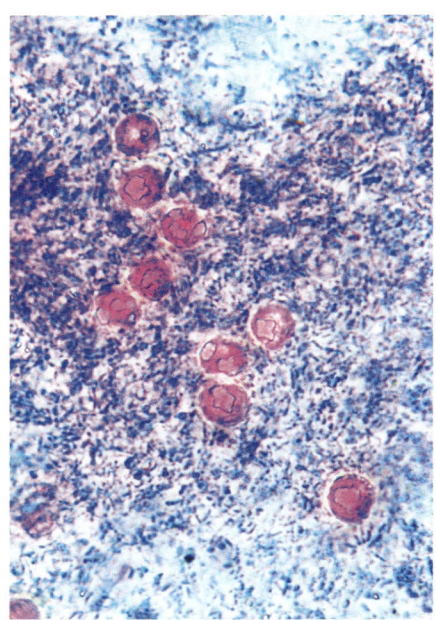

Cryptosporidium is a protozoan that occurs naturally in surface waters such as streams and lakes. Its presence in groundwater indicates that surface water is gaining access to the aquifer, either naturally or induced through pumping. *Cryptosporidium* may cause serious intestinal illness.

✷ The U.S. Environmental Protection Agency's drinking-water regulations can be found on the Internet at <http://www.epa.gov/safewater>.

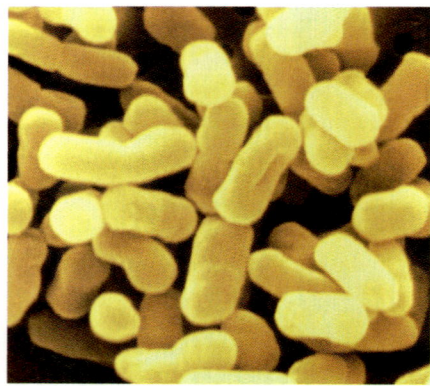

Escherichia coli is a bacteria found in the intestinal tract of warm-blooded animals. Its presence in groundwater indicates that a pathway exists between the land surface and the aquifer, and may indicate that the water is not safe to drink.

Pathogen survival depends on a number of factors, including soil moisture content (longer survival when moist), soil temperature (longer survival at low temperatures), soil acidity (shorter survival in acid soils), amount of nutrients (longer survival of bacterial pathogen if high), and the activity of microbes that normally live in the soil. Normal soil bacteria are fully adapted to their environment and should easily outcompete bacterial pathogens for available nutrients. Some will also release one or more chemicals (toxins) that may kill the pathogens.

In most cases, only a small portion of the original pathogen density will reach a well intake unless a rapid pathway exists (such as fractured rock). Despite this, most reported waterborne disease outbreaks reported in the United States each year are associated with groundwater. Between 1971 and 2000, almost 60 percent of the approximately 700 reported outbreaks, with more than 90,000 cases of illness, were associated with water systems using groundwater.

Protecting Public Health

How can the public be protected from fecally contaminated groundwaters? Drinking-water supplies using groundwater can be protected from fecal contamination of the groundwater in several ways. Most importantly, suppliers can ensure that their well meets the state well-construction code. They can identify any sources of fecal contamination near the well intake and take proper corrective measures. Suppliers can also have an expert conduct a periodic on-site inspection, including an assessment of how vulnerable the source water is to nearby sources of fecal contamination.

Another means by which a water supply can be protected is to periodically test its source-water quality. To test for all possible pathogens, however, is not practical; the wide variety of potential waterborne pathogens and the limitations of analytical methods would make testing for all such pathogens extremely difficult, time-consuming, and expensive. Instead, inexpensive and easily measured indicators of fecal contamination are used rather than testing for specific pathogens. These include microbes that are extremely common in the gut of humans and other warm-blooded animals, but not elsewhere. If such indicators are absent, the pathogens whose normal habitat is the intestinal tract of humans and other warm-blooded animals should also be absent. Among the fecal indicators used for environmental waters, including groundwater, are *Escherichia coli* (a bacterium), fecal streptococci (a small group of bacteria), enterococci (very similar to fecal streptococci), and coliphage (a group of viruses that infect and kill *E. coli*).

Water disinfection provides another barrier to pathogen entry and survival in the pipe network that distributes water (distribution system), especially if the source water is contaminated or vulnerable to fecal contamination. Disinfection kills pathogens and other microbes. Commonly used disinfectants for groundwaters include chlorine and ultraviolet light.

The U.S. Environmental Protection Agency develops regulations that apply to every water supplier that serves at least 25 people for at least 60 days during a year (defined as a public water system).✷ One regulation that protects groundwater systems from pathogens is the Total Coliform Rule (TCR). The TCR requires every public water system to monitor its tap water for a group of closely related bacteria, called total coliforms, at a frequency that depends on the number of people a system serves. While total coliforms are not a fecal indicator, they are used to detect problems in water treatment and in the distribution system. If any sample is total coliform positive, the system must then test for a fecal indicator (normally *E. coli*).

Another regulation, the Surface Water Treatment Rule (SWTR), in addition to surface waters, also covers groundwater systems that are directly under the influence of surface waters. Among other requirements, the SWTR requires all such systems to disinfect. An important element in both the TCR and the SWTR, as amended, is a requirement for systems to have a periodic on-site inspection (sanitary survey) by a state official or a state-approved agent.

As of 2002, the Environmental Protection Agency was developing another regulation, the Groundwater Rule, which will require groundwater systems to determine whether the source water is fecally contaminated or is vulnerable to such contamination. If so, the system will be required to take corrective measures. SEE ALSO DRINKING-WATER TREATMENT; ECOLOGY, FRESH-WATER; ENVIRONMENTAL PROTECTION AGENCY, U.S.; GROUNDWATER; HUMAN HEALTH AND WATER; KARST HYDROLOGY; LAND USE AND WATER QUALITY; LANDFILLS: IMPACT ON GROUNDWATER; LIFE IN EXTREME WATER ENVIRONMENTS; MICROBES IN LAKES AND STREAMS; POLLUTION OF GROUNDWATER; POLLUTION OF GROUNDWATER: VULNERABILITY; POLLUTION SOURCES: POINT AND NONPOINT; SAFE DRINKING WATER ACT; SEPTIC SYSTEM IMPACTS; STREAM, HYPORHEIC ZONE OF A; SUPPLIES, PROTECTING PUBLIC DRINKING-WATER; SUPPLIES, PUBLIC AND DOMESTIC WATER; WASTEWATER TREATMENT AND MANAGEMENT; WELLS AND WELL DRILLING.

Paul S. Berger

Bibliography

Atlas, Ronald M., and Richard Bartha. *Microbial Ecology*. Redwood City, CA: The Benjamin/Cummings Publishing Company, 1993.

Bitton, Gabriel, and Charles P. Gerba, eds. *Groundwater Pollution Microbiology*. New York: John Wiley & Sons, 1984.

Chapelle, Francis H. *Ground-Water Microbiology and Geochemistry*. New York: John Wiley & Sons, 1993.

Frederickson, James K., and Tullis C. Onstott. "Microbes Deep Inside the Earth." *Scientific American* 275, no. 4 (1996):68–73.

Microbes in Lakes and Streams

Many sights and sounds attract people to the water, from the waves lapping at the shore of a lake, to the fish and turtles that break the surface now and again. But other components of the fresh-water **ecosystem** cannot be seen with the naked eye: the **microorganisms**, or microbes, that constitute the foundation of the aquatic abundance that people see and enjoy. Microbes include bacteria, bacteria-like organisms called archaea, viruses, protozoa, helminths, and protists. Microbes are natural and vital members of all aquatic communities, and are the foundation of lake and stream ecology—without them the natural water worlds would not be possible. Certain microbes, however, when present in excessive numbers, pose a threat to human health.

ecosystem: the community of plants and animals within a water or terrestrial habitat interacting together and with their physical and chemical environment

microorganism: a microscopic organism

Microbes in Lakes and Streams

Leaves and branches that fall into lakes and streams are an important source of organic matter. Bacteria and other microbes break down these materials, making their nutrients available for other aquatic creatures.

photosynthesis: the process by which plants manufacture food from sunlight; specifically, the conversion of water and carbon dioxide to complex sugars in plant tissues by the action of chlorophyll driven by solar energy

cyanobacteria: also known as blue-green algae, primitive single-celled organisms structurally similar to bacteria, sometimes joined in colonies or filaments

food chain: the levels of nutrition in an ecosystem, beginning at the bottom with primary producers, which are principally plants, to a series of consumers—herbivores, carnivores, and decomposers

organic: pertaining to, or the product of, biological reactions or functions

Ecological Roles of Fresh-Water Microbes

Like all ecosystems, fresh-water ecosystems require energy inputs to sustain the organisms within. In lakes and streams, plants and also certain microbes conduct **photosynthesis** to harvest the Sun's energy. Microbial photosynthesizers include protists (known as algae) and **cyanobacteria**. Other protists and animals feed on these organisms, forming the next link in the **food chain**. Plant material from the land also enters lakes and streams at their edges, providing an important nutrient source for many waterbodies.

Decomposers form an especially important part of fresh-water ecosystems because they consume dead bodies of plants, animals, and other microbes. These microbial agents of decay are an important part of the ecosystem because they convert detritus (dead and decaying matter) and **organic** materials into needed nutrients, such as nitrate, phosphate, and sulfate. Decomposers and other microbes are thus essential to the major biogeochemical cycles by which nutrients are exchanged between the various parts of the ecosystem, both living and nonliving. Without microbial decomposers, minerals and nutrients critical to plant and animal growth would not be made available to support other levels of the fresh-water food chain.

Aerobes and Anaerobes. Aerobic decomposers in water need oxygen to survive and do their work. The lapping waves and babbling brook help increase the level of dissolved oxygen that is crucial to so many creatures in lake and stream ecosystems, none more so than the bacteria. If there is not enough oxygen in the water, many parts of the system suffer: the aerobic decomposers cannot digest plant matter, insects cannot develop and mature, and the fish cannot play their part, whether browsing for small food particles or eating other fish. Eventually, the stream or pond will be changed, starting at the microbial level.

Human interaction can jeopardize parts of this system in a variety if ways. One principal way is through the runoff of fertilizers or sewage into a waterbody. Both contain nutrients that plants, algae, and cyanobacteria can use to grow; and excessive nutrient amounts can lead to very rapid growth. Interconnected sequences of physical, biological, and chemical events may eventually deplete the water's dissolved oxygen supply, leading to changes in the aquatic ecosystem. If the conditions become severe enough, only a few species (known as anaerobes) tolerant of low-oxygen conditions will survive. This process, called cultural eutrophication, can have profound and lasting consequences on the waterbody. ✳

✳ See "Lake Management Issues" for a schematic of cultural eutrophication.

Microbes and Human Health

Fresh water is host to numerous microorganisms that affect human health directly.✳ Polluted drinking water is a major source of illness and death throughout the world, particularly in developing countries. In almost all cases, the organisms responsible cycle from the waterbody through the digestive tract of humans or other animals. Released in fecal waste of the infected host, they enter the water again to complete their life cycle. Most infections derived this way cause diarrhea, abdominal cramping, and potentially more serious symptoms, including fever, vomiting, and intestinal bleeding. Some common microbes in lakes and streams that are responsible for disease include:

✳ See "Human Health and Water" for a summary table of common waterborne pathogens.

- The protist *Giardia lamblia*, found in fresh-water bodies throughout the world. Giardia infection is a common waterborne illness in the United States.

- The bacterium *Vibrio cholerae*, while rare in the United States, remains a significant source of disease and death in countries without advanced sewage treatment and with no potable water supplies. For example, a cholera epidemic in 1991 killed more than a thousand people in Peru (South America), where more than 150,000 cases of the illness were confirmed.

- The bacterium *Escherichia coli*, a very common waterborne pollutant. Humans have a large and harmless population of *E. coli* in their lower, large intestines, and bacteria make up a large fraction of the volume of human feces. When released into drinking water or recreational water sources, *E. coli* can be ingested and enter the upper small intestine, causing diarrhea. Other fecal bacteria known as "coliform" bacteria cause similar symptoms. The level of fecal coliform bacteria in pools, ponds, and other waterbodies is frequently measured during the summer months to assess the safety of recreation in these waters.

Treating and Preventing Microbial Pollution

Sewage treatment is the single most important strategy for preventing waterborne microbial pollution. While the earliest treatment techniques simply settled the solids and released the liquids into a much larger waterbody (such as a river or bay), modern sewage treatment relies on multistage facilities that use a sequence of physical, chemical, and biological treatments

Microbes in the Ocean

Outdoor enthusiasts must plan ahead to avoid waterborne illnesses linked to contaminated streams and lakes. These campers are pumping water from what appears to be a pristine lake, but microbes in the water can make it unsafe for drinking. Various treatment options are available to make natural surface waters safe for consumption.

✳ See "Wastewater Treatment and Management" for a photograph of a sewage treatment plant.

and filters to treat sewage until it often is cleaner than the original water source from a river, lake, or reservoir.✳

Hikers and other outdoor enthusiasts can protect themselves by never drinking untreated water from lakes or streams. Water can be made safe in several ways:

- Boiling;
- Filtering (e.g., using small portable filters); and
- Adding chemical tablets, often containing iodine.

SEE ALSO ALGAL BLOOMS IN FRESH WATER; CLEAN WATER ACT; DRINKING-WATER TREATMENT; ECOLOGY, FRESH-WATER; FRESH WATER, NATURAL CONTAMINANTS IN; HUMAN HEALTH AND WATER; LAKES: BIOLOGICAL PROCESSES; LAKES: CHEMICAL PROCESSES; MICROBES IN GROUNDWATER; PLANKTON; POLLUTION SOURCES: POINT AND NONPOINT; SEPTIC SYSTEM IMPACTS; STREAM, HYPORHEIC ZONE OF A; WASTEWATER TREATMENT AND MANAGEMENT.

Neil Clark and Richard Robinson

Bibliography

Atlas, Ronald M. and Richard Bartha. *Microbial Ecology: Fundamentals and Applications*, 3rd ed., Redwood City, CA: Benjamin/Cummings, 1993.

Ford, Timothy E., ed. *Aquatic Microbiology*. 1993. Boston, MA: Blackwell Science Inc., 1993.

Microbes in the Ocean

The oceans teem with microorganisms such as bacteria, viruses, and protists. Many of these microbes fundamentally influence the ocean's ability to sustain life on Earth. Some microbes living and transported in ocean water, however, threaten human health.

Occurrence and Role in the Ocean

In the open ocean, far from the influences of coastal human habitation, sea water still contains huge numbers of microbes. Coastal areas can contain even greater concentrations. Vast numbers of bacteria and **plankton** occur both at the surface and in deep ocean waters. Viruses are entities that require bacteria or other cells in order to make copies of their genetic material and to construct new casings that house the genetic material. Scientific studies have shown that 10 to 100 million viruses can be present in a teaspoonful of sea water.

Plankton. More plankton exist in sea water than any other organism. Microscopic forms include protists and bacteria. Phytoplankton are photosynthetic organisms, including algae. By harvesting the energy of the Sun and converting it to their tissues, phytoplankton form the basis of the **food chain** in the ocean. All ocean organisms depend on phytoplankton either directly or indirectly. Eventually, humans consume ocean creatures such as fish. Even human life, therefore, is tied to the presence of phytoplankton.

Microbes such as plankton also have other benefits. In the ocean, they help make some nutrients available to other living marine creatures. Elemental iron, for example, is important for living creatures but is scarce in the ocean. Sunlight can change iron to a form that can be taken up by plankton and other microbes. The microorganisms are used as food by other organisms, such as fish and ocean mammals, making the iron available to other creatures in the food chain.

Viruses. Marine viruses can be both detrimental and beneficial to the ocean's health. Some viruses attack and kill plankton, eliminating the base of the ocean food chain in a particular area. At the same time, this dead plankton can become a source of carbon that is not otherwise readily available to other sea life. Scientists estimate that up to 25 percent of all living carbon in the oceans is made available through the action of viruses. When these aspects remain in proper balance, the ocean functions normally.

Bacteria. Bacteria are single-celled organisms without cell nuclei. They are found in all portions of the water column, the sediment surface, and the sediments themselves. Some are aerobic (requiring oxygen), whereas others are anaerobic (not requiring oxygen). Most bacteria are free-living, but some live as partners (symbionts) within other organisms. For instance, many deep-sea fish harbor symbiotic bacteria that emit light, which the fish use to signal other members of their species. The bacteria's ability to emit light is called bioluminescence. Bioluminescence causes water to glow, a phenomenon most noticeable at the surface but present at all depths.

Cyanobacteria, a type of bacteria, played an important role in the history of Earth and in ocean processes, including the development of stromatolites (see photograph on page 80). Living in colonies, the cyanobacteria produced oxygen during the process of photosynthesis, which generated the oxygen in the Earth's atmosphere that many living beings require today. Although cyanobacteria also are called "blue-green algae," it is important to remember that cyanobacteria are relatives of bacteria and not algae. They are, however, related to the chloroplasts within algae; the chloroplasts used by some plants to produce food are actually cyanobacteria living within plants' cell walls.

Photobacteria research is one of many subspecialties of marine microbiology. The study of bioluminescence (the biological generation of light) provides insights into fundamental aspects of cell biology and has far-ranging applications, including ocean dynamics and naval operations.

plankton: an assemblage of small, often microscopic aquatic organisms encompassing aquatic plants (phytoplankton) and aquatic animals (zooplankton) that float or drift passively with water currents, having no or very limited powers of locomotion

food chain: the levels of nutrition in an ecosystem, beginning at the bottom with primary producers, which are principally plants, to a series of consumers—herbivores, carnivores, and decomposers

Microbes in the Ocean

Stromatolites are layered, mushroom-shaped structures formed by lime secreted by cyanobacteria. Fossilized stromatolites have provided scientists with critical information about the earliest development of life on Earth. Living stromatolites are among the unique forms of life in the ocean.

organic: pertaining to, or the product of, biological reactions or functions

Some marine bacteria can interact with diatoms, another type of marine microbe, in such a way that influences the cycling of silicon in the ocean. Diatoms, a group of unicellular algae, are characterized by their highly ornate two-part shell-like structures made from silica.

One species of bacteria, *Thiomargarita namibiensis*, plays a critical role in hydrogen sulfide eruptions from diatomaceous sediments off Africa's Namibia coast. Known as the "sulfur pearl of Namibia," this anaerobic species digests **organic** matter under low-oxygen (or no-oxygen) conditions that are caused by high rates of phytoplankton growth in the Benguela upwelling zone, and the subsequent decay of large masses of dead phytoplankton that have fallen to the seafloor.

The anaerobic activity leads to the formation of hydrogen sulfide gas (H_2S) in the sediment. Over time, the gases build up and are periodically released into the water column in a "sulfide eruption." At the water surface, the H_2S oxidizes to microgranules of sulfur, discoloring it a milky green. These surface features can be observed by satellites (see photograph on page 81).

Recent Discoveries: Archaea. Knowledge of the diversity of microbial life in the oceans continues to grow. Until the 1990s, knowledge of microbial populations was determined using assays that relied on the growth of the microbe. Now, detection and identification of microbes are possible by the examination of their genetic material. These molecular assay techniques have revealed much larger numbers and types of microbes in the ocean than scientists previously suspected.

The bacteria-like microbes known as Archaea represent one example of research surprising to marine microbiologists. Archaea are one of the major domains of life on Earth. Since their discovery in 1970, these microorganisms have been found in many extreme environments on Earth, including hydrothermal vents on the ocean floor. Recently, scientists determined that Archaea also exist in the open sea. Moreover, these microbes may comprise up to half the mass of life in the oceans, and so must play an important role in the processes that occur in the oceans.

This 2002 satellite image by SeaWiFS (Sea-Viewing Wide Field-of-View Sensor) shows microcrystalline particles of sulfur off the coast of Namibia in southwestern Africa. Once thought to be a periodic phytoplankton bloom, the highly visible occurrence (shown here as wispy areas near the coast) is now known to be the result of periodic hydrogen sulfide gas eruptions from the diatomaceous sediments underlying the highly productive waters of the northern Benguela upwelling zone. Anaerobic microbial breakdown of decaying phytoplankton produces hydrogen sulfide gas in the seafloor sediment, and the gas buildup is periodically released, rising and oxidizing as it approaches the water surface. The hydrogen sulfide released from the surface waters causes the Namibian coast to smell like rotten eggs, and it often drives away lobsters.

Since 2001, the examination of sediment from the sea bottom has revealed the presence of another type of Archaea that exist by using methane, an important gas that is contributing to the warming of the Earth's atmosphere. In addition, bacteria that exist by using the rocks of the sea bottom as food have been discovered. The release of material from the sediment by the action of the rock-using bacteria may influence the chemistry of the oceans.

Animal and Human Health Impacts

Some ocean microorganisms can cause unhealthy effects in both land and sea animals. This is particularly true in coastal regions, where the influence of humans is more evident. For example, runoff or the deliberate release of

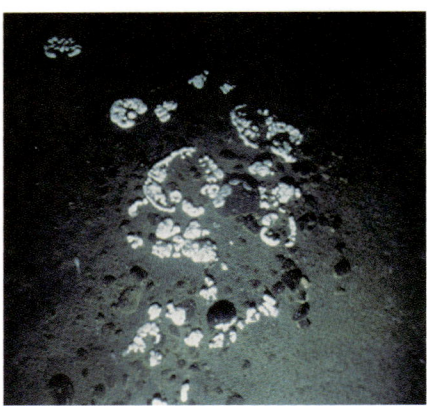

A bacteria colony on the ocean floor illustrates the ubiquitous nature of microorganisms. Marine bacteria play a major role in the ocean's nutrient cycles.

ecosystem: the community of plants and animals within a water or terrestrial habitat interacting together and with their physical and chemical environment

estuary: a tidally influenced coastal area in which fresh water from a river mixes with sea water, generally at the river mouth; the resulting water is brackish, which results in a unique ecosystem

sewage into the oceans releases huge numbers of bacteria and viruses into the water. These microbes normally live in the intestinal tracts of humans and other warm-blooded animals. The water that is contaminated by these microbes can be the source of diseases.

Just as humans are susceptible to microbial infections, so too can marine animals (e.g., mammals) develop infections. It is believed that infections are to blame for at least some cases of marine mammal "beachings," in which whales or dolphins become stranded on the shore. It is thought that human pollution may exacerbate this problem by increasing the likelihood of infection and decreasing the quality of the water.

Another potentially harmful microbe found naturally in the ocean is a protist called a dinoflagellate. At certain times and under certain conditions, some dinoflagellate species and other algal species can undergo population explosions called blooms, sometimes in response to human-caused pollution. These blooms often are called "red tides" because the algal pigments color the water. Further, some of the bloom-causing algae may produce natural poisons known as biotoxins. These biotoxins are transferred to ocean animals that feed on the toxin-containing algae, and also are released into the water as the dead algae decay. These biotoxins can bioaccumulate in the ocean food chain, sickening or killing higher-order animal consumers and tainting fisheries and shellfisheries used by humans.

Human Pathogens. The ocean has long been used as a means of disposal of human wastes, and still is used this way in many areas of the world. The enormous size of the oceans historically and incorrectly was assumed able to dilute any noxious material to concentrations low enough to render them harmless. While small amounts of sewage can dissipate quickly, the routine dumping of large amounts of human waste has caused long-term harm to many shoreline **ecosystems**, and rendered many coastal waters unfit for recreational activities.

For example, high numbers of disease-causing viruses and a bacterial species called *Escherichia coli* can occur in coastal waters influenced by human wastes (e.g., sewage). If ingested, these microbes can cause intestinal upset and organ damage that varies from inconvenient to life-threatening. Furthermore, the viruses that cause hepatitis A, hepatitis E, and polio, along with protozoan pathogens that cause giardiasis and cryptosporidiosis, are sometimes found in coastal waters and **estuaries**.

Another example related to improperly treated or untreated sewage is *Vibrio cholerae*, the bacterium that causes the intestinal infection cholera. Huge outbreaks of cholera have occurred throughout recorded history. In recent years, severe outbreaks of cholera occurred in Bangladesh and in Peru. Some evidence suggests that the cyclical temperature increase associated with El Niño currents increases the likelihood of such outbreaks, by increasing the ability of the bacterium to reproduce in polluted waters.

Vibrio cholerae and other infectious bacteria and viruses also are commonly present in high numbers in the ballast water of ocean-going ships, and can be widely spread by these ships. Ballast water is water that is pumped into the hull of a ship in order to stabilize it against the rough conditions of an ocean voyage. When the ship reaches port, millions of liters of ballast water are pumped out into the harbor, releasing the microbes into the new environment. Currents can act as highways for microbes, carrying them along in the water flow. Viruses, which can live in the ocean for days or weeks, can be transported great distances via ocean currents. SEE ALSO ALGAL BLOOMS, HARMFUL; ALGAL BLOOMS IN THE OCEAN; CARBON DIOXIDE IN THE OCEAN AND ATMOSPHERE; EL NIÑO AND LA NIÑA; FOOD FROM THE SEA; LIFE IN EXTREME WATER ENVIRONMENTS; HUMAN HEALTH AND THE OCEAN; HUMAN HEALTH AND WATER; OCEAN BIOGEOCHEMISTRY; OCEAN CURRENTS; PLANKTON; POLLUTION BY INVASIVE SPECIES; POLLUTION OF THE OCEAN BY SEWAGE, NUTRIENTS, AND CHEMICALS.

Brian D. Hoyle and Richard Robinson

Bibliography

Fuhrman, Jed A. "Marine Viruses and Their Biogeochemical and Ecological Effects." *Nature* 399 (1999):541–548.

Karner, Markus B., Edward F. DeLong, and David M. Karl. "Archaeal Dominance in the Mesopelagic Zone of the Pacific Ocean." *Nature* 409 (2001):507–509.

Orphan, Victoria J. et al. "Methane-Consuming Archaea Revealed by Directly Coupled Isotopic and Phylogenetic Analysis." *Science* 293 (2001):484–487.

Ruiz, Gregory M. et al. "Global Spread of Microorganisms by Ships." *Nature* 408 (2000):49–50.

Suttle, Curtis A. "The Significance of Viruses to Mortality in Aquatic Microbial Communities." *Microbial Ecology* 28 (1994):237–243.

Mid-Ocean Ridges

The mid-ocean ridge is an interconnected system of undersea volcanoes that meander over the Earth like the raised seams on a baseball. It is a continuous 40,000-mile (60,000-kilometer) seam that encircles Earth and bisects its oceans. The mid-ocean ridge represents an area where, in accordance with **plate tectonic** theory, **lithospheric plates** (also called tectonic plates) move apart and new crust is created by magma (molten rock) pushing up from the **mantle**. The mid-ocean ridge system is an example of a divergent (rather than a convergent or transform) plate boundary.

The mid-ocean ridge system has been understood only since the development and acceptance of plate tectonic theory in the 1960s. Four major scientific developments spurred the formation of the theory: (1) demonstration of the young age of the ocean floor; (2) confirmation of repeated reversals of Earth's magnetic field in the geologic past; (3) emergence of the seafloor-spreading hypothesis and associated recycling of the oceanic crust; and (4) precise documentation that Earth's earthquake and volcanic activity is concentrated along subduction zones and mid-ocean ridges.

Ridge Characteristics

Mid-ocean ridges have different shapes (morphology) depending on how fast they are spreading, how active they are magmatically and volcanically, and how much tectonic stretching and faulting is taking place. Scientists believe that the most likely reason for the different morphologies is due to the strength of the ocean crust at these different sites, and how cold and brittle the upper part of the tectonic plate is.

Ridge Types. There are two types of mid-ocean ridges: fast-spreading and slow-spreading. Fast-spreading ridges like the northern and southern East

plate tectonics: the theory that the Earth's lithosphere can be divided into a few large plates that are slowly moving relative to one another; plate sizes change and intense geologic activities occur at plate boundaries (e.g., earthquakes, volcanism, mountain-building)

lithospheric plate: a section of the lithosphere that acts as a single mass and interacts with other plates; lithospheric plates are created at spreading centers and destroyed at subduction zones

mantle: the region of the Earth between the molten core and the outer crust, composed mainly of silicate rock, and around 2,900 kilometers (1,800 miles) thick; also the interior of another planet, moon, or large asteroid between the core and the crust

Mid-Ocean Ridges

A composite map showing the topography (relief) of the ocean floor clearly reveals the mid-ocean ridge system, which appears here as dark "seams" extending through the oceans. In the Atlantic Ocean, the mid-ocean ridge is called the Mid-Atlantic Ridge. It curves from the Arctic Ocean southward, through Iceland, down the center of the Atlantic, and around the bottom of Africa. There it divides, with one branch running up the Red Sea and the other going around Australia into the Pacific and north back to the Arctic. A branch off South America is called the East Pacific Rise.

Pacific Rise have smoother topography at the ridge crest, and look somewhat like domes. They have relief of 100 to 200 meters (328 to 656 feet). The East Pacific Rise moves at an average of 15 centimeters (5.9 inches) per year.

Slow-spreading ridges like the Mid-Atlantic Ridge have large, wide, rift valleys, sometimes as wide as 10 to 20 kilometers (6 to 12 miles) and very rugged terrain at the ridge crest that can have relief of up to 3.2 kilometers (2 miles). The Mid-Atlantic Ridge moves at an average of 2.5 centimeters (1 inch) per year.

Fast-spreading ridges are "hotter," meaning that more magma is present beneath the ridge axis, and that more volcanic eruptions occur. Because the plate under the ridge crest is hotter, scientists think that the plate responds to the divergent spreading process more fluidly, and that the ridge behaves like hot taffy being pulled apart. In this scenario, the ridge crest does not have a chance to subside (sink or settle).

At slower spreading ridges, the seafloor behaves more like a cold chocolate bar—when pulled, it cracks and breaks to form ridges and valleys. As the sheets of oceanic crust move away from the mid-ocean ridge, the rock is cooled and thus becomes heavier. After about 200 million years, the cooled lithospheric plate has become heavier than the **asthenosphere** that it rides over, and it sinks, thereby producing a subduction zone.

Fracture Zones. Mid-ocean ridges do not form straight lines but are instead offset in many places by fracture zones, or transform faults. Fracture

asthenosphere: the zone inside the Earth beneath the lithosphere to a depth of less than 700 kilometers (approximately 400 miles), containing a low percentage of molten rock, and constituting the source of seafloor basalts

zones are thought to occur due to zones of weakness in the pre-existing continent before it was rifted apart. Most mid-ocean ridges are divided into hundreds of segments by fracture zones. Along the Mid-Atlantic Ridge, fracture zones occur at an average interval of 55 kilometers (34 miles). As the Mid-Atlantic Ridge is some 16,000 kilometers (10,000 miles) long, it is divided by fracture zones into about 300 distinct segments. The ridge crest and its associated faults are the locus of nearly all shallow earthquakes occurring in mid-ocean areas.

Water and Minerals. Ocean water is constantly percolating through fissures (cracks) at the mid-ocean ridge. Downward-convecting cold ocean water meets the hot new crust far below the surface, and many types of metals such as sulfur, copper, zinc, gold, and iron are transferred to the water. This hot, mineral-laden water gushes back up through the cracks, forming hydrothermal vents. As the hot water, which can reach temperatures of 371°C (700°F), escapes from the vents and comes in contact with the near-freezing water of the ocean bottom, the metals quickly **precipitate** out of solution. The results are surging black clouds of particle-rich water called black smokers, which often erupt out of tall chimneys of previously deposited solidified mineral.✱ Because so much metal is spewed out, hydrothermal vents have been responsible for many of the world's richest ore deposits. These unique features also are found to harbor a diverse array of deep-ocean life. SEE ALSO HOT SPRINGS ON THE OCEAN FLOOR; LIFE IN EXTREME WATER ENVIRONMENTS; MINERAL RESOURCES FROM THE OCEAN; OCEAN-FLOOR BATHYMETRY; PLATE TECTONICS; VOLCANOES, SUBMARINE.

Larry Gilman and K. Lee Lerner

precipitate: (verb) in a solution, to separate into a relatively clear liquid and a solid substance by a chemical or physical change; (noun) the solid substance resulting from this process

✱ See "Hot Springs on the Ocean Floor" for a photograph of a black smoker.

Bibliography

Coulomb, J. *Seafloor Spreading and Continental Drift.* Dordrecht, Netherlands: D. Reidel Publishing Co., 1972.

Nicolas, A. *The Mid-Oceanic Ridges.* Berlin, Germany: Springer Verlag, 1995.

Thurman, Harold, and Elizabeth Burton. *Introductory Oceanography,* 9th ed. Upper Saddle River, NJ: Prentice Hall, 2001.

Mineral Resources from Fresh Water

Rivers and streams transport water and **sediment** downslope to lakes and oceans. Along the way, sediment may be deposited, only to be eroded later during floods, and transported farther along the system. Dynamic processing of sediment in the ever-changing river environment separates grains according to density, size, shape, and resistance to **weathering**. Higher density and larger grains are found in the high-energy environment near the river channel. Lower density, smaller, and plate-shaped sediment grains are found in low-energy regions such as meander cutoffs and the **floodplain**, where water velocity is minimal or flow is absent.

This natural processing, which sorts sediments and minerals according to their physical properties, serves as a mechanism by which potentially valuable earth deposits are concentrated to a point where it is economically feasible to extract them for human use. These resources may be metallic, as in tin, gold and platinum placer deposits, or nonmetallic, such as deposits of sand and gravel. Sand and gravel, though not as spectacular as the precious metals, are among the most valuable mineral deposits.

sediment: rock particles and other earth materials that are transported and deposited over time by geologic agents such as running water, wind, glaciers, and gravity; sediments may be exposed on dry land and are common on ocean and lake bottoms and river beds

weathering: the decay or breakdown of rocks and minerals through a complex interaction of physical, chemical, and biological processes

floodplain: the low-lying land adjoining a river that is sometimes flooded; generally covered by fine-grained sediments (silt and clay) deposited by the river at flood stage

Mineral Resources from Fresh Water

Aggregate companies, such as this sand and gravel plant in Holland, locate where mineral deposits can be economically mined. These sediments were originally deposited in freshwater environments such as stream channels in ancient river valleys.

Aggregate Minerals

The well-sorted sand and gravel deposits in ancient and modern-day river channels are important sources of aggregate materials. Sand and gravel are used in the construction industry for roadbed, concrete, and so on. The highest quality aggregate, used for making concrete, often is found near active river channels. High-quality aggregate may also be present in old river **terraces** where river channels once were located.

Instream mining extracts the sand and gravel from the modern channel system. The legal definition of instream mining varies from state to state, and in many states it is not allowed. Some states define instream mining as occurring within the limits of the 2-year floodplain (i.e., the area inundated by a flood having a 50-percent chance of occurrence in any year).

Off-channel mining extracts sand and gravel from older deposits often associated with floodplain terraces, which are remnants of an older floodplain level. Off-channel mining is regulated to protect floodplain resources within the 100-year floodplain.

Instream and off-channel mining can damage habitat for fish, and change the relationships between the channel and floodplain. Carefully designed and regulated mining may become a tool to help maintain or even increase the diversity of channel and floodplain habitat.

terrace (river): an old alluvial floodplain, ordinarily flat or undulating, bordering a river, but at a higher level than the current floodplain; result from a river's downcutting ability being accelerated, leaving remnants of the former floodplain perched above the new stream level as terraces; stream terraces are frequently called second bottoms (as contrasted to floodplains) and are seldom subject to overflow

Diatomaceous Earth

Diatoms are common unicellular algae whose different species live in freshwater, **brackish** water, and salt-water environments. Each species is distinguished by the shape and ornamentation of the frustule, the shell-like structure composed of silica. The frustules of dead diatoms accumulate with mineral and rock grains to form sediment known as diatomaceous earth. When lithified into a rock, diatomaceous earth is called diatomite. Although diatoms are widespread, relatively pure and thick deposits composed primarily of diatom frustules are rare because under normal conditions, the frustules are overwhelmed by other sediment.

brackish: describes water having a salinity from 0.05 to 17 parts per thousand; typically a mixture of sea water and fresh water (e.g., as found in an estuary)

Diatomite has many human uses, such as in the production of paint, paper, and plastics; as a nontoxic alternative to chemical pesticide dusts; and as a filtration media for liquids, including drinking water, fruit juices, and medicine.

The filtering efficiency of diatomite is controlled by the physical characteristics and relative abundance of the different diatom species that comprise the deposit. Each deposit has a different species grouping that depends on the aquatic environment in which it formed, and the time in the geologic past when it formed. Thus, a diatomite mine operator works closely with the client to optimize the properties of the excavated product.

Peat

In swamp environments, organic material composed mostly of plant remains may become submerged in oxygen-poor, stagnant water. The water protects the plant debris from oxidation and decomposition. The resulting sedimentary deposits may be organic-rich sediment, or in special cases, peat. Under average peat-forming conditions, only about 10 percent of the organic material produced in a swamp survives as peat. If peat is compacted and heated during burial, it becomes coal.

The poorly developed drainage systems in recently glaciated landscapes of the Northern Hemisphere are favorable settings for modern peat formation. In these settings, the swamps are relatively isolated from streams that carry mineral and rock debris. Similarly, coastal swamps are sites of peat formation.

Peat that has experienced little decomposition and is composed mostly of moss is used in gardening and agricultural production as peat moss. If peat is more highly decomposed and compressed, it can be used as a fuel after it is dried. However, the heat released is low relative to other common fuels such as wood. Yet new technologies that convert peat to methane gas by either bacterial digestion or by heating to 400°C to 500°C (752°F to 932°F) are changing the way peat is used.

Salt Deposits

Sea water contains almost 3.5 percent dissolved salt. When it evaporates, the dissolved solids form minerals such as halite, or rock salt (NaCl), gypsum ($Ca_2SO_4 \cdot 2H_2O$), anhydrite (Ca_2SO_4), and potash (KCl) salts. In the geologic past there have been a number of times when shallow and/or marginal seas have gone through periodic flooding and evaporation, and have led to the accumulation of a significant thickness of these materials.

Approximately 400 million years ago the evaporation of a shallow sea resulted in the accumulation of hundreds of meters of predominantly halite in the area now occupied by Michigan, Ohio, West Virginia, Pennsylvania, and New York. Additional deposits are found in India, Algeria, and China. Thick salt deposits in the Gulf Coast area have lead to the occurrence of salt domes and have been instrumental as petroleum traps. Halite is used primarily in cooking and as a food preservative, but also in the manufacture of soda ash for the glass industry and other sodium compounds.

The largest gypsum deposits are of Permian age.✱ Notable deposits are now found in Greece, Austria, Italy, France, Spain, England and Mexico. In

> **PANNING FOR GOLD**
>
> Gold is a placer mineral that historically sparked mass migrations of people to sites of newly discovered deposits, often changing the course of cultural and economic development. Panning for gold mimics stream actions to separate grains of gold from other mineral and rock grains. To accomplish this separation, the panner takes advantage of the difference in the specific gravity of gold (19.3 grams per cubic centimeter) and common rocks and minerals (generally about 3 grams per cubic centimeter).
>
> Sediment to be tested for gold is placed in the pan, which is tilted while the watery mixture is swirled. Swirling allows the low-density minerals to be washed out while the heavy minerals and gold remain behind. Continued careful swirling separates the heavy minerals from the gold, which can then be removed from the pan.

✱ See the frontmatter of this volume for a geologic timescale.

the United States, deposits are found in Kentucky, Ohio, Michigan, South Dakota, and Utah. Gypsum commonly is mined to be used in plaster, Portland cement, paper filler, as a soil conditioner (to add sulfur) and in the manufacture of sulfuric acid. Wallboard is the largest single user of gypsum.

Heavy Minerals

Included in some river sediments are minerals resistant to weathering and abrasion, and that have higher density than common rocks and minerals. Because of their higher density, these so-called "heavy minerals" become concentrated, whereas the lower density minerals are removed by the water current. Locally, these heavy minerals may become aggregated in placer deposits, and may attain concentrations of economic importance. Gold, tin, and platinum are examples of placer minerals.

Where rivers carry heavy minerals to the ocean, the minerals become concentrated on the beach by wave action. The high-density and resistant minerals are left behind as the waves winnow away the lower density and less resistant rocks and minerals. Gold, diamonds, garnet, chromium, tin, iron, and titanium are examples of heavy minerals that may be extracted from beach placer deposits. Nearshore underwater mining, and mining of uplifted marine **terraces** are important sources of these valuable mineral resources. SEE ALSO BRINES, NATURAL; FRESH WATER, NATURAL COMPOSITION OF; FRESH WATER, PHYSICS AND CHEMISTRY OF; KARST HYDROLOGY; LAKES: PHYSICAL PROCESSES; MINERAL RESOURCES FROM THE OCEAN; STREAM CHANNEL DEVELOPMENT; STREAM HYDROLOGY.

Michael Cummings

terrace (marine): an ancient beach area perched above the current beach level; often flat and gently sloping toward the sea; has been elevated relative to current beach level by a lowering of sea level or uplift of the coastal area

Bibliography

Angier, Bradford. *Looking for Gold: The Modern Prospector's Handbook.* Mechanicsburg, PA: Stackpole Books, 1980.

Barnes, John W. *Ores and Minerals: Introducing Economic Geology.* Hoboken, NJ: John Wiley & Sons, Inc., 1991.

Evans, Anthony M. *Introduction to Economic Geology.* Malden, MA: Blackwell Publishers, 1997.

Internet Resources

Prospecting for Gold in the United States. U.S. Geological Survey. <http://pubs.usgs.gov/gip/prospect2/prospectgip.html>.

Mineral Resources from the Ocean

Oceans cover 70 percent of Earth's surface, host a vast variety of geological processes responsible for the formation and concentration of mineral resources, and are the ultimate repository of many materials eroded or dissolved from the land surface. Hence, oceans contain vast quantities of materials that presently serve as major resources for humans. Today, direct extraction of resources is limited to salt; magnesium; placer gold, tin, titanium, and diamonds; and fresh water.

Ancient ocean deposits of sediments and **evaporites** now located on land were originally deposited under marine conditions. These deposits are being exploited on a very large scale and in preference to modern marine resources because of the easier accessibility and lower cost of terrestrial

evaporites: sediments that form as the result of the precipitation of minerals during the evaporation of water, primarily sea water, and that may form sedimentary rock; principal minerals are gypsum and halite

Mineral Resources from the Ocean

These mounds of sea salt were mined from deeply buried beds deposited when sea water evaporated in an ancient environment. The beds were preserved by being covered and then uplifted in a modern terrestrial setting. Mining accounts for most of the annual salt production, even though it also can be obtained by evaporating ocean water.

resources. Yet the increasing population and the exhaustion of readily accessible terrestrial deposits undoubtedly will lead to broader exploitation of ancient deposits and increasing extraction directly from ocean water and **ocean basins**.

Principal Mineral Resources

Resources presently extracted from the sea or areas that were formerly in the sea range from common construction materials to high-tech metals to water itself. Chemical analyses have demonstrated that sea water contains about 3.5 percent dissolved solids, with more than sixty chemical elements identified. The limitations on extraction of the dissolved elements as well as the extraction of solid mineral resources are nearly always economic, but may also be affected by geographic location (ownership and transport distance) and hampered by technological constraints (depth of ocean basins).

basin (ocean): the topographic low area occupied by oceans; the floor of ocean basins consists of basaltic crust that is more dense than typical continental rocks

The principal mineral resources presently being extracted and likely to be extracted in the near future are briefly considered here.

Salt. Salt, or sodium chloride, occurs in sea water at a concentration of about 3 percent and hence constitutes more than 80 percent of the dissolved chemical elements in sea water. The quantity available in all the oceans is so enormous that it could supply all human needs for hundreds, perhaps thousands, of years. Although salt is extracted directly from the oceans in many countries by evaporating the water and leaving the residual salts, most of the nearly 200 million metric tons of salt produced annually is mined from large beds of salt. These beds, now deeply buried, were left when waters from ancient oceans evaporated in shallow seas or marginal basins, leaving residual thick beds of salt; the beds were subsequently covered and protected from solution and destruction.

Potassium. Like the sodium and chlorine of salt, potassium occurs in vast quantities in sea water, but its average concentration of about 1,300 parts per million (or 0.13 percent) is generally too low to permit direct economic extraction. Potassium salts, however, occur in many thick evaporite se-

Mineral Resources from the Ocean

quences along with common salt and is mined from these beds at rates of tens of millions of metric tons per year. The potassium salts were deposited when sea water had been evaporated down to about one-twentieth of its original volume.

Magnesium. Magnesium, dissolved in sea water at a concentration of about 1,000 parts per million, is the only metal directly extracted from sea water. Presently, approximately 60 percent of the magnesium metal and many of the magnesium salts produced in the United States are extracted from sea water electrolytically. The remaining portion of the magnesium metal and salts is extracted from ancient ocean deposits where the salts precipitated during evaporation or formed during **diagenesis**. The principal minerals mined for this purpose are magnesite ($MgCO_3$) and dolomite ($CaMg[CO_3]_2$).

diagenesis: the process of chemical and physical changes that occur within sediments after their accumulation; includes the processes of compaction, the cementation of minerals to one another, recrystallization of minerals, and replacement of one mineral by another

Sand and Gravel. The ocean basins constitute the ultimate depositional site of sediments eroded from the land, and beaches represent the largest residual deposits of sand. Although beaches and near-shore sediments are locally extracted for use in construction, they are generally considered too valuable as recreational areas to permit removal for construction purposes. Nevertheless, older beach sand deposits are abundant on the continents, especially the coastal plains, where they are extensively mined for construction materials, glass manufacture, and preparation of silicon metal. Gravel deposits generally are more heterogeneous but occur in the same manner, and are processed extensively for building materials.

Limestone and Gypsum. Limestones (rocks composed of calcium carbonate) are forming extensively in the tropical to semitropical oceans of the world today as the result of **precipitation** by biological organisms ranging from mollusks to corals and plants. There is little exploitation of the modern limestones as they are forming in the oceans. However, the continents and tropical islands contain vast sequences of limestones that are extensively mined; these limestones commonly are interspersed with dolomites that formed through diagenetic alteration of limestone. Much of the limestone is used directly in cut or crushed form, but much is also calcined (cooked) to be converted into cement used for construction purposes. Gypsum (calcium sulfate hydrate) forms during evaporation of sea water and thus may occur with evaporite salts and/or with limestones. The gypsum deposits are mined and generally converted into plaster of paris and used for construction.

precipitation: the separation of a solid phase (precipitate) from solution (dissolved state)

Manganese Nodules. The deep ocean floor contains extremely large quantities of nodules ranging from centimeters to decimeters in diameter (that is, from less than an inch to several inches). Although commonly called manganese nodules, they generally contain more iron than manganese, but do constitute the largest known resource of manganese.

Despite the abundance and the wealth of metals contained in manganese nodules (iron, manganese, copper, cobalt, and nickel), no economic way has yet been developed to harvest these resources from the deep ocean floor. Consequently, these rich deposits remain as potential resources for the future. Terrestrial deposits of manganese are still relied on to meet human needs.

precipitate: (verb) in a solution, to separate into a relatively clear liquid and a solid substance by a chemical or physical change; (noun) the solid substance resulting from this process

Phosphorites. Complex organic and inorganic processes constantly **precipitate** phosphate-rich crusts and granules in shallow marine environments.

These are the analogs (comparative equivalents) of the onshore deposits being mined in several parts of the world, and represent future potential reserves if land-based deposits become exhausted.

Metal Deposits Associated with Volcanism and Seafloor Vents.

Submarine investigations of oceanic rift zones have revealed that rich deposits of zinc and copper, with associated lead, silver, and gold, are forming at the sites of hot hydrothermal emanations commonly called black smokers. These metal-rich deposits, ranging from chimney- to pancake-like, form where deeply circulating sea water has dissolved metals from the underlying rocks and issue out onto the cold seafloor along major fractures. The deposits forming today are not being mined because of their remote locations, but many analogous ancient deposits are being mined throughout the world.

Placer Gold, Tin, Titanium, and Diamonds.

Placer deposits are accumulations of resistant and insoluble minerals that have been eroded from their original locations of formation and deposited along river courses or at the ocean margins. The most important of these deposits contain gold, tin, titanium, and diamonds.

Today, much of the world's tin and many of the gem diamonds are recovered by dredging near-shore ocean sediments for minerals that were carried into the sea by rivers. Gold has been recovered in the past from such deposits, most notably in Nome, Alaska. Large quantities of placer titanium minerals occur in beach and near-shore sediments, but mining today is confined generally to the beaches or onshore deposits because of the higher costs and environmental constraints of marine mining.

Water.

The world's oceans, with a total volume of more than 500 million cubic kilometers, hold more than 97 percent of all the water on Earth. However, the 3.5-percent salt content of this water makes it unusable for most human needs.

The extraction of fresh water from ocean water has been carried out for many years, but provides only a very small portion of the water used, and remains quite expensive relative to land-based water resources. Technological advances, especially in **reverse osmosis**, continue to increase the efficiency of fresh-water extraction. However, geographic limitations and dependency on world energy costs pose major barriers to large-scale extraction. SEE ALSO DESALINATION; MINERAL RESOURCES FROM FRESH WATER; PETROLEUM FROM THE SEA.

James R. Craig

reverse osmosis: process in which dissolved substances are removed from water by forcing water, but not dissolved salts, through a semipermeable membrane under high pressure; commonly used to treat contaminated drinking water or process water; in desalination, reverse osmosis is used to extract fresh water from salty water

Bibliography

Craig, James R., David J. Vaughtan, and Brian J. Skinner. *Resources of the Earth. Origin, Use, Environmental Impact*, 3rd ed. Upper Saddle River, NJ: Prentice Hall, 2001.

Lahman, H. S., and J. B. Lassiter III. *The Evolution and Utilization of Marine Mineral Resources.* Books for Business, 2002.

Internet Resources

USGS Minerals Information: Mineral Commodity Summaries. U.S. Geological Survey. <http://minerals.usgs.gov/minerals/pubs/mcs/>.

USGS Minerals Information: Minerals Yearbook. U.S. Geological Survey. <http://minerals.usgs.gov/minerals/pubs/myb.html/>.

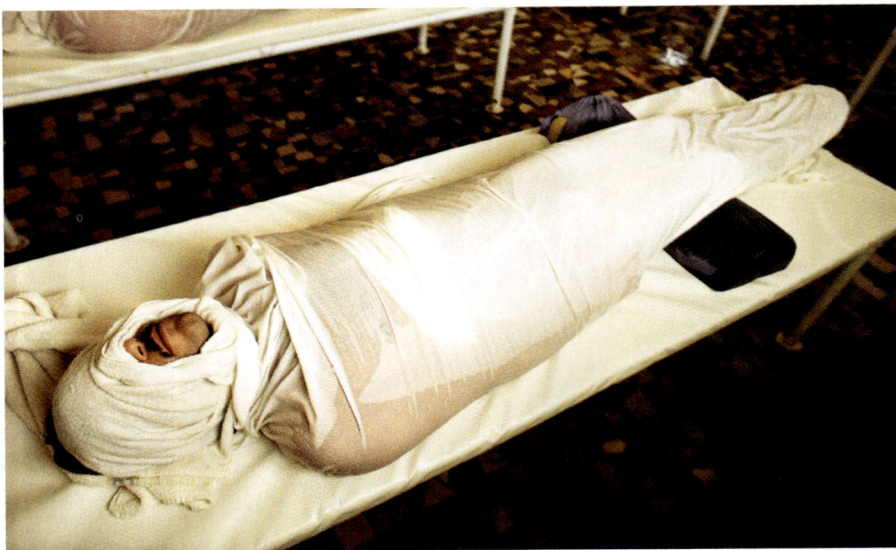

A spa patron has a cool herbal wrap after a mud bath and hot mineral soak at Murrieta Hot Springs near Temecula, California.

Mineral Waters and Spas

Mineral waters are those that contain some dissolved minerals in sufficient concentration to change the taste or perceived health effects of the water. Mineral water originates as **groundwater** and flows along the local groundwater gradient, dissolving available geologic material until it reaches the surface as a spring, or until the water is pumped to the surface from a well.

Types of Mineral Waters

The chemical and physical characteristics of mineral water depend on the nature of the geologic materials through which the groundwater has moved. Common soluble minerals include calcium carbonate, iron compounds, sodium and magnesium salts, sulfur compounds, and metals.

Mineral waters are usually classified as alkaline, saline, ferrous, sulfurous, acidulous, or soda, and may be either cold (down to about 4.5°C or 40°F) or hot (up to about 100°C or 212°F). Some mineral waters may originate from deep within the Earth, being heated by **geothermal** sources of either volcanic or **tectonic** nature. Hotter water typically dissolves more materials, making the mineral concentration higher.

Health Benefits

Mineral water has been sought after for various health-related benefits since ancient times. Soda (naturally carbonated) waters have been used as table beverages and to dilute spirits or wines. Saline waters are typically ingested for their perceived medicinal effects.

Mineral Water Baths. Mineral water used for bathing can be enjoyed by complete submersion, or by wrapping the body in wet sheets or towels. Baths at skin temperature (about 37°C or 98.6°F) are relaxing. The hot bath stimulates, relieves pain (particularly of cramps and sometimes of arthritis), controls convulsions, and induces sleep. Quickening the pulse and respiration, it also increases perspiration, thereby relieving the kidneys of part of their

groundwater: generally, all subsurface (underground) water, as distinct from surface water, that supplies natural springs, contributes to permanent streams, and can be tapped by wells; specifically, the water that is in the saturated zone of a defined aquifer

geothermal: describes terrestrial heat, usually associated with water, as around hot springs

tectonic: formed by tectonism, the shaping by deformation of the crust of a planet or moon

MINERAL SPRINGS IN THE UNITED STATES

Mineral springs with significant flows can be found in Saratoga, New York; Berkeley Springs, Virginia; and White Sulfur Springs, West Virginia. Well-known springs are also located in Kentucky, Tennessee, Arkansas, Indiana, Wisconsin, Nevada, New Mexico, California, and Washington. Many smaller or lesser-known mineral springs and spas can be found throughout the country.

work and temporarily decreasing weight. Hot packs are helpful in muscular disorders. The cold bath can be stimulating as well, and also is helpful in reducing high fever and limiting inflammation.

Although thousands of people suffering a variety of ailments frequent mineral baths in search of the cures attributed to local waters and mud, physicians generally doubt their medical value. Among the most popular medicinal baths are those in which the waters of natural warm mineral springs can be found. Resorts, sometimes called spas, have grown up near such springs.

Spas, by definition, are located where natural sources of mineralized water are used for drinking or bathing to enhance good health. The term "spa" was named after Spa, a town in eastern Belgium, which is a popular resort known for its baths and mineral springs. The waters have been frequented since ancient times. Such spas remain popular worldwide. SEE ALSO Geothermal Energy; Hot Springs and Geysers; Human Health and Water; Senses, Fresh Water and the.

Rick Graff and Kari Salis

Pluto Water, the advertising name given to highly mineralized water from French Lick Springs, Indiana, was world-famous for its perceived health benefits. Bottled since 1848 and once sold in drugstores throughout the Western Hemisphere, Pluto Water was especially known for its laxative effects.

Bibliography

Ahman, Nathaniel. *Healing Springs: The Ultimate Guide to Taking the Waters*, Rochester, VT: Healing Arts Press, 2000.

Hem, John D. *Study and Interpretation of the Chemical Characteristics of Natural Water*, 3rd ed. Alexandria, VA: U.S. Geological Survey, Water-Supply Paper W2254 (1985).

Minorities in Water Sciences

As with so many other scientific fields, African Americans, Hispanics, Native Americans, and Pacific Islanders are poorly represented in the ranks of practicing water science professionals. Minority students interested in becoming oceanographers, marine biologists, fisheries scientists, hydrologists, ecologists, aquatic chemists, or limnologists have few role models to emulate. In part, this reflects a history of a lack of minority science and mathematics teachers in the K–12 schools. A number of institutions, such as Northern Arizona University and State University of New York at Oswego, are trying to correct that by offering programs that are directed towards elementary and secondary environmental education. Students can see profiles of minorities in aquatic sciences at <http://www.aslo.org/mas/> and <http://www.marinecareers.net>.

However, there is also good news for minority students who are interested in becoming water scientists. Reflecting the views of many, Dan Goldin, former administrator of the National Aeronautics and Space Administration, emphasized the importance of opening educational programs in science and mathematics to all levels and cultures in order for the United States to be competitive in these fields in the twenty-first century. Federal and state agencies, professional associations of water scientists, and colleges and universities all recognize the need to improve minority representation. As such, a number of specific programs have been created to help promote minority participation. In addition to these special programs targeted toward minority students, there are numerous other programs open to all students, and

Hispanic scientists collect water samples from a canal near the U.S.–Mexico border in Texas. The canal receives industrial waste, then runs through a residential area. Water pollution is suspected of causing serious illnesses in local residents.

many of these make special efforts to recruit minority students. Nearly all institutions are trying to increase the diversity of their students.

Academic Focus

Many students, whether minorities or not, become attracted to a career in water science while they are still in high school. Others make that decision after starting college. For those who know early that they want to study to become water scientists, there are a number of colleges to which they can apply that offer specialty undergraduate degrees in the field. For others who make the decision to study water science only after they have begun their college career, there are several options available. They can transfer to an institution that offers the program at the undergraduate level, they can take a summer or a semester away at a special program, or they can apply to study aquatic sciences in graduate school. Regardless of the path taken, students must master certain essentials in order to succeed in their studies.

The successful water scientist must have a strong background in basic academics. The abilities to quantify, think critically, and communicate are essential. High school and undergraduate students should take very seriously their studies of mathematics, science, and English composition. In high school, they should work toward the highest level of math offered, generally calculus or precalculus, and then complete a sequence of calculus and statistics in college. Although desirable, high school calculus is not a prerequisite to pursue a career in water sciences. A solid background in algebra, **trigonometry**, and geometry is important for success in the aquatic sciences, and those high school courses will provide the foundation for calculus at the university level. High school students should also take as much basic science as they can. This generally includes biology, chemistry, physics, earth or physical science, and whatever advanced-placement science course might be available.

In college, students need to follow the same path, taking courses in all the basic sciences. This includes biology, 2 years of chemistry, and a year of physics. This strong foundation will prepare the student to take courses

trigonometry: a branch of mathematics dealing with trigonometric functions, triangles, and solutions of plane or spherical triangles

in the highly interdisciplinary field of water science. For example, a course in marine biology draws from elements of chemistry, physics, botany, zoology, microbiology, and ecology, while a course in hydrology may draw from chemistry, biology, physics, mathematics and geology.

Institutions

The minority high school or community college student seeking an undergraduate degree in water science can choose from both minority and majority institutions. Leading programs at historically black colleges and universities (HBCUs), such as Hampton University in Virginia and Savannah State University in Georgia, and Hispanic-serving institutions (HSI), such as the University of Puerto Rico at Mayaguez, include both public and private institutions. Many other institutions with a high population of minorities, too numerous to mention here, have programs that can prepare students for successful careers in water sciences. These programs differ in their areas of emphasis, but are tailored to meet the needs of all students, including minorities. Prospective minority students of aquatic science should search the internet for the terms "water resources," "aquatic science," "university," "college," and "minorities" to obtain listings for many of these institutions. Students may also search for specialty terms such as "hydrology," "groundwater," "oceanography," and "ecology" to reflect their specific interests.

Increasing numbers of minority scientists in water resources hold a variety of jobs and are making important contributions to scientific understanding of how aquatic systems function. This aquarium curator is active in many sea turtle rescues.

Minority students may, of course, attend majority institutions with undergraduate water science programs. Most states have at least one large university that has undergraduate programs in fisheries, ecology, or natural sciences. Generally, but not exclusively, inland schools specialize in freshwater lakes and rivers, while programs located in coastal states tend to be oriented toward marine systems.

Off-Campus Programs

While earning an undergraduate degree in aquatic sciences can provide excellent preparation for a career in the field, most professionals in the field earn degrees in one of the basic sciences (biology, chemistry, geology, or physics) before beginning their aquatic science studies in graduate school. However, many of these scientists participated in special (generally off-campus) programs as undergraduates. Many spent summers at marine or fresh-water stations to gain hands-on experience in specialized courses. Others participated in special undergraduate research programs or worked as assistants to professors conducting research.

A growing pathway to a career in water sciences is through intern programs. Many high schools and universities encourage their students to seek an association with a professional aquatic scientist outside the academic environment. Although most of these are unpaid positions, they are invaluable as a source of experience. Interested students may begin this process by talking with their instructors or contacting individuals directly; state and federal water agencies may provide a rich opportunity to experience the "real world" applications. Most water scientists cite such an undergraduate experience as the true beginning of their careers.

Special off-campus programs for minority students are offered at a variety of institutions. Like the special degree programs at minority-serving institutions, special off-campus programs are tailored to meet the needs of

Minorities in Water Sciences

Tribal fisheries biologists monitor populations of fresh-water and marine fish on which many Native peoples depend for sustenance, economic benefit, and cultural preservation. These scientists on the Tulalip Indian Reservation in Washington state measure salmon smolts (juveniles).

minority students. Many are built on the Research Experience for Undergraduates (REU) model developed and supported by the National Science Foundation. These programs emphasize the student as part of a research team, conducting his or her own research project under the supervision of scientists and graduate students. The students earn a stipend that is comparable to what they would make at a typical summer job. Some (e.g., MIMSUP and NAMSS, below) involve earning credit at another institution during the regular school year. Some examples of extra-institutional minority programs for undergraduates follow.

Alabama A & M University, Center for Hydrology, Soil Climatology, and Remote Sensing (HSCARS)
<http://hscars1.saes.aamu.edu/>

Arkansas State University, Research Internships in Science of the Environment (RISE)
<http://www.cas.astate.edu/rise/>

California State University, Center for Environmental Analysis-NSF. Centers of Research Excellence in Science and Technology (CEA-CREST)
<http://cea-crest.calstatela.edu>

Colorado University (Boulder), Cooperative Institute for Research in Environmental Sciences (CIRES)
<http://cires.colorado.edu/outreach.html>

Hampton University, Multicultural Students At Sea Together (MAST)
<http://www.Hamptonu.edu/science/marine/mast>

Hampton University, American Society of Limnology and Oceanography (ASLO), Special Program for Minority Students Interested in Aquatic Sciences
<http://www.hamptonu.edu/science/ASLO.htm>

Montana State University, College of Engineering Minority Program (Wetlands)
<http://www.coe.montana.edu/coe/emp/emp2.htm>

Northern Arizona University, Environmental Studies (Applied Indigenous Studies)
<http://www.ais.nau.edu/>

Oregon State University, Native Americans in Marine and Space Sciences (NAMSS)
<http://www.oce.orst.edu/native/>

South Carolina Department of Natural Resources, Marine Resources Division, Minority Internship Program
<http://www.dnr.state.sc.us/marine/minority>

University of New Orleans, Coastal Research Laboratory, Department of Geology and Geophysics, Minority Field Camp Program (Wetlands)
<http://www.coastal.uno.edu>

College of William and Mary, Virginia Institute of Marine Science (VIMS), Summer Intern Program (SIP)
<http://www.vims.edu/education/interns.html>

Western Washington University, Minorities in Marine Science Undergraduate Program (MIMSUP)
<http://www.ac.wwu.edu/~mimsup/>

ORGANIZATIONS OF INTEREST

Below is a partial list of professional and scientific organizations whose activities are at least in part devoted to aquatic sciences. Each organization provides information that is career-related and often has programs that target minority students. Students may join these organizations.

American Association for the Advancement of Science
<http://www.aaas.org/>

American Fisheries Society
<http://www.fisheries.org/>

American Geophysical Union
<http://www.agu.org/>

American Indian Science & Engineering Society
<http://www.aises.org/>

American Institute of Hydrology
<http://www.aihydro.org/>

American Institute of Marine Studies
<http://www.aimsamerica.org/>

American Society of Limnology and Oceanography
<http://www.aslo.org/>

American Water Resources Association
<http://www.awra.org>

American Water Works Association
<http://www.awwa.org/>

Association for Women in Science
<http://www.awis.org/>

Center for the Advancement of Hispanics in Science and Engineering Education
<http://www.cahsee.org/>

Council on Undergraduate Research, Geosciences Division
<http://www.cur.org/geology.html>

Ecological Society of America
<http://www.esa.org>

Environmental and Engineering Geophysical Society
<http://www.eegs.org/>

Institute of Marine Engineering, Science and Technology
<http://www.imarest.org/>

International Association of Hydrogeologists
<http://www.iah.org/>

International Association of Theoretical and Applied Limnology
<http://www.limnology.org/>

Geological Society of America
<http://www.geosociety.org/>

Groundwater Foundation
<http://www.groundwater.org/>

Marine Technology Society
<http://www.mtsociety.org/>

National Association for Black Geologists and Geophysicists
<http://nabgg.com/>

National Groundwater Association
<http://www.ngwa.org/>

North American Lake Management Society
<http://www.nalms.org/>

Oceanic Engineers Society
<http://www.oceanicengineering.org/>

The Oceanography Society
<http://www.tos.org/>

Society for Environmental Geochemistry and Health
<http://www.segh.net/>

Society of Wetland Scientists
<http://www.sws.org/>

Soil Science Society of America
<http://www.soils.org/>

Xavier University of Louisiana, Office of Environmental Education (African American Focus)
<http://www.xula.edu/Academic/oee/>

Minority students also can take advantage of a variety of nontargeted programs that are offered by a multitude of institutions. A list of those with an REU-type structure can be accessed on the World Wide Web at <http://

www.geo.nsf.gov/oce>. For example, the Sea Education Association offers students a semester aboard a tall sailing ship, where they study marine science and participate in a program that is keen on increasing the diversity of its participants (see <http://www.sea.edu>).

Minority students may obtain much information regarding the aquatic science programs that are being made available at the ASLO website (<http://aslo.org/mas>). At this website, links to a number of useful sites can be found: for example, a student directory (<http://aslo.org/mas/directory.html>), profiles of aquatic scientists (<http://aslo.org/mas/profiles.html>), and resources (<http://aslo.org/mas/resources.html>). Students can register on-line with the MAS program at <http://aslo.org/mas/reg_form.html>.

Another useful source of information to minority students regarding the water science profession is through professional organizations and scientific societies. These organizations are open to students and often provide minority students with information targeted for them. The box on page 97 lists several such organizations and their web addresses.

In addition to the professional organizations listed in the box, there are many state and international organizations that students can explore, either through direct membership or via the internet. Most of these organizations have sections that are devoted to different career paths within the discipline. SEE ALSO WOMEN IN WATER SCIENCES.

Benjamin Cuker and Dennis O. Nelson

Bibliography

Bolster, W. J. *Black Jacks: African American Seaman in the Age of Sail.* Cambridge, MA: Harvard University Press, 1997.

Cuker, Benjamin. "Minorities at Sea Together (MAST): A Model Interdisciplinary Program for Minority College Students." *Current: The Journal of Marine Education* 18:45–51.

Cuker, Benjamin. "Steps to Increasing Minority Participation in the Aquatic Sciences: Catching Up With Shifting Demographics." *Bulletin of the American Society of Limnology and Oceanograpy* 10:17–21.

Day, J. C. "Population Projections of the United States by Age, Sex, Race, and Hispanic Origin: 1995 to 2050." U.S. Bureau of the Census, *Current Population Reports, P25-1130.* Washington, D.C.: U.S. Government Printing Office, 1996.

Gilligan, M. R. "Promoting Diversity in the Fisheries Profession: The Role of Historically Black Colleges and Universities." *Fisheries* 21 (1996): 26–29.

Huang, G., N. Taddese, and E. Walter. *Entry and Persistence of Women and Minorities in College Science and Engineering Education.* NCES 2000-601. Washington, D.C.: National Center for Education Statistics, 2000.

National Science Foundation. *Women, Minorities and Persons with Disabilities in Science and Engineering: 2000.* NSF 00-327. Arlington, VA: National Science Foundation, 2000.

Mississippi River Basin

discharge: the volume of water or a watery solution flowing past a point per unit time; common units are cubic feet per second or cubic meters per second

delta: an alluvial deposit of sediment deposited at the mouth of a river where it enters quieter or deeper water, such as a lake or ocean

The Mississippi River is North America's longest and largest river in terms of **discharge**, and the fifth largest discharge river worldwide, at an average of 17,330 cubic meters per second (811,530 cubic feet per second). The Mississippi flows 3,763 kilometers (2,333 miles) from Lake Itasca in northern Minnesota to its **delta** in southern Louisiana (see map).

Mississippi River Basin

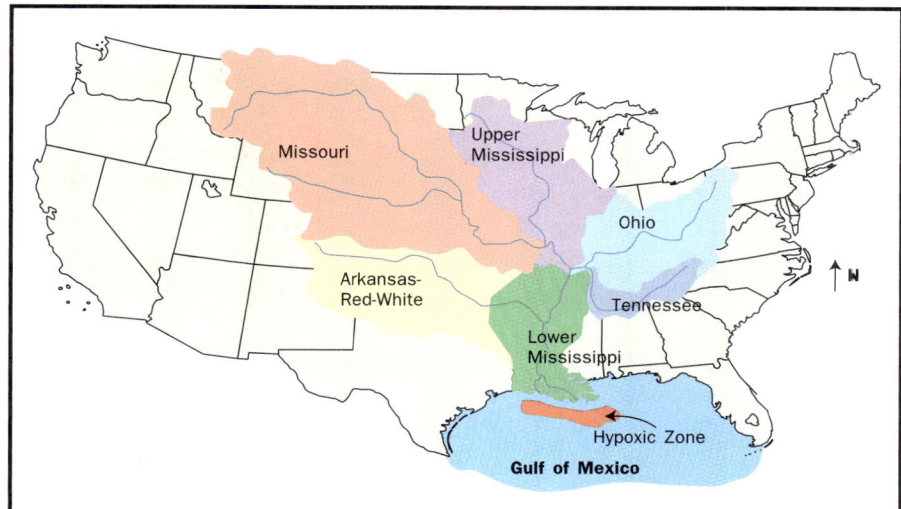

The Mississippi River Basin encompasses more than 40 percent of the U.S. land area. Its diverse features, rich history, and heavy human demands challenge water resource managers who must balance complex issues.

The Mississippi River **drainage basin** is the world's second largest, draining 4.76 million square kilometers (1.83 million square miles), including **tributaries** from thirty-two U.S. states and two Canadian provinces. The Mississippi River **watershed** encompasses 40 percent of the contiguous United States. Major tributaries include the Missouri, Ohio, Arkansas–Red–White, and Tennessee Rivers.

River Vitality

The Mississippi River is vital to the people of the basin. Over 18 million people rely on the Mississippi for water supply. Communities up and down the Mississippi use the river to discharge industrial and municipal wastes (as permitted by the U.S. Environmental Protection Agency). Commercial navigation on the Mississippi River allows Midwestern farmers to ship grains to world markets, with over 60 percent of U.S. grain exports shipped via the Mississippi. The river generates close to $2 million annually from commercial fishing and over $1 billion from Upper Mississippi River recreation alone.

Owing to the environmental diversity of the watershed, the river system has developed extraordinary **species** richness. The river provides habitat for 241 fish species and 37 mussel species, while its bluffs and bottomlands support 45 amphibious and reptile species and 50 mammal species. The **floodplain** includes the largest continuous system of **wetlands** in North America, which are used by up to 40 percent of North America's waterfowl and wading birds.

History of River Management

The Mississippi River has undergone tremendous natural change. It has carried the **meltwater** of **continental ice sheets** in the recent geologic past; it has carved and filled valleys; and it has responded to floods and droughts. However, one of the greatest changes to the Mississippi River is neither natural nor ancient—it is the engineering of the river for commercial navigation and flood control.

Navigation. The history of commercial navigation began in the 1830s, with the U.S. Army Corps of Engineers removing rock from the Des Moines

drainage basin: the land area drained by a river and its tributaries; also called catchment, drainage area, river basin, or watershed

tributary: a smaller stream that flows into a larger stream

watershed: the land area drained by a river and its tributaries; also called river basin, drainage basin, catchment, and drainage area

species: the narrowest classification or grouping of organisms according to their characteristics; members of a species can reproduce only with others of that group

floodplain: the low-lying land adjoining a river that is sometimes flooded; generally covered by fine-grained sediments (silt and clay) deposited by the river at flood stage

wetland: an area that is periodically or permanently saturated or covered by surface water or groundwater, that displays hydric soils, and that typically supports or is capable of supporting hydrophytic vegetation

meltwater: water resulting from the melting of snow, ice, or glacial ice

continental ice sheet: a relatively permanent layer of ice (a large ice cap) covering an extensive tract of land; also called continental glacier, as that of the Antarctic continent

Mississippi River Basin

anastomotic: describes the character of individual stream paths within a braided stream; stream paths appear to interconnect with one another around sand bars

(Iowa) and Rock Island (Illinois) rapids. At this time, the river still retained much of its natural **anastomotic** character: that is, its waters naturally split and rejoined in countless side channels, producing a dynamic mosaic of islands, floodplains, and waters.

Later in the nineteenth century, steamboat traffic increased and the U.S. Congress authorized a series of channel improvements, beginning in 1866 with the establishment of a 4-foot-deep channel. Further regional growth prompted channel deepening to 4.5 feet in 1878, 6 feet in 1907, and 9 feet in 1930. The 9-foot channel project began a new era of river development, entailing construction of twenty-six locks and dams from Minneapolis (Minnesota) to St. Louis (Missouri). The U.S. Army Corps of Engineers maintains the 9-foot channel and the U.S. Coast Guard monitors watercraft safety.

levee: a natural or artificially-made earthen obstruction along the edge of a stream, lake, or river; also, a long, low embankment usually built to restrain the flow of water out of a river bank and to protect land from flooding

Flood Control. Individuals began building agricultural **levees** before the Civil War, with organized levee districts developing soon after the war. By 1914, fifty-two levee and drainage districts had been established between Rock Island, Illinois and Cape Girardeau, Missouri, laying the foundation for flood control far into the future. In 1917, Congress passed the country's first flood control act, authorizing the Corps of Engineers to work on levees from Illinois to Louisiana. After the devastating flood of 1927, these efforts were reinforced in the 1928 Flood Control Act.

In the 1920s, environmentalists began to warn that river management for navigation and agriculture was threatening river ecology. In 1924, Congress established the Upper Mississippi River Wildlife and Fish Refuge, which grew to include 78,900 hectares (195,000 acres). The U.S. Fish and Wildlife Service manages this refuge and many other refuges within the watershed.

Contemporary Management Issues

While the lessons of the past provide insights into the Mighty Mississippi, river managers are constantly faced with new challenges. The many perspectives on the values of the Mississippi River and how it should be managed illustrate the significance of this important resource.

Preventing Natural Channel Change. Like all rivers, the Mississippi naturally changes course over time. As lower reaches of the river flow through the delta, sediment deposition slowly reduces the river's slope until the channel breaches (breaks) its banks and establishes a shorter, steeper route to the Gulf of Mexico.

In recent decades, the Mississippi River has been drawn toward the Atchafalaya River, seeking a more direct route to the Gulf (see map). Because of the port facilities and extensive industrial corridor in southern Louisiana, a shift of the river channel to this new route would be economically devastating. Thus the Corps of Engineers wages a continuous battle against the forces of nature to maintain the existing channel. Although severely damaged in the flood of 1973, the "Old River Control Structure" continues to divert a small percentage of the flow into the Atchafalaya River, yet most of the discharge continues to the Gulf via the Mississippi.

ZEBRA MUSSELS IN THE MISSISSIPPI RIVER BASIN

In 1991, the zebra mussel, a species not indigenous to the Mississippi River Basin, moved from Lake Michigan into the Illinois River. This rapidly reproducing, invasive mollusk is now found throughout the Mississippi River and its major tributaries. In addition to increasing operating costs of river infrastructure, the mussel threatens the ecological diversity of native river species.

Major Floods. The Great Flood of 1993 devastated the Upper Mississippi River and Missouri River Basins, causing over $12 billion in damages. (The Upper Basin is above the confluence with the Ohio River at Cairo, Illinois.)

Sandbags afford only minor protection from the Mississippi River's flood waters. Buildings and infrastructure located in the floodplain (such as this water-supply pumping station in Hannibal, Missouri) are vulnerable to direct damage from flowing water, debris, and saturated soils, as well as indirect damage from contamination. For shallow unconfined aquifers, floodwaters may seep directly down to the water table.

In some places, the flood magnitude was statistically described as one that is expected to recur only an average of once every 500 years. This extreme event reminded residents of their vulnerability and generated change in the philosophy of flood protection. Some entire communities—such as Valmeyer, Illinois—were moved out of the floodplain to higher ground nearby. Certain floodplain lands once occupied by farms, houses and other structures were restored to natural uses as wetlands and wildlife refuges. But despite these preventive measures and the known risks of continued floodplain development, such development nonetheless continues in many areas, where future floods will once again cause great economic damages and personal hardship.

"The Dead Zone." Perhaps the greatest concern of the late twentieth and early twenty-first centuries focuses on the effects of the Mississippi River on the Gulf of Mexico. Extensive agriculture throughout the watershed is responsible for elevated river concentrations of nutrients such as nitrogen and phosphorus. The Mississippi provides 90 percent of the fresh water flowing into the Gulf, and the nutrient load it delivers has generated a serious problem: namely, a critical lack of dissolved oxygen in thousands of square kilometers of Gulf waters known as the "Dead Zone." Given the importance of Gulf marine life (for example, a multi-million dollar shrimp fishery), this **hypoxia** problem is receiving a great deal of attention. Proposed solutions include improved nutrient management on farms and reconstruction of **riparian** forests and wetlands capable of removing excess nutrients. SEE ALSO ALGAL BLOOMS IN FRESH WATERS; FLOODPLAIN MANAGEMENT; POLLUTION BY INVASIVE SPECIES; RIVER BASIN PLANNING; RIVERS, MAJOR WORLD.

Christopher J. Woltemade

hypoxia: a condition in which natural waters have a low concentration of dissolved oxygen (about 2 milligrams per liter as compared with a normal level of 8 to 10 milligrams per liter); most game and commercial species of fish avoid such waters

riparian: pertaining to the banks of a river, stream, or waterway

Bibliography

Changnon, Stanley A., ed. *The Great Flood of 1993: Causes, Impacts, and Responses.* Boulder, CO: Westview Press, 1996.

Galat, David L., and Ann G. Frazier, eds. "Overview of River-Floodplain Ecology in the Upper Mississippi River Basin," Vol. 3. Kelmelis, John A., ed. *Science for Floodplain Management into the 21st Century.* Washington, D.C.: U. S. Government Printing Office, 1996.

Scientific Assessment and Strategy Team. "Science for Floodplain Management into the 21st Century." *Report of the Interagency Floodplain Management Review Committee to the Administration Floodplain Management Taskforce.* Washington, D.C.: U. S. Government Printing Office, 1994.

Theiling, Charles H. "Habitat Rehabilitation on the Upper Mississippi River." *Regulated Rivers: Research and Management* 11(1995):227–238.

Wiener, James G., Richard V. Anderson, and David R. McConnville, eds. *Contaminants in the Upper Mississippi River: Proceedings of the 15th annual meeting of the Mississippi River Research Consortium.* Boston, MA: Butterworth Publishers, 1984.

Internet Resources

Mississippi River Basin. Office of Water, U.S. Environmental Protection Agency. <http://www.epa.gov/msbasin/index.htm>.

Modeling Groundwater Flow and Transport

The main purpose for numerically modeling flow and transport in **groundwater** systems is to solve a variety of problems. For example, a city may use groundwater to supply its public water system; city officials use a model to plan where wells should be located. In another case, wastes may leak into the ground from an industrial facility. A model is used to determine where the contamination is moving and guide scientists in how best to clean it up. Models draw from mathematics, geology, chemistry, and biology.

Overview of the Modeling Process

Modeling flow and transport begins by gathering information about the problem and conceptualizing how the natural system functions based on available data. When the data are gathered and interpreted in a conceptual model, the computer model may then be run. The modeling process begins by setting up the computer model to match the conceptual model, then inputting all the data. Inputs represent the best estimate of how the system functions, but in reality, the result is uncertain. Normally the input values change as the model is calibrated (i.e., adjusted to fit the actual groundwater system).

Once calibration of the model is achieved, results are generated for the specific problem to be solved. In models of water-supply wells, results may be generated to show how the groundwater elevation changes at the well over time when the well is pumped at different rates. Alternately, results may be generated to show how pumping the well affects groundwater discharge into a critical **wetland** area or stream. In contamination studies, results normally are generated to show the groundwater area that has been contaminated, or the time period that will be required before a chemical moves to the area of a water-supply well. The modeler will decide which results should be generated in order to best present the solution to the problem.

---

groundwater: generally, all subsurface (underground) water, as distinct from surface water, that supplies natural springs, contributes to permanent streams, and can be tapped by wells; specifically, the water that is in the saturated zone of a defined aquifer

wetland: an area that is periodically or permanently saturated or covered by surface water or groundwater, that displays hydric soils, and that typically supports or is capable of supporting hydrophytic vegetation

Modeling Groundwater Flow and Transport

Input Data: Flow and Transport

As noted above, groundwater flow models first require input data, which describe the geometry or shape of the system to be modeled. Questions to ask could be: How thick, wide, and long is the area to be modeled? What types of soils and underlying geological materials are present? How easily can groundwater move through the geologic material in the project area? The person operating the computer model assigns the characteristics of the geologic materials (e.g., permeability and thickness) in the computer program to match the real situation.

The modeler needs information about where the groundwater originates and where it is going. Recharge areas correspond to locations where water enters the groundwater system, either by precipitation that percolates downward through the soil, or where rivers, streams, and lakes leak water into the subsurface. Wastewater disposal and stormwater runoff from city streets also can enter a groundwater system as part of the recharge. Aquifer storage and recovery systems are yet another means by which a groundwater system is recharged.

Discharge areas correspond to locations where water leaves the groundwater system, either naturally via seeps, springs, and streams, or artificially via pumping at wells. In modeling groundwater flow, recharge and discharge areas become the boundary conditions that constrain the model.

Groundwater flows in the direction of the hydraulic gradient. The hydraulic gradient of a river is the slope of the water surface; similarly, the hydraulic gradient of a groundwater system is the slope of the water level in the **aquifer**. The general direction of groundwater flow can be estimated by determining the direction of the hydraulic gradient. Measurements of groundwater hydraulic head at three or more locations provide this information (see box on this page).

Transport. When preparing to model the transport of a chemical, the modeler must determine the supply, or source, of the chemical to the groundwater, and where it occurs. The source term for modeling describes the mass (amount) per unit of time (mass loading) entering the groundwater system. For example, it could be determined that **nitrate** enters the groundwater at a wastewater pond, and it enters at a rate of 9 kilograms per day.

When a chemical moves in groundwater, it disperses within the pore spaces in the aquifer. It also may stick, or adsorb, onto soil particles. A parameter called the dispersivity is used to characterize the dispersion behavior of the chemical. The net effect of dispersion is a dilution of the concentration of the chemical. Another parameter, called the retardation factor, is used to describe the tendency for a chemical to adsorb onto the soil or aquifer particles. The retardation factor has the effect of slowing down the rate of transport.

Many chemical compounds change their identity in groundwater when they undergo a chemical reaction. One chemical compound can turn into another. This process is addressed in transport models by a decay parameter, or decay rate. The model predicts a decrease in the original chemical with time. Some models may track the new chemical(s) formed by decay.

---

aquifer: a water-saturated, permeable, underground rock formation that can transmit significant quantities of water under ordinary hydraulic gradients to wells and springs

nitrate: the highly leachable form of soil nitrogen taken up by most plants through their roots; it is a common groundwater contaminant, especially in agricultural areas and locations with a high density of septic systems, that is regulated by the U.S. Environmental Protection Agency with a drinking water standard of 10 ppm (parts per million) of nitrogen in the nitrate form

---

DETERMINING GROUNDWATER FLOW DIRECTION

Groundwater hydraulic head is the elevation of the groundwater in a well. By determining the hydraulic head at three or more locations in a groundwater flow system (e.g., by measuring the static water level in wells), hydrologists can draw a map showing lines (contours) of equal hydraulic head. These lines are much like topographic contours showing the land surface elevation, but instead of land, they show the elevation of groundwater hydraulic head. The general direction of groundwater flow under typical conditions occurs at 90-degree angles to the contour lines, and from higher to lower hydraulic head.

Model Calibration

Calibration is an important procedure in modeling. The computer uses all input data (e.g., recharge, discharge, geologic characteristics) to calculate how the groundwater occurs and how it moves as a function of the input data. During calibration, the model is fit to the actual groundwater system.

In order to judge how well the model fits the actual groundwater system, calibration targets are needed. For example, groundwater flow models calculate the elevation of the groundwater at many locations. When the groundwater elevation is measured in a well at some or all of these locations, then a comparison can be made. The model calculations can be compared to the field measurements; thus, the field measurements become the calibration targets. Calibration targets are then developed that show flow rates (e.g., the discharge of groundwater at a spring or into a river). Calibration targets also can be developed for chemical concentrations in groundwater (e.g., the concentration of a chemical in water pumped from one or more wells).

Calibration traditionally was performed by the trial-and-error method. The modeler simply changed an input value to the model (e.g., hydraulic conductivity or a boundary condition). The model was then executed and the new results were again compared to the calibration targets. This procedure was repeated, perhaps dozens (or even hundreds) of times, until it was no longer possible to obtain a better match to the calibration targets. Today, new computer tools have become available that use sophisticated **algorithms** to make the calibration changes automatically. SEE ALSO AQUIFER CHARACTERISTICS; ARTIFICIAL RECHARGE; FRESH WATER, NATURAL COMPOSITION OF; FRESH WATER, PHYSICS AND CHEMISTRY OF; GROUNDWATER; HYDROGEOLOGIC MAPPING; POLLUTION OF GROUNDWATER; SUPPLIES, PUBLIC AND DOMESTIC WATER.

Mark Cunnane

algorithm: a formalized, step-by-step procedure or set of equations designed for the purpose of solving a particular type of problem

Bibliography

Freeze, R. Allan, and John A. Cherry. *Groundwater.* Upper Saddle River, NJ: Prentice Hall, 1979.

Spitz, Karlheinz, and Joanna Moreno. *A Practical Guide to Groundwater and Solute Transport Modeling.* New York: John Wiley & Sons, 1996.

Modeling Streamflow

Hydrologic systems are complex, with processes occurring over different geographic areas characterized by highly variable parameters. In general, numerical models used for **surface-water** studies simulate the processes of interest as equations. Model results are difficult to confirm and often rely on the experience and judgment of the analyst; hence, it is the responsibility of the analyst to ensure that the model used produces results that are reasonable for a given watershed. The analyst also is faced with decisions regarding the length of time to be simulated. These modeling decisions require significant experience, not only as a hydrologist, but also as a user of a particular model.

Users of surface-water hydrologic modeling have two basic needs: (1) to determine the magnitude and frequency of flood flows; and (2) to deter-

surface water: water found above ground and open to the atmosphere, such as the oceans, lakes, ponds, wetlands, rivers, and streams

mine the long-term availability of water for consumption. The two needs require different modeling approaches.

Flood Runoff

If flood runoff is important (e.g., to determine drainage design), then only rainfall and runoff need to be modeled. In this case, relatively simple models can be used to determine a design discharge (i.e., the streamflow to be used for engineering design) and the runoff hydrograph (i.e., the plot of streamflow against time).

Two models that can be applied are the rational method and the unit hydrograph method. In both cases, short-term flow rates are the primary interest. Short-term flow rates are those that occur over a period of hours to days. For short-term times, hydrologic processes such as soil infiltration, percolation, and **evapotranspiration** can be ignored. Therefore, modeling approaches for short-term flows focus on the hydrograph of runoff from a single precipitation event.

evapotranspiration: water discharged to the atmosphere as a result of evaporation from the soil and surface-water bodies and by plant transpiration

Moreover, modeling of flood flow focuses on the maximum flow (discharge), or peak discharge, for an event with a particular exceedence probability. The exceedence probability is the probability of a particular event being equaled or exceeded over a given period of time, usually 1 year. The designer chooses the exceedence probability based on the perceived risk to human life or property damage if the event is exceeded.

Small design problems (those dealing with watershed drainage areas less than about 200 acres) typically use exceedence probabilities in the range of 10 percent to 20 percent because the consequences of failure are limited. For example, a structure like a bridge would be designed to withstand a stream discharge that would be exceeded less than 20 percent of the time. For such problems, it is acceptable to use simplified computational approaches that use approximate equations based on watershed properties.

Methods of Analysis.
In the simple computations procedures known as the rational method, the rate of maximum runoff is related to a runoff coefficient, a rainfall rate, and the drainage area through a basic mathematical formula. But if the drainage area exceeds about 200 acres, or if ponds or lakes complicate drainage, then a more complicated approach using a computer model is required. The most common method uses a unit hydrograph, which represents the runoff from the watershed as a unit pulse of runoff.

What complicates this procedure is that both a precipitation distribution (in time) as well as the watershed response (the unit hydrograph) must be determined. No longer can a single precipitation event (as represented by the rainfall intensity in the rational method) be considered. This is because there are many possible precipitation events for a given exceedence probability, each of which differs by the length of time the storm lasts. Therefore, to determine the maximum volumetric runoff intensity (the peak discharge), the designer must analyze several, perhaps many, different storm durations. This is why a computer model generally is used for this type of analysis.

In addition to the rational method and the unit hydrograph method, two other approaches are commonly used for flood analysis. If a stream gage is present at or near the location of interest, then the principles of statistics can be applied to determine the runoff and exceedence probability. This approach is called frequency analysis and can be used to estimate the flow rate. But if a stream gage is not available, then measurements at other sites can be used to create an estimate of the flow at a site of interest. The procedure used is called regionalization, and usually is accomplished by using statistical methods at several stream gages, then applying the results at the site of interest.

Water Supply

If water supply is important, the analyst must focus on longer time periods (e.g., years or decades) compared to the short intervals used for flood analysis. Because a specific storm event is no longer the focus, all hydrologic processes come into play, and the models used to study long-term water quantity must reflect this complexity. The physical processes are represented by simplified submodels, yet the entire model must yield a reasonable solution.

The basic concept is the hydrologic budget, an accounting of the water as it moves along various pathways at the Earth's surface (in particular on the watershed of interest) in a manner that preserves (conserves) the mass of water. Components are precipitation, runoff, evapotranspiration, and movement to groundwater. The formula is $P - E - R - G = \Delta S$, where P is incoming precipitation, E is evapotranspiration, R is runoff from the watershed, G is percolation to groundwater, and ΔS (delta S) is the change in water storage in the active part of the soil profile. The terms in this equation can be expressed as either time rates or by depth or volume over some convenient period (such as 1 day). Because there are so many terms involved in the hydrologic budget, computer models are most often used for these analyses. SEE ALSO HYDROLOGIC CYCLE; RUNOFF, FACTORS AFFECTING; STREAM HYDROLOGY.

David B. Thompson

Bibliography

Viessman, Jr., Warren, and Gary L. Lewis. *Introduction to Hydrology*, 4th ed. New York: HarperCollins, 1996.

Mollusks See Bivalves; Cephalopods.

Moorings and Platforms

Moorings and platforms are structures that allow water scientists to position instruments, collect samples, and take long-term measurements in waterbodies. A mooring typically consists of a flexible cable that is tethered to the ocean floor by a weight or anchor and suspended from the sea surface by a buoy. A platform has a flat workspace for instruments and people, and can be free-floating, moored to the seafloor, or rigidly fixed in place.

Water scientists, including physical and chemical oceanographers, marine biologists and geologists, **meteorologists**, and lake scientists called limnologists collect many types of data from instruments installed on moorings

meteorology: the science that deals with the atmosphere, especially with regard to climate and weather

Moorings and Platforms

Instruments that measure physical and chemical aspects of the oceans and atmosphere can mounted on or suspended from a buoy. Together with shipboard and satellite data, the global network of measuring sites provides further understanding of the ocean–atmosphere system and its influence on global climate and biological productivity.

and platforms. Oceanographers measure physical and chemical properties such as temperature, salinity, current velocity and turbidity—the amount of suspended sediment and other particles in the water—using equipment suspended beneath the sea surface on mooring cables. Marine biologists collect water and specimen samples, conduct dives, and make observations of marine flora and fauna from platforms. They also use oceanographic information to study the distribution of marine **ecosystems**. Marine geologists collect seafloor samples and sub-seafloor cores that hold clues to the Earth's geologic history and past climate, as well as clues to the location of **petroleum reservoirs**, from fixed platforms.

ecosystem: the community of plants and animals within a water or terrestrial habitat interacting together and with their physical and chemical environment

petroleum reservoir: a porous and permeable rock in which petroleum accumulates; primarily marine sedimentary rocks such as sandstone and limestone

107

Equipment and Instrumentation

Moorings and platforms often accommodate numerous pieces of measuring equipment. This economical approach allows scientists to collect several types of data at a site, to compare data from the site, and to observe data patterns that affect a number of natural systems. Changes in current direction or temperature within the water column, for example, may correlate to changes in biological productivity or weather patterns.

A typical oceanographic mooring, like one deployed in the northwest Atlantic Ocean by the Global Ocean Ecoystems Dynamics (GLOBEC) program, holds a large array of instrumentation: seven current meters, seven temperature gauges, three optical turbidity scanners, four salinity/conductivity/pressure meters, and one Acoustic Doppler Current Profiler (ADCP) that records surface ocean current patterns around the mooring. The GLOBEC mooring buoy houses instruments that record wind speed and direction, air temperature, relative humidity, and amount of sunlight reaching the buoy at the sea surface.

Deployment. Tethered moorings and platforms are deployed from ships and left at their sites to record data for extended periods of time. The GLOBEC moorings were deployed for 5 years and serviced every 6 months. In the past, ocean scientists had to retrieve data from recorders at such "stand alone" moorings during return visits to the sites. Modern moorings, however, now include transmitters that beam data to scientists' computers via satellite. They also carry global positioning system (GPS) instruments that measure their precise location over time, and transmitters that relay information about mechanical difficulties at moorings. Some systems can even receive signals to adjust their instruments. These technological advances allow continuous data monitoring at land stations, improve data quality, and reduce the number of expensive visits to mooring sites.

An international initiative called the Global Ocean Observing System (GOOS) is a global scientific collaboration that seeks to collect data from moorings and platforms all over the planet, and to compile those data into a global database that would be available to scientists for comparison and analysis.

Specialized Design. Tethered moorings and platforms are designed to reliably collect, record, and transmit accurate data while submerged in corrosive saltwater and exposed to extreme weather conditions for long periods of time. Their buoys have heavy-duty steel or aluminum flotation collars, resilient solar panels, batteries for backup power, safety devices such as beacons and radio transmitters, and insulated, reinforced wells that protect sensitive scientific instruments.

Mooring cables, while flexible, are extremely strong. Electronic cables that carry data from the submerged instruments to the sea surface are encased in an insulating sheath that surrounds the tether. Anchor weights, which are designed to hold moorings in place indefinitely, can be mechanically or electronically signaled to release when it is time to retrieve the mooring.

Platforms

Platforms provide flat workspaces to accommodate scientists and large equipment at the study site and serve as field laboratories for water scientists. In addition to collecting the types of data provided by unmanned moor-

Moorings and Platforms

The data collected at moorings and platforms give scientists the information they need to pursue some of the most pressing scientific and environmental questions of the twenty-first century. Here, oceanographers with the National Oceanic and Atmospheric Administration prepare floats for a mooring deployment from the deck of a research vessel in Antarctica's Ross Sea.

ings, staffed platforms allow scientists to collect biological specimens, sea water, and sediment samples, to observe marine organisms, to conduct diving expeditions, and to deploy to manned and unmanned submersible vehicles. Instruments installed on drifting platforms measure oceanographic, atmospheric, and biological data along the path of an ocean current.

Geological studies that seek to collect deep sediment cores, or to drill into crystalline rock, require platforms that are stable in shifting seas. Some platforms float around a central "moon pool" that isolates the drilling equipment from platform motion. Other drilling platforms stand above the sea surface on legs that are rooted in the seafloor.

Most drilling platforms are too expensive for publicly funded scientific studies, but they are common equipment in the petroleum industry. Oil platforms over major petroleum reservoirs may house hundreds of staff members who live and work on the rig for weeks or months at a time. The petroleum industry also maintains smaller, unmanned platforms that house oil and gas pumps once an offshore field is established. SEE ALSO CARBON DIOXIDE IN THE OCEAN AND ATMOSPHERE; CLIMATE AND THE OCEAN; GEOSPATIAL TECHNOLOGIES; OCEAN CURRENTS; OCEAN-FLOOR SEDIMENTS; OCEANOGRAPHY, BIOLOGICAL; OCEANOGRAPHY, CHEMICAL; OCEANOGRAPHY, GEOLOGICAL; OCEANOGRAPHY, PHYSICAL; PETROLEUM FROM THE OCEAN; SUBMARINES AND SUBMERSIBLES.

Brian D. Hoyle and Laurie Duncan

Bibliography

Irish, James D. et al. "The Next Generation Ocean Observing Buoy in Support of NASA's Earth Science Enterprise." *Sea Technology* 40, no. 5 (1999):37–43.

Open University Course Team. *Ocean Circulation.* Oxford, U.K.: Pergamon Press, 1993.

Weller, Robert, et al. "Outposts in the Ocean." *Oceanus* 42, no. 1 (1999):20–23.

Internet Resources

Global Ocean Ecosystem Dynamics. <http://www.pml.ac.uk/globec/main.htm>.

The Global Ocean Observing System. Intergovernmental Oceanographic Commission (IOC). <http://ioc.unesco.org/goos/>.

oceanography: the broad category of science that deals with oceans

Nansen, Fridtjof

Norwegian Arctic Explorer and Oceanographer
1861–1930

The science of **oceanography** was in its infancy in 1893 when Fridtjof Nansen, a 32-year-old Norwegian, purposely allowed his ship, the *Fram*, to be captured in an Arctic ice pack. Through this deliberate act, Nansen hoped to prove his theory that the Arctic current flowed from Siberia towards the North Pole and then southward to Greenland. The way to show this to be true, Nansen reasoned, was to create a specially built ship that would not be crushed by ice so that he could drift in the ship wherever the ice pack moved. If the *Fram* (meaning "Forward") was carried close enough, Nansen also hoped to become the first person to reach the North Pole.

Greenland Ice Cap

Nansen thought in original ways. While a university student in Christiana (now Oslo, Norway), he took a voyage in 1882 to Arctic regions aboard a seal-hunting ship. Nansen kept records of winds, ice movements, and animal life. On this voyage, he became intrigued with the idea of ocean currents while observing the eastern coast of Greenland. He decided to attempt to cross Greenland's inland **ice cap** in order to study more closely continental **glaciers**.

With a team of five, Nansen accomplished this trek in 1888. One of the more unusual aspects of the expedition was the direction he chose to travel; the team went from east to west across Greenland instead of the more usual west-to-east direction. Nansen decided that because it was impossible for a ship to wait off the inhospitable east coast once his team had disembarked, he and his men would be forced to move toward the inhabited west coast, where their ship could safely meet them.

It was a plan that worked. Nansen had become the first person to cross Greenland's ice cap. His observations on this trip demonstrated that continental glaciers are thick and heavy enough to depress the Earth's crust beneath their weight. His work provided support for the theory of **isostatic** rebound—the idea that when the Earth's crust sinks under a heavy weight, it will slowly return to its original position when that weight is removed.

The *Fram* Expedition

By the time he returned home to Norway, Nansen was famous. Yet people thought him foolish in planning his next voyage—one in which he would purposely let his ship be caught up in the Arctic ice. They reasoned that the *Fram* would be crushed; or, if by some miracle the vessel escaped destruction, then his crew would become insane by virtue of being trapped on a ship for several years without any contact with the outside world.

But attention to detail was Nansen's strength. He found a shipbuilder who designed the *Fram* as an iron-clad, three-hulled ship able to be thrust up and not crushed by the pressure of ice. The propeller and rudder would be built into the ship to protect them.

Nansen knew his team would be constantly busy making scientific observations. But for the infrequent off-hours, he put a library, musical instruments, and games onboard to entertain the crew. To captain the ship, Nansen chose Otto Sverdrup, one of the men who had crossed Greenland with him.

Successful Journey.

Nansen's meticulous planning paid off. His trip was successful, and had a most unusual finale. Because the *Fram* did not drift as close to the North Pole as he had hoped, Nansen decided to make an overland dash for it. With one companion, dogs, sleds, and kayaks, he left the *Fram* under the capable care of Sverdrup and, in March 1895, headed for the Pole. Conditions forced the two men to turn back, but they had gotten closer than anyone as of that date. After a demanding journey of 483 kilometers (300 miles), they made it to a safe wintering place, an island in Franz Josef Land.

After breaking camp in the spring, Nansen and his companion headed south and had a wonderful piece of luck. On the ice and in the middle of nowhere, they met the leader of a British scientific expedition who sent them home to Norway in one of his ships. They arrived in their country on August 13, 1896, the same day the *Fram* finally became free of pack ice and also headed home. The ship had done exactly what Nansen predicted—drifted with the Arctic current. His work provided enormous knowledge about the currents, climate, and marine life of the Arctic Ocean.

Fridtjof Nansen, shown here in 1929, completed daring Arctic expeditions that provided pioneering knowledge of ocean currents, glaciers, sea ice, and climate.

ice cap: an extensive perennial accumulation of snow and ice that forms when glaciers completely fill their subglacial valleys and coalesce (join together); ice caps are smaller than ice sheets

glacier: a huge mass of ice, formed on land by the compaction and recrystallization of snow, which moves slowly downslope or outward owing to its own weight

isostasy: describes the concept that the elevation of the Earth's surface (over tens of millions of years) seeks a balance between the weight of lithospheric rocks and the buoyancy of asthenospheric fluid (hot, plastic, partially molten rock); a mountain range where erosion has moved a significant amount of rock material may rise isostatically, whereas the basin that receives this eroded sediment may sink under the added weight; "isostatic" means pertaining to or related to isostasy

The ship used in Fridtjof Nansen's 1895–1896 Arctic expedition was built to withstand sea-ice conditions. More than a century later, specially designed research vessels (like the one shown here) still explore the vast polar oceans.

Later in Life

From 1896 to 1917, Nansen served as a professor at the University of Christiania, where he performed research in oceanography. He published a six-volume report of the *Fram* expedition that remains a major reference on the Arctic Ocean. He also designed several instruments still in common use, including the Nansen bottle, a device used to collect pure samples of sea water at given depths. Several scientific voyages to the North Atlantic added to his knowledge.

As a statesman, Nansen worked for the peaceful separation of Norway from Sweden. His efforts were rewarded by an appointment as Norway's first minister to Great Britain from 1906 to 1908.

Toward the end of World War I, Nansen became internationally known for his service to famine-stricken Russia as well as for his work in returning prisoners of war to their homes. His contributions won him the Nobel Peace Prize in 1922. SEE ALSO GLACIERS AND ICE SHEETS; ICE AGES; ICE AT SEA; OCEAN CURRENTS; OCEANS, POLAR.

Barbara Johnston Adams

Bibliography

Huntford, Roland. *Nansen: The Explorer as Hero*. New York: Barnes and Noble, 1997.

Nansen, Fridtjof. *Farthest North*. New York: Modern Library, 1999.

habitat: the environment in which a plant or animal grows or lives; the surroundings include physical factors such as temperature, moisture, and light, together with biological factors such as the presence of food and predators

environment: all of the external factors, conditions, and influences that affect the growth, development, and survival of organisms or a community; commonly refers to Earth and its support systems

National Environmental Policy Act

As concern for environmental problems grew during the 1960s, the need for better project planning to prevent environmental degradation became obvious. Too many projects (e.g., dams, power plants, highways) had been built without regard for their negative impacts on water quality, soil erosion, aquatic and terrestrial **habitat**, or negative economic and social impacts on nearby communities. Many of these projects caused irreversible harm to the **environment**. Congress recognized that federal projects were

National Environmental Policy Act

The National Environmental Policy Act is intended to promote environmental considerations into federal agencies' decision-making processes. Impacts on wetland systems (shown here) often are the subject of NEPA-related processes.

culprits of environmental degradation, and that federal and state agencies would continue to support such projects unless Congressional statute forced them to consider their negative environmental impacts.

The National Environmental Policy Act (NEPA) of 1969 requires all federal agencies to use an interdisciplinary approach in their project planning and decision-making. Because federal agencies had no history or desire to work together on projects, tried to protect their "turf," and did not want the expense of the environmental impact assessment process to affect their budgets, they initially fought with Congress to prevent NEPA from being enacted. But since that time, every federal agency has developed a set of NEPA guidelines for its personnel to follow.

NEPA Components and Requirements

The National Environmental Policy Act has several major requirements. First, it provides a comprehensive approach to environmental protection by requiring that federally funded or federally permitted projects must perform environmental impact assessments and write an environmental impact statement (EIS) if that project would have a potentially significant environmental impact. An environmental assessment (EA) is conducted to determine whether impacts are potentially significant.

The EIS must describe:
- The proposed action;
- Various environmental impacts of the proposed action;
- Unavoidable adverse environmental effects of the action;
- Alternatives (and their negative impacts) to the proposed action; and
- Long-term environmental damages.

Alternatives often give decisionmakers methods to mitigate some negative project impacts or carry out actions entirely different than those originally planned. For example, the city of Denver, Colorado proposed the Two Forks Dam to obtain additional water supplies. In 1989, the U.S Environmental Protection Agency denied approval when one EIS alternative

> **COUNCIL ON ENVIRONMENTAL QUALITY**
>
> The National Environmental Policy Act (NEPA) created the Council on Environmental Quality and placed it in the Executive Office of the President. The council oversees implementation of NEPA, advises the president on environmental issues, and writes an annual report evaluating national environmental conditions and trends, and regulatory adequacy to protect the environment.

suggested making Denver meter all buildings and houses in order to stop wasting the water it already had.

Research for environmental impact assessments is done in categories such as: land use and development; social and economic effects; relocations and neighborhood effects; noise; traffic; transportation; environmental health and public safety; historical and archaeological resources; visual resources; air quality; water resources (including **wetlands**); wildlife and aquatic species and their habitats; floodplains; and coastal areas. Because of its complexity and comprehensive nature, the research is expensive. EISs for large projects typically cost tens of millions of dollars and take years to complete.

To ensure public participation, the public is notified when the draft EIS (DEIS) is published; all notices regarding NEPA processes are published in the Federal Register. The public and special interest groups typically have 45 to 60 days to read and comment on it, either in writing or at public hearings. Federal agencies must incorporate public comment into the final EIS (FEIS). The FEIS states which of the alternatives has been chosen for implementation, and the justifications for selecting that alternative over the others.

Visionary Scope. NEPA is important to water resources management because the assessment done for the water resources portion of an EIS helps decisionmakers balance the numerous and complex variables when considering a project. They must take into account how their project can have a ripple effect on water resources, wildlife, habitat, and local communities. An EIS helps illustrate that water resources must be managed not only for immediate human uses, but for long-term **ecosystem** sustainability. For example, a recent controversy over how to manage the Missouri River is pitting the interests of upstream recreational economies and their necessary habitat preservation with downstream economic interests of barge traffic and flood control. SEE ALSO BALANCING DIVERSE INTERESTS; ENVIRONMENTAL MOVEMENT, ROLE OF WATER IN THE; LAND-USE PLANNING; LEGISLATION, FEDERAL WATER; PUBLIC PARTICIPATION.

Laurel E. Phoenix

wetland: an area that is periodically or permanently saturated or covered by surface water or groundwater, that displays hydric soils, and that typically supports or is capable of supporting hydrophytic vegetation

ecosystem: the community of plants and animals within a water or terrestrial habitat interacting together and with their physical and chemical environment

Bibliography

Marriott, Betty Bowers. *Environmental Impact Assessment: A Practical Guide.* New York: McGraw-Hill, 1997.

Reimold, Robert J. *Watershed Management: Practice, Policies, and Coordination.* New York: McGraw Hill, 1998.

Rogers, Peter. *America's Water: Federal Roles and Responsibilities.* Cambridge, MA: MIT Press, 1993.

Internet Resources

Council on Environmental Quality. <http://www.whitehouse.gov/ceq/>.

Federal Register. U.S. Government Printing Office. <http://www.access.gpo.gov/su_docs/aces/aces140.html>.

U.S. Government's NEPAnet. <http://ceq.eh.doe.gov/nepa/nepanet.htm>.

National Oceanic and Atmospheric Administration

The National Oceanic and Atmospheric Administration (NOAA) was officially established under the U.S. Department of Commerce in 1970. It arose from the same reorganization plan that formed the Environmental Protec-

National Oceanic and Atmospheric Administration

Scientists with the National Oceanic and Atmospheric Administration collect data and conduct research that supports the agency's regulatory missions. Here, fisheries biologists with the National Marine Fisheries Service study populations of groundfish off the Oregon coast.

tion Agency. During this time, a new approach to environmental regulation was beginning in which separate agencies were being consolidated in recognition of the interrelations existing in the **environment**.

NOAA is charged with the mission of collecting scientific information, predicting changes in the environment, and protecting life and property. In the years since its formation, NOAA's efforts have provided a better understanding of the behavior of natural systems, and how to effectively manage resources to allow for economic development while protecting environmental quality.

environment: all of the external factors, conditions, and influences that affect the growth, development, and survival of organisms or a community; commonly refers to Earth and its support systems

Major Divisions

NOAA gathers information and conducts research primarily through its five major organizational units. Together, these organizational units work to achieve the agency's goal of ensuring effective management and stewardship

National Oceanic and Atmospheric Administration

weather: the condition of the atmosphere at any given time and location, including the temperature, pressure, and humidity of the air; wind direction and speed; and phenomena such as clouds, rain, and snow

climate: the long-term average of weather conditions at a given location

meteorology: the science that deals with the atmosphere, especially with regard to climate and weather

biodiversity: a measure of the variety of the Earth's species, of the genetic differences within species, and of the ecosystems that support those species

habitat: the environment in which a plant or animal grows or lives; the surroundings include physical factors such as temperature, moisture, and light, together with biological factors such as the presence of food and predators

sustainable: as in "sustainable development," describes efforts that guide economic growth in a manner that meets current needs without compromising the ability of future generations to meet their needs; in terms of natural resources, also encompasses development conducted in an environmentally sound manner, with an emphasis on natural resource conservation, including water and aquatic life

of natural resources. NOAA collects much of the data on water in the oceans and atmosphere. These data are used by an array of decisionmakers, scientists, special interest groups, and the general public, ranging from applications as diverse as ocean-related policy-making, to investigations of climate change, to the planning of daily personal activities.

National Weather Service. Perhaps the most well-known organization within NOAA is the National Weather Service (NWS). The NWS collects **weather** data and provides forecasts and warnings to protect the nation's residents and to provide communities with the information needed to plan and prepare for weather events. Commercial weather organizations also use information provided by the NWS to provide the public with weather forecasts.

National Environmental Satellite, Data, and Information Service. The National Environmental Satellite, Data, and Information Service (NESDIS) operates the nation's environmental satellites, providing valuable information about weather and climate used by the NWS for short-term forecasting and long-term monitoring and prediction of **climate** variation. The organization also maintains large databases for **meteorology**, oceanography, geophysics, and solar–terrestrial sciences used by scientists working to better understand Earth systems.

National Ocean Service. The National Ocean Service (NOS) provides information to coastal communities and those navigating ocean waters to allow for the safe use of ocean resources while also preserving and protecting those resources for future use. The NOS also administers the National Marine Sanctuary Program, which, like the U.S. Department of Interior's National Parks Program, protects areas identified for their **biodiversity**, ecological integrity, and cultural legacy. The sanctuary program was created in 1972 by the Marine Protection, Research, and Sanctuaries Act.

National Marine Fisheries Service. NOAA Fisheries, or the National Marine Fisheries Service (NMFS), studies and manages marine fisheries, conserves fishery **habitats**, and enforces federal statutes, such as the Endangered Species Act, the Marine Mammal Protection Act of 1994, and the Magnuson–Stevens Act of 1996. This organization's research and management efforts support the exploitation of living marine resources while working to achieve **sustainability** of resource use.

Office of Oceanic and Atmospheric Research. The Office of Oceanic and Atmospheric Research (OAR) is NOAA's primary research arm. OAR maintains several research laboratories, including oceangoing vessels and undersea research centers. It also works with academic institutions, providing research grants to generate new knowledge about the oceans.

NOAA's plan for the future includes the strong promotion of sustainable development. As a bureau of the U.S. Department of Commerce, NOAA collects information and conducts research to allow for economic growth and the wise use of natural resources while maintaining the quality of the environment. SEE ALSO CLIMATE AND THE OCEAN; EL NIÑO AND LA NIÑA; ENDANGERED SPECIES ACT; FISH AND WILDLIFE SERVICE, U.S.; FISHERIES, MARINE: MANAGEMENT AND POLICY; GEOSPATIAL TECHNOLOGIES; NATIONAL PARK SERVICE; WEATHER AND THE OCEAN.

Vincent G. McGowan

Bibliography

U.S. Department of Commerce. *NOAA Strategic Plan: A Vision for 2005—Executive Summary*. Washington, D.C.: U.S. Department of Commerce, 1998.

Internet Resources

National Marine Sanctuaries. <http://www.sanctuaries.nos.noaa.gov>.

NOAA History. U.S. Department of Commerce. <http://www.publicaffairs.noaa.gov/grounders/noaahistory.html>.

NOAA Home Page. U.S. Department of Commerce. <http://www.noaa.gov>.

National Park Service

The U.S. Congress created the National Park Service (NPS) in 1916 as a bureau of the Department of the Interior. Its purpose was to coordinate the administration for an increasing number of national parks. NPS objectives have evolved to include (1) preserving the natural and cultural resources and values of the parks, and (2) providing for public enjoyment of these areas while leaving them unimpaired.

The diverse areas managed by the NPS—employing such professionals as foresters, naturalists, engineers, biologists, historians, geologists, archaeologists, rangers, and guides—are collectively known as the National Park System. As of 2002 it encompassed 384 areas, totaling about 33.8 million hectares (83.6 million acres) in 49 states, the District of Columbia, American Samoa, Guam, Puerto Rico, Saipan, and the Virgin Islands.

Activities, Services, and Administration

The NPS engages in such activities as control of water **pollution**; fire prevention and control; wildlife **conservation**; and protection of natural, historic, or prehistoric features. In addition, it provides public services such as lectures, guided tours, and informational programs.

The National Park Service administers over twenty types of areas that are categorized as natural, historical and cultural, and recreational. Areas dedicated to the preservation of natural features include national parks, rivers, preserves, and many of the monuments. Areas preserved for historic and cultural purposes include national monuments, military parks, battlefields, historical parks, and historic sites. Recreational areas include national seashores and lakeshores, and national recreation areas.

The administration of natural resources includes the preservation of:

- Physical resources, such as water, soil, air, **topographic** and geologic features, and paleontological resources;
- Physical processes, such as weather, erosion, cave formations, and wildland fires;
- Biological resources, such as native animals, plants, and communities;
- Biological processes, such as **photosynthesis**, ecosystem succession, and evolution;
- **Ecosystems**; and
- Highly valued associated characteristics, such as scenic views.

pollution: any alteration in the character or quality of the environment, including water in waterbodies or geologic formations, which renders the environmental resource unfit or less suited for certain uses

conservation: the organized management and planned use of living and nonliving natural resources; "water conservation" refers to strategies that increase the efficiency of water use, reuse, recycling, production, or distribution, or that decrease demand

topography: the shape and contour of a surface, especially the land surface or ocean-floor surface

photosynthesis: the process by which plants manufacture food from sunlight; specifically, the conversion of water and carbon dioxide to complex sugars in plant tissues by the action of chlorophyll driven by solar energy

ecosystem: the community of plants and animals within a water or terrestrial habitat interacting together and with their physical and chemical environment

---

LARGE AND SMALL PROPERTIES

The largest National Park Service area is Wrangell–St. Elias National Park and Preserve in Alaska, containing 5.3 million hectares (13.2 million acres), or 16.3 percent of the entire National Park System. The smallest unit is Thaddeus Kosciuszko National Memorial in Pennsylvania, with 0.008 of a hectare (0.020 of an acre).

National Park Service

Environmental and historical education are offered by National Park Service rangers, naturalists, guides, and historians. Here a ranger involves two young visitors while explaining the Florida Everglades' water resource system.

surface water: water found above ground and open to the atmosphere, such as the oceans, lakes, ponds, wetlands, rivers, and streams

groundwater: generally, all subsurface (underground) water, as distinct from surface water, that supplies natural springs, contributes to permanent streams, and can be tapped by wells

point source: a pollutant release or discharge originating from one specific location (e.g., an outfall pipe from a factory) rather than over a wide land area (e.g., water runoff from a farm field)

nonpoint source: a pollutant release or discharge originating from a land use active over a wide land area (e.g., agriculture) rather than from one specific location (e.g., an outfall pipe from a factory)

wetland: an area that is periodically or permanently saturated or covered by surface water or groundwater, that displays hydric soils, and that typically supports or is capable of supporting hydrophytic vegetation

Water Resources

As part of its administration, the NPS maintains the quality of its **surface waters** and **groundwaters** because they are critical components of aquatic and terrestrial ecosystems. Any pollution incidents from **point sources** and **nonpoint sources** are quickly resolved, knowing that such pollutants reduce visitor appeal to park waters. NPS regularly works with various government bodies in order to maintain or restore water quality as laid out under the federal Clean Water Act and other applicable federal, state, and local laws and regulations. Oftentimes, it is activities that take place outside park boundaries that have profound effects on the Service's ability to protect natural resources inside parks.

In order to maintain the high standard of the park's natural resources many water-related systems are important: for example, floodplains, **wetlands**, watershed (drainage basin) and stream processes, and shorelines and barrier islands.

Floodplains and Wetlands. The NPS is committed to protect, preserve, and restore the functions of floodplains. It manages wetlands in order to prevent destruction, loss, or degradation in compliance with NPS mandates and the requirements of Executive Order 11990 (Wetland Protection), the Clean Water Act, and the Rivers and Harbors Act of 1899, among others.

Watershed and Stream Processes. The NPS manages watersheds as complete hydrologic systems, and minimizes human disturbance to the natural upland processes that deliver water, sediment, and debris to streams. These processes include runoff, erosion, and disturbance to vegetation and soil caused by fire, insects, weather events, and mass movements of earth materials. The NPS manages streams to protect processes that create habitat features such as floodplains, riparian systems, and natural pools.

Shorelines and Barrier Islands. On NPS properties, natural shoreline processes such as erosion, dune formation, and inlet formation are allowed to evolve without interference. Where human activities or structures have altered the nature or rate of natural processes, the NPS investigates alternatives for mitigating the effects of such activities or structures and for restoring natural conditions.✶

The National Park Service is an important partner in water management, especially given its responsibilities for many of the country's most precious wilderness areas. Water and aquatic resources are valued as an important natural resource in the NPS's planning and management activities. Scientists and resource managers are increasingly called upon to address disruptions of water resources that threaten the quality of life and environmental sustainability in U.S. national parks. SEE ALSO BEACHES; CLEAN WATER ACT; FISH AND WILDLIFE SERVICE, U.S.; FLOODPLAIN MANAGEMENT; NATIONAL OCEANIC AND ATMOSPHERIC ADMINISTRATION; POLLUTION SOURCES: POINT AND NONPOINT; WETLANDS.

William Arthur Atkins

Brandon, Katrina et al. *Parks in Peril: People, Politics, and Protected Areas.* Washington, D.C.: Island Press, 1998.

Buccino, Sharon et al. *Reclaiming Our Heritage: What We Need To Do To Preserve America's National Parks.* Washington, D.C.: Natural Resources Defense Council Press, 1997.

Internet Resources

2001 NPS Management Policies, Chapter 4: Natural Resource Management. National Park Service. <http://www.nps.gov/policy/mp/chapter4.htm>.

History of the National Park Service. National Park Service. <http://www.cr.nps.gov/history/hisnps/NPSHistory.htm>.

National Park Service: ParkNet. <http://www.nps.gov>.

National Park Service, Water Resources Division. <http://www.nature.nps.gov/wrd>.

Water Management: Everglades National Park. National Park Service. <http://www.nps.gov/ever/eco/h2omgmt.htm>.

> **NATIONAL MARINE SANCTUARIES**
>
> In 1972, exactly 100 years after the first national park was created, the U.S. Congress made a similar commitment to protecting selected marine areas by establishing the National Marine Sanctuary Program. Since then, thirteen sanctuaries have been designated, representing a variety of ocean environments and one Great Lakes (fresh-water) area. The National Marine Sanctuary Program is administered by the National Oceanic and Atmospheric Administration, part of the U.S. Department of Commerce.
>
> Marine sanctuaries encompass deep ocean gardens, nearshore coral reefs, whale migration corridors, deep-sea canyons, and even underwater archaeological sites. They range in size from about 65 hectares (160 acres or 0.25 square mile) in Fagatele Bay, American Samoa, to roughly 1.37 million hectares (about 3.39 million acres or 5,300 square miles) in Monterey Bay, California, one of the largest marine protected areas in the world. For more information, see <http://www.sanctuaries.nos.noaa.gov>.

✶ See "Coastal Ocean" for a photograph of a newly formed barrier island.

Navigation at Sea, History of

The first Western civilization known to have developed the art of navigation at sea were the Phoenicians, about 4,000 years ago (c. 2000 B.C.E.). Phoenician sailors accomplished navigation by using primitive charts and observations of the Sun and stars to determine directions.

Navigation at Sea, History of

Maps, compasses, astrolabes, and calipers are among the early tools used by ocean navigators. In the modern era, these tools have been largely replaced by electronic and technological equivalents.

Despite these early beginnings, it would take many centuries before global navigation at sea became possible. Until the fifteenth century, mariners were essentially coastal navigators. Sailing on the open sea was limited to regions of predictable winds and currents, or where there was a wide **continental shelf** to follow. Farther ventures were enabled by the development of scientifically and mathematically based methods and tools.

Early Navigational Tools

Determining **latitude** can be accomplished relatively easily using celestial navigation. In the Northern Hemisphere, mariners could determine the latitude by measuring the **altitude** of the North Star above the horizon. The angle in degrees was the latitude of the ship.

Mariner's Compass. One of the earliest human-made navigational tools used to aid mariners was the mariner's compass, which was an early form of the magnetic compass. Early mariners thought the mariner's compass was often inaccurate and inconsistent because they did not understand the concept of magnetic variation, which is the angle between true north (geographic) and magnetic north. It was primarily used when the Sun was not visible to help identify the direction from which the wind was blowing.

Nautical Charts. During the mid-thirteenth century, mariners began realizing that maps could be helpful and began keeping detailed records of their voyages. Thus, the first nautical charts were created. These first charts were not very accurate, but were considered valuable and often kept secret from other mariners. There was no latitude or **longitude** labeled on the charts, but between major ports there was a **compass rose** indicating the direction to travel. (The term "compass rose" comes from the figure's compass points, which resemble rose petals.)

Astrolabe, Sextant, and Chip Log. Some of the early instruments used to assist sailors in determining latitude were the cross-staff, astrolabe, and quadrant. The astrolabe dates back to ancient Greece, when it was used by astronomers to help tell time, and was first used by mariners in the late

---

continental shelf: the relatively flat, submerged natural platform, about 1-degree slope, that extends seaward from the beach for about 70 kilometers (45 miles), with water depth up to 130 meters (425 feet) maximum, and ending where the slope and water depth increase

latitude: the angular distance north or south of the Earth's or another planet's equator, measured in degrees along a meridian

altitude: in navigational usage, the angle measured from the horizon to a celestial body, such as the Sun, Moon, or stars; in general usage, the height above the ground surface, called elevation if referring to height above sea level

longitude: the angular distance measured east or west from the prime meridian (which runs through Greenwich, England), to the meridian passing through a position; expressed in degrees (or hours), minutes, and seconds

compass rose: a circular diagram printed on a chart or map to show the direction of north and other principal directional points of the compass

Tall ships keep alive the history of ocean navigation. Today, ships such as these call to mind images of merchant ships from long ago and pirates in their heyday during the eighteenth century.

fifteenth century. It was used to measure the altitude of the Sun and stars to determine latitude.

Around 1730, an English mathematician, John Hadley (1682–1744), and an American inventor, Thomas Godfrey (1704–1749), independently invented the sextant. The sextant provided mariners with a more accurate means of determining the angle between the horizon and the Sun, moon, or stars in order to calculate latitude.

During the sixteenth century, the chip log was invented and used as a crude speedometer. A line containing knots at regular intervals and weighted to drag in the water was let out over the stern as the ship was underway. A seaman would count the number of knots that went out over a specific period of time and the ship's speed could then be calculated.

Longitude and the Chronometer. Throughout the history of navigation, latitude could be found relatively accurately using celestial navigation. However, longitude could only be estimated, at best. This was because the measurement of longitude is made by comparing the time-of-day difference between the mariner's starting location and new location. Even some of the best clocks of the early eighteenth century could lose as much as 10 minutes per day, which translated into a computational error of 242 kilometers (150 miles) or more.

In 1764, British clockmaker John Harrison (1693–1776) invented the seagoing chronometer. This invention was the most important advance to marine navigation in the three millenia that open-ocean mariners had been going to sea.

In 1779, British naval officer and explorer Captain James Cook (1728–1779) used Harrison's chronometer to circumnavigate the globe. When he returned, his calculations of longitude based on the chronometer proved correct to within 13 kilometers (8 miles). From information he gathered on his voyage, Cook completed many detailed charts of the world that completely changed the nature of navigation.

In 1884, by international agreement, the Prime Meridian (located at 0° longitude) was established as the meridian passing through Greenwich, England.

Modern Navigation

The twentieth century brought important advances to marine navigation, with radio beacons, radar, the gyroscopic compass, and the global positioning system (GPS). Most oceangoing vessels keep a sextant onboard only in the case of an emergency.

Gyroscopic Compass. The gyroscopic compass (or gyro compass) was introduced in 1907. The primary benefit of the gyro compass over a magnetic compass is that the gyro is unaffected by the Earth's, or the ship's, magnetic field, and always points to true north.

Radar. The first practical radar (short for "radio detection and ranging") system was produced in 1935. It was used to locate objects beyond range of vision by projecting radio waves against them. This was, and still is, very useful on ships to locate other ships and land when visibility is reduced.

Loran. The U.S. navigation system known as Long Range Navigation (Loran) was developed between 1940 and 1943, and uses pulsed radio transmissions from so-called "master" and "slave" stations to determine a ship's position. The accuracy of Loran is measured in hundreds of meters, but only has limited coverage.

GPS. In the late twentieth century, the global positioning system (GPS) largely replaced the Loran. GPS uses the same principle of time difference from separate signals as Loran, but the signals come from satellites. As of 2002, the system consisted of 24 satellites, and gave the mariner a position with accuracy of 9 meters (30 feet) or less. SEE ALSO COOK, CAPTAIN JAMES; GEOSPATIAL TECHNOLOGIES; TRANSPORTATION.

Amy J. Bratcher

Bibliography

Rosenbach Company. *The Sea: Books and Manuscripts on the Art of Navigation, Geography, Naval History, Shipbuilding, Voyages, Shipwrecks, and Mathematics, Including Atlases and Maps.* Storrs-Mansfield, CT: Maurizio Martino Publishers, 2003.

Sobel, Dava. *Longitude: The True Story of a Lone Genius Who Solved the Greatest Scientific Problem of His Time.* New York: Penguin USA, 1996.

Thurman, Harold V. *Introductory Oceanography,* 7th ed. New York: Macmillian Publishing Company, 1994.

Toghill, Jeff E. *Celestial Navigation.* New York: W. W. Norton & Co., 1988.

Internet Resources

Boat Safe Kids: The History of Navigation. International Marine Educators, Inc. <http://www.boatsafe.com/kids/navigation.htm>.

The Age of Exploration. The Mariners' Museum, Newport News, Virginia. <http://www.mariner.org/age/index.html>.

Nutrients in Lakes and Streams

When considering the water quality of lakes and streams, two important questions come to mind: "What are nutrients?" And: "Why are nutrients a problem in lakes and streams?"

Nutrients are chemical elements critical to the development of plant and animal life. In healthy lakes and streams, nutrients are needed for the growth of **algae** that form the base of a complex **food web** supporting the entire aquatic **ecosystem**. The most common nutrients in lakes and streams are nitrogen and phosphorus.

Under the right conditions, including abundant nutrients, algae and aquatic plants will continue to grow and multiply well beyond the amount needed to support the food web. The excess growth then dies, and **microorganisms** break it down, consuming dissolved oxygen from the water in the process. Dissolved oxygen, which aquatic organisms need just as humans need oxygen from the air, can be completely used up by the breakdown process. When this happens, aquatic organisms die from lack of oxygen. Extensive fish kills can result.

algae: (singular, alga) simple photosynthetic organisms, usually aquatic, containing chlorophyll, and lacking roots, stems, and leaves

food web: a complex food chain, with several species at each level, so that there is more than one producer and more than one consumer of each type

ecosystem: the community of plants and animals within a water or terrestrial habitat interacting together and with their physical and chemical environment

microorganism: a microscopic organism

Eutrophication and its Impacts

Eutrophication is the process of enrichment of lakes and streams with nutrients, and the associated biological and physical changes. Eutrophication is a natural process, but human activity has dramatically increased its rate in many waterbodies.✳ Lakes and ponds are particularly vulnerable to eutrophication because the nutrients carried into them continue to buildup; in contrast, the nutrients can be carried away in moving water.

✳ See "Lake Management Issues" for a schematic showing natural and cultural (human-induced) eutrophication in a lake.

Some results of excessive eutrophication are visible: thick mats of algae in the water; scum and foam; odor and taste problems; and death and disease of fish and other aquatic organisms. Other effects, such as the reduction in dissolved oxygen, cannot be seen directly, although the conditions often produce visible results such as dead fish.

An increase in the water's **pH** as a result of the increased growth of algae is another impact that is not directly visible. High pH can be toxic to fish and other organisms, and it can also make other substances, such as ammonia, even more toxic than they are otherwise.

Excess nutrients not only affect stream health but also may impact human health and livestock. Although phosphorus is not toxic to human adults in moderate concentrations, high levels of **nitrate** in drinking water (10 milligrams per liter or greater) can injure or kill livestock or human infants. Nuisance species of algae, such as some cyanobacteria (also called blue-green algae), produce toxins that affect the nervous system and liver, posing a threat to animals and humans who ingest them.

The worldwide increase in **red tides** and other blooms of toxic algae in coastal ocean waters has been linked to nutrient enrichment coming from coastal rivers. Nuisance species such as these, in fresh water as well as coastal oceans, can increase and force out less tolerant species, resulting in a loss of aquatic **biodiversity**.

pH: a measure of the acidity of water; a pH of 7 indicates neutral water, with values between 0 and 7 indicating acidic water (0 is very acidic), and values between 7 and 14 indicating alkaline (basic) water (14 is very alkaline)

nitrate: the highly leachable form of soil nitrogen taken up by most plants through their roots; it is a common groundwater contaminant, especially in agricultural areas and locations with a high density of septic systems, that is regulated by the U.S. Environmental Protection Agency with a drinking water standard of 10 ppm (parts per million) of nitrogen in the nitrate form

red tide: a visible coloration of the sea caused by the excessive growth (bloom) of microscopic algae, commonly dinoflagellates; the red, brown, green, purple, or yellow tint in the water is a result of the high concentration of algal pigments; some red tide events are harmful to marine animals and/or humans

biodiversity: a measure of the variety of the Earth's species, of the genetic differences within species, and of the ecosystems that support those species

An extreme example of overenrichment by nutrients is this canal in Bangkok, Thailand. Untreated sewage and rubbish discharged into waterways contribute to the proliferation of algae and water weeds, which clog small canals and float in the larger rivers. In the United States, environmental regulations are designed to prevent extreme eutrophication caused by human influences.

Sources of Nutrients

Several sources of nutrients are found in lakes and streams. Some are from natural sources, but many stem from human activities.

Natural Sources. Nutrients are present naturally in lakes and streams, but human activity has greatly increased the amounts going into surface waters. Background levels of nitrogen and phosphorus are generally quite low and are normally measured in milligrams per liter. Background nitrate concentrations in streams are usually less than 0.6 milligrams per liter, whereas background phosphorus rates in streams are even lower, less than 0.1 milligrams per liter. Soil and rocks are the primary natural sources of phosphorus, usually in the form of **phosphates**. Natural nitrogen sources include leaves and other organic debris from **riparian** vegetation.

Sewage Treatment Plants. Wastewater (or sewage) treatment plants are **point sources** of nutrients by virtue of the **effluent** which they discharge directly to rivers and streams. Unless the effluent has received tertiary treatment, or treatment to remove nutrients, it can be a significant contributor.

In the United States, treatment plants are regulated under the federal Clean Water Act. Under the act, if nutrients are a problem, then more stringent controls can be imposed. Tertiary treatment is expensive, and requires new systems, so some plants spread their effluent on land during times when nutrients could cause water quality problems.

Household Detergents. In the past, household detergents brought high loads of phosphorus to treatment plants, which then were discharged with the effluent. In the United States, however, laws restricting the phosphorus content of detergents have produced markedly reduced phosphate levels.

Septic Systems. Septic systems may contribute large amounts of nutrients, particularly if located close to the water. Standard septic systems do not remove nitrates; however, special systems like sand filters that remove nutrients are now becoming more common.

phosphate: the general term for phosphorus-containing derivatives of phosphoric acid (H_3PO_4); phosphates can be environmentally harmful when phosphate-rich wastewaters reach waterbodies; in surface waters, phosphates can act as a primary nutrient source for algae, whose accelerated growth and subsequent death and decay can deplete the oxygen needed for aquatic organisms

riparian: pertaining to the banks of a river, stream, waterway, or other, typically, flowing body of water as well as to plant and animal communities along these waterbodies

point source: a pollutant release or discharge originating from one specific location (e.g., an outfall pipe from a factory) rather than over a wide land area (e.g., water runoff from a farm field)

effluent: a liquid that flows out of or away from an area of waste processing or containment; includes treated wastes from municipal sewage plants, brine wastewater from desalinization operations, and coolant waters from a nuclear power plant

Nutrients in Lakes and Streams

These South Korean entrepreneurs have developed a washing machine that needs very little or no detergent, thereby saving consumers money and reducing the harmful effects of detergent wastewater on receiving lakes and streams. Under certain circumstances, the phosphorus content in household and commercial detergents may enhance eutrophication in these waterbodies.

Sediment. Sediment from excessive erosion is a nonpoint source that transports phosphorus in particles attached to soil. Construction sites lacking effective erosion control systems can dramatically increase the amount of sediment reaching lakes and streams, bringing in large phosphorus loads.

Animal Manure. Manure is a significant source of nutrient pollution in lakes and streams. Manure from livestock, if not properly managed, can reach streams through runoff or from direct deposits by animals in the water. The U.S. Geological Survey estimates that more than 7 million metric tons (nearly 16 billion pounds) of nitrogen and more than 2 million metric tons (more than 4 billion pounds) of phosphorus are applied to agricultural lands as manure each year. In the same way, pet waste is a nutrient source in urban and suburban areas, and aggregations of ducks, geese, and other waterfowl have also caused problems.

Commercial Fertilizers. Commercial fertilizers are a major source of both phosphorus and nitrogen. According to the U.S. Geological Survey, about 12 million metric tons (26 billion pounds) of nitrogen and 2 million metric tons (4 billion pounds) of phosphorus are applied annually in commercial

SO ALIVE IT WAS DYING

The Tualatin River, located southwest of Portland, Oregon, has a history of pollution problems dating back 100 years, when industry and sewage treatment plants dumped waste into the stream. These point sources of pollution were cleaned up and regulated under the federal Clean Water Act (enacted in 1972), yet pollution problems stemming from overenrichment continued.

Studies identified various nonpoint sources of nutrients, including soil erosion and manure from livestock. The state of Oregon helped landowners improve runoff control, while sewage plants upgraded their phosphorus removal technology.

Despite major progress, officials could not identify all the sources of phosphorus until continued study revealed groundwater as a nutrient source. Releasing more water from upstream dams in the summer to dilute the phosphorus has helped keep the Tualatin River from again being so alive it was dying.

fertilizer in the United States. Depending on the composition of the soil in an area, irrigation amounts and application methods, and the amount of rainfall, nutrients not needed by crops either run off the land into lakes and streams, build up in the soil, or seep down into groundwater. Groundwater can seep into a stream and be a source of nutrients.

Atmospheric Nitrogen. Atmospheric nitrogen comprises about 78 percent of the air that humans breathe. The burning of **fossil fuels** forms oxidized nitrogen compounds, which then reach the Earth when it rains or snows. In some parts of the United States, in particular the Northeast and the Upper Midwest, the so-called "acid rain" associated with these processes conveys large nitrogen loads to lakes and streams. The U.S. Geological Survey estimates that more than 3.5 million metric tons (nearly 7 billion pounds) of atmospheric nitrogen are deposited in the United States each year. SEE ALSO ACID RAIN; ALGAL BLOOMS IN FRESH WATER; ALGAL BLOOMS IN THE OCEAN; CHEMICALS FROM AGRICULTURE; CLEAN WATER ACT; ESTUARIES; FRESH WATER, NATURAL COMPOSITION OF; GROUNDWATER; LAKE MANAGEMENT ISSUES; LAKES: BIOLOGICAL PROCESSES; LAKES: CHEMICAL PROCESSES; LAKES: PHYSICAL PROCESSES; POLLUTION OF LAKES AND STREAMS; POLLUTION OF THE OCEAN BY SEWAGE, NUTRIENTS, AND CHEMICALS; SEPTIC SYSTEM IMPACTS; STREAM HEALTH, ASSESSING.

Roberta J. Lindberg

fossil fuel: substance such as coal, oil, or natural gas, found underground in deposits formed from the remains of organisms that lived millions of years ago

Bibliography

Carpenter, S. et al. "Nonpoint Pollution of Surface Waters with Phosphorus and Nitrogen." *Issues in Ecology* 3 (1998).

Mueller, D. K., and D. R. Helsel. *Nutrients in the Nation's Waters—Too Much of a Good Thing?* U.S. Geological Survey Circular 1136 (1996).

Puckett, L. J. *Nonpoint and Point Sources of Nitrogen in Major Watersheds of the United States.* Water-Resources Investigations Report 94-4001 (1994).

Ocean Basins

Ocean basins can be described as saucer-like depressions of the seabed. They vary in size from relatively minor features of the **continental margin** to vast structural divisions of the deep ocean. The largest ocean basins are 3 to 5 kilometers (2 to 3 miles) deep and stretch from the outer margins of the continents to the mid-ocean ridges.

continental margin: region where continental crust meets oceanic crust; extending from the shoreline to the deep-ocean basin, this feature includes the continental shelf, continental slope, and continental rise

Ocean basins cover approximately 71 percent of Earth's surface or about 361 million square kilometers (140 million square miles). Their average depth is 5,000 meters (16,000 feet), and the total volume is about 1.35 billion cubic kilometers (322 million cubic miles). There are five major subdivisions of the world ocean: the Pacific Ocean, Atlantic Ocean, Indian Ocean, Southern Ocean, and Arctic Ocean. The Pacific, Atlantic, and Indian Oceans are conventional ocean basins and are bounded by the continental masses or by ocean ridges and currents; they merge below 40° South latitude in the Antarctic Circumpolar current, or west Wind Drift, at the Southern (or Antarctic) Ocean. In the North Polar Region, the nearly circular Arctic Ocean, almost landlocked except between Greenland and Europe, is considered the fifth ocean subdivision.

Pacific Ocean Basin

The Pacific Ocean is bounded on the east by the North and South American continents; on the north by the Bering Strait; on the west by Asia, the Malay Archipelago, and Australia; and on the south by the Southern Ocean. In the Southeast it is arbitrarily divided from the Atlantic Ocean by the Drake Passage along 68° West longitude. It is by far the largest and deepest of the world's oceans and contains more than half of its free water. In area, this represents about 155 million square kilometers (59 million square miles). In comparison, the area of the continental United States, Hawaii, and Alaska is about 4 million square miles, sixteen times less in surface area. All the continents could fit into the Pacific basin.

The Pacific is the oldest of the existing ocean basins, its oldest rocks having been dated at about 200 million years. The major features of the basin have been shaped by the phenomena associated with **plate tectonics**. The coastal shelf, which extends to depths of about 180 meters (600 feet), is narrow along North and South America, but is relatively wide along Asia and Australia.

The East Pacific Rise, a mid-ocean ridge, extends from the Gulf of California to a point west of the southern tip of South America, and rises an average of 2,130 meters (7,000 feet) above the ocean floor. Along the East Pacific Rise, molten rock (magma) upwells from Earth's **mantle**, adding crust to the plates on each side of the rise. These plates are thus forced apart, causing them to collide with the continental plates adjacent to their outer edges. Under this tremendous pressure, the continental plates fold into mountain ranges and the oceanic plates are forced downward, forming deep **trenches** called subduction zones. The stresses at these areas of subduction are responsible for the earthquakes and volcanoes that give the Pacific basin the name "Ring of Fire."✱

plate tectonics: the theory that the Earth's lithosphere can be divided into a few large plates that are slowly moving relative to one another

mantle: the region of the Earth between the molten core and the outer crust, composed mainly of silicate rock, and around 2,900 kilometers (1,800 miles) thick

trench: an elongated surface feature on the sea floor that marks the location where the lithosphere bends downward at a subduction zone

✱ See "Volcanoes, Submarine" for the general location of the "Ring of Fire."

Atlantic Ocean Basin

The Atlantic Ocean is the second largest of Earth's five oceans, the most heavily traveled, and the most intensely studied, principally because of its importance in ship traffic between Europe and North America. This ocean's name is derived from Atlas, one of the Titans of Greek mythology.

The Atlantic Ocean occupies about 20 percent of Earth's surface, representing approximately 75 million square kilometers (29 million square miles). This includes its marginal seas: the Baltic Sea, Black Sea, Caribbean Sea, Davis Strait, Denmark Strait, part of the Drake Passage, Gulf of Mexico, Mediterranean Sea, North Sea, Norwegian Sea, and almost all of the Scotia Sea.

The Atlantic is divided into two nominal sections: the part north of the equator is called the North Atlantic; the part south of the equator, the South Atlantic. The South Atlantic is arbitrarily separated from the Indian Ocean on the east by the 20° East meridian and from the Pacific on the west along the line of shallowest depth between Cape Horn and the Antarctic Peninsula. The ocean is essentially an S-shaped north–south channel, extending from the Arctic Ocean in the north to the Southern Ocean in the south, and situated between the eastern coast of the American continents and the western coasts of Europe and Africa.

Ocean Basins

The world ocean with five major subdivisions covers nearly three-fourths of Earth's surface. From left to right are the easily recognizable eastern Pacific, Atlantic, Indian, and western Pacific Oceans, whereas the Arctic Ocean (top) and Antarctic (Southern) Ocean (bottom) are less apparent.

✷ See the frontmatter of this volume for a geologic timescale.

continental shelf: the relatively flat, submerged natural platform, about 1-degree slope, that extends seaward from the beach for about 70 kilometers (45 miles), with water depth up to 130 meters (425 feet) maximum, and ending where the slope and water depth increase

The Atlantic Ocean has an average depth of 3,926 meters (12,881 feet). At its deepest point, in the Milwaukee Deep in the Puerto Rico Trench, the bottom is 8,605 meters (28,231 feet) below the surface.

The Atlantic began to form during the Jurassic period, about 150 million years ago, when a rift opened up in the supercontinent of Gondwana, resulting in the separation of South America and Africa.✷ The separation continues today at a rate of approximately 2.5 centimeters (approximately 1 inch) a year along the Mid-Atlantic Ridge.

Along the American, African, and European coasts are the **continental shelves** of the Atlantic basin. These are areas of debris washed from the continents. Submarine ridges and rises extend roughly east-west between the continental shelves and the Mid-Atlantic Ridge, dividing the eastern and western ocean floors into a series of basins, also known as abyssal plains. The three subbasins on the American side of the Mid-Atlantic Ridge are more than 5,000 meters (16,400 feet) deep: the North American basin, the Brazil basin, and the Argentina basin. The European–African side is marked by several basins that are smaller but just as deep: the Iberia, Canaries, Cape Verde, Sierra Leone, Guinea, Angola, Cape, and Agulhas basins.

The break in the ridge at the equator, called the Romanche furrow, is important because it gives the deep ocean water a gap to flow through, which influences the currents and temperature of the Atlantic Ocean.

Indian Ocean Basin

The Indian Ocean is the third largest of the five oceans. It is bounded on the west by Africa, on the north by Asia, on the east by Australia and the Australasian islands, and on the south by the Southern Ocean. No natural boundary separates the Indian Ocean from the Atlantic Ocean, but a line about 4,020 kilometers (2,500 miles) long, connecting Cape Agulhas at the southern end of Africa with Antarctica, is generally considered to be the boundary.

It has a total area of 68 million square kilometers (26 million square miles), which includes its marginal seas: Adaman Sea, Arabian Sea, Bay of Bengal, Great Australian Bight, Gulf of Aden, Gulf of Oman, Mozambique Channel, Persian Gulf, Red Sea, and the Strait of Malacca. The average depth of the basin is about 4,210 meters (13,800 feet), although the Java Trench reaches a depth of over 7,258 meters (23,812 feet). The Indian basin also is divided in half by the Mid-Indian Ocean Ridge.

Southern Ocean

A decision by the International Hydrographic Organization in the spring of 2000 delimited a fifth world ocean, the Southern Ocean. Although not a true ocean basin, it extends from the coast of Antarctica north to 60° South latitude, which coincides with the Antarctic Treaty Limit. The Southern Ocean is now the fourth largest of the world's five oceans. The Southern Ocean has the unique distinction of being a large circumpolar body of water totally encircling the continent of Antarctica. Its area is 20 million square kilometers (7 million square miles) and includes the Amundsen Sea, Bellingshausen Sea, part of the Drake Passage, Ross Sea, a small part of the Scotia Sea, and Weddell Sea.

Arctic Ocean Basin

The smallest of Earth's five ocean basins is the Arctic. The Arctic Ocean extends south from the North Pole to the shores of Europe, Asia, and North America. The surface waters of the Arctic Ocean mingle with those of the Pacific Ocean through the Bering Strait, by way of a narrow and shallow channel, which has a depth of about 55 meters (180 feet). More importantly, the Arctic waters mix with those of the Atlantic Ocean across a system of submarine sills (shallow ridges) that reach from Scotland to Greenland and from Greenland to Baffin Island at depths of about 500 to 700 meters (1,640 to 2,300 feet).

The total surface area of the Arctic Ocean—including its major subdivisions, the North Polar Sea (the main portion), the Norwegian Sea, the North Sea, and the Barents Sea—is about 14 million square kilometers (5.4 million square miles).

Approximately one-third of the Arctic Ocean is underlain by continental shelf, which includes a broad shelf north of Eurasia and the narrower shelves of North America and Greenland. Seaward of the continental shelves lies the Arctic Basin proper, which is subdivided into a set of three parallel ridges and four basins (also known as deeps). These features have been discovered and explored only since the late 1940s. The Lomonosov Ridge, the major ridge, cuts the North Polar Sea almost in half, extending as a submarine bridge 1,700 kilometers (1,060 miles) from Siberia to the northwestern tip of Greenland. Parallel to it are two shorter ridges: the Alpha Ridge on the North American side, defining the Canada and Makarov basins, and the mid-ocean ridge on the Eurasian side, defining the Nansen and Fram basins.

The average depth of the Arctic Ocean is only about 1,500 meters (4,900 feet) because of the vast shallow expanses on the continental shelves. The deepest point in the Arctic Ocean is 5,450 meters (17,880 feet).

Unlike the other oceans, the Arctic Ocean is ice-covered. At lower latitudes, the ice melts during the summer months. At polar latitudes, however, the ice cover is permanent. Another distinctive feature of the Arctic Ocean is the presence of islands composed of ice. These ice islands move, as does much of the ice cover, in the currents that exist in the underlying ocean water. Scientists have used the islands as research bases to study the arctic ice movements and other aspects of the far North. SEE ALSO COASTAL OCEAN; MARGINAL SEAS; OCEAN CURRENTS; OCEAN-FLOOR BATHYMETRY; OCEANS, POLAR; OCEANS, TROPICAL.

Brian D. Hoyle and K. Lee Lerner

Bibliography

Lebow, Ruth, and Tom S. Garrison. *Oceanus: the Marine Environment.* Belmont, CA: Wadsworth Publishing, 1989.

Internet Resources

Earth's Oceans: An Introduction. EnchantedLearning. <http://www.enchantedlearning.com/subjects/ocean/>.

Physiography of the Ocean Basins. Okanagan University College. <http://www.geog.ouc.bc.ca/physgeog/contents/10p.html>.

The World Factbook 2002. U.S. Central Intelligence Agency. <http://www.cia.gov/cia/publications/factbook/index.html>.

Ocean Biogeochemistry

Biogeochemistry is the study of the interactions of the biology, chemistry, and geology of the Earth. In the case of a large body of water such as the ocean, biogeochemistry can be thought of as a huge experiment or set of reactions. Instead of happening in a clean glass beaker, the reactions have the ocean floor as the container.

Marine snow is a critical component of the ocean biogeochemical cycled because it transports organic and inorganic material from the water column to the ocean floor. The oceanographers shown here use bongo nets to collect the falling debris, which is faintly visible and somewhat resembles falling snow on land.

The surface of the water is open to the air, and every day more dust and dirt from land blows over the ocean and falls in. Moreover, the surface of the water contains many small plant forms that are continually growing and being consumed by animals that are themselves consumed by other animals.

As this life and death drama continues, the scraps and leftovers drift downward towards the ocean floor like a snowfall; hence the name "marine snow." (See the photograph on page 130.) Around the edges of the ocean, rivers empty water and **sediment**. Deep in the ocean, mud-dwelling creatures await the arrival of their next meal from the falling biological debris (marine snow). These events are linked to each other, to the history of life on Earth, and to variations in Earth's climate.

sediment: rock particles and other earth materials that are transported and deposited over time by geologic agents such as running water, wind, glaciers, and gravity; sediments may be exposed on dry land and are common on ocean and lake bottoms and river beds

Cycles

Scientists who study biogeochemistry usually consider the cycling of materials through the different parts of the system. To do this, they deal with reservoirs of materials and the fluxes of a substance from one reservoir to another. For example, they examine reservoirs such as the surface ocean water versus the deep ocean water, or the transfer of masses of materials per unit time (fluxes). An example of this kind of approach to biogeochemical cycles in the ocean can be seen in the following figure, where the reservoirs represented are the atmosphere, **lithosphere**, terrestrial (land-based) biosphere, surface ocean, **phytoplankton**, and deep ocean. The figure shows the global carbon cycle, a network of interrelated processes that transports carbon between different reservoirs on Earth.

lithosphere: the rigid outer layer of Earth made up of the crust and the uppermost mantle

phytoplankton: microscopic floating plants, mainly algae, that live suspended in bodies of water and that drift about because they cannot move by themselves or because they are too small or too weak to swim effectively against a current

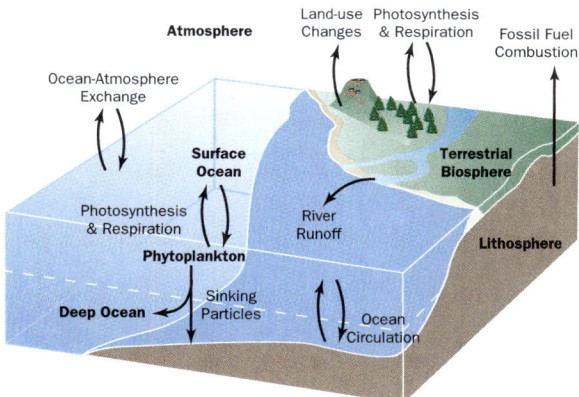

Most scientific study has focussed on the carbon cycle. Carbon, after all, is the basis of life on Earth, and its gaseous form, carbon dioxide, is linked to the **greenhouse effect** and changes in Earth's climate over time. For these reasons, understanding the carbon cycle has been the focus of several large research programs supported by the U.S. government. Three examples include:

- the U.S. Global Change Research Program (USGCRP): a joint project to design a carbon cycle research program; funded by the Department of Energy; the National Aeronautic and Space Administration, the National Oceanic and Atmospheric Administration, National Science Foundation, and U.S. Geological Survey;

greenhouse effect: the phenomenon whereby a planetary body's atmosphere traps solar radiation; caused by the presence in the atmosphere of gases such as carbon dioxide, water vapor, and methane, that allow incoming sunlight to pass through but trap heat radiated back from the body's surface

Ocean Biogeochemistry

- Global Ocean Ecosystems Dynamics (GLOBEC): a major research program funded by the National Science Foundation to determine how global change affects the marine ecosystem and what the feedbacks to the physical climate system will be; and

- the Global Carbon Program (GCP): a study funded by the National Oceanic and Atmospheric Administration to improve scientists' ability to predict the fate of human-derived carbon dioxide and future concentrations of atmospheric carbon dioxide.

Other substances also have well-studied cycles. Water, of course, is constantly moving into, through, and out of the ocean. Some of the atmospheric gases such as oxygen and carbon dioxide are vitally important to life. Nutrient elements such as nitrogen, phosphorus, and silicon are necessary to the phytoplankton, and form the basis for the oceanic **food web**.

A Cycling Example. The presence of life forms on Earth is tremendously important in the cycling of elements through the major reservoirs. Consider the ocean as an example: If one focuses on the impact of a single **diatom** on the ocean, the following story emerges.

Diatoms are a group of **algae** living by the millions in each cubic centimeter of surface ocean water. There each alga has access to the sunlight needed for photosynthesis; the CO_2 (carbon dioxide), N (nitrogen), and P (phosphorus) needed to make its soft tissue; the Si (silicon) needed for its shell-like covering; and a number of rare or trace substances in sea water, including Cu (copper) and Fe (iron). To reproduce, it undergoes cell division. Its life processes produce O_2 (oxygen) that can be used by other organisms; **organic** tissue that becomes food for the next higher creatures in the food web; and often an exudate or slime.

Once the diatom has been consumed by an animal (a copepod, for example), its life is over, but its effect on the ocean is not. The copepod digests and derives energy from the diatom's soft tissue, then packages the remains into a fecal pellet that is discharged as waste to become part of the falling debris (marine snow) headed for the ocean floor.

The pellet lands on the ocean floor, forming a site for bacteria to live as well as food for them to consume. The **inorganic** part of the diatom that remains (the silica shell) will begin to dissolve on the way to the ocean bottom, and Si taken out of the surface water is returned to deeper water as the shell dissolves. Decomposition of sinking organic matter by bacteria returns N, C, and P to the water and removes **dissolved** O_2.

Carbon. Ocean water itself is changed by life processes. During the growth of diatoms and the consumption of diatoms by **zooplankton**, carbon is removed from ocean water and in turn from the atmosphere as the diatoms use it to grow. The transfer of this carbon toward the ocean floor and its partial burial in the sediments is often referred to as the carbon pump; it is one of the processes that slow the accumulation of CO_2 in the atmosphere.

Silicon. The silicon (Si) used in the diatom shell enters the ocean from rivers, from the hot springs along mid-ocean ridges and by diffusion from deep-sea sediments. Diatoms remove Si so efficiently from the ocean surface water that it is a very scarce element there, and mixing and upwelling processes are necessary to redistribute enough Si back to the surface to pro-

---

food web: a complex food chain, with several species at each level, so that there is more than one producer and more than one consumer of each type

diatom: any of the microscopic unicellular or colonial algae constituting the class *Bacillarieae* that have a silicified cell wall, which persists as a skeleton after death

algae: (singular, alga) simple photosynthetic organisms, usually aquatic, containing chlorophyll, and lacking roots, stems, and leaves

organic: pertaining to, or the product of, biological reactions or functions

inorganic: an element, molecule, or substance that did not form as the direct result of biologic activity

dissolved: describes the chemical breakdown of a solid in a solution into individual atoms or molecules and their dispersement in the fluid medium; for example, describes the dissolved solids or dissolved gases in water

zooplankton: microscopic animals that live suspended in bodies of water and that drift about because they cannot move by themselves or because they are too small or too weak to swim effectively against a current; composed primarily of protozoans, microcrustacea (copepods, cladocera, rotifers) and larval stages of certain invertebrates

Ocean Biogeochemistry

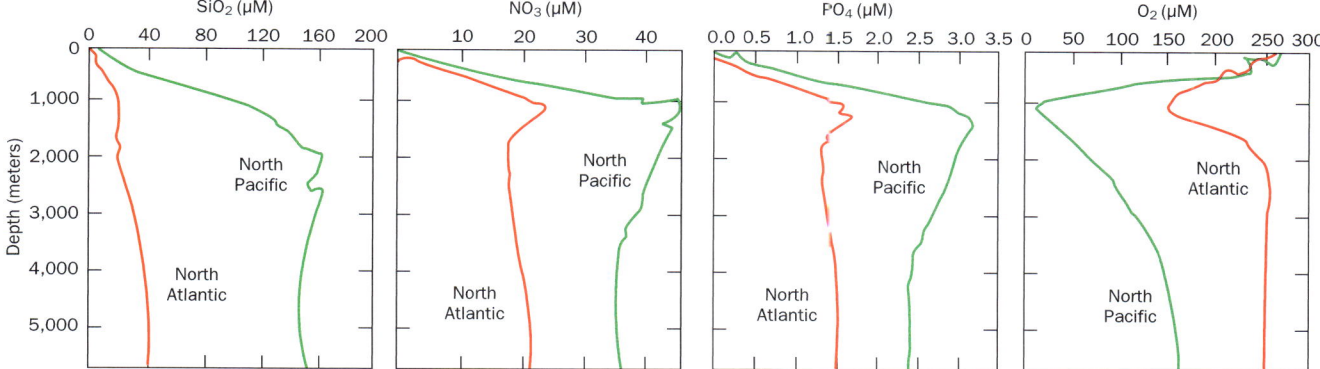

vide for diatom growth. For that reason, Si as well as N, P, and other biologically important elements are in low concentration in surface water of the ocean, and increase with depth, as shown in the figure above.

Oxygen. Another consequence of ocean biogeochemistry can be seen in the distribution of O_2 (oxygen) with depth (see figure above). The oxygen content at the surface is relatively high (about 6 milliliters per liter) and is replenished from the air. Deeper in the water, the O_2 content begins to decrease with depth, until at about 1,000 meters (3,082 feet), the value reaches a minimum. The reason for the decrease is the consumption by bacteria of the rain of organic debris (marine snow) falling through the water. The process requires O_2, and below the surface there is no immediate source to return the O_2 being used up.

The exact amount of O_2 at the O_2 minimum varies with location in the ocean; below the minimum, O_2 content begins to increase again with depth. The increase is related to water circulation in the ocean. The deep water in the ocean starts out at the surface in polar regions, where it becomes very dense because of the extreme cold, and sinks to great depths in the ocean, carrying with it dissolved oxygen from the surface waters. This cold, dense, deep water flows along the ocean floor close to the bottom, well beneath the depths of the O_2 minimum. These factors combine to give the observed shapes of O_2 profiles in the ocean.

Hydrothermal Processes. There are other processes that play a role in determining the nature of the ocean. For example, hydrothermal activity at mid-ocean ridges results in significant changes in the chemistry of ocean water. The water that comes out of these hot springs comes from normal deep-ocean water that runs down into deep cracks on the ocean floor alongside the ridges. As the water penetrates into the oceanic crust, it becomes heated to very high temperatures, and reacts with the rocks. The water that comes out of the vents is very hot; contains sulfide (S^-) instead of sulfate (SO_4^{2-}); contains no Mg (magnesium) or O_2 (oxygen); and contains large amounts of Si (silicon). Because the entire volume of the ocean circulates through the mid-ocean ridge system every 10 million years, these changes are of great significance to the oceans and the organisms that live in them. SEE ALSO HOT SPRINGS ON THE OCEAN FLOOR; MID-OCEAN RIDGES; OCEAN CHEMICAL PROCESSES; OCEANOGRAPHY, BIOLOGICAL; OCEANOGRAPHY,

Vertical profiles illustrate the changing concentration, with depth, of common constituents in the ocean: silicon dioxide (SiO_2), nitrogen as nitrate (NO_3^-), phosphorus as phosphate (PO_4^{3-}), and oxygen (O_2).

Chemical; Oceanography, Geological; Sea Water, Physics and Chemistry of; Volcanoes, Submarine.

Martha R. Scott

Bibliography

Libes, Susan. *An Introduction to Marine Biogeochemistry.* New York: John Wiley & Sons, 1991.

Thurman, Harold V., and Elizabeth A. Burton. *Introductory Oceanography*, 9th ed. Upper Saddle River, NJ: Prentice Hall, 2001.

Ocean Chemical Processes

Why is the sea salty? Sea water contains about 35 grams per kilogram of dissolved salt. The most obvious source for the salt is river water, which can easily be observed **weathering** rocks (from which the water derives minerals), carrying **sediment**, and flowing continually into the ocean.✷ Because the water added to the ocean evaporates but the dissolved salts do not, it seems reasonable to suggest that river water brings salt to the ocean.

But a closer look shows that the process must be more complicated. Table 1 compares the major substances dissolved in river water and ocean water. If sea water is simply concentrated river water, these elements should be present in the same ratios in both types of water. For both water types, the Cl/Cl ratio is 1 because that is the chosen standard of comparison. Notice that the ratio patterns for most components for the two water types are quite different. This pattern means that simple evaporation of water cannot change river water into sea water.

Addition–Removal Processes and Considerations

The composition of sea water is controlled by many different processes, all acting at the same time, and adding and removing substances at different rates (see figure). The sum of all the processes, a kinetic (changing) balance, determines sea-water chemistry (see Table 2 on page 136).

When sea water dries up completely, it leaves behind a salt deposit called an **evaporite**. Evaporites of greatly different ages on Earth all are similar,

weathering: the decay or breakdown of rocks and minerals through a complex interaction of physical, chemical, and biological processes; water is the most important agent of weathering; soil is formed through weathering processes

sediment: rock particles and other earth materials that are transported and deposited over time by geologic agents such as running water, wind, glaciers, and gravity; sediments may be exposed on dry land and are common on ocean and lake bottoms and river beds

✷ See "Oceanography from Space" for a photograph of a river in Borneo adding water and sediment to the coastal ocean.

evaporite: sediments that form as the result of the precipitation of minerals during the evaporation of water, primarily sea water, and that may form sedimentary rock; principle minerals are gypsum and halite

TABLE 1. COMPARISON OF RIVER WATER AND SEA WATER COMPOSITION

Ions	Average river water (mM/l)	Average sea water (mM/l)	River water ratio to Cl	Sea water ratio to Cl
HCO_3^-	0.86	2.38	5.375	0.0044
SO_4^{2-}	0.069	28.2	0.43125	0.0517
Cl^-	0.16	545	1	1
Ca^{2+}	0.33	10.2	2.0625	0.0187
Mg^{2+}	0.15	53.2	0.9375	0.09761
Na^+	0.23	468	1.4375	0.8587
K^+	0.03	10.2	0.1875	0.0187

Ocean Chemical Processes

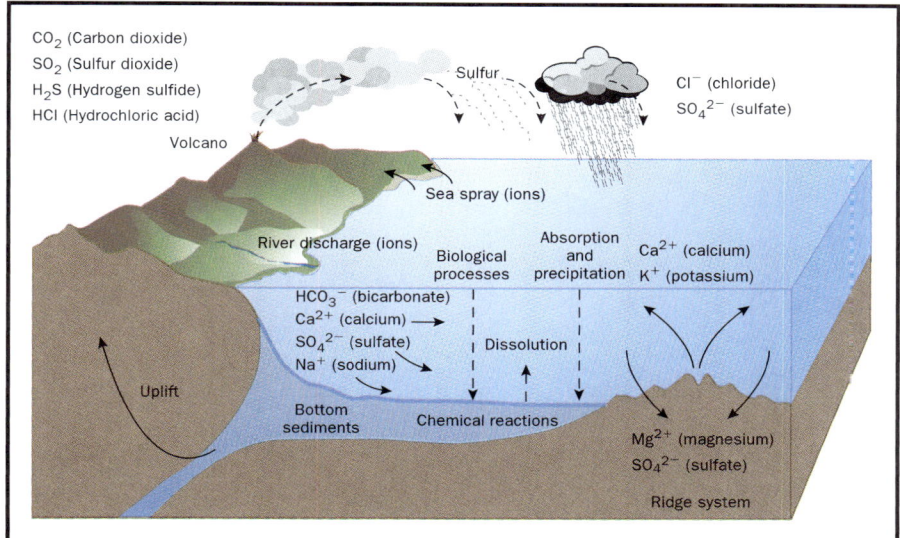

This schematic summarizes the processes that distribute and regulate the major constituents in sea water. For example, river water liberates ions from earth materials and adds them to the ocean. Sea spray from waves removes ions from the ocean when it deposits a film of salts on the land. Other methods of ion addition and removal also are depicted.

so scientists reached the conclusion that sea water must have had roughly the same chemistry over hundreds of millions of years.

If this is true, then all the processes affecting sea-water chemistry must be at steady state—that is, operating so that the input of salt equals the output. For a steady-state ocean, it is possible to find out how long a particular element stays in the ocean (i.e., its residence time) before it is removed. The ocean is at steady state for a particular element if that element is added and removed at the same rate.

For example, Na (sodium) is added to the ocean at the rate of about 7.9×10^{12} **moles** per year (3.4×10^5 tons per year) The whole ocean contains 6.4×10^{20} moles (2.8×10^{13} tons) of Na, and the total amount of Na present divided by the rate of addition gives 80×10^6 years, the residence time of Na. In other words, a sodium atom entering the ocean in river water will stay in the ocean for 80×10^6 years before it is removed from sea water.

mole: a quantity of a given element or compound, defined as the formula weight (atomic or molecular weight) expressed in grams; for example, the formula weight for the water molecule (H_2O) is 18, so a mole of water is a quantity of water having a mass of 18 grams; a mole of a substance comprises 6.023×10^{23} atoms or molecules

Residence Time. The concept of residence time is informative in several ways. Elements with long residence times in the ocean tend to be very soluble in sea water and to be evenly mixed throughout the ocean. Thus, Na, Cl, and other elements (Table 2) have long residence times and are known as conservative elements, occurring in the same ratio to one another throughout the ocean regardless of the total salinity.

Elements with short residence times (such as iron and aluminum) are relatively reactive, or insoluble in sea water; they are easily removed and are unevenly distributed throughout the ocean. This makes sense in comparison with the mixing time for the whole ocean, which is about 1,000 years. An element that remains dissolved for millions of years will have been mixed through the ocean many times over, and hence should be evenly distributed. In contrast, an element with a residence time of only 100 years will not be able to make it around once without being removed.

Estuaries. Rivers add huge amounts of dissolved materials to the ocean each year as well as many tons of soil and rock particles. At the boundary between the land and the sea are estuaries, or bodies of water that are

Ocean Chemical Processes

TABLE 2. CHEMISTRY OF SEA WATER
(salinity S = 35.000)

Major Constituents

Constituent	g/kg (ppt)	mmol/kg	Residence time (10^6 years)
Na^+	10.781	469.0	80
K^+	0.400	10.21	12
Mg^{2+}	1.284	52.83	13
Ca^{2+}	0.412	10.28	1
Sr^{2+}	0.008	0.091	5
Cl^-	19.353	545.9	
SO_4^{2-}	2.712	28.23	10
HCO_3^-	0.126	2.06	
Br	0.844	130	

Some Trace Constituents

Constituent	mmol/kg	Residence time (years)
Al	~0.03	600
Fe	~1×10^{-3}	50
Co	~3×10^{-5}	340
Ni	~8×10^{-3}	8200
Cu	~4×10^{-3}	970
Zn	~6×10^{-3}	500

pH: a measure of the acidity of water; a pH of 7 indicates neutral water, with values between 0 and 7 indicating acidic water (0 is very acidic), and values between 7 and 14 indicating alkaline (basic) water (14 is very alkaline)

precipitate: (verb) the process by which a solution separates into a relatively clear liquid and a solid substance by a chemical or physical change; (noun) the solid substance resulting from this process

continental shelf: the relatively flat, submerged natural platform, about 1-degree slope, that extends seaward from the beach for about 70 kilometers (45 miles), with water depth up to 130 meters (425 feet) maximum, and ends where the slope and water depth increase

✷ See "Ocean-Floor Sediments" for a photograph of dust blowing into the Red Sea.

chemical and physical transitions between rivers and oceans. As salinity and **pH** increase seaward, some dissolved substances, such as Fe (iron), may **precipitate** to form solids and then remove other dissolved elements, such as Mn (manganese), onto their surfaces. Other elements that arrive in the estuary adsorbed (bound) to river particle surfaces may be desorbed (unbound) by the influence of the higher salt content they encounter in the ocean. The dissolved substances enter the ocean, but as much as 90 percent of river-borne particles are trapped in the estuary and on the **continental shelf**.

Aerosols. Particles carried through the air are known as aerosols. They come from a variety of sources. Natural aerosols include sea-spray residues, windblown soil particles, volcanic particles, smoke from forest fires, and particles condensed from natural gases. Anthropogenic (human-derived) aerosols, often considered pollutants, include direct emissions such as from smokestacks and particles from conversion of anthropogenic gases.

Because most of the river-derived sediment load is trapped in estuaries and on the continental shelf, a large fraction of particles reaching the ocean from land consists of aerosols. The global mineral dust source is 100 $\times 10^{12}$ grams per year to 800×10^{12} grams per year [(3.5×10^{12} ounces per year to 28.2×10^{12} ounces per year)], compared to a river discharge of 15.5×10^{15} grams per year [(0.6×10^{15} ounces per year)].✷ Because most of the continental land mass is in the Northern Hemisphere, most of the natural and human-derived aerosols also are generated in the Northern Hemisphere. The amounts in air vary over time and tend to be concentrated in latitude zones.

Aerosols can be delivered to the ocean as dry fallout or as wet fallout if they are **entrained** into falling rain; in fact, most of the chloride (Cl^-) and sulfate (SO_4^{2-}) in sea water is believed to have come from volcanic gases that were dissolved in rain and delivered to the ocean over Earth's history.

Hydrothermal Processes.
Sea water continually reacts with its "container"—the **basalt** rocks that underlie the ocean. The most intense of these reactions occur in hot springs along mid-ocean ridges.✱ The entire volume of ocean water gradually circulates into the ocean floor, reacts with hot basalt, and returns greatly changed in chemical composition.

In this transit, sea water loses all its O_2 and all its Mg; further, all its SO_4^{2-} is stripped of oxygen. At the same time, the water gains Ca, Si, Fe, and Mn. When the hot, altered hydrothermal fluids mix with cold, normal sea water, Fe, Mn, and Si precipitate out in huge plumes that look like smoke: hence the names "**black smoker**" and "**white smoker**".✱ The solids formed in this precipitation remove trace metals (such as Cr, V, Mo, U) from solution. (Trace elements occur at concentrations of less than 1 part per million (ppm), or 1 milligram per kilogram of water.)

It is difficult to estimate how fast ocean water circulates through the mid-ocean ridge axis areas; a reasonable estimate of once every 10 million years suggests the importance of these reactions in establishing the composition of the ocean.

Biological Processes.
Life processes have an important effect on ocean chemistry. Certain surface-dwelling phytoplankton (e.g., coccolithophores and diatoms) remove calcium or silicon from sea water to make calcium carbonate shells or opalline silica shells, respectively. These hard particles eventually fall toward the ocean floor, where they are buried in the sediment and thus removed from sea water for many millions of years. The carbon removed by these same organisms to make their soft tissue becomes food for both higher organisms and for bacteria.

The carbon plus the Sun's energy make up the basic fuel for the biogeochemical cycling of material in the ocean. It may be difficult to imagine how an organism as small as a coccolithophore can affect the Ca content of the whole ocean until one considers the enormous numbers of these phytoplankton that occur in the surface ocean.✱

In addition to the biological debris that falls through the ocean as discrete particles and fluffy bits of marine snow, many inorganic particles from soils are caught up in the fecal pellets and other biogenic materials.✱ All these particles have surfaces that may adsorb dissolved material from sea water and cause them to be buried in ocean floor sediments, thus removed from the ocean.

Evaporation.
Evaporation of isolated bodies of sea water accounts for removal of large amounts of salt from the ocean. Salt deposits occur through the geologic record as layers of gypsum, rock salt, sylvite, and other minerals. These minerals represent dissolved salts that have been removed from sea water by evaporation and are mined for table salt and other uses.✱

Pore-Water Interactions.
Interaction with pore water can affect the chemistry of the ocean. Pore water is sea water that has been trapped between

---

entrain: to draw in and transport (as solid particles or gas) by the flow of a fluid; for example, water droplets may become entrained in rising air currents

basalt: a dark, volcanic rock with abundant iron and magnesium and relatively low silica common on all of the terrestrial planets

✱ See "Mid-Ocean Ridges" for a composite image of the mid-ocean ridge system.

✱ See "Hot Springs on the Ocean Floor" for a photograph of a black smoker.

black smoker: a sea-floor vent in which hot mineralized water from below the sea floor discharges into cold sea water; black color arises from the precipitation of dark sulfide minerals

white smoker: a sea-floor vent in which hot mineralized water from below the sea floor discharges into cool sea water; white color arises from the precipitation of minerals rich in barium, calcium, and silica

✱ See "Algal Blooms in the Ocean" for a photograph of a coccolithophore bloom.

✱ See "Ocean Biogeochemistry" for a photograph of marine snow.

✱ See "Mineral Resources from the Ocean" for a photograph of mined stacks of sea salt.

sediment grains. The chemistry of the pore water is susceptible to change by biological processes. For example, bacteria in the sediment consume organic tissue, at the same time using up much or even all the oxygen in the pore water.

In the absence of oxygen (O_2), various members of the bacterial population continue to metabolize their food using other energy sources; common examples are reduction of SO_4^{2-} to S^-, reduction of $Fe(OH)_3$ to Fe^{2+}, and reduction of Mn^{4+} to Mn^{2+}. Changed forms of the elements (such as Mn^{2+}) in some cases become soluble and diffuse out of the sediment into the overlying sea water; O_2 diffuses from sea water into the sediment. These processes are slow but continuous, and in the long term affect the chemistry of sea water.

Conclusion. Many questions concerning the chemistry of sea water remain unanswered. In recent years, work on the composition of tiny sea-water bubbles included in salt crystals from ancient evaporites suggests that the composition of sea water may have changed over the past 700 million years. The changes amount to about a factor of two for several of the major elements; these observations will lead to much future research on the processes that establish ocean chemistry and their changes over time. SEE ALSO ALGAL BLOOMS IN THE OCEAN; BRINES, NATURAL; CARBON DIOXIDE IN THE OCEAN AND ATMOSPHERE; MID-OCEAN RIDGES; MINERAL RESOURCES FROM THE OCEAN; OCEAN BIOGEOCHEMISTRY; PLANKTON; SEA WATER, FREEZING OF; SEA WATER, PHYSICS AND CHEMISTRY OF; TRACERS OF OCEAN-WATER MASSES.

Martha R. Scott

Bibliography

Broecker, Wallace. S., and T. H. Peng. *Tracers in the Sea*. Palisades, NY: LDGO Press, 1982.

Chester, Roy *Marine Geochemistry*, 2nd ed. London, U.K.: Blackwell Publishers, 2000.

Dasch, E. Julius, ed. *Encyclopedia of Earth Sciences*. New York: Macmillan Reference USA, 1996.

Libes, Susan. *An Introduction to Marine Biogeochemistry*. New York: John Wiley & Sons, 1991.

Pilson, Michael E. Q. *An Introduction to the Chemistry of the Sea*. Upper Saddle River, NJ: Prentice Hall, 1998.

Ocean Currents

Mariners have known for many centuries that the ocean contains currents that flow along generally consistent paths. The Spanish galleons transporting gold and silver from Mexico to Spain made use of the Gulf Stream to help them return home, while Benjamin Franklin used ships' log books to draw a map of this current in 1772 (see illustration on page 139 based on his original map). Since then, scientists have gained much more information on both where currents flow and why.

Why Do Currents Flow?

Currents exist at all depths in the ocean; in some regions, two or more currents flow in different directions at different depths. Although the current system is complex, ocean currents are driven by two forces: the Sun and the rotation of the Earth.

The Gulf Stream is one of the strong ocean currents that carries warm water from the tropics to the higher latitudes. In contrast to the nontechnological methods used to produce early maps of the Gulf Stream, today's remote sensing technology on satellites allows scientists to delineate the current's features and follow changes in its position. (See "Geospatial Technologies" for a satellite image of the Gulf Stream.)

The Sun affects the ocean in two ways. First, it heats the atmosphere, creating winds and moving the sea surface through friction. This tends to drag the water surface along as the wind blows over it. Although the wind strongly affects the surface layer, its influence does not extend much below about 100 meters (325 feet) in depth.

The second effect of the Sun is to alter the density of the ocean surface water directly by changing its temperature and/or its **salinity**. If water is cooled or becomes saltier through evaporation, it becomes denser. This can result in the water column becoming unstable, setting up density-dependent currents, also known as the thermohaline circulation.

salinity: the concentration of dissolved materials carried in an aqueous (watery) solution; typically expressed in grams per liter (parts per thousand) or milligrams per liter (parts per million)

The rotation of the Earth also affects the currents through the Coriolis force. This force causes water to move to the right in the Northern Hemisphere and to the left in the Southern Hemisphere. It exists because moving ocean water is affected by friction with the Earth only at the seafloor, and because the eastward linear velocity of the earth decreases from a maximum at the equator to zero at the poles (the rotational velocity, however, does not change). A parcel of water at the equator is moving at the same speed as the Earth. If it starts to move north, with no friction, it is then going faster than the Earth beneath it. To conserve momentum (the product of mass and velocity), it consequently moves more to the east as it gets farther from the equator. The Coriolis force therefore increases away from the equator.

Ekman's Theory. The first reasonable theory of how the wind affects surface currents was derived by Swedish oceanographer and mathematician Valfrid Ekman in 1890. Ekman considered an infinitely wide and deep ocean of constant density, divided into an infinite number of horizontal layers. The top layer is affected by the wind and by friction with the layer below it. The second layer is also affected by friction at top and bottom, and so on. The Coriolis force also affects the layers because they are moving.

Balancing the friction and the Coriolis force led Ekman to conclude that the resulting currents decreased exponentially with depth, that the surface current moved at a 45-degree angle to the wind direction, and that deviations from the surface wind direction increased with depth, forming a spiral (to become known as the Ekman spiral). Adding the movement at all depths gives a mean (average) current that moves at right angles to the wind direction, to the right in the Northern Hemisphere, and to the left in the Southern Hemisphere. Such Ekman spirals are rare, but have been observed in the ocean.

Because the oceans are neither infinitely wide nor of constant density, as Ekman assumed, complications arise at the boundaries, where water tends to "pile up." The surface of the ocean is then no longer flat, but has a slope, which sets up a horizontal pressure gradient. The combination of the Coriolis force and the horizontal pressure gradient produces a current that flows at right angles to the pressure gradient; when the two forces are equal, this is known as a geostrophic current. All major current systems in the ocean can essentially be considered geostrophic.

Because the density of the ocean varies both horizontally and vertically, scientists can use the density structure of the ocean to calculate the pressure field and hence the pressure gradient. The result is a map showing how the height of the sea surface at any given point varies relative to a particular pressure level; relative changes in these dynamic heights are used because the seafloor is neither flat nor of constant depth. These maps can be used to show where currents flow (the steeper the gradient, the faster the current). Scientists can also use the observed density distribution in the ocean to calculate current speeds.

Patterns of Surface Currents

The general pattern of surface currents in the ocean is shown in Figure 1. This shows a series of quasi-circular gyres, or large eddies, in each ocean **basin**. The gyres tend to be biased towards the western side of the basin, where strong, narrow flows are found in the Gulf Stream, Kuroshio, Agulhas, Brazil, and East Australian Currents, which flow at speeds up to 2.5 meters per second (6 miles per hour). The volumes involved in these western boundary currents are large; the Agulhas, for instance, carries about 80 **Sverdrups** of water in the upper 1,000 meters (about 0.6 mile). Current speeds away from the western boundaries of the gyres are generally much lower, and currents on the eastern boundaries of ocean basins are much wider.

The gyres rotate clockwise in the Northern Hemisphere and counterclockwise in the Southern Hemisphere because of the Coriolis force; in this they follow the prevailing wind patterns. The northern Indian Ocean is an exception because of the intense atmospheric **monsoon** system that changes its direction twice a year, but an intense circular current system, the Great Whirl, develops off Somalia during the southwest (summer) monsoon.

The gyres in both ocean and atmosphere help to transport heat away from the equator towards the poles (the Gulf Stream keeps northern Eu-

---

basin (ocean): the topographic low area occupied by oceans; the floor of ocean basins consists of basaltic crust that is more dense than typical continental rocks

Sverdrup: abbreviated as Sv, 1 Sverdrup is 1 million cubic meters of water per second

monsoon: a wind system that influences large climatic regions and reverses direction seasonally; best known as a wet, warm-season wind carrying drenching rains; also can describe the wintertime wind shift that carries dry, cooler air

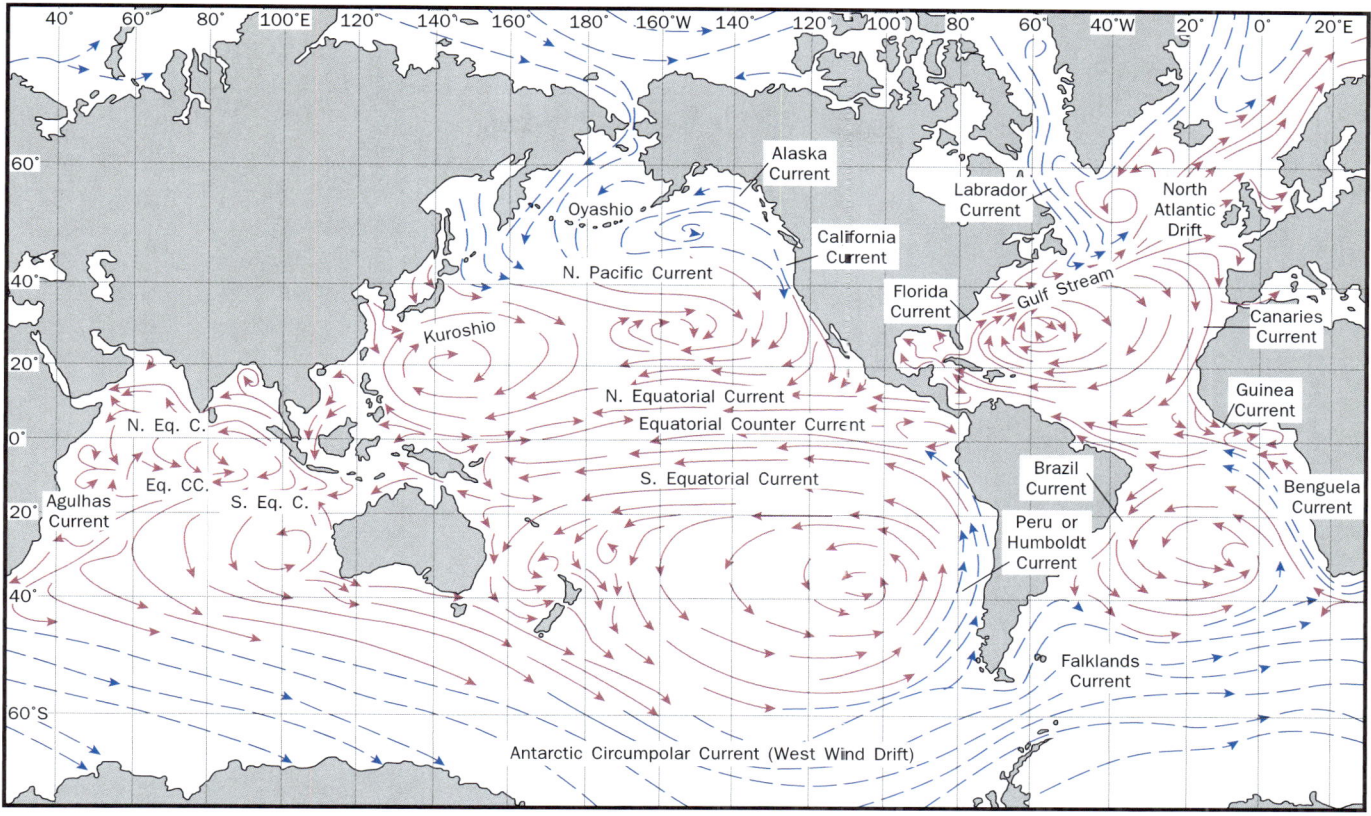

Figure 1. This map shows the global surface current system under average conditions for winter months in the Northern Hemisphere. Warm currents are shown as solid red arrows, and cold currents as dashed blue arrows.

rope considerably warmer than its latitude suggests). This northward flow of warm water in the North Atlantic and a similar flow (the Kuroshio) in the North Pacific are partly balanced by southward flow in the East Greenland and Labrador Currents and in the Oyashio, respectively, while additional southward flow occurs at greater depths. The surface equatorward flow along the eastern edges of the gyres is also considerably cooler than the poleward flow found on the western boundaries. This results from wind-driven upwelling; the equatorward wind stress caused by the trade winds "pushes" water away from the coast and cooler subsurface water upwells to replace it.

In two regions of the globe the surface currents do not form gyres. The Southern Ocean is a region of continuous westerly winds and is the only place on the globe where a continuous circumglobal current (the Antarctic Circumpolar Current) can form. This current therefore is a major region for mixing water between different ocean basins. It is the largest current on earth, having a total volume flux of at least 120 Sv.

Gyral motion is also absent near the equator. Equatorial currents are also driven by the wind, in particular by the trade winds—from the northeast in the Northern Hemisphere and from the southeast in the Southern Hemisphere. However, the Coriolis force here is zero, even though it becomes significant within one degree of latitude north or south. The boundary between the two sets of trade winds is usually slightly north of the equator. The trade winds set up two westward-flowing currents north and

Ocean Currents

south of the equator (the North and South Equatorial Currents), but because the southeast trades blow across the equator, this causes a divergence (upwelling) along the equator itself because of the change in direction of the action of the Coriolis force.

Between the two trade-wind belts is a region of generally light winds, known as the Doldrums. This allows water, which would otherwise pile up against the western boundary of the ocean in the Equatorial currents, to flow back eastwards in the surface Equatorial Countercurrent. There is also an eastward-flowing Equatorial Undercurrent, which forms a jet within the **thermocline**, driven by the horizontal pressure **gradient**. This system of eastward-flowing and westward-flowing currents is found in the upper 200 meters (650 feet) in all three oceans, although their distribution may change seasonally depending on the wind forcing. The equatorial undercurrents are much stronger than the surface currents at the equator, and can have flows of more than 50 Sv.

Deep Currents

Currents in the deep ocean exist because of changes in the density of sea water occurring at the surface. These density changes give rise to specific water masses, which have well-defined temperature and salinity characteristics, and which can be traced for long distances in the ocean.

When sea water freezes, much of the salt that it contains is frozen out, so that a layer of cold **brine** forms at the ocean surface. Being denser than the water below it, the brine sinks, **entraining** water as it does so, until it reaches a level where it has the same density as the surrounding sea water. This process takes place in several regions of the world's oceans, the most important being in the Greenland, Norwegian, and Labrador Seas in the Northern Hemisphere, and close to the Antarctic continent in the Weddell and Ross Seas in the south.

It is the dense waters formed by this process that set up the deep ocean current patterns. Water formed in the Weddell and Ross Seas spreads eastwards and northwards around Antarctica under the influence of the Coriolis force. As is the case with surface waters, most of the flow is concentrated on the western sides of ocean basins, but in this case movement is towards the north. In the South Atlantic Ocean, for instance, bottom water from the Weddell Sea can be identified flowing through the Argentine and Brazil Basins below 4,000 meters (2.5 miles) in depth; it eventually crosses the equator into the North Atlantic off the coast of Brazil. Similar effects are found in both the South Indian and South Pacific oceans, with the bottom waters being forced to follow the bottom **topography**.

Water masses formed in the Northern Hemisphere similarly flow southward. The deep water from the Greenland and Norwegian Seas fills up these basins until it spills over the ridges between Greenland, Iceland, and Scotland. From here it is forced to the right by the Coriolis force and follows the topography around the coast of southern Greenland and the Labrador Sea, eventually crossing under the Gulf Stream and flowing south along the east coast of the U.S. at depths between about 2,000 to 4,000 meters (1.3–2.5 miles). This water, known as North Atlantic Deep Water (NADW), continues south until it joins the Antarctic Circumpolar Cur-

---

thermocline: in a thermally stratified waterbody, the water layer of rapid temperature change over a short vertical interval; it serves as a barrier to water-column mixing

gradient: a measure of the change in magnitude of a parameter (e.g., temperature, elevation, chemical concentration) with distance; when a gradient exists, there is a tendency for a transfer to take place from the area of greater magnitude to the area of lesser magnitude

brine: water containing a higher concentration of dissolved salts than normal sea water (which contains approximately 35 parts per thousand); produced in oceans through the evaporation or freezing of sea water, or in groundwater through extensive reaction with bedrock minerals

entrain: to draw in and transport (as solid particles or gas) by the flow of a fluid; for example, water droplets may become entrained in rising air currents

topography: the shape and contour of a surface, especially the land surface or ocean-floor surface

---

STOMMEL MODEL

In 1948, Henry Stommel explained how friction, a rotating Earth, and a varying Coriolis force produced the strongest surface currents along the western boundaries of the ocean. A decade later, he modeled the deep circulation. He assumed that cold, dense water sinks only in restricted regions of the North Atlantic and Weddell Sea, but that the rising water replacing it does so throughout the tropics and subtropics. This theory requires strong deep flows towards the equator along the western boundaries of ocean basins, rather than away from it as at the surface. Recent observations have shown that these flows can exist east of mid-ocean ridges as well as along the western boundaries of each ocean.

rent, from where it supplies much of the salt to the deep waters of the southern hemisphere.

The Labrador Sea is also a source of dense water. Since, however, the winter conditions are not as severe here as in the Greenland and Norwegian Seas, the brine produced is less dense and does not sink as deep (only 1,500–2,000 meters, or 0.9–1.3 miles). However, the Labrador Sea Water can be traced southward in the North Atlantic as far as the equator on top of the NADW layer.

Role of Water Masses. All these water masses help to transfer oxygen from the atmosphere into the deep ocean. The sinking water is very cold and contains high concentrations of dissolved oxygen acquired at the surface, because cold water can hold more oxygen than warm water. During their flow, they mix with "older" water that has been away from the surface for a longer time, thus ensuring that the bottom waters of the ocean are supplied with oxygen. Additional oxygen is supplied in the southern hemisphere by Antarctic Intermediate Water, formed in a band near 50° S to 55° S latitude. In this region, water does not freeze in winter, but it does cool forming a low-salinity layer that sinks to about 1,000 meters (0.6 mile) depth and moves north in all three oceans.

Water need not be cooled to change its density. Large density changes also can be produced in areas where evaporation is more important than precipitation. Examples of regions where this occurs are the Mediterranean Sea, the Red Sea, and the Persian Gulf. Although the waters here are warm (in the Persian Gulf temperatures can exceed 30°C, or 86°F), the density can increase so that water leaving these enclosed basins sinks as it mixes into its surroundings. Mediterranean Water can be traced across the North Atlantic because of its high salinity, while Red Sea Water can be followed moving south along the east coast of Africa to the Agulhas Current.

An idealized version of the current patterns throughout the whole ocean is shown in Figure 2. This shows clearly that although the surface and deep current patterns may appear separate, they are actually closely linked. Deep water sinking in the northern North Atlantic is replaced at the surface by warmer water from nearer the equator. Similarly, the dense water forming off Antarctica is replaced by upwelling of deep water derived originally from the North Atlantic. Thus, there is a global thermohaline circulation that converts surface water in high latitudes into deep water that moves away from its source, mixing with the water into which it flows.

This flow can be traced from the northern North Atlantic, through the South Atlantic into the Circumpolar Current, and then back again via upwelling in the Pacific and Indian Oceans to the surface layers. Water flows from the Pacific to the Indian Ocean through the Indonesian passages, and the circuit is completed by warm water in the Agulhas Current south of Africa, which enters the South Atlantic and moves northward, crossing the equator again and merging into the Gulf Stream. Although this pathway can be traced in Figure 2, it is clearly much more complicated than stated here. The currents do not flow continuously, as there are many small gyres where water gets "stuck" on its journey and is forced to recirculate one or more times before it can continue around the globe.

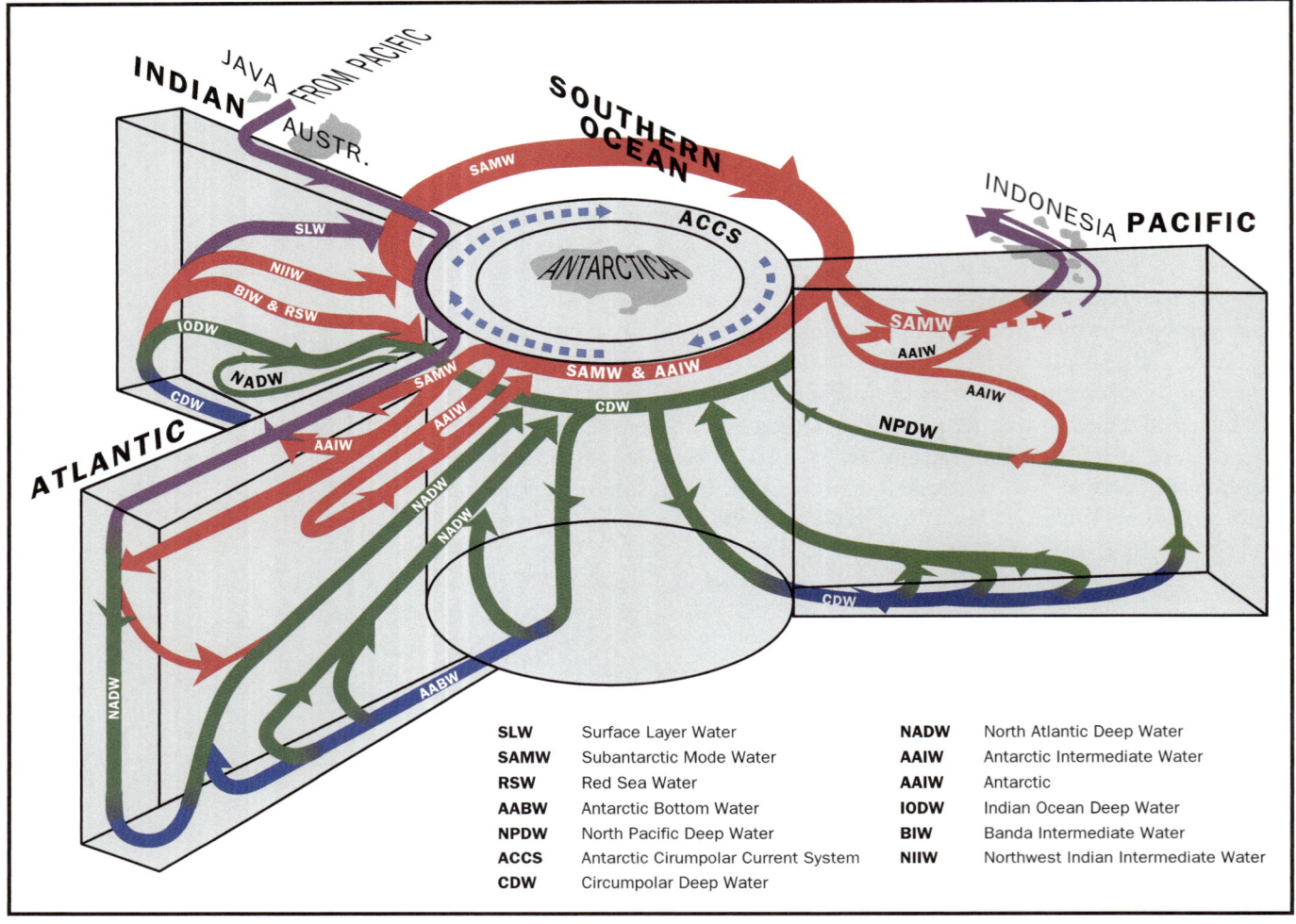

SLW	Surface Layer Water		**NADW**	North Atlantic Deep Water
SAMW	Subantarctic Mode Water		**AAIW**	Antarctic Intermediate Water
RSW	Red Sea Water		**AAIW**	Antarctic
AABW	Antarctic Bottom Water		**IODW**	Indian Ocean Deep Water
NPDW	North Pacific Deep Water		**BIW**	Banda Intermediate Water
ACCS	Antarctic Cirumpolar Current System		**NIIW**	Northwest Indian Intermediate Water
CDW	Circumpolar Deep Water			

Figure 2. This schematic shows generalized interbasin flow for the indicated oceans, and their horizontal connections in the Southern Ocean and the Indonesian Passages. The surface layer circulations are in purple, intermediate and SAMW are in red, deep in green, and near-bottom in blue.

Similarly, there is considerable variation in the paths traced out by the different currents. While the general path of a particular current is the same from one year to another, the actual path it takes can vary widely on scales of a few weeks. All the western boundary currents show considerable movement about their mean position. Quite why these occur is not really known, but they may be driven by changes in the wind stress upstream, or the shedding of eddies. SEE ALSO Climate and the Ocean; Coastal Ocean; Energy from the Ocean; Navigation at Sea, History of; Ocean Mixing; Oceanography, Physical; Oceans, Polar; Sea Water, Gases in; Weather and the Ocean.

Piers Chapman

Bibliography

Open University Course Team. *Ocean Circulation.* Oxford, U.K.: Pergamon Press, 1989.

Pond, Stephen, and George L. Pickard. *Introductory Dynamical Oceanography.* Oxford, U.K.: Pergamon Press, 1983.

Schmitz, William J. *On the World Ocean Circulation*, Vols I and II. Woods Hole Oceanographic Institution Technical Reports WHOI 96-03 and 96-08M. Woods Hole, MA: WHOI, 1996.

Tomczak, Matt, and J. Stuart Godfrey. *Regional Oceanography: An Introduction.* Oxford, U.K.: Pergamon Press, 1994.

Ocean Health, Assessing

The ocean has long been thought to have both a limitless bounty and ability to absorb human impacts. Its sheer volume supported the observation that "dilution is the solution" to **point-source** pollution, as tides and currents removed almost anything that entered the sea. However, increased pressures on the marine environment resulting from increases in human population, industry, and agriculture have led to concern that the ocean's health is being negatively affected by human activities.

Evaluation Methods

The ability of scientists to monitor the ocean's health is hindered by the ocean's complexity. Although the basic chemistry of sea water has been stable for millions of years, components that directly affect plant and animal life (e.g., nutrients and dissolved oxygen) and the living populations themselves vary naturally due to interactions between oceanic and atmospheric processes. Evaluating the health of the oceans therefore requires that human impacts must be distinguished from a natural, changing background.

Methods for evaluating the ocean's health include estimates of commercial fishery stocks (populations) and localized studies of plant and animal species (including estimates of **estuarine** productivity and coral bleaching). Impacts from **contaminants** and adverse water quality ideally are monitored through long-term baseline studies. In the United States, this approach has been followed by the National Oceanic and Atmospheric Administration's Mussel Watch program, for example, that focuses on organic and inorganic contaminants in mussels and oysters in U.S. coastal waters. The success of this approach has led to similar studies on an international scale.

Historical trends of contaminant input have been evaluated through "dated core" programs in which contaminants are measured in marine sediment layers and compared with estimates of when they were deposited. Larger spatial scales are evaluated by **remote sensing** to measure such variables as temperature, **plankton** populations, and sediment load of surface waters.

Ocean Threats

At present, pollution, habitat alteration, and overfishing are considered the primary threats to the ocean's health. Pollutants includes chemicals (including compounds that may affect animal development during sensitive life stages), sewage, floating debris such as plastic and trash, and nutrient elements (nitrogen and phosphorus) that are largely released to coastal areas either directly, via rivers, or via the atmosphere.

Ocean margins are impacted almost everywhere by alteration or destruction of critical habitat. These changes include erosion and loss of salt marshes; drainage of wetlands; siltation of estuarine areas after deforestation and erosion; alteration of fresh-water inputs; and restriction of fish migration routes by dams. Dredging, boating, and pressure from tourism have affected coral reefs.

Overfishing target species (e.g., whales, sharks) and the accidental removal of nontarget species (as bycatch) have damaged overall ocean health. Many fishery stocks (populations) have declined dramatically as a result of

Chemical spills and chemical-laden runoff from land contribute to ocean pollution. Here, a French Navy firefighter takes a water sample after the Italian tanker *Levoli*, carrying 6,000 metric tons of toxic chemicals, sunk in 2000 off the northwestern coast of France. Through controlled international efforts to remove the ship's cargo, little environmental damage was incurred.

point-source: describes a pollutant release or discharge originating from one specific location (e.g., an outfall pipe from a factory) rather than over a wide land area (e.g., water runoff from a farm field)

estuary: a tidally influenced coastal area in which fresh water from a river mixes with sea water, generally at the river mouth; the resulting water is brackish, which results in a unique ecosystem

contaminant: as defined by the U.S. Environmental Protection Agency, any physical, chemical, biological, or radiological substance in water, including constituents that may not be harmful to the environment or human health

remote sensing: the collection and interpretation of information about an object without being in physical contact with the object; most often, it refers to satellite-based collection of data to map and monitor the environment and resources on Earth

plankton: an assemblage of small, often microscopic aquatic organisms encompassing aquatic plants (phytoplankton) and aquatic animals (zooplankton) that float or drift passively with water currents

overfishing, and the annual catch has remained high only as a result of switching to new target species. At present, the seventeen major ocean fishery areas are fished to capacity, overfished, or depleted.

Looking to the Future

In the future, although human population and development in coastal areas will continue to expand, releases of chemical contaminants and nutrients will continue to be regulated in developed countries. Lessons learned in developed countries will assist developing nations in improving their economies and infrastructure while preserving their natural environments.

Unfortunately, the popularity of coastal areas and their importance in trade will exert continued pressure on habitat. Difficult policy decisions will be required if habitat is to be conserved or restored to earlier conditions. Similarly, the dependence of expanding human populations on fisheries as protein sources, particularly in developing countries, coupled with the cultural importance of fishing as a lifestyle, will continue.

Economic considerations may ultimately provide the impetus for conserving and improving marine environmental conditions. Increasing costs per unit catch will ultimately restrict the expansion of fishing, especially if government subsidies are reconsidered. Appreciation of the economic importance of ocean-based tourism and coastal habitat for renewing valuable living resources will provide additional incentives to maintaining ocean health. SEE ALSO COASTAL OCEAN; COASTAL WATERS MANAGEMENT; CORALS AND CORAL REEFS; EROSION AND SEDIMENTATION; ESTUARIES; FISHERIES, MARINE: MANAGEMENT AND POLICY; POLLUTION OF THE OCEAN BY PLASTIC AND TRASH; POLLUTION OF THE OCEAN BY SEWAGE, NUTRIENTS, AND CHEMICALS; POLLUTION SOURCES: POINT AND NONPOINT; SALMON DECLINE AND RECOVERY.

Robert J. Taylor

Bibliography

Thurman, Harold V., and Elizabeth A. Burton. *Introductory Oceanography*, 9th ed. Upper Saddle River, NJ: Prentice Hall, 2001.

Internet Resources

"Chemical Contaminants in Oysters and Mussels." *State of the Coastal Environment*. National Oceanic and Atmospheric Administration. <http://state-of-coast.noaa.gov/bulletins/html/ccom_05/ccom.html>.

Global Coral Reef Monitoring Network. National Oceanic and Atmospheric Administration. <http://coral.aoml.noaa.gov/gcrmn/>.

Ocean Mixing

Mixing in the ocean occurs on several scales, the smallest scale being molecular. If a layer of warm, salty water lies above a layer of colder, fresher water, the heat and salt will tend to diffuse (spread out) downwards to make a single layer with intermediate temperature and **salinity** values. However, because heat diffuses faster than salt, the process can lead to local instabilities in the density structure which cause mixing within a layer many meters thick. The best-known example of this process, known as salt fingering, occurs where very salty water from the Mediterranean outflow mixes into the North Atlantic.

salinity: the concentration of dissolved materials carried in an aqueous (watery) solution; typically expressed in grams per liter (parts per thousand) or milligrams per liter (parts per million)

Ocean Mixing

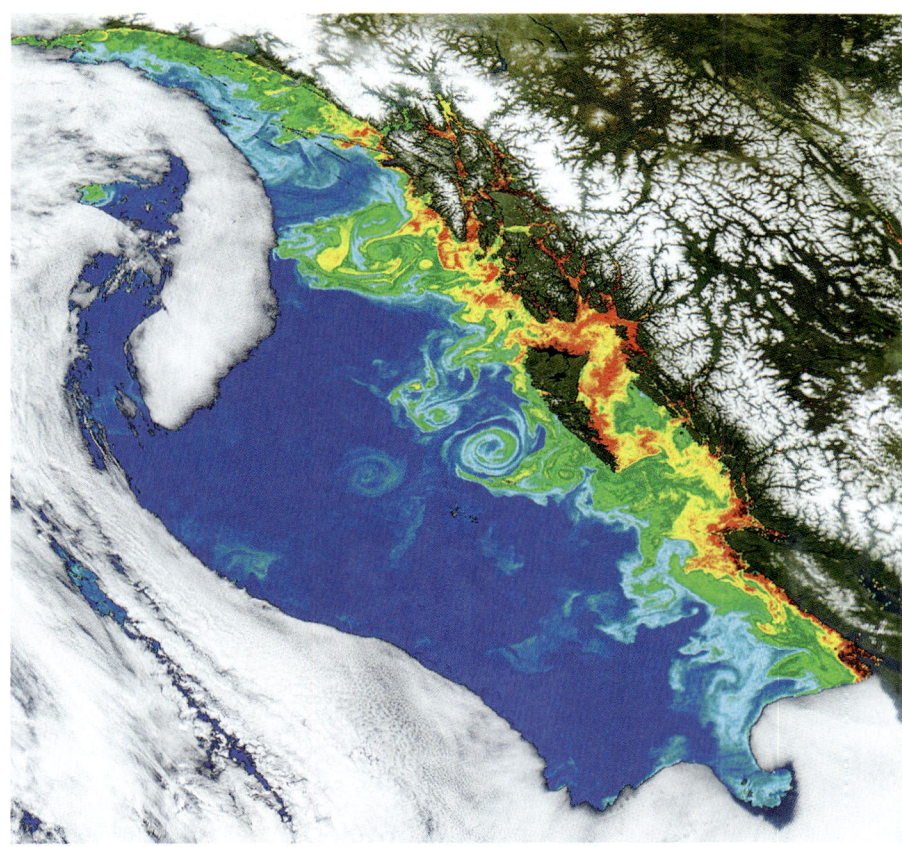

Several eddies in the Pacific Ocean are visible in this pseudo-color satellite view of British Columbia's Queen Charlotte Islands and Alaska's Alexander Archipelago. The eddies are formed by strong outflow currents from coastal rivers that are rich in nutrients from spring snowmelt. The nutrient-rich water helps stimulate phytoplankton blooms, whose chlorophyll concentrations yield the colored swirls.

Most mixing, however, takes place on larger scales in response to forcing by the wind (in the upper layers), by tides (particularly close to shore or in confined areas or those with rough topography), or by currents. Mixing by winds and tides often results in a surface mixed layer having homogeneous temperature and salinity. This layer may be separated from the water below it by a jump in temperature or salinity, known as a thermocline or pycnocline, respectively.✴

✴ See "Sea Water, Physics and Chemistry of" for a depth profile illustrating a thermocline and pycnocline.

Flow in the ocean is not smooth but turbulent, and this increases mixing at the edge of currents by causing eddies to form. Such turbulent mixing can often be seen in rivers immediately downstream of bridges, where the fast water flow is affected by the bridge piers.

Currents and Eddies

Because ocean currents are many kilometers wide, they have a correspondingly large effect on mixing in the oceans. As may be expected from the positions of ocean currents, most mixing in the upper layers of the ocean takes place on the western boundaries of ocean **gyres** where the current speeds are greatest. Considerable mixing also occurs along the **fronts** which border the Antarctic Circumpolar Current. The mixing is evident in temperature or color images of the sea surface because the edge of the current is drawn out into filaments (strands), which mix into the quieter waters outside the current.

Frequently, discrete parcels of water break off from the current as eddies, which may be up to about 200 kilometers (125 miles) across and last

gyre: a circular pattern of currents in an ocean basin

front (ocean): a region in the ocean where a sudden change in temperature, velocity, or other parameter causes a sharp line of demarcation at the surface, often visible to the eye; they may be small-scale and narrow, or larger-scale and extend across several kilometers; usually they denote a region of convergence, where water tends to sink

basin (ocean): the topographic low area occupied by oceans; the floor of ocean basins consists of basaltic crust that is more dense than typical continental rocks

brine: water containing a higher concentration of dissolved salts than normal sea water (which contains approximately 35 parts per thousand); produced in oceans through the evaporation or freezing of sea water, or in groundwater through extensive reaction with bedrock minerals

for many months. They drift across ocean **basins** at speeds of about 5 centimeters per second, which is equivalent to 4.3 kilometers (2.5 miles) per day, or roughly 0.2 kilometer (0.1 mile) per hour. In comparison, atmospheric eddies (tropical cyclones) are larger (up to 2,000 kilometers [1,250 miles] across) and move faster (up to 30 kilometers [18 miles] per hour). Depending on their source, oceanic eddies may contain water either warmer or colder than that into which they mix, or of very different salinity, as occurs where water from the Mediterranean mixes into the North Atlantic.

As the eddies revolve they lose water through mixing into their surroundings. This occurs most rapidly along surfaces of equal density, which tend to be almost flat in the ocean, so the mixing is generally horizontal rather than vertical. Vertical mixing occurs only in a few places, such as zones of divergence, or upwelling, along the eastern boundaries of oceans and near the equator. It occurs also in the Arctic and Antarctic where the freezing of water during the winter causes dense **brines** to be produced. As they sink, they cause additional mixing within the water column. SEE ALSO OCEAN CURRENTS; OCEANOGRAPHY FROM SPACE; OCEANOGRAPHY, PHYSICAL; SEA WATER, PHYSICS AND CHEMISTRY OF.

Piers Chapman

Bibliography

Open University Course Team. *Ocean Circulation*. Oxford, U.K.: Pergamon Press, 1989.

Pond, Stephen, and George L. Pickard. *Introductory Dynamical Oceanography*. Oxford, U.K.: Pergamon Press, 1983.

Ocean–Floor Bathymetry

The term bathymetry is defined as the depth of water relative to sea level. Thus bathymetric measurements can determine the topography of the ocean floor, and have shown that the sea floor is varied, complex, and ever-changing, containing plains, canyons, active and extinct volcanoes, mountain ranges, and hot springs. Some features, such as mid-ocean ridges (where oceanic crust is constantly produced) and subduction zones, also called deep-sea trenches (where it is constantly destroyed), are unique to the ocean floor.✱

✱ See "Plate Tectonics" for an illustration showing types of plate convergence.

Bathymetric mapping involves the production of ocean and sea maps based upon bathymetric data (see historic map). Bathymetric maps represent the ocean depth as a function of geographical coordinates in the same way topographic maps represent the altitude of Earth's surface at different geographic points. The most popular type of bathymetric maps are ones on which lines of equal depths (called isobaths) are represented.

Bathymetric Techniques

For hundreds of years, the only way to measure ocean depth was the sounding line, a weighted rope or wire that was lowered overboard until it touched the ocean floor. Not only was this method time-consuming, it was inaccurate; ship drift or water currents could drag the line off at an angle, which would exaggerate the depth reading. It was also difficult to tell when the sounding line had actually touched bottom.

England's Sir John Murray compiled this bathymetric (depth) chart of the North Atlantic in 1911. Murray's chart went far beyond American naval officer Matthew Maury's first attempt at bathymetric mapping in 1855. In addition, Murray's map gave birth to the idea of the Telegraphic Plateau, a submarine land formation from Canada to the British Isles, across which the first transatlantic cable was laid.

In the twentieth century, sounding lines were entirely replaced by sonar systems. Sonar (*so*und *na*vigation *r*anging), invented during World War II (1939–1945) measures distances by emitting a short pulse of high-frequency sound and measuring the time until an echo is heard. After the war, ships with sonar units attached to their hulls crisscrossed the world's oceans systematically, measuring depth. The data collected made possible complete bathymetric maps of the world's oceans. For the first time, scientists knew what 70 percent of Earth's surface really looked like (radar, which produces images by bouncing radio waves rather than sound waves off distant objects, cannot be used for bathymetry because water absorbs radio waves).

Many sonar techniques have been developed for bathymetry. When high-resolution images are desired, an underwater unit may be towed behind a ship, scanning to the left and right with multiple sonar beams (side-scan sonar).✷ Furthermore, orbiting visible-light cameras image the bottoms of some shallow waters, while satellite radar maps deep-sea topography by detecting the subtle variations in sea level caused by the gravitational pull of undersea mountains, ridges, and other masses.

The Ocean Floor in Cross-Section

The oceans begins, of course, at the shore, the irregular boundary where the surface of a continent descends first to sea level and then beneath it. If the depth of the ocean is measured along a line drawn straight out from a continental shore, the following sequence of bottom features are typically seen.

Continental Shelf. For many miles out, the ocean is only a few hundred feet deep and gets deeper quite slowly (i.e., slopes at an angle of 0.1°, or 1.7 meters per kilometer [9 feet per mile]). This flat, wide margin is found around every continent and is known as the continental shelf. The average width of a continental shelf is 70 kilometers (43 miles).

✷ See "Sound Transmission in the Ocean" for a photograph of a modern sonar device.

Continental Slope and Rise.
The continental shelf ends at a sudden drop-off called the shelf break. Beyond the shelf break, the slope of the ocean floor becomes much steeper, typically a 4° slope, or 70 meters per kilometer (370 feet per mile). This steep embankment is called the continental slope and is grooved by submarine canyons and gullies. The continental slope is about 16 kilometers (10 miles) wide, on average, and descends to a depth of about 2.4 kilometers (1.5 miles). There it ends as the slope moderates to a mere degree or two from horizontal. This gradual zone, which may be several hundred miles wide, is called the continental rise. It is composed of fine-grained continental sediments (silt and clay) washed down the many submarine canyons that notch the continental slope.

Abyssal Plain.
The abyssal plain, which is the deepest, most level part of the ocean, is found where the continental rise ends, at a depth of about 4 kilometers (2.5 miles). The abyssal plain is dotted with thousands of small, extinct volcanoes called abyssal hills. The abyssal plains of the Atlantic appear smooth because its abyssal hills are buried under a thick blanket of continental sediment, but in the Pacific Ocean basin, which is ringed by trenches that trap sediments before they can spread over the ocean floor, tens of thousands of unburied abyssal hills have been observed. Abyssal hills more than 1 kilometer (0.6 mile) high are called seamounts, and seamounts with flat tops are called guyots or tablemounts. Guyots are drowned volcanic islands that become submerged due to subsidence of the oceanic **lithosphere**.

lithosphere: the rigid outer layer of Earth made up of the crust and the uppermost mantle

Mid-Ocean Ridge.
Beyond the abyssal plain, which may be several hundred kilometers wide, the ocean floor begins to ascend again with a gentle slope. This area is the flank of the mid-ocean ridge, a long, undersea mountain chain that usually extends down the middle of the ocean. The Mid-Atlantic Ridge, for example, snakes down the middle of the Atlantic most of the way from the North Pole to Antarctica.✳ As the ocean floor climbs slowly toward the center of the mid-ocean ridge its sediment blanket gets thinner and its surface more irregular. Here, the ocean floor is marked by thousand-mile cracks called fracture zones that lie at right angles across the mid-ocean ridge.

✳ See "Mid-Ocean Ridges" for an image showing the Mid-Atlantic Ridge and other extensions of the ridge system.

Rift Valley.
Along the center of the mid-ocean ridge is the rift valley, a deep V-shaped notch. From this valley, new oceanic crust is constantly being extruded from Earth's **mantle** by processes not yet fully understood. Twin sheets of fresh, mile-thick crust emerge from the mantle along the sides of the rift valley and flow slowly away from it in opposite directions. In the case of the Mid-Atlantic rift valley, one sheet flows east and the other west, each moving at about half an inch per year. The older, more distant parts of these growing sheets of crust are gradually covered by sediments, eventually becoming the abyssal plains and continental rises described above. These growing sheets of crust have the effect of forcing the distant continents farther apart, a process called sea-floor spreading.

mantle: the region of the Earth between the molten core and the outer crust, composed mainly of silicate rock, and around 2,900 kilometers (1,800 miles) thick; also the interior of another planet, moon, or large asteroid between the core and the crust

Subduction Zones.
Beyond the rift valley, the depth sequence described above is observed in reverse: a gently declining slope of ridges and fractures, an abyssal plain, a continental rise, a steep continental slope, a somewhat flat continental shelf, and finally dry land again. This would be the case in crossing the Atlantic Ocean. Elsewhere, however, such neat symmetry is not always found. Some continental margins are characterized by subduction zones, whereby oceanic crust is destroyed beneath the edge of a continental plate.

Ocean–Floor Bathymetry

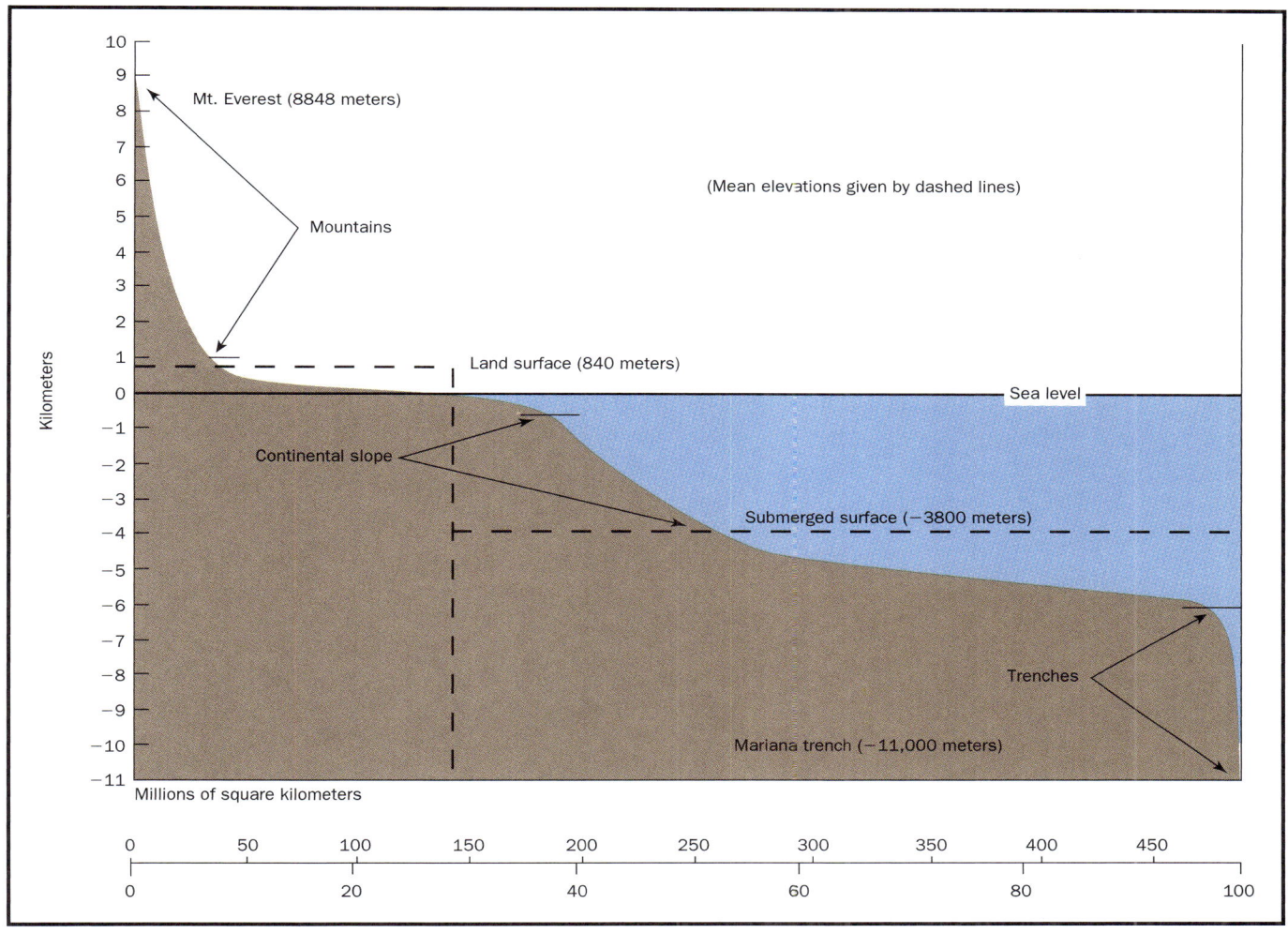

The hypsographic curve shows the amount of Earth's surface at various elevations and depths. Horizontal dashed lines indicate average height of the continents at 840 meters (2,750 feet) above sea level, and average depth of the oceans at 3,800 meters (12,460 feet) below sea level. The vertical dashed line marks the division between land and sea at present-day sea level.

The disappearance of oceanic crust into subduction zones is occurring primarily along the western, northern, and eastern edges of the Pacific Ocean. These subduction zones are the deepest places in the ocean, varying from 7 to 11 kilometers (4 miles to 6.8 miles) in depth. Along the length of a typical subduction zone or deep-sea trench, a sheet of old oceanic crust is forced beneath continental crust. The oceanic crust that is forced to submerge or subduct in this manner slides down into the mantle at an angle.

The Hypsographic Curve

The elevation features of the ocean floor—shelves, slopes, rises, plains, ridges, and trenches—are quantitatively summarized by oceanographers, along with the distribution of dry land at different altitudes, in a graph called the hypsographic curve (see the figure). The hypsographic curve shows what percentage of Earth's surface rises above present-day sea level to a given height, or sinks below it to a given depth. The curve shows that a small percentage of Earth's surface consists of high-altitude continental mountains, 30 percent of the surface consists of continental lowlands. Continental shelves and slopes account for about another 15 percent, abyssal plains and mid-ocean ridge systems for almost 50 percent, and deep-sea trenches account for a small fraction. The fact that Earth's surface is comprised mostly of flat plates of continental and oceanic crust, with mountains and deep-sea trenches occurring only along their edges where the plates collide, gives the hypsographic curve its characteristic shape. SEE ALSO GEOSPATIAL TECHNOLOGIES; MARGINAL SEAS; MID-OCEAN RIDGES; OCEANOGRAPHY FROM SPACE; PLATE TECTONICS; SOUND TRANSMISSION IN THE OCEAN; VOLCANOES, SUBMARINE.

Larry Gilman and K. Lee Lerner

Bibliography

Medwin, Herman, and Clarence Clay. *Fundamentals of Acoustical Oceanography.* New York: Academic Press, 1998.

Neumann, Gerhard, and Willard Pierson. *Principles of Physical Oceanography.* Englewood Cliffs, NJ: Prentice Hall, 1966.

Sabins, Floyd. *Remote Sensing: Principles and Interpretation,* 2nd ed. New York: W. H. Freeman and Company, 1987.

Thurman, Harold, and Elizabeth Burton. *Introductory Oceanography,* 9th ed. Upper Saddle River, NJ: Prentice Hall, 2001.

Ocean-Floor Sediments

Sediment on the seafloor originates from a variety of sources, including **biota** from the overlying ocean water, eroded material from land transported to the ocean by rivers or wind, ash from volcanoes, and chemical **precipitates** derived directly from sea water. A very small amount of it even originates as **interstellar** dust. In short, the particles found in sediment on the seafloor vary considerably in composition and record a complex interplay of processes that have acted to form, transport, and preserve them.

Geological oceanographers have coined the terms "terrigenous" to describe those sediments derived from eroded material on land, "biogenic" for those derived from biological matter, "volcanogenic" for those that include significant amounts of ash, "hydrogenous" for those that precipitate directly from sea water, and "cosmogenic" for those that come from interstellar space.

The seafloor, however, is not a random arrangement of these different sediment types. Oceanographers have painstakingly mapped the distribution of sediment around the globe and have learned that at any given location the sediments provide important information regarding the history of the ocean as well as the overall state of climate on the Earth's surface. By studying how the heterogeneous composition of sediment varies as a function of geographic location and age, oceanographers are able to document the geologic and climatic conditions that are responsible for that sediment.

Oceanographers study sediment by taking long cylindrical cores, which individually can be as long as 18 to 30 meters (60 to 98 feet). Because the bottom of the ocean is extremely cold (only 1 to 3 degrees above freezing), the cores are stored in refrigerators onboard the research ship prior to being stored in large refrigerated repositories at shore-based laboratories. In their laboratories, scientists study the physical, chemical, and biological makeup of the sediment.

Regardless of which type of sediment, there are three processes that are responsible for its final composition: namely, the production of the sedi-

---

biota: the plant and animal life of a region or ecosystem, as in a stream, lake, or ocean

precipitate: (verb) the process by which a solution separates into a relatively clear liquid and a solid substance by a chemical or physical change; (noun) the solid substance resulting from this process

interstellar: describes the region of space that occurs between individual stars, occupied by gas and dust as well as isolated molecules, including hydroxyl ions, water, sulfur oxide, as well as carbon-based molecules

Ocean-Floor Sediments

A core sample of sediment from Chesapeake Bay can tell scientists about the oceanographic history of that particular location, including climate change, pollution, and past changes in erosion.

ment; its transport; and its preservation. It is important to differentiate between these three processes. For example, if a sedimentary particle is produced, but not preserved, there will be no resulting sedimentary record. Thus, only if material is produced and transported and preserved will marine sediment result.

The different combinations of each process' effectiveness result in a commensurate variety of sedimentation rates. Sediment can accumulate as slowly as 0.1 millimeter (0.04 inch) per 1,000 years (in the middle of the ocean where only wind-blown material is deposited) to as fast as 1 meter (3.25 feet) per year along **continental margins**. More typical deep-sea rates are on the order of several centimeters per 1,000 years.

continental margin: region where continental crust meets oceanic crust

Production of Sediment

The production of marine sediment is more complex than it may seem. Terrigenous sediment is produced by an interplay of chemical and physical **weathering** processes, which collectively serve to create small grains of material ranging in size from thousandths of millimeters to 1 or 2 millimeters (0.04 or 0.08 inch). (The larger grains of coarse sand, gravel, and boulders are too large to be transported to the deep sea and therefore are not discussed here.)

weathering: the decay or breakdown of rocks and minerals through a complex interaction of physical, chemical, and biological processes; water is the most important agent of weathering; soil is formed through weathering processes

Physical weathering is caused by mechanical fracturing of rocks, such as that due to the freezing of water in cracks, and results in finer grained, compositionally similar examples of the original rock. On the other hand, chemical weathering, caused by the weak acid produced by the interaction of rainwater and atmospheric carbon dioxide, degrades the rock slowly and often produces fine-grained minerals that are compositionally distinct from the original rock.

Biogenic (biologically derived) sediment is produced by marine plankton, which are small, often microscopic, unicellular plants and animals that float in the surface waters of the ocean. The shells of these organisms are

Ocean-Floor Sediments

made of either calcium carbonate ($CaCO_3$) or silica (SiO_2). Although ubiquitous, particularly elevated concentrations of such organisms are most commonly found in biologically productive waters such as the Equatorial Pacific, or the Southern Ocean ringing the continent of Antarctica.

Volcanic ash is produced during volcanic eruptions, as can be seen in the billowing ejected material from many volcanos. Cosmogenic material is the remains of primordial material left over from the creation of the solar system (and perhaps from beyond) and, although very low in abundance, is ubiquitously distributed.

The production of hydrogenous sediment is most difficult to visualize, but involves either the slow precipitation of dissolved chemicals from sea water or the leaching of chemical elements from rocks that have extremely hot sea water (greater than 300°C [572°F]) circulating through them along mid-ocean ridges. When these hot solutions are injected into the cold sea water the leached chemical elements precipitate from the cooling water, leading to hydrothermal sediments near the mid-ocean ridge that are enriched in iron, manganese, copper, zinc, and other metals.

Transport of Sediment

The transport of sediment depends on its grain size and the original location where it was produced. Terrigenous sediment can be transported to the deep sea via rivers or by wind. Material transported by rivers most commonly ends up deposited on the continental margin, the shallow portions of the ocean that are within several hundred kilometers of land. When continental margin deposits accumulate fast and get overly steep, or when an earthquake or storm causes the sediment to be resuspended, **turbidity currents** provide additional transport out to the deep sea. The resuspension of the sediment into the bottom water causes it to be more dense than the overlying water, and thus these turbidity currents flow downslope to the more distant ocean basin.

The transport of sediment by wind is also extremely significant, and is particularly relevant to studies of Earth's climate in the past. When the Earth's climate is relatively dry (arid), such as during glacial periods, the land surface tends to be more dusty, and thus during such periods there will be more windblown terrigenous material delivered to the deep ocean. Also, during such time periods the wind speed tends to be higher, and thus terrigenous grains that are slightly larger than usual are preferentially transported. Thus, by examining the amount of dust, as well as its grain size, in the different layers of a sediment core, oceanographers learn how arid the land surface was at a given time, as well as how fast the average wind speeds were.

Although such dust is essentially invisible to the human eye, its transport is still an important and long-ranging process. For example, dust derived from the Sahara in North Africa is easily observed in Miami, Florida and even in the eastern Pacific Ocean. Moreover, volcanic ash ejected tens of kilometers into the atmosphere during the largest eruptions can be transported by winds all around the globe.

The microscopic shells of the plankton do not just simply fall to the seafloor. In fact, because they are so small, the plankton may not be able to

turbidity current: a gravity current resulting from a density increase brought about by increased water turbidity; possibly initiated by some sudden force, such as an earthquake, the turbid mass continues under the force of gravity down a submarine slope

MANGANESE NODULES

Manganese nodules are concretions enriched in manganese, iron, cobalt, copper, zinc, and other metals that are found laying on seafloor in regimes of extremely low sedimentation rate. Manganese nodules nucleate or begin to grow on a previously existing particle (commonly a small fish bone or shark's tooth) and derive their chemicals from the extremely slow precipitation of the metals directly from sea water. Because their accumulation rates are on the order of 10 millimeters (0.04 inch) per 1 million years, oceanographers interested in past ocean history can study the composition of their individual layers and derive a record of ocean change that can be as long as 50 million years in a single nodule of a 5-centimeter (2-inch) radius.

Wind-swept desert sands not only produce a cooling effect due to deflection of incoming solar radiation, but they also deposit sand, silt and dust on the ocean surface and ultimately on the ocean floor. (Shown here is a dust storm over the Red Sea.) Through mineralogical and chemical analysis, scientists can recreate historical patterns in climate and geological development.

fall individually. Oceanographers learn how such sediment is delivered to the seafloor by suspending sediment traps in the ocean. These traps are essentially large funnels, up to 1 or 2 meters (3.3 to 6.6 feet) in diameter, that collect the material as it falls through sea water.

By examining the material trapped by these instruments, it was discovered that plankton shells are delivered to seafloor by "biopackaging" via fecal pellets. In other words, various microorganisms that eat other plankton excrete their shells in fecal pellets. These "biopackaged" fecal pellets are large enough (0.2-1.5 millimeters, or 0.008-0.059 inch) and dense enough to sink to the seafloor, where they become part of the sediment.

Hydrothermal sediment is largely localized to within less than 10 kilometers (6.2 miles) of the mid-ocean ridge. The concentration of metals in these sediments decreases with distance from a ridge, yet small amounts can be found up to 500 to 1000 kilometers (300-600 miles) away.

Preservation of Sediment

Terrigenous sediment, whether it be delivered by rivers or wind, is not altered significantly on the seafloor and thus is well-preserved. During very deep burial (e.g., 5 kilometers, or 3 miles, below the seafloor), the terrigenous grains can be altered into different minerals, but this does not occur while the grains are lying on the seafloor and is generally a more important process for geologists rather than oceanographers.

Biogenic sediment, on the other hand, is very poorly preserved on the seafloor. The degradation of biogenic sediments is a complex, largely chemical suite of processes. The preservation of these sediments is a field of study that has captivated oceanographers for over 100 years, dating from when they were discovered in the mid-1800s during the first oceanographic research cruise by the ship *HMS Challenger*.

For example, significantly less than 1 percent of the siliceous plankton that are biopackaged to the seafloor are preserved. This is because sea water is undersaturated with respect to silica. Therefore, the siliceous plankton are living in an environment that is corrosive to their shells.

organic: pertaining to, or the product of, biological reactions or functions

While the plankon is alive, the shell is surrounded by **organic** protoplasm that protects it from the corrosive sea water. After death, however, even if biopackaged, this organic coating will degrade, exposing the shell of the siliceous plankton. When exposed to sea water, the shell will dissolve.

This process occurs over all depth and temperature ranges throughout the global ocean. Thus, the only regions of the seafloor where biogenic silica appreciably accumulates is where the production of biogenic silica is so enormous that it overwhelms the amount that is dissolved. In the modern oceans, this occurs at high latitudes in the North Pacific and Southern Ocean and the Equatorial Pacific Ocean.

hydrostatic: referring to the pressure exerted by water at a point, related to the weight of the water above the point

Plankton with shells made of calcium carbonate also commonly dissolve, but not as commonly as siliceous plankton. The dissolution of carbonate plankton is controlled by water depth and water temperature. Water depth and **hydrostatic** pressure correlate with each other—at greater depths there is greater pressure. At greater pressures, the solubility of carbon dioxide gas increases. An excellent analogy of this process is observed in a bottled carbonated beverage that is under pressure until opened—when the pressure is released the carbon dioxide comes out of solution and bubbles form. Similarly, at the great depths of the deepest seafloor, the solubility of carbon dioxide increases so much that calcium carbonate sediment may dissolve. This dissolution is also facilitated at the lower temperatures of the deep sea.

photosynthesis: the process by which plants manufacture food from sunlight; specifically, the conversion of water and carbon dioxide to complex sugars in plant tissues by the action of chlorophyll driven by solar energy

The converse is also true. At shallow water depths (that is, lower pressure) the carbonate does not dissolve and the warmer water temperatures (along with the increased light for **photosynthesis**) each serve to enhance the construction and preservation of coral reefs and other carbonate-producing biota. Thus, there is both a depth and latitudinal effect on the distribution of carbonate sediments due to their influence on temperature and pressure. SEE ALSO ALGAL BLOOMS IN THE OCEAN; MINERAL RESOURCES FROM THE OCEAN; OCEAN BIOGEOCHEMISTRY; OCEAN CURRENTS; OCEANS, POLAR; PLATE TECTONICS; PLANKTON; RADIONUCLIDES IN THE OCEAN.

Richard W. Murray

Bibliography

Charnock, H., J. M. Edmond, I. N. McCave, A. L. Rice, and T. R. S. Wilson. *The Deep Sea Bed: Its Physics, Chemistry, and Biology.* London, U.K.: The Royal Society, 1990.

Cronan, D. S., *Underwater Minerals.* London, U.K.: Academic Press, 1980.

Ernst, W. G., and J. G. Morin, eds. *The Environment of the Deep Sea.* Upper Saddle River, NJ: Prentice-Hall, 1982.

Lisitzin, A. P. *Oceanic Sedimentation, Lithology, and Geochemistry.* Washington, D.C.: American Geophysical Union, 1996.

Morse, J. W., and F. T. Mackenzie. *Geochemistry of Sedimentary Carbonates.* Amsterdam, Netherlands: Elsevier, 1990.

Oceanography, Biological

Biological oceanography is a field of study that seeks to understand what controls the distribution and abundance of different types of marine life, and how living organisms influence and interact with processes in the oceans.

Biological oceanographers study all forms of life in the oceans, from microscopic plants and animals to fish and whales. In addition, biological oceanographers examine all forms of oceanic processes that involve living organisms. These include processes that occur at molecular scales, such as **photosynthesis**, respiration, and cycling of essential **nutrients**, to large-scale processes such as effects of ocean currents on marine productivity.

A distinction is often made between the fields of biological oceanography and marine biology. Although there is considerable overlap between the two disciplines, the field of marine biology traditionally deals with the study of individual organisms, including their taxonomy, behavior, physiology and other aspects of their biology. In contrast, the emphasis of biological oceanography is the ocean and organisms as a system. As such, biological oceanographers tend to utilize a multidisciplinary approach, drawing on knowledge from various fields in addition to biology including, for example, physics, chemistry, and geology.

Tools and Technology

Biological oceanographers rely on a variety of tools and use a variety of approaches to aid them in their study of life in the sea. Some studies involve laboratory experiments with individual organisms. In other cases, the oceanographer must go into the water to directly sample and observe certain types of organisms such as **zooplankton**.

Other approaches involve underwater submersible vehicles to gain access to biological communities deep in the ocean, such as those associated with deep-sea hydrothermal vents. Many oceanographers use research vessels from which they lower instruments and specialized water sampling gear into the water. Biological oceanographers employ methods derived from various fields, including molecular biology, immunology, physiology, biochemistry, ecology, and many others.

In addition to making scientific observations, the biological oceanographer uses a variety of models to study the biology of the oceans. Theoretical models are used to examine problems in biological oceanography that cannot be answered through direct observation and measurement. Heuristic

---

photosynthesis: the process by which plants manufacture food from sunlight; specifically, the conversion of water and carbon dioxide to complex sugars in plant tissues by the action of chlorophyll driven by solar energy

nutrients: a group of chemical elements or compounds needed for all plant and animal life

zooplankton: microscopic animals that live suspended in bodies of water and that drift about because they cannot move by themselves or because they are too small or too weak to swim effectively against a current; composed primarily of protozoans, microcrustacea (copepods, cladocera, rotifers) and larval stages of certain invertebrates

Oceanography, Biological

This oceanography graduate student repairs a deep-water instrument in the biological research lab aboard a research vessel. Deployment of such devices allows biological oceanographers to conduct measurements at depth.

models are used to help to understand and explain an existing set of observations. Finally, some models are used to predict changes in biological processes that may occur because of natural and human-induced changes to the ocean environment.

Advances in technology have given biological oceanographers new insights about the living oceans. Lasers, fiber optics, high-speed digital video imaging and DNA microarrays are some of the high-tech "gadgets" that are used to study biological processes in the oceans. Robotic underwater vehicles reduce the risk and expense of manned submersibles while providing spectacular views of undersea communities. Other types of instruments are allowed to drift freely with ocean currents, towed behind a ship, or anchored at specific locations to provide detailed information over time and space. Among the most powerful tools available to biological oceanographers are satellite and airborne sensors, which provide large-scale views of the ocean and have greatly enriched the scientific understanding of biological processes and their relationship to physical phenomena.

Areas of Research

Major research programs in biological oceanography examine cycles of carbon and other biologically critical elements, such as nitrogen, phosphorus, silicon and iron. These biogeochemical cycles are key in understanding large-scale phenomena such as global warming. Living organisms, particularly phytoplankton (single-celled microscopic plants that utilize photosynthesis), bacterioplankton (marine bacteria), and small animals (zooplankton), play a critical role in biogeochemical cycles.

Other important areas of study include understanding linkages between different levels of the marine **food web**, from phytoplankton all the way up to fish and marine mammals. Biological oceanographers also study factors

food web: a complex food chain, with several species at each level, so that there is more than one producer and more than one consumer of each type

that influence biological diversity within the oceans, and the importance of diversity in maintaining biological function. Understanding and mitigating the decline in **biodiversity**, such as has occurred with losses of highly diverse coral reef communities, is a primary concern of biological oceanographers. Researchers also may deal with issues that affect society such as water pollution, overexploitation of fisheries, and harmful algal blooms.

biodiversity: a measure of the variety of the Earth's species, of the genetic differences within species, and of the ecosystems that support those species

Funding Sources. Biological oceanographers compete for a limited pool of funds to do their research by submitting proposals or bidding on contracts to various scientific agencies. Research is supported by federal agencies such as the National Science Foundation, Department of Commerce, Department of Defense, National Aeronautics and Space Administration, Environmental Protection Agency, Department of Energy, Minerals Management Service, and National Research Council, as well as many other government and private agencies. Funded programs strive to advance a basic understanding of the oceans and life within, provide strategic information required for national defense, and preserve and protect the valuable resources of the oceans. SEE ALSO ALGAL BLOOMS IN THE OCEAN; BIODIVERSITY; CEPHALOPODS; CORALS AND CORAL REEFS; CRUSTACEANS; ECOLOGY, MARINE; FISH; FISHERIES, MARINE; FISHES, CARTILAGINOUS; FOOD FROM THE SEA; GEOSPATIAL TECHNOLOGIES; LIFE IN WATER; MARINE MAMMALS; OCEAN BIOGEOCHEMISTRY; OCEAN HEALTH, ASSESSING; PLANKTON; REPTILES; SUBMARINES AND SUBMERSIBLES.

Steven E. Lohrenz

Bibliography

Kunzig, Robert. *Mapping the Deep: The Extraordinary Story of Ocean Science.* New York: W. W. Norton & Co., 2000.

Lalli Carol M., and Timothy R. Parsons. *Biological Oceanography: An Introduction*, 2nd ed. Woburn, MA: Butterworth Heinemann, 1997.

Prager, Ellen J., with Sylvia A. Earle. *The Oceans.* New York: McGraw-Hill, 2000.

Sumich, James, L., and Sneed Collard. *An Introduction to the Biology of Marine Life*, 7th ed. New York: McGraw-Hill Higher Education, 1998.

Thorne-Miller, Boyce, and Sylvia A. Earle. *The Living Ocean: Understanding and Protecting Marine Biodiversity*, 2nd ed. Covelo, CA: Island Press, 1998.

Oceanography, Chemical

Oceanography is the scientific discipline that studies Earth's oceans. Chemical oceanography is concerned with the study of the dissolved elements in sea water and the ocean's numerous chemical and biochemical cycles. Topics of study include the origin and evolution of sea water, the origin of the sediment that covers the seafloor, the relationships between the myriad of chemical constituents of sea water, and the significance of changes in ocean chemistry (i.e., the influence of changing geology, including biological activity, and human-induced pollution).

Chemical oceanography can be further divided into focused areas of study. For example, marine chemistry is concerned with the composition of sea water. Marine geochemistry is additionally concerned with the chemistry of the precipitated rocks and sediment found on the ocean floor. Additionally, marine biogeochemistry is concerned with the role of organisms (particularly microorganisms) in the alteration or formation of geological features in the oceans.

Oceanography, Chemical

Water and sediment sampling are at the heart of chemical oceanography. This deep-water sampling device is one of many designs that allows oceanographers to study processes occurring in the ocean's depths.

The study of pollutants holds a high priority among many chemical oceanographers. Runoff of sewage, oil, fuel, and agricultural chemicals into the oceans decreases sea-water quality, particularly along the coast. At a local scale (i.e., beach or ocean/estuary interface), the decreasing water quality is more easily detected. In contrast, the global effect of ocean pollution is more difficult to determine. The full scope and the significance of the pollution-related changes are currently not clear, although chemical oceanographers are involved in clarifying the interaction between the ocean water with various pollutants, and with the ocean surface and the sea floor.

Another increasingly important aspect of chemical oceanography research concerns the study of the role of oceans in the global carbon cycle. The oceans are a major source and reservoir of carbon dioxide. Too much carbon dioxide in the atmosphere traps the escape of heat, leading to increasing global temperatures ("greenhouse effect"). The role of the oceans in potential global warming remains to be clarified.

Many elements are soluble in sea water, the oceans are major reservoirs of these elements. The study of trace elements (such as mercury and arsenic that are usually present in nature at very low levels) in sea water is important in the understanding of cycling of these elements between inorganic and organic processes. For example, naturally occurring elements such as mercury and arsenic are toxic to humans in high concentrations, and a deeper understanding of how the oceans contribute to potential human exposure (e.g., through consumption of mercury-laden fish) is gaining research importance.

Precise elemental studies may require sophisticated equipment and ultra-clean sampling containers. Obtaining high-quality results can be time-consuming and difficult, but the results have proved significant. Chemical oceanographers, for example, were among the scientists who first discovered and unraveled the unique ecosystem of hydrothermal vents that are present at the extremely cold, lightless bottom of the ocean floor.

A related area of chemical oceanography is concerned with the speciation of trace metals in ocean water. Some metals exist in a number of different forms, or species. Metals such as manganese, iron, nickel, and zinc form certain chemical species when organisms utilize the metals. Thus, the discovery of an abundance of such metal species is a clue to the presence of life in the ocean system under study. SEE ALSO CARBON DIOXIDE IN THE OCEAN AND ATMOSPHERE; HOT SPRINGS ON THE OCEAN FLOOR; OCEAN BIOGEOCHEMISTRY; OCEANOGRAPHY, GEOLOGICAL; RADIONUCLIDES IN THE OCEAN; TRACERS OF OCEAN-WATER MASSES.

Brian D. Hoyle and K. Lee Lerner

Bibliography

Donat, John R., and Kenneth W. Bruland. "Trace Elements in the Oceans." In *Trace Elements in Natural Waters*. Boca Raton, FL: CRC Press, 1995.

Libes, Susan M. *An Introduction to Marine Biogeochemistry*. New York: John Wiley & Sons, 1992.

Oceanography from Space

The use of space satellite data for ocean observations allows marine scientists to view biological, chemical, and physical interactions within the oceans on regional and global scales. Satellite studies have revolutionized our ideas of how the ocean works. Satellite sensors measure a myriad of different phenomena including: sea surface temperature, surface wind, ocean color and productivity, ocean height, tides, and currents.

Altimeter Data

Several different instruments are used to collect oceanographic measurements. For example, the TOPEX/Poseidon satellite measures the height of the sea surface using an **altimeter**. The altimeter flies aboard the satellite at approximately 1,335 kilometers (830 miles) above Earth and emits a radar pulse that reflects off the sea surface. Because the speed of the pulse and location of the satellite are known, scientists can calculate the height of the sea surface.

In addition, the strength of the radar signal depends on the size of the surface ripples, which are in turn related to the wind speed, allowing the wind speed to be calculated. Because currents are detectable as slopes in the sea surface, the world's ocean currents can also be identified and monitored.

Altimeter data is also used for identifying the **topography** of the seafloor. For example, when there is a topographic high, such as a mountain on the seafloor, then there is a related topographic high or "mountain" in the sea level. This seafloor feature can be a subsurface **seamount**, or it may be a local increase in density in the Earth's crust.

Radiometer Data

A radiometer can be used to collect surface temperature data and ocean color data. A radiometer measures the amount of the Sun's visible light and infrared radiation reflected off the ocean. Examples of radiometers include the Coastal Zone Color Scanner (CZCS) and SeaWiFS (Sea-viewing Wide Field-of-view Sensor).

The temperature of the sea surface can be calculated using the infrared portion of the data. Because currents and water masses vary considerably in

altimeter: an instrument that determines height above ground surface, especially one mounted in an aircraft or satellite and incorporating a barometer or radar device

topography: the shape and contour of a surface, especially the land surface or ocean-floor surface

seamount: an isolated conical submarine mountain rising 1,000 meters (3,280 feet) or more above the sea floor; most form as submarine volcanoes at spreading centers and are transported to the deep ocean by plate movement

SATELLITE PHOTOGRAPHS

This encyclopedia contains several photographs taken from satellites. Entries with notable images include:

- "Algal Blooms in the Ocean"
- "Bays, Gulfs, and Straits"
- "Geospatial Technologies"
- "Marginal Seas"
- "Microbes in the Ocean"
- "Ocean Mixing"
- "Ocean-Floor Sediments"

Oceanography from Space

This view from the space shuttle *Columbia* shows silt running into the sea from the Mahakam River in Borneo, Indonesia, and the delta that has formed as a result. The delta is the roughly triangular-shaped landmass extending from where the river branches into many distributaries, out to the coastal area. The feathery areas seaward of the delta are very fine-grained sediments that are being transported away by coastal ocean currents.

plankton: an assemblage of small, often microscopic aquatic organisms encompassing aquatic plants (phytoplankton) and aquatic animals (zooplankton) that float or drift passively with water currents, having no or very limited powers of locomotion; the term "planktonic" describes any floating or drifting organism, including large plants and animals

temperature, this data is particularly useful in observing currents and circulation processes.

Eddies are one feature in particular that can be identified using an infrared radiometer. These are generated by large-scale currents, such as the Gulf Stream. Eddies can affect the distribution of marine life and can last for many years before dissipating. Locating such eddies and studying their dynamics can help researchers track pollution such as oil spills and determine where marine life may be located.

Ocean Color. Ocean color can be determined by measuring the portions of the visible spectrum reflected from the ocean surface. It can indicate a number of things to an oceanographer, such as amount of **plankton** and amount of vegetation. The color of the ocean changes slightly, from a bright blue to a dark blue or black when plankton float freely or concentrate in areas. These concentrations are called blooms. These colors can indicate to scientists the productivity of the oceans and potential for greater amounts of wildlife since plankton are the basis of the marine food web and without plankton all marine life would suffer.

Satellite data have become accurate and dependable enough that they is now integrated with other forms of marine data collection. In addition, satellite data provide a large-scale view of ocean dynamics that otherwise would be unavailable. What has emerged is exciting new information about vast areas of previously unstudied open water. SEE ALSO ALGAL BLOOMS IN the Ocean; Ocean Currents; Ocean-Floor Bathymetry; Ocean Mixing; Weather and the Ocean.

Alison Cridland Schutt

Bibliography

Mellor, George L. *Introduction to Physical Oceanography*. New York: Springer Verlag, 1996.

Ross, David A. *Introduction to Oceanography*, 5th ed. New York: HarperCollins College Publishers, 1995.

Sandwell, David T. "Geophysical Applications of Satellite Altimetry." *Reviews of Geophysics*. supplement (1990):132-137.

Oceanography, Geological

Geological oceanography is the study of Earth beneath the oceans. A geological oceanographer studies the topography, structure, and geological processes of the ocean floor to discover how the Earth and oceans were formed and how ongoing processes may change them in the future. Geological oceanography is one of the broadest fields in the Earth Sciences and contains many subdisciplines, including geophysics and **plate tectonics**, petrology and sedimentation processes, and micropaleontology and stratigraphy. Geological oceanographers study many features of the oceans such as rises and ridges, trenches, seamounts, abyssal hills, the oceanic crust, sedimentation (clastic, chemical, and biological), erosional processes, **volcanism**, and seismicity.

plate tectonics: the theory that the Earth's lithosphere can be divided into a few large plates that are slowly moving relative to one another

volcanism: the activity and phenomena of volcanoes on Earth

Tools and Techniques

Many different tools are used by geological oceanographers. For example, the structure and topography of the ocean floor are studied through the use of satellite mapping, which measures the level of the ocean surface to estimate the shape of the ocean floor. A detailed study in 1978 by the National Oceanic and Atmospheric Administration (NOAA) involved the Seasat satellite which produced frequent, short pulses of microwave radiation to measure the level of the surface of the sea with great accuracy (within 3 centimeters or 1 inch). Underwater mountains and valleys cause subtle variations in Earth's gravitational field. The stronger gravity near high massive formations attracts more water molecules, raising the level of the ocean slightly. Similarly valleys on the ocean floor produce weaker areas of gravity, so the level of the ocean will be lower. Using this technique, a complete survey of the ocean floor was accomplished.

Seismic techniques are used to measure the subsurface structure. This type of study is carried out by teams of two ships: one fires an explosive in the water and the other uses sensitive instruments to record the sound waves as they reach the second ship. Some waves travel directly to the second ship; others travel to the ocean floor, are refracted (bent) within the layers of sediment, and then travel to the second ship. By measuring the time it takes for the energy to arrive and the distance between the boats, the thickness of sediments and other features can be determined. Structures may also be analyzed by studying natural earthquake waves that travel through deeper oceanic rocks and may be recorded at stations around the world.

Oceanography, Geological

Geological oceanography and coastal geology are closely related yet distinct disciplines. Geological oceanographers study the rocks and structure of the ocean floor, the ocean-floor sediment that covers them, and the processes that formed them. Coastal geologists focus on these structures and processes in a coastal environment (shown here).

Other ways oceanic sediments are studied are by dredging processes and deep-sea exploration projects such as the Ocean Drilling Program (ODP), which obtains samples of seafloor sediment from the entire world. The ocean sediments are found to consist of rock particles and organic remains whose compositions depend on depth, distance from continents, and local variations such as submarine volcanoes or high biological activity. For example, clay minerals, which are formed by the weathering of continental rocks, are carried out to sea by rivers and are usually abundant in the deep sea. Thick deposits of weathered rock material, commonly coarser than deep-sea clays, are often found near the mouths of rivers and on the continental shelves.

Programs. Universities and private individuals, as well as governments, have established institutions and programs for the study of the ocean; today about 250 such entities exist. Examples include the Ocean Drilling Program, funded by the U.S. National Science Foundation and 22 international partners to conduct research into the history of ocean basins and the nature of the crust beneath the ocean floor; Sea Grant, a university-based program that receives support from the U.S. Department of Commerce through NOAA and that studies all aspects of the ocean, including geological, chemical, and physical processes; and RIDGE (Ridge Interdisciplinary Global Experiments), a program funded by the National Science Foundation whose the primary objective is to understand the geological, chemical, biological, and physical oceanographic interactions between the oceans and the mid-ocean ridge system, which forms the plate tectonic boundaries between diverging plates.

Notable U.S. oceanographic institutions include: Scripps Institution of Oceanography in San Diego, California; Woods Hole Oceanographic Institute at Woods Hole, Massachusetts; Lamont–Doherty Geological Observatory of Columbia University in Palisades, New York; the University of Miami's Rosenstiel School of Marine and Atmospheric Science in Miami, Florida; the University of Washington in Seattle, Washington; and Oregon State University in Corvallis, Oregon. SEE ALSO COASTAL OCEAN; HOT SPRINGS ON THE OCEAN FLOOR; MID-OCEAN RIDGES; OCEAN BASINS; OCEAN-FLOOR BATHYMETRY; OCEAN-FLOOR SEDIMENTS; PLATE TECTONICS; TSUNAMIS; VOLCANOES, SUBMARINE.

Alison Cridland Schutt

Bibliography

Ross, David A., *Introduction to Oceanography*. Boston, MA: Addison-Wesley Educational Publishers, 1997.

Oceanography, Physical

Although oceanography is the scientific study of the ocean, the subdiscipline of physical oceanography is principally concerned with the study of the structure and movement of water in the oceans. Physical oceanographic studies utilize a number of scientific specialties, and studies can encompass a diversity of technologies—from echo-sounding determinations of seafloor structure and seismic studies of movements in oceanic crust to satellite estimations of current flow based on radar reflections and thermal imaging.

Physical oceanography studies the many factors that influence the movement of ocean waters. Wind can push surface water, and the gravity fields of the Sun and Moon continually exert gravitational tugs that push and pull massive amounts of water in tidal cycles. Earth's rotation also contributes to the physical movement of water, as do density and temperature differences between oceans or between layers of water within the same ocean.

Understanding ocean movement is important as some 70 percent of the planet's surface is covered by ocean water; increased oceanographic knowledge will lead to better understanding of such diverse topics as flooding, current movement, fish migration, and **remediation** of ocean damage such as oil spills. Physical oceanography also is important in the

remediation: the cleanup, through a variety of methods, to remove or contain a toxic spill or hazardous materials from a contaminated site

Oceanography, Physical

These physical oceanographers studying the Ross Sea in Antarctica derived a makeshift depth profiler by mounting a recording conductivity–temperature–depth (CTD) device on a trainwheel.

understanding of the climate of Earth, the erosion of coastlines, and how the world's oceans both provide and store vital nutrients and compounds such as carbon dioxide.

Instrumentation is vital to the study of physical oceanography. Determining wave height, water temperature, or current flow is impossible without a means of measurement. For example, the height of the waves pounding onto the shore can be measured using conventional measuring instruments. Away from the shore, the wave height in deep water is measured using an instrument called a tide gage. The gage is immersed in the water and measures the weight of the water on top. A higher weight, for example, means more water and a higher wave that—depending on other physical factors—might eventually produce a higher wave at the shore.

Temperature determination interments vary depending on the location and the length of time the measurement is being recorded. Dangling a thermometer from a boat manually accomplishes a one-time surface measurement. Measurement of deeper-water temperature or the temperature over time, however, often utilizes a thermometer attached to a deep-water device or buoy, makes indirect measurements based on the speed of sound in water. Buoys, for example, are used in the Pacific Ocean to chart the water temperature fluctuations associated with the **El Niño** phenomenon.

Physical oceanographers routinely use satellites to gain measurements over large distances and areas. Instruments on satellites can measure ocean height and thereby allow estimations of ocean surface temperature. Spectral studies can detect the presence of surface organic material such as algae. Other instruments can allow estimates of wave height and wind speed.

Measurements in physical oceanography occur over a large range of scales. For instance, measurements of ocean current can vary from a few centimeters to the entire globe, and measurements of current variability show movement occurring over a few seconds to estimates spanning thousands of years.

Funding for physical oceanographic research typically comes from governments: directly, through agencies like the National Oceanic and Atmospheric Association (NOAA), or indirectly, through funding of university or other institutional research. Many universities have departments where physical oceanography research is carried out (e.g., San Diego's Scripps Institution of Oceanography, which is affiliated with the University of California). Additionally, entire institutions devoted to oceanographic research exist around the world (e.g., Woods Hole Oceanographic Institution in Woods Hole, Massachusetts). Funding also comes from private sources, such as oil companies. SEE ALSO MOORINGS AND PLATFORMS; OCEAN CURRENTS; OCEAN MIXING; OCEANOGRAPHY FROM SPACE.

Brian D. Hoyle and K. Lee Lerner

El Niño: an occasional warming of sea-surface temperatures in the equatorial Pacific off the coast of South America

SATELLITE OCEANOGRAPHY

The launching in 1978 of Seasat, the first oceanographic satellite, revolutionized measurements of physical properties of the ocean. Within a few years, sea-surface temperature, wave height, variations in sea surface contours, ice cover, chlorophyll content, and other parameters could be measured and reported almost instantly from satellites.

Bibliography

Pickard, George L. and William J. Emery. *Descriptive Physical Oceanography: An Introduction*, 5th ed. New York: Pergamon Press, 1990.

Ross, David A. *Introduction to Oceanography*. New York: HarperCollins College Publishers, 1995.

Oceans, Polar

The Arctic Ocean and the Southern Ocean (the ocean around Antarctica) have different characteristics than the rest of the world's oceans in terms of circulation, formation of bottom water, convergent and divergent water masses, productivity, ice cover, and biological diversity. In addition, polar oceans are inextricably linked to climate change, and hence continued to be studied in the context of global warming.

Antarctic Ocean

The water surrounding the Antarctic continent is often called the Antarctic or Southern Ocean. The Antarctic Circumpolar Current (ACC), at 0 to 200 meters (0 to 650 feet) in depth, is the dominant surface-water circulation pattern in this polar region. ✸ The ACC, a **geostrophic current**, is influenced by the prevailing Antarctic wind patterns and is constrained by the adjacent landmass, the Antarctic continent. The ACC, also called the West Wind Drift, flows from west to east around the Antarctic landmass.

✸ See "Ocean Currents" for depictions of the ocean surface currents and deep currents.

geostrophic current: an ocean current that flows along a line of equal dynamic topography on the sea surface, oriented so that the high topographies are on the right in the Northern Hemisphere and on the left in the Southern Hemisphere

In polar oceans, seasonal melting of sea ice creates gaps in the ice cover, thereby allowing more sunlight to penetrate the surface waters. The favorable light conditions stimulate phytoplankton growth and yield surges in primary productivity.

To the north of the ACC is the warmer, more saline (salty) Subantarctic Surface Water. Between the ACC and the Antarctic landmass is the East Wind Drift, a counter-current that flows from east to west.

Antarctic Water Masses. Antarctic Bottom Water (AABW) has average salinity, temperature, and density values of 34.65, $-0.5°C$ (31°F), and 1.03 gram per cubic centimeter, respectively. The majority of the AABW (the Earth's densest water mass) is formed in the Weddell Sea during the winter when the sea-ice formation process produces cold **brine** that sinks and mixes with Antarctic Circumpolar Current water. This water mixture subsequently settles along the **continental shelf**, slowly spreading along the seafloor in the form of a northward-flowing sheet below the North Atlantic Deep Water. The Antarctic Deep Water (AADW) which is formed in less extreme latitudes and is less salty and warmer than AABW, flows northward near the surface until it reaches the Antarctic Polar Front Zone, where the AADW sinks and continues to flow northward beneath the warmer, less dense North Atlantic Deep Water.

The Antarctic Polar Front Zone encircles the Antarctic continent between 50°S and 60°S latitude. Antarctic Intermediate Water (AAIW) includes water from the Antarctic and is formed by mixing below the surface in the Antarctic Polar Front Zone. The AAIW flows northward beneath the South Atlantic Central and the South Pacific Central Waters and above

brine: water containing a higher concentration of dissolved salts than normal sea water (which contains approximately 35 parts per thousand); produced in oceans through the evaporation or freezing of sea water, or in groundwater through extensive reaction with bedrock minerals

continental shelf: the relatively flat, submerged natural platform, about 1-degree slope, that extends seaward from the beach for about 70 kilometers (45 miles), with water depth up to 130 meters (425 feet) maximum, and ends where the slope and water depth increase

the North Atlantic Deep Water until it converges with the Arctic Intermediate Water and North Pacific Intermediate Water.

Arctic Ocean

The Arctic Ocean is divided into the Eurasian and Canadian basins by the Lomonosov Ridge, a **bathymetry** feature that extends from Greenland past the North Pole to Siberia. Arctic Ocean surface water (0–200 meters or 0–650 feet) predominately flows counterclockwise in the Eurasian basin and clockwise in the Canadian basin. The general direction of Arctic Ocean water movement below the surface is counterclockwise.

The primary site of water export from the Arctic to the Atlantic Ocean is the Fram Strait, which has a **sill** depth of 2,600 meters (8,530 feet) and is located between Greenland and Spitsbergen. Water from the Atlantic is introduced into the Arctic by the Norwegian current that flows northeastward passing between Norway and Spitsbergen.

The Bering Strait with a shallow sill depth of 45 meters is located between Alaska and Siberia and serves as the water exchange site for the Arctic Ocean and the Bering Sea. The amount of water exchanged through the Bering Strait is approximately one-tenth of that which is exchanged through the Atlantic Ocean pathways.

Arctic Intermediate Water, which is actually formed in the subarctic region, flows southward beneath the North Atlantic Central Water, meeting the Antarctic Intermediate Water. Deep (Bottom) Water exchange between the Arctic, Atlantic, and Pacific Oceans is restricted by seafloor bathymetry.

Arctic Water Masses. There are three distinct marine water masses located within the Arctic Ocean: the Arctic Surface Water (0–200 meters); the Atlantic Water (200–900 meters or 650–2,950 feet); and the Arctic Deep Water (900 meters–seafloor). The Arctic Surface Water is divided into three layers: the surface, subsurface, and lower surface layers. Each of these water layers has distinct salinity and temperature characteristics.

The Atlantic Water (AW) is located below the Arctic Surface Water (ASW) and above the Arctic Deep Water (ADW). The average temperature (3°C [37.4°F]) of the AW is warmer than both that of the ASW (−1.9°C to −1.0°C [28.6°F to 30.2°F]) and the ADW (−0.8°C to 2.0°C [30.6°F to 35.6°F]). The AW has a higher salinity range (34.8–35.1) than that of the ASW (28.0–34.0). The ADW, with a salinity range of 34.9 to 34.99, represents approximately 60 percent of the Arctic Ocean total water volume and is comprised of the Norwegian Sea, Greenland Sea, Eurasian basin, and the Canadian basin deep waters.

Ice and Productivity

In the polar oceans, ice exists primarily in the form of either icebergs (glacier fragments) or **sea ice**. Sea ice, the major form of ice in the polar oceans, is formed by a sequence of events that occur once suitably cold (−1.8°C [28.8°F]) conditions exist to freeze sea water.

After sea water begins to freeze, frazil ice (small ice crystals) is formed. Frazil ice eventually accumulates to form grease ice (surface ice slicks), which in turn accumulates to form small ice chunks and **floes** that aggregate

bathymetry: the science of measuring the depths and underwater topography of seas, oceans, lakes, and reservoirs

sill: the shallow area that separates coastal bays or marginal seas from the adjacent oceans or that separates two basins from one another

sea ice: a general term for any form of ice found at sea which has originated from the freezing of sea water; includes types of ice such as grease ice, frazil ice, pancake ice, and pack ice

floe: a contiguous piece of ice on the surface of water (e.g., rivers, seas)

together to form a solid ice cover. Over time this solid ice cover will thicken into sea ice. Sea ice is present in the Arctic Ocean in three forms: the Polar Ice Cap, pack ice, and fast ice. Unlike Antarctica, the Arctic Ocean has no central landmass.

Polynyas. Polynyas are large areas of open water surrounded by sea ice. Polynyas can range in size from a few square kilometers to more than 50,000 square kilometers (more than 19,000 square miles). Polynyas, which are of biological and physical interest, are produced by either the removal of sea ice or by the prohibition of sea-ice formation.

Nearshore polynyas are generally formed by strong surface winds blowing sea ice offshore, leading to sea-ice removal, surface-water exposure, and, in some cases, new production of sea ice. Polynyas also exist in the open ocean as a result of convection, a process that allows warmer subsurface waters to rise above sinking colder surface waters.

Recurrent polynyas play significant roles in the marine **ecosystem** by triggering early and intense **phytoplankton** production. Additionally, polynyas serve as wintering grounds for marine mammals.

Primary Productivity. **Primary productivity** is affected by the availability of sunlight, carbon dioxide, and **inorganic** nutrients (nitrates, phosphates, and trace elements). In the marine environment, nutrients are recycled from phytoplankton to animals to decomposers (such as bacteria) before returning to phytoplankton. One of the most effective pathways for nutrients to be re-incorporated into phytoplankton is through the **upwelling** of nutrient-rich deep waters in which the bodies of marine plants and animals have previously decomposed.

In the polar oceans, phytoplankton blooms (explosive population growth) occur during the summer months as a result of favorable light conditions which lead to short-term increased primary productivity. During these months, the Antarctic Ocean's upwelling zone exhibits some of the Earth's highest primary productivity.

In the Arctic and Antarctic Oceans, the sea-ice formation and melt processes also play important roles in primary productivity. Frazil ice is mixed with surface and subsurface water, entrapping phytoplankton between ice crystals that are eventually incorporated into pack ice. The phytoplankton (mainly **diatoms**) will grow within the sea-ice brine channels, causing the pack ice to appear greenish-brown. During the yearly ice melt process, the diatoms are released back into the water, resulting in local increased primary productivity.

Biological Diversity and Biota

Marine **biodiversity** generally decreases towards high latitudes, reaching a minimum in the polar oceans. The few marine species that exist in the polar oceans tend to grow at slower rates, live longer lives, attain larger sizes, and have fewer offspring than their tropical counterparts. In addition, the few marine species that are able to withstand the relatively harsh polar conditions tend to exist as larger populations than their more diverse tropical counterparts.

Unlike the Antarctic, the Arctic Ocean is dominated by shallow **marginal seas**, a major factor that has resulted in the different **biota** spatial distributions in these two oceans. While the majority of the Antarctic Ocean

---

ecosystem: the community of plants and animals within a water or terrestrial habitat interacting together and with their physical and chemical environment

phytoplankton: microscopic floating plants, mainly algae, that live suspended in bodies of water and that drift about because they cannot move by themselves or because they are too small or too weak to swim effectively against a current

primary productivity: the rate at which biomass is produced by photosynthetic and chemosynthetic organisms in the form of organic substances

inorganic: an element, molecule, or substance that did not form as the direct result of biologic activity

upwelling: in marine environments, the movement of nutrient-rich water from great depths to the ocean surface

diatom: any of the microscopic unicellular or colonial algae constituting the class *Bacillarieae* that have a silicified cell wall, which persists as a skeleton after death and forms kieselguhr (loose or porous diatomite)

biodiversity: a measure of the variety of the Earth's species, of the genetic differences within species, and of the ecosystems that support those species

marginal sea: a semi-closed sea attached to a continent and formed during rifting and early spreading

biota: the plant and animal life of a region or ecosystem, as in a stream, lake, or ocean

Polar habitats support populations of diving birds such as penguins and puffins, and marine mammals such as whales, seals, and polar bears. These animals are more visible than the invertebrate and microscopic communities found in the water column, on the seafloor, and in sea ice.

biota reside and feed within the water column, the Arctic Ocean biota reside and feed throughout the water column and along the seafloor.

Among the larger populations of biota present in the Antarctic are at least five types of seals (crabeater, elephant, leopard, ross, weddell), six varieties of penguins (adelie, chinstrap, emperor, gentoo, macaroni, king), and five whale species (blue, sperm, orca, mink, southern bottlenosed) in addition to various seabirds, squid, fish, krill, copepods, and diatoms.

The major sea mammals associated with the Arctic Ocean are whales (beluga, orca, bowhead, California gray, narwhal), polar bears, sea otters, seals (ringed, ribbon, bearded, spotted), and walruses. Arctic birds such as the tufted puffin, laysan albatross, and spectacled eider rely on the Arctic Ocean as a primary food source. These birds feed by diving into the water for fish (e.g., arctic cod), crustaceans, and/or mollusks.

In the Arctic Ocean, the dominant types of phytoplankton and **zooplankton** are diatoms and copepods, respectively. Water-column productivity of the shallow Arctic marginal seas encourages the growth of productive benthic (bottom) communities that include mollusks, polychaetes, brittle stars, and amphipods, which support bottom feeding by the spectacled eider, walrus, bearded seal, sea otter, and the California gray whale. SEE ALSO CLIMATE AND THE OCEAN; GLOBAL WARMING AND THE OCEAN; ICE AT SEA; MARINE MAMMALS; MARGINAL SEAS; OCEAN CURRENTS; OCEAN-FLOOR SEDIMENTS; RADIONUCLIDES IN THE OCEAN; SEA WATER, FREEZING OF.

zooplankton: microscopic animals that live suspended in bodies of water and that drift about because they cannot move by themselves or because they are too small or too weak to swim effectively against a current; composed primarily of protozoans, microcrustacea (copepods, cladocera, rotifers) and larval stages of certain invertebrates

Ashanti Johnson Pyrtle

Bibliography

Pickard, George L., and William J. Emery. *Descriptive Physical Oceanography: An Introduction*, 5th ed. New York: Pergamon Press, 1990.

Internet Resources

Polar Programs. National Science Foundation. <http://www.nsf.gov/home/polar/>.

Teachers Experiencing Antarctica and the Arctic. Rice University. <http://tea.rice edu/>.

Oceans, Tropical

Tropical oceans encircle Earth in an equatorial band between the Tropic of Cancer (23.5° North latitude) and the Tropic of Capricorn (23.5° South latitude). *See "Climate and the Ocean" for an illustration of the tropical zones.* The central portions of the Pacific and Atlantic Oceans and most of the Indian Ocean lie in the tropics. The warm tropical oceans play a critical role in regulating Earth's climate and large-scale weather patterns. Much of the planet's biological diversity resides in the tropics, and the global distribution of species and ecosystems depends on oceanographic and atmospheric processes that occur in the equatorial oceans.

Heat from the Sun drives global circulation of Earth's oceans and atmosphere. Much of that critical solar radiation initially falls on the tropics, where the Sun lies almost directly overhead for the entire year. The water temperature of tropical oceans thus typically exceeds 20°C (68°F) and stays relatively constant throughout the year. Particularly intense radiation directly over the equator evaporates seawater and forms a mass of very warm, humid tropical air that subsequently rises and cools as it flows north and south. Because cool air holds less moisture than warm air, the water vapor quickly condenses into clouds and falls as precipitation. Heavy, warm, year-round rains are a hallmark of Earth's tropical regions. Fragile, biologically diverse ecosystems such as rainforests and coral reefs thrive in the warm, wet tropics.

Uneven heating of the sea surface between the tropics and the poles creates heat-driven convection currents in the atmosphere and oceans. *See "Climate and the Ocean" for an illustration of circulation zones.* Vertical circulation in the tropical oceans also affects the distribution of heat and biological nutrients throughout the global ocean. In general, surface water sinks at downwellings where surface currents flow toward a continental coastline, or where two surface currents converge. Deep ocean water rises to the surface at upwellings where surface currents flow away from land, or where surface currents diverge. Tropical downwellings transfer heat and nutrients to the deep-ocean circulation system. At tropical upwellings,

White sand, palm trees, and warm, shallow water comprise the classic image of a tropical beach. The brilliant turquoise hue of clear tropical waters is largely the result of the selective scattering and absorption of visible light.

cool, oxygen-rich and nutrient-rich deep water supports abundant marine life. Because normal tropical currents flow from east to west, downwellings often occur along the east coasts of tropical continents, and upwellings are common along their west coasts. In the Pacific Ocean, for example, an upwelling off the west coast of South America usually feeds extremely productive fisheries of coastal Peru and Ecuador, and a downwelling in Polynesia forces warm, oxygen-depleted water into the deep ocean.

Tropical upwellings support huge populations of microscopic plants and animals called phytoplankton and zooplankton. Plankton, in turn, feed many species of fish and other marine life, and humans who depend on fish for food. Tropical fisheries account for about half of the world's fish catch, even though tropical oceans represent only 0.01 percent of Earth's ocean volume.

Coral reefs are another well-recognized feature of tropical oceans. The seas surrounding tropical islands and low-latitude continental shelves away from major river deltas are ideal for coral reef formation. Over millennia, very large reefs have formed in the Caribbean Sea, and especially in the southwest Pacific Ocean. For example, the Great Barrier Reef of northeastern Australia covers thousands of square kilometers. SEE ALSO CARBON DIOXIDE IN THE OCEAN AND ATMOSPHERE; CLIMATE AND THE OCEAN; CORALS AND CORAL REEFS; EL NIÑO AND LA NIÑA; OCEAN CURRENTS; OCEAN MIXING; PLANKTON; WEATHER AND THE OCEAN.

Brian D. Hoyle and Laurie Duncan

Bibliography

Open University Course Team. *Ocean Circulation.* Oxford, U.K.: Pergamon Press, 1993.

Press, Frank, and Raymond Siever. *Understanding Earth.* New York: W.H. Freeman and Company, 2001.

Ross, David A. *Introduction to Oceanography.* New York: Harper Collins College Publishers, 1995.

Turk, Daniela, Michael J. McPhoden, Antonio J. Busalacchi, and Marlon R. Lewis. "Remotely Sensed Biological Production in the Equatorial Pacific." *Science* 293 (July 20, 2000): 471–474.

Ogallala Aquifer

The Ogallala **Aquifer** occupies the High Plains of the United States, extending northward from western Texas to South Dakota. The Ogallala is the leading geologic formation in what is known as the High Plains Aquifer System. The entire system underlies about 450,000 square kilometers (174,000 square miles) of eight states. Although there are several other minor geologic formations in the High Plains Aquifer System, such as the Tertiary Brule and Arikaree and the Dakota formations of the Cretaceous, these several units are often referred to as the Ogallala Aquifer.

Characteristics of the Ogallala

The Ogallala is composed primarily of unconsolidated, poorly sorted clay, silt, sand, and gravel with **groundwater** filling the spaces between grains below the **water table**. The Ogallala was laid down about 10 million years ago by **fluvial** deposition from streams that flowed eastward from the Rocky Mountains during the Pliocene epoch.✳ Erosion has removed the deposits

aquifer: a water-saturated, permeable, underground rock formation that can transmit significant quantities of water under ordinary hydraulic gradients to wells and springs

groundwater: generally, all subsurface (underground) water, as distinct from surface water, that supplies natural springs, contributes to permanent streams, and can be tapped by wells; specifically, the water that is in the saturated zone of a defined aquifer

water table: the upper surface of the zone of saturation in an unconfined aquifer below which all voids in rock, sediment, and other geologic materials are saturated (completely filled) with water

fluvial: pertaining to the action of a river, stream, or flood flow, as in fluvial processes of erosion or the deposition of alluvium

✳ **See the frontmatter of this volume for a geologic timescale.**

Ogallala Aquifer

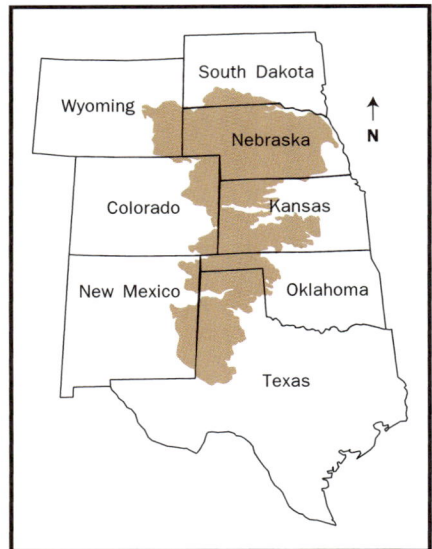

The Ogallala Aquifer (shaded area) is in a state of overdraft owing to the current rate of water use. If withdrawals continue unabated, the aquifer could be depleted in only a few decades.

recharge: the process by which precipitation infiltrates below the surface and replenishes an aquifer

unconfined: describes an aquifer whose upper surface is the water table that is free to fluctuate under atmospheric pressure

semiarid: a climate or region where moisture is normally greater than under arid conditions but still limits the growth of most crops; either dryland farming methods or irrigation generally are required for crop production

saturated thickness: the thickness of the portion of the aquifer in which all pores, or voids, are filled with water; specifically, in a confined aquifer, this is generally the aquifer thickness, whereas in an unconfined aquifer, this is the distance between the water table and the base of the aquifer

groundwater mining: the withdrawal of groundwater at a rate exceeding natural recharge, resulting in a permanent lowering of the water table

between the mountains and the existing western boundary of the Ogallala, so there is no longer water **recharge** being received from the Rockies.

The Ogallala is an **unconfined** aquifer, and virtually all recharge comes from rainwater and snowmelt. As the High Plains has a **semiarid** climate, recharge is minimal. Recharge varies by amount of precipitation, soil type, and vegetational cover and averages less than 25 millimeters (1 inch) annually for the region as a whole. In a few areas, recharge from surface water diversions has occurred. Groundwater does flow through the High Plains Aquifer, but at an average rate of only 300 millimeters (12 inches) per day.

The depth to the water table of the Ogallala Aquifer varies from actual surface discharge to over 150 meters (500 feet). Generally, the aquifer is found from 15 to 90 meters (50 to 300 feet) below the land surface. The **saturated thickness** also varies greatly. Although the average saturated thickness is about 60 meters (200 feet), it exceeds 300 meters (1,000 feet) in west-central Nebraska and is only one-tenth that in much of western Texas. Because both the saturated thickness and the areal extent of the Ogallala Aquifer is greater in Nebraska, the state accounts for two-thirds of the volume of Ogallala groundwater, followed by Texas and Kansas, each with about 10 percent.

Using and Protecting the Ogallala

The Ogallala Aquifer, whose total water storage is about equal to that of Lake Huron in the Midwest, is the single most important source of water in the High Plains region, providing nearly all the water for residential, industrial, and agricultural use. Because of widespread irrigation, farming accounts for 94 percent of the groundwater use. Irrigated agriculture forms the base of the regional economy. It supports nearly one-fifth of the wheat, corn, cotton, and cattle produced in the United States. Crops provide grains and hay for confined feeding of cattle and hogs and for dairies. The cattle feedlots support a large meatpacking industry. Without irrigation from the Ogallala Aquifer, there would be a much smaller regional population and far less economic activity.

Because of the Ogallala, the High Plains is the leading irrigation area in the Western Hemisphere. Overall, 5.5 million hectares (nearly 13.6 million acres) are irrigated in the Ogallala region. The leading state irrigating from the Ogallala is Nebraska (46%), followed by Texas (30%) and Kansas (14%).

Slowing the Rate of Depletion.
The Ogallala Aquifer is being both depleted and polluted. Irrigation withdraws much groundwater, yet little of it is replaced by recharge. Since large-scale irrigation began in the 1940s, water levels have declined more than 30 meters (100 feet) in parts of Kansas, New Mexico, Oklahoma, and Texas. In the 1980s and 1990s, the rate of **groundwater mining**, or overdraft, lessened, but still averaged approximately 82 centimeters (2.7 feet) per year.

Increased efficiency in irrigation continues to slow the rate of water-level decline. State governments and local water districts throughout the region have developed policies to promote groundwater conservation and slow or eliminate the expansion of irrigation. Generally, management has emphasized planned and orderly depletion, not sustainable yield. Depletion re-

Center-pivot sprinklers are among the irrigation methods used in the High Plains. Large quantities of groundwater pumped from the Ogallala Aquifer allows these semiarid western lands to yield abundant harvests.

sults in reduced irrigation in areas with limited saturated thickness and increased energy cost in all areas as the depth to water increases.

The average **specific yield** for the High Plains Aquifer is about 0.15. This means that only 15 percent of all the water available in the aquifer can be recovered using irrigation pumps, while the rest remains unused and locked up in the **unsaturated zone**. Groundwater depletion problems could be forestalled if this presently nonrecoverable water could be forced to the **saturated zone**. One experimental means of accomplishing this is by injecting air into the unsaturated zone, which breaks down **capillary action** and permits the movement of water down to the saturated zone. Air injection experiments have shown positive results for very localized areas. However, the widespread applicability of this technology has not yet proven effective.

Reducing Contamination. Groundwater contamination in the Ogallala became an issue in the 1990s. In its natural state, the High Plains Aquifer is, for the most part, of high quality. The water is generally suitable for domestic use, stock watering, and irrigation without filtration or treatment. Surveys of groundwater samples have detected traces of pesticides and nitrates. Sources include irrigated agriculture and confined livestock feeding operations. The **percolation rates** of contaminants from the surface to the water table have not been established in the areas where polluted water has been found.

specific yield: the volume of water released by gravity from a unit area (square meter or square foot) of an unconfined aquifer when the water table drops a unit length (meter or foot); expressed as a fraction, for example, if 0.2 cubic meter of water will drain from 1 cubic meter of aquifer sand, the specific yield is 0.2 (20 percent)

unsaturated zone: the zone between the ground surface and the water table that contains both air and water

saturated zone: an area where pore spaces within the soil are entirely filled with water

capillary action: the action by which water is drawn around and into soil particles due to the forces of adhesion, cohesion, and surface tension acting in a liquid that is in contact with a solid

percolation rate: the rate, usually expressed as a velocity, at which water moves through saturated granular material; also applies to quantity per unit of time of such movement

Managing for the Future. The future economy of the High Plains depends heavily on the Ogallala Aquifer, the main source of water for all uses. The Ogallala will continue to be the lifeblood of the region only if it is managed properly to limit both depletion and contamination. SEE ALSO AGRICULTURE AND WATER; AQUIFER CHARACTERISTICS; GROUNDWATER; IRRIGATION MANAGEMENT; SUSTAINABLE DEVELOPMENT.

David E. Kromm

Bibliography

Kromm, David E., and Stephen E. White, eds. *Groundwater Exploitation in the High Plains.* Lawrence: University Press of Kansas, 1992.

Sophocleous, Marios, ed. *Perspectives on Sustainable Development of Water Resources in Kansas.* Lawrence: Kansas Geological Survey, 1998.

White, Stephen E., and David E. Kromm. "Local Groundwater Management Effectiveness in Colorado and Kansas Ogallala Region." *Natural Resources Journal* 35 (1995):275–307.

Oil Spills: Impact on the Ocean

Oil wastes that enter the ocean come from many sources, some being accidental spills or leaks, and some being the results of chronic and careless habits in the use of oil and oil products. Most waste oil in the ocean consists of oily stormwater drainage from cities and farms, untreated waste disposal from factories and industrial facilities, and unregulated recreational boating.

It is estimated that approximately 706 million gallons of waste oil enter the ocean every year, with over half coming from land drainage and waste disposal; for example, from the improper disposal of used motor oil. Offshore drilling and production operations and spills or leaks from ships or tankers typically contribute less than 8 percent of the total. The remainder comes from routine maintenance of ships (nearly 20 percent), **hydrocarbon** particles from onshore air pollution (about 13 percent), and natural seepage from the seafloor (over 8 percent).

hydrocarbon: a chemical compound that consist entirely of carbon and hydrogen, such as petroleum, natural gas, and coal

Prevalence during Drilling versus Transportation

Offshore oil spills or leaks may occur during various stages of well drilling or workover and repair operations. These stages can occur while oil is being produced from offshore wells, handled, and temporarily stored; or when oil is being transported offshore, either by flowline, underwater pipeline, or tanker. Of the approximately 706 million gallons of waste oil in the ocean each year, offshore drilling operations contribute about 2.1 percent, and transportation accidents (both ships and tankers) account for another 5.2 percent. The amount of oil spilled or leaked during offshore production operations is relatively insignificant.

Oil waste from offshore drilling operations may come from disposal of oil-based drilling fluid wastes, deck runoff water, flowline and pipeline leaks, or well failures or blowouts. Disposal of offshore production waste can also pollute the ocean, as can deck runoff water, leaking storage tanks, flowline and pipeline leaks, and the wells themselves. Oil spilled from ships and tankers includes the transportation fuel used by the vessels themselves or their cargos, such as crude oil, fuel oil, or heating oil.

Oil Spills: Impact on the Ocean

Over half the ocean's waste oil comes from land-based sources and from unregulated recreational boating. The heavy development in this busy California port illustrates one potential source of petroleum contamination in coastal waters. (Note dark plume in left foreground.)

Oil Spill Behavior

When oil is spilled in the ocean, it initially spreads in the water (primarily on the surface), depending on its relative density and composition. The oil slick formed may remain cohesive, or may break up in the case of rough seas. Waves, water currents, and wind force the oil slick to drift over large areas, impacting the open ocean, coastal areas, and marine and terrestrial habitats in the path of the drift.

Oil that contains **volatile organic compounds** partially evaporates, losing between 20 and 40 percent of its mass and becoming denser and more viscous (i.e., more resistant to flow). A small percentage of oil may dissolve in the water. The oil residue also can disperse almost invisibly in the water or form a thick **mousse** with the water. Part of the oil waste may sink with suspended particulate matter, and the remainder eventually congeals into sticky tar balls. Over time, oil waste weathers (deteriorates) and disintegrates by means of photolysis (decomposition by sunlight) and biodegradation (decomposition due to microorganisms). The rate of biodegradation depends on the availability of nutrients, oxygen, and microorganisms, as well as temperature.

Oil Spill Interaction with Shoreline.
If oil waste reaches the shoreline or coast, it interacts with sediments such as beach sand and gravel, rocks and boulders, vegetation, and terrestrial habitats of both wildlife and humans, causing erosion as well as **contamination**. Waves, water currents, and wind move the oil onto shore with the surf and tide.

volatile organic compounds: organic compounds that can be isolated from the water phase of a sample by purging the water sample with inert gas, such as helium, and subsequently analyzed by gas chromatography

mousse: thick, foamy oil-and-water mixture formed when petroleum is subjected to mixing with water

contamination: impairment of the quality of water or the environment by natural or human-made substances to a degree that is considered undesirable for certain uses; this term usually implies a human or environmental health threat, but some types of contamination are merely nonaesthetic rather than harmful

Oil Spills: Impact on the Ocean

Crude oil from the *Sea Empress* tanker spill coats a beach at Pembrokeshire, Wales in 1996. Although marine transportation accidents can result in such oil spills, they account for only about 5 percent of the waste oil that enters the ocean annually.

biomass: the total mass of living organic matter in a defined location; generally expressed as grams per unit volume or per unit area

mortality: for a particular animal population, the number of deaths in a given area or period, or from a particular cause

Beach sand and gravel saturated with oil may be unable to protect and nurture normal vegetation and populations of the substrate **biomass**. Rocks and boulders coated with sticky residue interfere with recreational uses of the shoreline and can be toxic to coastal wildlife.

Examples of Large Spills. The largest accidental oil spill on record (Persian Gulf, 1991) put 240 million gallons of oil into the ocean near Kuwait and Saudi Arabia when several tankers, port facilities, and storage tanks were destroyed during war operations. The blowout of the *Ixtoc I* exploratory well offshore Mexico in 1979, the second largest accidental oil spill, gushed 140 million gallons of oil into the Gulf of Mexico. By comparison, the wreck of the *Exxon Valdez* tanker in 1989 spilled 11 million gallons of oil into Prince William Sound offshore Alaska, and ranks fifty-third on the list of oil spills involving more than 10 million gallons.

The number of large spills (over 206,500 gallons) averaged 24.1 per year from 1970 to 1979, but decreased to 6.9 per year from 1990 through 2000.

Damage to Fisheries, Wildlife, and Recreation

Oil spills present the potential for enormous harm to deep ocean and coastal fishing and fisheries. The immediate effects of toxic and smothering oil waste may be mass **mortality** and contamination of fish and other food species, but long-term ecological effects may be worse. Oil waste poisons the sensitive marine and coastal organic substrate, interrupting the food chain on which fish and sea creatures depend, and on which their reproductive success is based. Commercial fishing enterprises may be affected permanently.

Wildlife other than fish and sea creatures, including mammals, reptiles, amphibians, and birds that live in or near the ocean, are also poisoned by oil waste. The hazards for wildlife include toxic effects of exposure or ingestion, injuries such as smothering and deterioration of thermal insulation, and damage to their reproductive systems and behaviors. Long-term ecological effects that contaminate or destroy the marine organic substrate and thereby interrupt the food chain are also harmful to the wildlife, so species populations may change or disappear.

Coastal areas are usually thickly populated and attract many recreational activities and related facilities that have been developed for fishing, boating, snorkeling and scuba diving, swimming, nature parks and preserves, beaches, and other resident and tourist attractions. Oil waste that invades and pollutes these areas and negatively affects human activities can have devastating and long-term effects on the local economy and society. Property values for housing tend to decrease, regional business activity declines, and future investment is risky.

Long-term Fate of Oil on Shore

The fate of oil residues on shore depends on the spilled oil's composition and properties, the volume of oil that reaches the shore, the types of beach and coastal sediments and rocks contacted by the oil, the impact of the oil on sensitive habitats and wildlife, weather events, and seasonal and climatic conditions. Some oils evaporate, disperse, emulsify, weather, and decompose more easily than others. The weather and seasonal and climatic conditions may accelerate or delay these processes.

Oil Spills: Impact on the Ocean

In 2000, several thousand penguins were affected by a fuel oil spill after the iron-ore carrier *Treasure* sank off South Africa. Many oil-soaked birds were cleaned and released.

Oil waste that coalesces into a tar-like substance or that saturates sediments above the surf and tide level is especially persistent. Efforts to remove the oil and clean, decontaminate, and remediate an oil-impacted shoreline may make the area more visibly attractive, but may be more harmful than helpful in terms of actual recovery.

Cleanup and Recovery

The techniques used to clean up an oil spill depend on oil characteristics and the type of environment involved; for example, open ocean, coastal, or **wetland**. Pollution-control measures include containment and removal of the oil (either by skimming, filtering, or *in situ* combustion), dispersing it into smaller droplets to limit immediate surficial and wildlife damage, biodegradation (either natural or assisted), and normal weathering processes. Individuals of large-sized wildlife species are sometimes rescued and cleaned, but micro-sized species are usually ignored.

wetland: an area that is periodically or permanently saturated or covered by surface water or groundwater, that displays hydric soils, and that typically supports or is capable of supporting hydrophytic vegetation

Oil spill countermeasures to clean up and remove the oil are selected and applied on the basis of many interrelated factors, including ecological protection, socioeconomic effects, and health risk. It is important to have contingency plans in place in order to deploy pollution control personnel and equipment efficiently.

Environmental Recovery Rates. The rate of recovery of the environment when an oil spill occurs depends on factors such as oil composition and

Workers clean up an oil refinery spill that polluted Anacortes Bay, Washington. The floating ring of absorbent pads trailing behind the boat is being used to contain some of the oil that has spilled.

properties and the characteristics of the area impacted, as well as the results of intervention and remediation. Physical removal of oil waste and the cleaning and decontaminating of the area assist large-scale recovery of the environment, but may be harmful to the substrate biomass. Bioremediation efforts—adding microorganisms, nutrients, and oxygen to the environment—can usually boost the rate of biodegradation.

Because of the type of oil spilled and the Arctic environment in which it spilled, it is estimated that the residue of the *Exxon Valdez* oil spill will be visible on the Alaskan coast for 30 years.

Costs and Prevention

The costs of an oil spill are both quantitative and qualitative. Quantitative costs include loss of the oil, repair of physical facilities, payment for cleaning up the spill and remediating the environment, penalties assessed by regulatory agencies, and money paid in insurance and legal claims. Qualitative costs of an oil spill include the loss of pristine habitat and communities, as well as unknown wildlife and human health effects from exposure to water and soil pollution.

Prevention of oil spills has become a major priority; and of equal importance, efforts to contain and remove oil that has spilled are considered to be prevention of secondary spills. The costs associated with oil spills and regulations governing offshore facilities and operations have encouraged the development of improved technology for spill prevention. The Oil Pollution Act of 1990 was enacted by the U.S. Congress to strengthen oil spill prevention, planning, response, and restoration efforts. Under its provisions, the Oil Spill Liability Trust Fund provides cleanup funds for oil pollution incidents.

Responsibility for the prevention of oil spills falls upon individuals as well as on governments and industries. Because the sources of oil waste in

---

RECOVERING FROM THE EXXON VALDEZ OIL SPILL

A large quantity of crude oil was deposited on beaches in Prince William Sound and along the shoreline of the Gulf of Alaska after the *Exxon Valdez* tanker wrecked in 1989. The oil waste has been closely monitored to determine its status and its effects in the ocean and along the coast.

Initial efforts to remove the oil from intertidal areas included flushing them with hot water applied with high pressure, which proved fatal for much of the marine life involved. Natural rates of biodegradation and recovery have been slower than anticipated, and visible residue may persist for up to 30 years.

the ocean are generally careless, rather than accidental, truly effective prevention of oil spills involves everyone. SEE ALSO BEACHES; COASTAL WATERS MANAGEMENT; CORALS AND CORAL REEFS; ENERGY FROM THE OCEAN; FISHERIES, MARINE; MARINE MAMMALS; OCEAN HEALTH, ASSESSING; PETROLEUM FROM THE OCEAN.

Carolyn Embach

Bibliography

American Petroleum Institute. *Fate of Spilled Oil in Marine Waters*. Publication Number 4691. Washington, D.C.: American Petroleum Institute, 1999.

Carls, Mark G. et al. "Persistence of Oiling in Mussel Beds after the Exxon Valdez Oil Spill." *Marine Environmental Research* 51, no. 2 (2001):167–190.

Raloff, Janet. "Valdez Spill Leaves Lasting Impacts." *Science News* no. 143 (February 13, 1993):102.

U.S. Environmental Protection Agency. *Understanding Oil Spills and Oil Spill Response*. Publication Number 9200.5–105. Washington, D.C.: U.S. Environmental Protection Agency, 1993.

Petroleum from the Ocean

Petroleum (literally, "rock oil") is a substance that has formed beneath the surface of the Earth over eons. The remains of ancient plants and animals have been buried and compressed beneath thousands of feet of sand, mud, and rock. The organic materials, under certain geological conditions, have been transformed by overburden pressure and subsurface heat into hydrocarbon compounds, such as **crude oil** and **natural gas**.

crude oil: naturally occurring liquid composed of mixtures or organic chemicals called hydrocarbons; can be distilled to produce gasoline and many other products

natural gas: naturally occurring gas composed of methane and other light hydrocarbons

Discovery and Use

Petroleum substances have been discovered and used as fuels for many centuries, originally at the sites of natural seeps and leaks from fissures in the Earth's surface. In the modern era, petroleum resources are sought and exploited from locations that are more difficult and hazardous, including offshore. The petroleum deposits discovered in offshore areas contain the same kind of hydrocarbon substances as those found onshore. For example, a form of natural gas called gas hydrate is available both onshore and offshore, but offshore resources are particularly of interest because of their volume and potential for future large-scale development.

The reason offshore petroleum exploration and development are so prominent in this century is that many onshore petroleum resources have either been exhausted, are not economically producible, or are unable to be fully developed because of restrictions of national ownership or geopolitical problems. The expansion of resource development offshore is inevitable because of the growing demand for petroleum and petroleum products.

Locations of Resources. Offshore locations where petroleum resources are being found and developed for the market occur worldwide, wherever prospects for economic development exist. Presently, the petroleum industry is investing in exploration and development of oil and gas fields in offshore locations such as Newfoundland, Alaska, California, the Gulf of Mexico, the Caribbean Sea, Mexico, Brazil, Venezuela, West Africa, the North Sea, the North Atlantic Ocean, the Caspian Sea, Western Australia, Malaysia, Indonesia, New Zealand, and the South China Sea. Prospective

> **GAS HYDRATE**
>
> Gas hydrate is an ice-like, crystalline solid consisting of gas molecules, each surrounded by a "cage" of water molecules. Gas hydrate may form as cement in the pore spaces of sediment as well as in layers and nodules of pure hydrate. Most marine gas hydrate contains methane.
>
> Pressure and temperature are the two major factors controlling the formation and stability of gas hydrate. Gas hydrate is stable at the temperatures and pressures that occur in ocean-floor sediments at depths greater than 500 meters (1,640 feet). Gas hydrates are also stable in association with permafrost in polar regions, both in offshore and onshore sediments.

Petroleum from the Ocean

This platform, located in the Java Sea, is producing petroleum from a reservoir under the seafloor.

continental shelf: the relatively flat, submerged natural platform, about 1-degree slope, that extends seaward from the beach for about 70 kilometers (45 miles), with water depth up to 130 meters (425 feet) maximum, and ending where the slope and water depth increase

reserves: referring to petroleum, the amount of petroleum that can be extracted, depending on economics and technology

petroleum reservoir: a porous and permeable rock in which petroleum accumulates; primarily marine sedimentary rocks such as sandstone and limestone

areas are being sought offshore in all of the onshore petroleum basins, and wherever correlations of continental geologic structures are found.

Technological Advances

The technology is presently available to explore, drill, produce, and transport petroleum in water up to 2,300 meters (7,500 feet) deep, and further innovations are predicted to extend the depth range of future operations. In addition, remote sensing and remote control technology are being applied in facilities that cannot be directly operated by personnel, who live and work either in onsite quarters or onshore.

Every country that has dominion over part of the **continental shelf** of the ocean is investigating the possibility that petroleum resources may be found and developed there, with the help of modern technology. The estimated petroleum **reserves** of a region provide a solid incentive for capital investment in its economic and social development.

Offshore Exploration. Exploration of offshore petroleum prospects involves various technologies. Seismic surveying uses seismic waves propagated in the crust of the Earth to identify and map rock structures that are likely to contain **petroleum reservoirs**, such as basins and traps. When likely structures are found, other technologies are applied to obtain additional information about the formations being studied. These may include core sampling of the rocks with geological and laboratory analysis; data log-

Petroleum from the Ocean

Offshore petroleum production yields tremendous economic benefits yet presents substantial risk to human life and the environment. In March 2001, the Brazilian-owned *Petrobras* oil rig collapsed after explosions destroyed its substructure, and several workers were killed. The company collected or dispersed all but 3,200 gallons of waste oil at the location, averting a major spill.

ging and interpretation of formation characteristics; geochemical sampling and analysis of reservoir rocks and fluids; and reservoir modeling to evaluate the potential for economic development and exploitation.

The exploration stage is increasingly being integrated with other stages in petroleum resource development and management, so that all information and data obtained are additive in understanding the geologic structures being characterized and exploited.

Offshore Drilling. If the prospects for reservoir and field development are economically feasible, based on estimates of recoverable reserves, current market prices, and possible environmental impact, exploratory drilling can begin. The drilling program is designed according to the characteristics of rocks and formations that have been identified during the exploration stage.

Offshore drilling involves the use of various technologies and types of equipment, continually being upgraded and improved for purposes of both cost efficiency and environmental protection. Drilling ships, semi-submersible drilling barges, or other mobile drilling units are preferentially used for offshore drilling. In some locations, permanent drilling platforms are installed offshore so that development wells can also be drilled at that location, though this was done more often in the past. Offshore wells being serviced or worked over or redrilled may also require reentry of the boreholes or addition of more boreholes to an existing well, requiring the use of specialized drilling rigs and equipment.

The evaluation of the reservoir continues as data are accumulated and as the primary resources are depleted, and additional boreholes may be installed in order to exploit fully the oil or gas deposit.

Petroleum from the Ocean

Tanker carriers are a major mode of export for petroleum resources. This tanker is assisted by a tugboat in the Panama Canal.

Offshore Production. Production can begin when the wells have been drilled and completed—that is, when all downhole and surface equipment has been installed and the startup of production is approved. As noted previously, offshore production involves the use of a wide variety of equipment and technology. Offshore production structures in the past were commonly installed as a permanent structure, but modern production structures include special-design production ships, mobile production units, columnar and compliant structures and buoys that are tethered or moored to the sea floor, as well as subsea production systems with underwater wellheads. If an underwater pipeline is not used to transport the well fluids onshore or to another production platform, facilities for offshore storage and offloading may also be installed for pickup by a tanker.

Offshore production wells may be vertically arranged with respect to the producing formation, or may drain reservoir fluids through inclined or horizontal boreholes. Some formations may best be drained with multilateral wells, with several branches entering the producing formation at different depths and in several flow units.

Production may be possible without well pumps in some locations, if there is sufficient reservoir pressure to drive the fluids toward the wells. However, artificial lift and pressure boosting of well fluids are common procedures in offshore wells. Enhanced recovery methods such as gas or water injection are also used to drive or control reservoir fluid flow toward the wells.

Offshore production and drilling structures that are not removed from their offshore locations when production ceases and instead are abandoned by operators can be converted into artificial reefs that provide a habitat for marine life. On the other hand, oil leaks and spills from the abandoned structures can be ecologically harmful.

Marine Transportation. The export of oil and gas produced from offshore fields to onshore markets is accomplished primarily by underwater pipelines or by tanker carriers. Pipelines of various diameters and design may be buried in the sea floor or lie on top of it; the pipelines may be connected into systems of flowlines from multiple wells and fields, or may feed the product they carry to nearby storage tanks or tanker carriers. The tankers may serve a single well or field, or may gather and deliver products at several ports and transportation terminals. Both pipeline and tanker designs are changing to better accommodate well fluid properties, distance and hazards of the transportation route, and structural standards.

The reliability, economic operation, and environmental protection aspects of both underwater pipelines and tankers are improving. Remote sensing, remote control, and deep-water maintenance and repair are both detecting and preventing pipeline leaks. New standards for construction and operation of tankers are being applied to avoid accidents and associated spills. New technology is being developed to treat and process well fluids offshore, in order to decrease the volume of product being transported and to decrease the risk of leaking or spilling well fluids in the ocean environment.

Economic Value

The value of energy-rich petroleum in all of its forms lies in the many products that can be made from it and the importance of their uses. Crude oil, natural gas, and other hydrocarbon compounds are the bases of the fuel products that are essential for modern modes of transportation, which are predominantly fueled by motor gasoline, jet fuel, and diesel fuel. Petroleum also provides fuels for heating, industrial–manufacturing processes, and generation of electricity. Petroleum can be converted into petrochemicals and derivative products, such as pharmaceutical ingredients, plastics, and building materials, which represent other portions of the petroleum-source market.

Both supply and demand factors drive the investments and operations that the petroleum industry makes in exploring and exploiting petroleum resources. When petroleum prices in the market are high, the industry can afford to produce resources that might otherwise be uneconomic and can invest in new technology to reach resources that would otherwise be unavailable. When petroleum prices in the market are low, even proven reserves may not be produced because the cost of development and production would not allow any profit for operators.

The definition of reserves depends upon economic factors as well as actual accumulations of petroleum. Limitations of the technology needed to exploit reserves may also affect the definition and the potential for present development.

Countries without domestic petroleum resources are increasingly dependent on imported oil and gas, derived fuels, and petrochemicals, a factor that affects their domestic security. Petroleum resources define modern trade patterns and stimulate both international marketing and the potential for war. As petroleum supplies decrease and their prices climb, the pressure to develop more expensive reserves and to develop alternative energy sources is expected to increase. SEE ALSO ENERGY FROM THE OCEAN; MINERAL RESOURCES FROM THE OCEAN; OIL SPILLS: IMPACT ON THE OCEAN; TRANSPORTATION.

Carolyn Embach

Bibliography

Energy Information Administration. *Annual Energy Outlook 2000: With Projections to 2020.* DOE–EIA–0383 (2000). Washington, D.C.: December 1999.

Glasner, David. *Politics, Prices, and Petroleum: The Political Economy of Energy.* San Francisco, CA: Pacific Research Institute for Public Policy, 1985.

Hyne, Norman J. *Geology for Petroleum Exploration, Drilling, and Production.* New York: McGraw-Hill, 1984.

Plankton

Awareness is growing regarding the importance of the oceans and the variety of life they support. Research in many branches of oceanography is discovering the vast unknown of the marine world, and has expanded interest in the understanding of the marine environment and the role each member plays in a complex community.

The free-floating organisms known as plankton, from the Greek "wandering," are the drifters of the ocean. Although most of these organisms are motile (moving), they cannot swim or move against currents, but they can move vertically in the water column.

Many marine plankton are found in the deep waters of the outer ocean, or pelagic waters, whereas others are found in the shallow waters known as the neritic zone. Many of the neritic plankton are known as meroplankton, and spend only a brief period of their life cycle in the planktonic category. Many pelagic forms, such as the holoplankton, are planktonic during their entire lifespan.

The size of plankton can also determine its general name.

- Picoplankton: Smaller than 2 μm; includes bacteria, prochlorophytes, and viruses

- Nanoplankton: 2 to 20 μm; includes diatoms, coccoliths, and silicoflagellates

- Microplankton: 20 to 200 μm; includes large diatoms, dinoflagellates, and small zooplankton, such as ciliates

- Macroplankton: 200 to 2,000 μm; includes large zooplankton, copepods, and invertebrate larvae

- Megaplankton: Larger than 2,000 μm; includes fish larvae and gelatinous zooplankton

Phytoplankton

Many kinds of marine and fresh-water organisms utilize **inorganic** carbon (as carbon dioxide) and fix it into **organic** compounds by **photosynthesis**.

μm: The abbreviation for micrometer, where μ means "micro," or 10^{-6}; hence, a micrometer is 0.000001 meter

inorganic: an element, molecule, or substance that did not form as the direct result of biologic activity

organic: pertaining to, or the product of, biological reactions or functions

photosynthesis: the process by which plants manufacture food from sunlight; specifically, the conversion of water and carbon dioxide to complex sugars in plant tissues by the action of chlorophyll driven by solar energy

Plankton

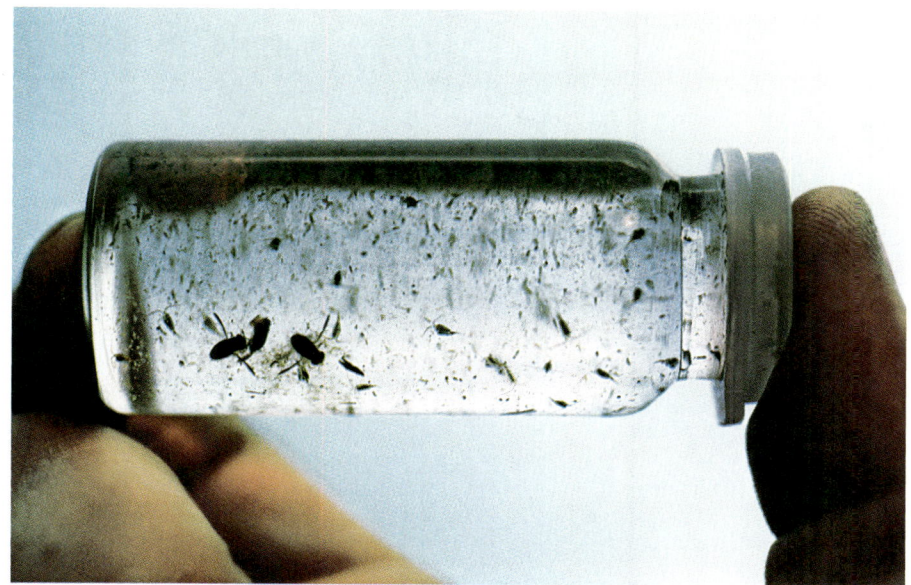

This vial of plankton illustrates the small sizes: some are barely visible to the naked eye, and some are visible only through a microscope. Plankton form the basis of aquatic food webs, including the ocean.

The principal taxa of microscopic planktonic producers, **primary producers**, are found over most of the world's oceans, lakes, rivers, and estuaries, and comprise the base of the **food web**. Phytoplankton consist primarily of diatoms, dinoflagellates, coccolithophorids, silicoflagellates, bacteria, and viruses. All of the organisms discussed below are key players in the microbial food web.

primary producer: an organism capable of using the energy from light or a chemical substance to manufacture organic compounds

food web: a complex food chain, with several species at each level, so that there is more than one producer and more than one consumer of each type

Diatoms. Diatoms have cell walls of silica and pectin, and float in the water column or attach to surfaces as single cells or chains. They are one of the major contributors to primary production in coastal waters, and occur everywhere in the ocean, but are most abundant in colder, nutrient-rich, nearshore waters. Cell division occurs by fission, which is accompanied by a reduction in cell size. They are one of the principal groups that fix carbon through photosynthesis, and this production is prominent during seasonal blooms of short duration.

Dinoflagellates. Dinoflagellates occur as single cells, either naked or within a cellulose cell wall, and many species use **flagella** to move. These organisms are sometimes classified as protozoa and algae because of their ability to photosynthesize and also absorb nutrients by being parasitic, or by ingesting organic particles. They are second to diatoms in contributing to primary production, and are widespread in the oceans, but are most abundant in nutrient-poor waters offshore. Reproduction is by cell division. Some species are bioluminescent (emitting a pale blue glow seen at night) Dinoflagellates often are the cause of red and brown tides, so named because the algal pigments give the water a colored tint.

flagellum: any of various elongated, threadlike or whiplike appendages of plants or animals; plural is flagella

Coccolithophorids. Coccolithophorids are single-celled organisms. Many are flagellated, and are protected by ornate calcareous plates, called coccoliths, embedded in a gelatinous sheath that surrounds the cell. These organisms may form cysts that produce spores to produce new individuals. They are most abundant in warm, open-ocean waters, and are sometimes found nearshore.✷ Coccolithophores can photosynthesize (autotrophic) and may also absorb organic matter (heterotrophic).

✷ See "Algal Blooms in the Ocean" for a photograph of a coccolithophore bloom near the coast.

Plankton

The microscopic phytoplankton shown here are mainly comprised of diatoms and dinoflagellates. Other phytoplankton in the microscopic ranges include coccolithophorids, silicoflagellates, bacteria, and viruses. Some phytoplankton are much larger.

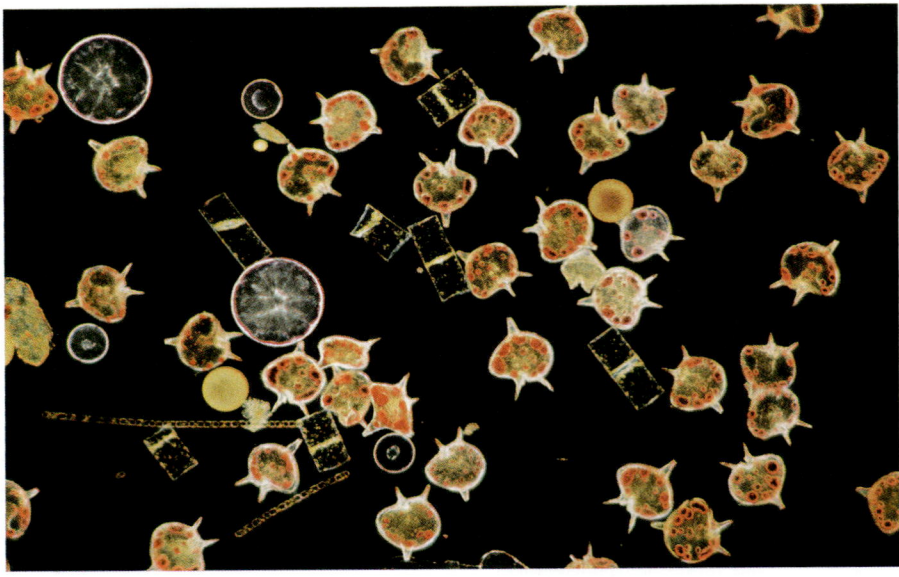

autotroph: an organism that is able to form nutritional organic substances from simple inorganic substances such as carbon dioxide

heterotroph: an organism deriving its nutritional requirements from complex organic substances; animals or microorganisms that live on producers

prokaryote: a microscopic, single-celled organism that has neither a distinct nucleus with a membrane nor other specialized organelles; includes organisms such as bacteria and cyanobacteria

chitin: a nitrogen-containing polysaccharide (carbohydrate whose molecules consist of a number of sugar molecules bonded together) forming a hard outer layer in many invertebrates, especially insects; also found in the cell walls of many fungi

metabolism: the sum total of biochemical processes that occur within a living organism, or a portion of it, in order to maintain life; the biochemical changes by which energy is provided to living cells and new material is assimilated

Silicoflagellates. Silicoflagellates occur as single flagellated cells and typically secrete a silicious outer skeleton. Like coccoliths, these organisms are both **autotrophic** and **heterotrophic**, and are most abundant in cold, nutrient-rich waters.

Bacteria. Bacteria are **prokaryotes** with cell walls made of **chitin**, and occur as single coccoid cells or long filaments. They often are restricted to waters with low oxygen, and are important in the **metabolism** of aquatic ecosystems. To support their metabolism, they obtain nutrients by the uptake of organic matter and the release of exoenzymes to lyse (distintegrate or dissolve) particulate organic matter, and attack diatoms, dinoflagellates, and flagellates. Blue-green algae, or cyanobacteria, are photosynthetic. Bacterial activity in marine waters is strongly affected by availability of nutrients and organic matter. Their productivity increases as phytoplankton productivity increases.

Viruses. Viruses play an important role in marine food webs. They infect a wide range of hosts, including bacteria and phytoplankton. They can potentially reduce phytoplankton and bacterial production by viral lysing of their cells and the releasing of dissolved organic carbon. This dissolved carbon can than be utilized by other phytoplankton cells.

Prochlorophytes. Prochlorophytes are a recently discovered group of extremely abundant producers that are barely visible by microscopy. They are most abundant at the lower layers of the illuminated region of the water column, and are now considered to be another major player in primary production.

Nanoflagellates. Nanoflagellates are both autotrophic and heterotrophic. They feed on viruses, bacteria, and some picoplankton and nanoplankton. Nanoflagellates are major consumers of bacteria; some experiments show that they may be able control their abundances when larger predators, such as dinoflagellates, are not present. However, this is less likely to occur in nature.

Protozoans. Nanoplanktonic and microplanktonic protozoan groups are mainly ciliates and heterotrophic dinoflagellates. They consume bacteria, nanoplankton, and microplankton. While these groups engulf their prey, they also release nutrients that stimulate the growth of these same prey.

Zooplankton

Zooplankton are planktonic free-floating animals in fresh and marine aquatic systems, and are the major consumers of the organisms in the microbial food web. These organisms possess a wide range of feeding strategies, from the nematocysts (stinging cells) of cnidarians (e.g., jellyfish) to the complicated mouthparts of copepods. Some are carnivorous (animal-eaters), some are herbivorous (plant-eaters), and some are omnivorous (eaters of plants and animals).

These animals can move by means of cilia, flagella, jointed appendages, jet propulsion, or tailed larvae (as in tunicates to larval fish). Reproduction varies from asexual, to fission and fragmentation, to sexual reproduction where some **gametes** are released into the water and fertilized, yet others are retained and fertilized internally.

gamete: a sex cell; in some of the simplest organisms, the gametes are not differentiated into egg and sperm

Zooplankton include many phylum, and not all can be discussed here. Some live their entire life cycle in the water (holoplankton), whereas only the larval stages of fish and other benthic organisms (such as starfish) live in the water column for a short time (meroplankton). All are considered zooplankton. An overview of the major zooplankton phyla follows.

Protozoa. Discussed previously, this group includes ciliates, dinoflagellates, foraminifera, and radiolarian.

Coelenterata (Cnidaria). Typically known as jellyfish, the major groups are Hydrozoa, Scyphozoa, and Anthozoa. The hydrozoans medusae are the prominent members in zooplankton, and the most common forms are aurelia, pelagia, and siphonophores. These gelatinous animals are major consumers of smaller zooplankton and some of the microbial food web.

Ctenophora. Best known as comb jellies, these possess eight "comb" rows of fused cilia. When they are abundant, these animals can consume phytoplankton and zooplankton, and can clear the water of food for other zooplankton.

Chaetognatha. Known as the arrow worm, this is a common member of deep-water plankton. Smaller species are found in coastal waters, whereas larger species are abundant offshore in blue water. They are predacious carnivores that grasp their prey and paralyze them before ingesting them.

Annelida. This includes many species of marine polychaetes. Many of these organisms can be seen on the surface at night, shedding gametes for sexual reproduction. Their larvae are abundant in the zooplankton community.

Mollusca. This includes marine gastropod larvae, pteropods, and cephalopods (commonly known as squid and octopus).✱ Mollusks are consumers of larger zooplankton.

✱ See "Cephalopods" for a photograph of an octopus's suction-cupped tentacles, and the *Nautilus*.

Echinodermata. This includes starfish, brittle stars, and sea cucumber. All these animals are meroplankton. Their larvae are a major presence in the zooplankton community.

* See "Crustaceans" for photographs of crustaceans, and "Ecology, Marine" for a photograph of a krill swarm.

estuary: a tidally influenced coastal area in which fresh water from a river mixes with sea water, generally at the river mouth; the resulting water is brackish, which results in a unique ecosystem

Arthropoda. These are the major members of zooplankton and include copepods, shrimp, crabs, lobsters, amphipods, crustaceans, and euphausids, or krill, which are the major source of nutrition for some whales. The most studied of crustacea are the copepods. These animals are found in all parts of the world's oceans, lakes, and **estuaries** and are considered the major consumers of most of the organisms in the microbial loop. Because they are holoplankton, spending their entire life in water, they can consume a wide range of food particles, from nanoplankton to microplankton, as they mature. Copepods are responsible for much of the carbon energy transferred from phytoplankton to larger zooplankton.

Chordata. Known as the urochordates (tunicates), this includes ascidians (or sea squirts) and are found on the coast, whereas larvaceans, oikopleura, thaliaceans, salps, and doliolids are pelagic and spend their entire life cycle in the water column. Tunicates are now realized to be major consumers of phytoplankton and smaller zooplankton, and can contribute to the entire food-web dynamics as much or even more than copepods.

Larvaceans have retained their notochord and tail as adults and produce a mucus net, or "house," around their bodies to capture food particles. The house is either ingested or abandoned.

The salps and doliolids are free-swimming tunicates with a cylindrical or barrel-shaped body with up to eight muscle bands to aid in swimming by jet propulsion and feeding with an internal mucus net. These animals have a complicated life cycle that includes a sexual stage and one or two asexual stages. They are known for their ability to create "blooms," or a rapid increase in their abundance, exceeding 1,000 animals in a cubic meter of water in a short period of time. With this rapid increase in population and their ability to filter feed a wide range of food sizes, they can outcompete copepods during these bloom events. Their role in the food web is being studied more intensely because of their production of large, fast-sinking fecal pellets that can transfer organic matter produced by primary producers to fish and benthic organisms. SEE ALSO ALGAL BLOOMS, HARMFUL; ALGAL BLOOMS IN THE OCEAN; CEPHALOPODS; CRUSTACEANS; ECOLOGY, FRESH-WATER; ECOLOGY, MARINE; EL NIÑO AND LA NIÑA; LAKES: BIOLOGICAL PROCESSES; LIFE IN WATER; MICROBES IN THE OCEAN; OCEAN BIOGEOCHEMISTRY.

Deidre M. Gibson

Bibliography

Barnes, R. S. K., and K. H. Mann. *Fundamentals of Aquatic Ecosystems.* Oxford, U.K.: Blackwell Scientific Publications, 1980.

Parsons, T. R., M. Takahashi, and B. Hargrave. *Biological Oceanographic Processes*, 3rd ed. Oxford, U.K.: Pergamon Press, 1984.

Smith, Deboyd L. *A Guide to Marine Coastal Plankton and Marine Invertebrate Larvae.* Dubuque, IA: Kendall/Hunt, 1977.

Valiela, Ivan. *Marine Ecological Processes*, 2nd ed. New York: Springer Verlag, 1995.

Planning and Management, History of Water Resources

Few governmental functions have been more important historically than managing water. Early in United States history, despite some questions over the constitutionality of the federal government getting involved in local and regional economic development, most people recognized that a partnership between government and the private sector was needed in order to plan for and manage the coastal and inland waterways of the new nation. The expenses involved in such activities as dredging rivers, constructing **levees**, and building canals were too great for private enterprise to undertake alone. Hence, Congress created federal agencies to assist with regional economic development projects and waterborne transportation systems.

levee: a natural or artifically-made earthen obstruction along the edge of a stream, lake, or river; also, a long, low embankment usually built to restrain the flow of water out of a river bank and to protect land from flooding

Major Water Management Agencies

U.S. Army Corps of Engineers.
An 1802 statute created the Army Corps of Engineers, whose original purposes were to improve navigation on existing waterways and to explore western water routes for an expanding nation. Throughout the nineteenth century, army engineers distinguished themselves in the exploration and mapping of the continent. The most famous expedition was one by Lewis and Clark, who set out from St. Louis, Missouri, in May 1804 to find a route to the Pacific Ocean.

An 1824 act of Congress established the Army Corps of Engineers as the nation's preeminent water resources manager. It appropriated funds for improving navigation on the Mississippi and Ohio Rivers, and it authorized the agency's engineers to use their expertise in determining the most practicable way of doing so.

Legislation passed in 1850 added water resources planning to the Corps' responsibilities, and in 1879 a Mississippi River Commission was established, with the Corps in charge of planning for an entire river basin. The agency's interest in planning and managing the nation's waterways continues to this day in the form of numerous activites, including **channelization** projects, dredge and fill activities, harbor improvements, **floodplain management**, and the construction and maintenance of a vast system of locks and dams on the nation's largest rivers.

channelization: any excavation and construction activities intended to widen, deepen, straighten, or relocate a natural river channel; the term does not include maintenance activities on existing channels, such as the clearing of debris or dredging of accumulated sediments

floodplain management: the societal process of decision-making to achieve the best use of floodplains, the low-lying land adjoining a river that is sometimes flooded

Bureau of Reclamation.
In 1902, a second federal agency, the U.S. Bureau of Reclamation, was created to deal with the physical and hydrological conditions peculiar to the western United States. As the celebrated explorer John Wesley Powell (1834–1902) pointed out, the western two-thirds of the continent was **arid**. What worked in the East could not work in the West, and so the federal government established a reclamation fund and a bureau for the express purpose of "reclaiming" desert lands for agricultural and municipal uses.

arid: describes a climate or region where precipitation is exceeded by evaporation; in these regions, agricultural crop production is impractical or impossible without irrigation

Whereas the Army Corps of Engineers was given authority to undertake projects throughout the entire United States, the Bureau of Reclamation was limited by statute to working in the sixteen westernmost states. Nevertheless, the Bureau, like the Corps, developed into a powerful planner and manager of water resources during the twentieth century.

Tennessee Valley Authority.
Another federal agency with a regional focus is the Tennessee Valley Authority (TVA). Created in 1933, the TVA was the brainchild of U.S. senator George Norris of Nebraska. Like president Franklin Roosevelt, Norris was a staunch advocate of comprehensive natural resources planning. That long-range, integrated approach, which today many call **ecosystems** management, was embodied in the legislation setting

ecosystem: the community of plants and animals within a water or terrestrial habitat interacting together and with their physical and chemical environment

The establishment of the Tennessee Valley Authority in 1933 was a bold experiment designed to integrate the use of all natural resources in a river basin. The TVA project, and other large federal water projects, also created jobs and contributed to the ongoing economic development of the regions in which they were built.

up the TVA. The work undertaken by the agency in the 1930s, during the worst years of the Great Depression, greatly assisted the economic and social rehabilitation of an entire region. The TVA today continues to be a significant component of the nation's water planning and management infrastructure.

Changing Goals and Values

Throughout most of the twentieth century, the U.S. Army Corps of Engineers and the Bureau of Reclamation, and the TVA, utilized the Progressive Era concepts of conservation and multipurpose development to guide their planning of water resources projects. Conservation at the turn of the century meant using a scarce resource such as water to the fullest extent possible. It dovetailed with the multipurpose idea, in that the construction of a dam, for example, not only would provide flood control but also would store waters behind the dam for use as drinking water, for recreation, and for irrigation of crops.

The 1909 Rivers and Harbors Act authorized the agencies to additionally consider the provision of hydroelectric power in their planning. At the same time, president Theodore Roosevelt (1858–1919) and other policymakers stressed the need for comprehensive river basin planning as the best approach to the conservation of the nation's resources.

During the depths of the Great Depression in the 1930s, president Franklin Roosevelt (1882–1945) used the federal water agencies to help jump-start a nearly prostrate economy. Huge construction projects involving the most sophisticated planning were undertaken, including Hoover Dam on the Arizona–Nevada border; Bonneville and Grand Coulee Dams in Washington and Oregon; Norris Dam and others in the Tennessee River Valley; and Shasta Dam in California. (The construction of dams on virtually every major river in America would continue for two more decades.) In the important 1936 Flood Control Act, Congress mandated the Army Corps of Engineers to employ a form of economic analysis known as benefit-cost analysis in its project planning. Other federal agencies quickly followed suit.

The Green Book. In 1950, policymakers in Washington, D.C. undertook a comprehensive review and analysis of water resources planning and management. Known as the Green Book for the color of its cover, the report presented the classic economic efficiency model as the standard for analysis.

In 1958, the Green Book was revised and published with the title, *Proposed Practices for Economic Analysis of River Basin Projects*. It covered the basic concepts of benefit-cost analysis, principles and procedures for project and program formulation, analysis of various project purposes, and cost allocation methods. Most of the report's findings were incorporated into the President's Bureau of the Budget guidelines for water planning and management known as *Circular A-47*. This document went into effect in 1953.

Both documents contained the prevailing philosophy of the era, which was that the federal government's role in water resources planning and management should reflect state-of-the-art economic analysis. In addition, opinion at that time dictated that federal projects should not be undertaken unless their net benefits exceeded their costs, and that national economic development was the primary goal of federal project planning.

During the 1960s, environmentalists challenged many of the report's basic assumptions. Although classical economic analysis was not abandoned by federal water agencies, it was significantly modified in ensuing decades.

Environmental Era. For more than 150 years, water planning and management in the United States had as its principal goals economic and social development. It was spectacularly successful. By the end of the Second World War (1939–1945), the United States was an economic superpower rivaled only by the former Soviet Union.

But beginning about mid-twentieth-century, the recognition spread in the United States that economic development, urbanization, and population growth came at a heavy cost to the natural environment. A major change in societal **values** occurred in the 1960s and 1970s, and these values became increasingly reflected in the water resources policy arena.

As the twenty-first century opened, for example, the Bureau of Reclamation no longer considered itself a construction agency, but instead a management and planning organization that employs watershed management and river basin planning to assist states and the private sector in meeting *all* water needs of an arid but highly populated West. For its part, the Corps continues to be a construction and engineering agency, but is also pursuing a number of more environmentally sensitive programs such as wetland protection, **mitigation banking**, floodplain management, and watershed planning.

values: abstract concepts of what is right and wrong, and what is desirable and undesirable

mitigation banking: a mitigation bank with respect to wetlands is a wetland area that has been restored, created, enhanced, or (in exceptional circumstances) preserved, which is then set aside to compensate for later conversions of wetlands for development activities

Responding to a public opinion which highly values environmental protection, both Congressional legislators and all recent presidents have encouraged the Corps and the Bureau, as well as other federal resource agencies, to pursue these new directions. SEE ALSO ARMY CORPS OF ENGINEERS, U.S.; BUREAU OF RECLAMATION, U.S.; CANALS; DAMS; DROUGHT MANAGEMENT; ECONOMIC DEVELOPMENT; FLOODPLAIN MANAGEMENT; HOOVER DAM; LEGISLATION, FEDERAL WATER; LEWIS, MERIWETHER AND WILLIAM CLARK; PLANNING AND MANAGEMENT, WATER RESOURCES; PORTS AND HARBORS; POWELL, JOHN WESLEY; REISNER, MARC; RIVER BASIN PLANNING; SUSTAINABLE DEVELOPMENT; TENNESSEE VALLEY AUTHORITY; TRANSPORTATION.

Jeanne Nienaber Clarke

Bibliography

Clarke, Jeanne Nienaber, and Daniel C. McCool. *Staking Out the Terrain: Power and Performance among Natural Resource Agencies*, 2nd ed. Albany: State University of New York Press, 1996.

Lilienthal, David E. *The TVA: Democracy on the March*. New York: HarperCollins, 1944.

National Research Council. *New Directions in Water Resources Planning for the U.S. Army Corps of Engineers*. Washington, D.C.: National Academies Press, 1999.

Reisner, Marc. *Cadillac Desert: The American West and Its Disappearing Water*. New York: Penguin, 1986.

Rogers, Peter. *America's Water: Federal Roles and Responsibilities*. Cambridge, MA: MIT Press, 1993.

Planning and Management, Water Resources

Most public utilities engage in some form of planning, although the extent and scope of planning vary greatly. Utility planning can be characterized by four general approaches: traditional supply planning, least-cost utility planning, integrated resource planning, and total water management. Each iteration retains the tried-and-true methods of the preceding approach while expanding the scope of the planning horizon to include new issues and potential solutions.

Traditional Supply Planning

Traditional planning for water utilities is not that different from traditional planning by electricity utilities, which can be characterized by its focus on utility ownership and control of all production resources (including central-station power plants), its reliance on system and financial planning processes internal to the utility, and its emphasis on the goals of minimizing electricity prices and maintaining a high level of system reliability. Risk avoidance also appears to be a paramount consideration in traditional utility planning. In the case of water, ownership and control of resources is more constrained (for example, by limits to **groundwater** withdrawals). However, the emphasis on utility ownership of the water delivery infrastructure (plants, pipes, and so on) prevails much like the case of electricity distribution.

Shortcomings. Three principal concerns about the traditional approach have been advanced. First, forecast demand is taken as a given, and virtually no attempt is made to integrate supply management and demand management options.

---

groundwater: generally, all subsurface (underground) water, as distinct from surface water, that supplies natural springs, contributes to permanent streams, and can be tapped by wells; specifically, the water that is in the saturated zone of a defined aquifer

Traditional supply-planning activities have tended to focus on providing new sources of supply to meet future water demands. Dam-building has been an important component of traditional water-supply planning, and civil engineers have played a major role in the construction and maintenance of the world's dams.

Second, the public-at-large, outside experts, and government regulators generally have little or no involvement in traditional utility planning. Demand analysis and the assessment of supply alternatives takes place within the utility (or a single planning unit within the utility); only the final product is made available for review or regulatory approval. Often, major investment decisions are made with little or no oversight.

Third, traditional planning also tends to be confined to individual utilities in virtual isolation. The decision-making process excludes parties who not only have a vested interest but also unique insights about resource options.

In sum, there is a tendency for traditional planning to be narrowly focused and exclusionary. Once a utility has committed itself to a plan, little room is left for altering the chosen course.

Supply-Driven Focus. Like other types of public utilities, the prevailing planning processes undertaken by water utilities have been internally driven and dominated by supply considerations. The result has been an emphasis on maintaining reliable water supplies and, accordingly, the engineering of facilities for source development, treatment and storage, and transmission and distribution of water.

Water-supply planning generally takes the form of forecasting future demand and developing and analyzing supply options to meet the projected demand level, plus a comfortable margin. The result is a disaggregated planning approach focusing only on new supply alternatives, while initiatives in the areas of demand management and conservation are consigned to separate programs. Moreover, conservation effects often are not reflected in revised demand and revenue forecasts.

Delivery, Not Demand. Engineering considerations have always been central to water utility planning, at times to the exclusion of other perspectives. Engineering does take into account the benefits and costs of supply projects.

Planning and Management, Water Resources

If water planners do not carefully consider current and future water needs for a given area, expansion of local and regional economies may outpace available water supplies; conversely, available water supplies could end up far exceeding local demands. Here a large water-supply pipe is en route to a construction site in the Colorado River Basin.

externality: the unintended or unwanted byproduct of production or consumption that must be borne by society in general; a negative externality arises from the detrimental effects of use or production (e.g., water pollution may represent a negative externality of watercraft operation); a positive externality arises from beneficial effects (e.g., decreased disease incidence arises from health vaccinations)

However, the emphasis on supply options can result in additions to capacity that outpace growth in demand, which is a problem familiar to the electricity industry.

Water's natural abundance in many areas may explain why the central water supply issue is that of engineering water delivery (getting water to where it is needed) rather than managing water demand. Historically, it was easy to presume that there always would be enough water to go around and that plentiful amounts would generally be so inexpensive that investments in demand management would not be cost-effective. The utility corporate culture, which understandably emphasizes selling more (not less) product, might also favor the supply side. At the very least, utilities traditionally have had little incentive to promote conservation.

Least-Cost Planning

The shortcomings of traditional planning, along with significant external economic and political forces, gave rise to the current interest in least-cost planning, or integrated resource planning (IRP). Least-cost planning emerged in the context of the energy industries during the 1980s as a response to rising costs, a poor record of forecasting, and growing concerns about environmental **externalities**. According to its proponents, planning allows regulators to be more proactive, that is, to actively affect utility decisions rather than simply react to them later on.

Least-cost planning emphasizes a balanced consideration of supply management and demand management options in identifying feasible least-cost alternatives for meeting future water needs. Compared with traditional planning, least-cost planning recognizes that water demand is malleable and that forecast demand does not have to simply be taken as a given in the planning process.

Which Cost Should Be Least? Different definitions emphasize the minimization of rates, customer bills, utility revenue requirements, and production (both capacity and operating) costs. The variety of available perspectives leaves open the question of whose costs are supposed to be least: existing customers, future customers, utility shareholders, or society at large.

Another unfortunate implication of the concept of least cost is that it suggests a single, optimal solution. Regulators and utility managers do not necessarily have identical views about the meaning of least-cost planning. According to one survey, "regulators tend to view least-cost planning with an emphasis on conservation, whereas utilities tend to regard least-cost planning as an integrated supply-and-demand analysis."

A Broader Context. Least-cost planning has come to be understood as the comprehensive evaluation of all supply and demand alternatives with the end result, in an attempt to minimize costs, of creating a flexible plan allowing for uncertainty and a changing economic environment. Cost minimization, resource diversity, risk management, and flexibility are the hallmarks of least-cost planning.

In practical terms, least-cost planning can be characterized by the consideration of a diverse set of resource options (including conservation, **load management**, pricing, and purchases from other producers); coordination among several departments within the utility; and the use of multiple resource selection goals, including those that address prices, costs, revenue requirements, utility financial condition, risk reduction, technological diversity, environmental quality, and economic development. Least-cost planning in no way abandons sound engineering practices, but it does place them in a wider context.

load management: steps taken to reduce water demand at peak load times, such as shifting some of it to off-peak times; may refer to peak hours or peak days

It follows that least-cost planning can facilitate regulatory review of supply management and demand management practices and regulatory approval of capacity expansion projects prior to construction. Least-cost planning also can be used to educate policymakers about the critical issues in utility supply decisions and induce utilities to aggressively engage in long-term strategic planning. Least-cost planning can focus on a particular utility or take the form of a statewide assessment of all utilities in a given utility sector, the latter of which is typically performed by a state resource management agency.

Demand versus Supply. In contrast to traditional supply planning, least-cost planning gives much weight to the distinction between demand management and supply management activities, both of which can be used to achieve reliability goals. Least-cost planning recognizes that demand is malleable rather than simply a fixed input to planning models. The demand side involves any strategy to eliminate or defer the need for an investment in new capacity by the utility, including load management, conservation, and pricing strategies. The supply side involves determining the most efficient method of meeting growing demand, including investments in new capacity.

All demand management activities that decrease the demand for utility services tend to affect supply management since existing system capacity is released for other customers and other uses. That is, the freed or redirected utility capacity can be compared to that provided by more traditional means. Thus emerged the concept of "negawatts," meaning electricity "produced" through conservation and efficient use of existing electricity supply resources.

As water management has advanced into integrated resource planning, many stakeholders now participate in water planning activities. Their presence has resulted in the representation of increasingly diverse views and more participatory decision-making processes.

A related idea is the trend to establish energy service companies, as compared with traditional electric utilities, who could market efficiency (or demand management) as well as electrical power.

Complexity, Risk, and Cost. Least-cost utility and planning is complicated by the lack of familiarly with demand management, barriers to coordination with nearby utilities, the use of broad definitions of costs (such as those associated with externalities), and the inclusion of goals (such as resource protection) not directly attributable to the utility. Traditional planners tend to view least-cost planning as more risky, as least in terms of supply reliability. Advocates of least-cost planning would counter that planning actually reduces some forms of risk, such as the revenue recovery risk associated with excess capacity.

While highly pertinent to the debate, these disagreements can be overcome. One issue that merits especially careful attention, however, is that of the cost of planning itself. Certainly, the benefits of planning should outweigh the costs. This calculation itself poses a rational decision-making problem. The benefits of increased awareness and understanding and reduced ignorance and uncertainty can be substantial but are not easily quantified.

Integrated Resource Planning

Integrated resource planning is a somewhat more encompassing term than least-cost utility planning, although the two are consistent and can be used interchangeably for many analytical purposes. In fact, the term "least-cost integrated resource planning" sometimes is used. The concept of integrated planning evolved in part to address the potential misconceptions and complexities arising from use of the term "least cost" as well as any unjustified bias against supply-side solutions.

Comprehensive and Participatory. Integrated resource planning encompasses the concept of least-cost planning, which emphasizes balancing supply and demand management considerations and identifying feasible planning alternatives that meet the test of least cost without unduly sacrificing other policy goals. Integrated resource planning is a more comprehensive evaluation system that goes further to emphasize the construction of various planning scenarios in which key variables and assumptions can be altered. These scenarios can be used to help utilities incorporate uncertainties, environmental externalities, and community needs into decision-making.

Integrated resource planning also emphasizes the importance of establishing a more open and participatory decision-making process and coordinating the many water **institutions** that govern water resources. Thus, IRP encourages the development of new institutional roles in addition to new analytical tools. It also promotes consensus building and alternative dispute resolution over conflict and litigation. Importantly, planning does not preclude the development and use of markets or market-like mechanisms (such as competitive bidding), which in fact may be essential for the purpose of identifying least-cost opportunities.

institution: an established organization, especially one dedicated to public service

Integrative Assessments. Like least-cost planning, IRP explicitly recognizes that demand management can be a cost-effective and viable resource option. In a somewhat broadened sense, IRP recognizes that demand management can help achieve multiple policy goals (such as cost control and pollution prevention). Advocates of IRP have long argued that better planning methods are needed to account for environmental and social externalities associated with expanding utility capacity. More recently, analysts have recognized that IRP can help utilities deal with uncertainties and risks as well.

The generalized concept of IRP also can be used to address short- and long-term community needs that span from environmental protection to economic development. Prescreening resource options and the construction of alternative planning scenarios can be used to evaluate the implications of a given resource mix on the utility, the environment, and the community. In addition to integrating resource options, IRP also seeks integration along temporal (short- and long-term) and spatial (local and regional) dimensions.

Internal and External Linkages. Integrated resource planning clearly entails new roles and responsibilities for water-supply utilities. Integration means that environmental, engineering, public health, financial, rate-making, social, and economic considerations all feed into the planning process. Planning data and information are linked internally to the other management activities of the water utility (physical facilities management, financial management, environmental management, research and development, economic development, and public involvement).

Integrated planning also links water utility planning with external planning processes (planning by other water, wastewater, and energy utilities; local and regional planning; river basin planning; and statewide, interstate, and federal water planning and policy). Some of these relationships are formal (as in permit processes involving state water resource or drinking-water regulators), while others are less so (as in the use of regional water planning data by the utility to develop forecasts).

For the water sector, a particularly important issue is the relationship of water utility planning to the activities of various government agencies whose policies may constrain utility planning choices. Clearly, better coordination mechanisms could be established among the three or more state agencies that regulate water systems from the public health, natural resource, and economic perspectives. If IRP does nothing else but facilitate coordination and improvement in the institutional structures governing water resources, it will be worth the effort.

Total Water Management

In the 1990s, total water management emerged as a potentially salient concept for water and wastewater utilities. Total water management reflects the philosophy that water resources should be managed for the greatest good of people and the environment with opportunities for participation in water policy by all segments of society.

While not strictly a planning model, total water management seems to encompass the basic principles of IRP. It also has considerable symbolic connotations. According to a white paper by the American Water Works Association, "Total water management recognizes the paradigm shift from considering water available in unlimited quantities to understanding water supply as a limited resource." Total water management seeks to inspire the water industry to embrace such ideas as sustainability, stewardship, unified water resource policies, watershed and ecosystem management, water conservation, and the importance of public and political support for water management decisions.

Total water management recognizes that water resources are a part of numerous complex systems, both natural and social. Advocates of integrates resources planning (encompassing, for example, water, energy, and land-use planning) make a similar point. These perspectives present numerous intellectual, analytical, and evaluative challenges.

The term "total water management" and its manifestations are more normative and prescriptive than the seemingly value-neutral realm of rational analysis. The explicit consideration of values in planning can be uncomfortable, particularly for members of the many science-based disciplines involved in water resource issues. But in making policy choices, trade-offs among competing values are inevitable. A paradigm that allows for a dialogue about how to make these trade-offs should be welcomed by the water industry.

Planning in Practice

The practice of planning requires attention to the potential barriers to success. Three key areas of concern are (1) access to analytical tools and adequate information, (2) the level of commitment of utilities and regulators to considering and pursuing new options, and (3) the consistency of approaches and methods within the real-world context of existing utility and regulatory practices. If these issues are relevant to least-cost planning for energy utilities, they are as much or more applicable to the case of water.

Information Base. For water systems, information resources vary substantially, and the need to develop data processing and analytical capabilities is

---

WHAT IS A PARADIGM?

A paradigm is a prevailing philosophical, theoretical, and analytical framework that guides how a discipline or area of inquiry is defined, explored, and modeled over time. Paradigms shape how academicians generalize, investigate, and experiment as well as how practitioners and policymakers interpret and understand the world around them. Disagreement and controversy can arise over the existence, application, or change of paradigm.

A paradigm shift, or the emergence of a new paradigm, represents a fundamental change in thinking and occurs relatively infrequently. In the water sector, an example is the shift from a focus on supply management to integrated resource planning that recognizes the joint relevance of supply and demand management.

clear. Attention should be paid to the design and implementation of planning strategies that minimize the effects of inadequate information.

Participants. The second barrier involves the attitudes and dispositions of those involved in resource planning. In part, success in IRP will depend on whether the prevailing corporate culture accepts conservation and related concepts as legitimate utility goals.

Realistic Approaches. Finally, successful IRP requires good practice as well as good theory. Water utility managers need practical methods for dealing with new forms of uncertainty, such as the revenue uncertainty associated with conservation. SEE ALSO CONSERVATION, WATER; DEMAND MANAGEMENT; DROUGHT MANAGEMENT; ETHICS AND PROFESSIONALISM; FLOODPLAIN MANAGEMENT; HYDROLOGIC CYCLE; INFRASTRUCTURE, WATER-SUPPLY; INTEGRATED WATER RESOURCES MANAGEMENT; MARKETS, WATER; PLANNING AND MANAGEMENT, HISTORY OF WATER RESOURCES; PRICING, WATER; RIVER BASIN PLANNING; UTILITY MANAGEMENT.

Janice A. Beecher

Bibliography

Dzurik, Andrew. *Water Resources Planning*, 2nd ed. Savage, MD: Rowman & Littlefield Publishers, Inc., 1990.

Reisner, Marc. *Cadillac Desert: The American West and Its Disappearing Water*, Rev. ed. New York: Penguin, 1993.

Viessman Jr., Warren, and Mark J. Hammer. *Water Supply and Pollution Control*. Menlo Park, CA: Addison Wesley Longman, Inc., 1998.

Internet Resources

"White Paper on Total Water Management." American Water Works Association. <http://www.awwa.org/Advocacy/govtaff/totwapap.cfm>.

Plate Tectonics

Plate tectonics is the unifying theory of geology that describes and explains that all earthquakes, volcanic activity, and mountain-building processes are caused by the gradual movement of rigid slabs of rock, called plates, that make up the Earth's surface layer. Given the expanse of geologic time, even modest movements—measured in centimeters or inches per year—result in substantial changes in the distribution of lands and oceans over millions of years.

Earth Structure

The Earth's internal structure can be viewed in two ways: either in terms of compositional layers, or in terms of layers of varying strength. There are three main compositional layers: the crust, mantle, and core. The crust, the outermost layer, is relatively buoyant and very thin compared to the mantle and core.

Beneath the oceans, the oceanic crust varies very little in thickness, generally extending only about 5 kilometers (3.1 miles), and is composed of **basalt**. The crust beneath the continents, however, is much more variable in thickness, averaging about 30 kilometers (18.6 miles); under large mountain ranges it can extend to depths of up to 100 kilometers (62.1 miles).

basalt: a dark, volcanic rock with abundant iron and magnesium and relatively low silica common on all of the terrestrial planets

Plate Tectonics

The San Andreas fault is the border between two tectonic plates—the North American Plate and Pacific Plate. Los Angeles is located on the Pacific Plate, and San Francisco is on the North American Plate. In a few million years, the two geographic areas will be right next to each other because the western side of the fault (the Pacific Plate) is moving northward with respect to the rest of the state. The fault is moving at about 2 centimeters (just under an inch) per year.

Continental crust is mostly formed of granite, which is less dense than basalt. This density difference is important in driving the motion of plates, as described below.

Below the crust is the mantle, a dense, hot layer approximately 2,900 kilometers (1,802 miles) thick. At the center of the Earth lies the core, which is composed of an iron–nickel alloy. It is divided into two regions—a liquid outer core and solid inner core. As the Earth rotates, the liquid inner core spins, creating the Earth's magnetic field.

Within the crust and mantle, there also are two important mechanical layers—the lithosphere and asthenosphere. The lithosphere is the outermost of these layers, and comprises the crust and uppermost mantle. The lithosphere is relatively cool, making the rock strong and resistant to deformation. The lithosphere is broken into the moving tectonic (or lithospheric) plates.

Below the lithosphere is a relatively narrow, mobile zone of the mantle called the asthenosphere. The asthenosphere is a weak zone, formed of mostly solid rock (with perhaps a little **magma** mixed in), and flows very slowly, in a manner similar to the ice at a bottom of a glacier. The rigid lithosphere is believed to "float" or move about on the slowly flowing asthenosphere.

magma: molten rock found in the mantle and crust of the Earth (also found on planets, moons, and asteroids); when forced toward the surface, it cools and solidifies to become igneous rock; when it erupts at the surface, it is called lava

Plate Tectonic Theory is Developed

The plate tectonic theory known today evolved in the 1950s, owing to four major scientific developments:

(1) Demonstration of the young age of the ocean floor;

(2) Confirmation of repeated reversals of the Earth's magnetic field in the geologic past;

(3) Emergence of the seafloor-spreading hypothesis and associated recycling of the oceanic crust; and

(4) Precise documentation that the Earth's earthquake and volcanic activity was concentrated along subduction zones and mid-ocean ridges.

Youthful Seafloor. Before the nineteenth century, the depth of the open ocean was a matter of speculation, although most scientists believed it to be flat and featureless. Only in 1855 did the first bathymetric maps reveal the first evidence of underwater mountains in the central Atlantic. In 1947, seismologists found that the sediment layer on the floor of the Atlantic was much thinner than previously thought. Scientists believed that the oceans were over 4 billion years old, and were perplexed by the distinct lack of sediment cover. The answer to this question would prove vital to advancing the theory of plate tectonics.

Magnetic Field Reversals. In the 1950s, scientists began recognizing magnetic variations in the rocks of the ocean floor. This was not entirely unexpected, since it was known that basalt contained the mineral magnetite, and this mineral was known to locally distort compass readings. In the early part of the twentieth century, geologists recognized that oceanic rocks had normal or reverse polarity (i.e., in normal polarity, the rocks have the same orientation of today's magnetic field). This can be explained by the ability of the magnetite grains to align themselves in the molten basalt with the Earth's magnetic field. When the rock cools, these grains are "locked" in, recording the magnetic orientation or polarity (normal or reversed) at the time of cooling. As more of the ocean floor was mapped, patterns of alternating stripes of normal and reverse polarity were noted; this became known as magnetic striping.

Seafloor Spreading. With the discovery of magnetic striping at mid-ocean ridges, scientists began to theorize that mid-ocean ridges mark structurally weak zones where magma from deep within the Earth rises and erupts at the surface. This theory, called seafloor spreading, quickly gained acceptance, but raised an additional question: If new crust is continually being formed at mid-ocean ridges, and the Earth is not increasing in size, what is happening to the old crust? Harry Hess and Robert Dietz postulated that the old crust must be destroyed in the deep canyon-like oceanic trenches, while new crust if formed at the mid-ocean ridges. This theory explained why the Earth is not expanding, there is little sediment on the ocean floor, and oceanic crust is much younger than continental rocks.

Subduction Zones. The final scientific discovery that cemented the theory of plate tectonics occurred with improvements in seismic detection in the 1950s. Seismologists identified regions of earthquake activity that coincided with Hess's predicted areas of ocean crust generation (mid-ocean ridges) and oceanic lithosphere destruction (subduction zones). Today scientists know that tectonic plates move, because they can measure their motion directly using the global positioning system (GPS).

Plate Tectonic Boundaries

Plate tectonic boundaries are regions where lithospheric plates meet. There are three types of plate tectonic boundaries: divergent, convergent, and transform.

Divergent. Divergent boundaries occur along spreading ridges where plates are moving apart and new crust is being created by ascending magma from the mantle. An example of a divergent plate boundary is the Mid-Atlantic Ridge. This submerged mountain chain extends from the Arctic to the southern tip of Africa, and is one part of the global ridge system that extends around the Earth.✶ The Mid-Atlantic Ridge spreads at a rate of approximately 2.5 centimeters (1 inch) per year.

✶ See "Mid-Ocean Ridges" for an image illustrating the mid-ocean ridge system.

Convergent. Convergent boundaries are regions where lithospheric plates collide. The type of convergence depends on the types of plates involved: namely, (1) oceanic–oceanic convergence; (2) oceanic–continental convergence; (3) continental–continental convergence (see figure).

- *Oceanic–Oceanic.* When two oceanic plates collide, one plate is subducted beneath the other. This occurs as one lithospheric plate becomes older, colder, and denser than the underlying hot, weak asthenosphere. As the lithosphere sinks slowly through the asthenosphere, the uppermost sediments are melted, and the resulting magma reaches the surface to form volcanoes. As a result, subduction zones are marked by an arc of volcanoes parallel to and about 150 kilometers (93 miles) from the plate margin. An example of oceanic-oceanic collision is the Marianas Trench and the Aleutian Islands in the Pacific Ocean.

- *Oceanic–Continental.* When oceanic and continental plates collide, the oceanic plate is the one that is subducted beneath the continental plate, because the continental crust is lighter and less dense. An example of oceanic-continental collision is seen at the Cascadia Subduction Zone, where the Pacific Plate is being subducted beneath the North American Plate.

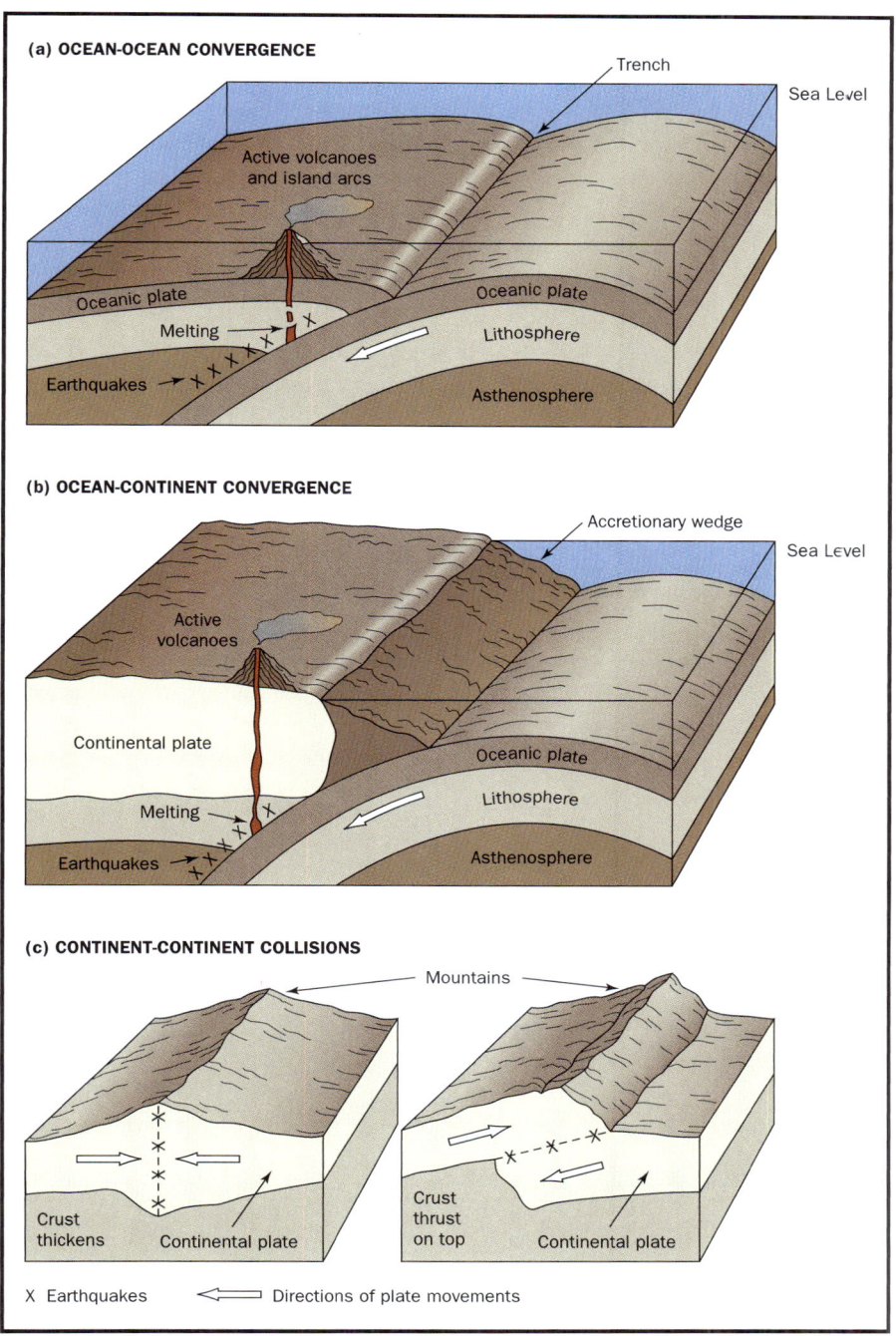

This schematic illustrates the three types of plate convergence.

- *Continental–Continental.* Continental–continental convergence results in spectacular mountain ranges such as the Himalayas, the Alps, and the Appalachians. Because continental crust is buoyant, neither plate will subduct, and a collision zone is the result.

Transform. Transform boundaries mark regions where plates slide past one another. Transform boundaries are great vertical fractures that extend down through the lithosphere. An example of a transform boundary is the San Andreas Fault in Southern California (see the photograph of the fault on page 202).

Plate Movements

The lithospheric plates do not randomly meander about the Earth's surface, but are driven by internal forces. The mantle is believed to move in circular motions rather like soup boiling on a stovetop, wherein the heated soup rises to the surface, cools, and sinks back to the bottom of the pot, where it is heated and rises again. This cycle is called convective flow, and it is the same process that occurs in the mantle today. However, the heat source within the Earth is radioactive decay of minerals and residual heat from the formation of the Earth.

Until the early 1990s, scientists believed that mantle convection, seafloor spreading, and magma intrusion at mid-ocean ridges (called "ridge push") were the predominant mechanisms that drove plate motion. However, in recent years, the significance of subduction mechanisms over mid-ocean ridge processes has taken precedence. The gravity-controlled sinking of a cold, dense, oceanic slab into a subduction zone (called "slab pull") now is considered the driving mechanism behind plate tectonics.

Although scientists know that forces deep within the Earth drive plate motion, they may never know the exact details, because no mechanism can be directly tested. The fact that lithospheric plates have moved in the past and are still in motion today is beyond dispute, but the exact mechanisms of how and why they move will continue to challenge scientists in the future. SEE ALSO GEOTHERMAL ENERGY; HOT SPRINGS ON THE OCEAN FLOOR; MID-OCEAN RIDGES; OCEAN BASINS; OCEAN-FLOOR BATHYMETRY; VOLCANOES, SUBMARINE.

Alison Cridland Schutt

Bibliography

Skinner, Brian J., and Stephen C. Porter. *This Dynamic Earth*. New York: John Wiley & Sons, 1992.

Internet Resources

This Dynamic Earth. U.S. Geological Survey. <http://www.pubs.usgs.gov/publications/text/>.

Policy-Making Process

Public **policies** are developed by officials within institutions of government to address public issues through the political process. When it comes to creating public policy, policymakers are faced with two distinct situations. The first situation, and the ideal one, is for policymakers to jointly identify a desirable future condition, and then create policies and take actions to move toward that desired future state, monitoring progress to allow for necessary adjustments. The alternative, and less desirable, situation occurs when policymakers are unable to reach a consensus regarding a desirable future condition. In this later instance, policymakers try instead to move away from present situations judged as undesirable, even though no consensus exists about the preferred alternative.

policy: a pattern of goal-oriented choice and action; a plan of action

Aspects of Policy-Making

The context for the public policy-making process in the United States reflects several important aspects, which are highlighted in the following paragraphs.

Policy-Making Process

Public policymakers evaluate complex aspects of water planning and management, then derive courses of action that attempt to balance competing interests. A hand vote can be a simple way to gage the degree of consensus.

Guidance for Policymakers. Ideally, policymakers are guided by core principles. Four examples follow.

1. Politicians and public servants are **accountable** to the public.
2. **Elites**, in politics and the private sector, do not have the right to pursue their interests without constraints.
3. Government bureaucratic and decision processes must be open, accessible, and **transparent**, as well as being responsive to public concerns.
4. Individuals and communities affected by projects have the right to information regarding proposed developments; the right to challenge the need for, and the design of, projects; and the right to be involved in planning and decision-making processes.

Public Demands. In addition to the guidance and associated constraints placed on policymakers, demands from the general public, or "bottom up" initiatives, can be as influential as "top down" directives. The general public is reasonably educated and informed, and can mobilize to demand and support desired initiatives.

Special Interest Groups. Powerful special interest groups can and do apply significant pressure on elected officials and public servants in order to achieve their ends, regardless of the public welfare. A special interest group is an organized group that exists primarily to advance its own specific interests. For example, a Chamber of Commerce usually advocates for interests of business, whereas the Sierra Club normally represents the views of people interested in the environment. Elites sometimes use questionable means in order to achieve their ends, and such influence is often exerted "behind closed doors."

Complex Issues. Public policy issues normally are complex, occur in rapidly changing and turbulent environments characterized by uncertainty, and involve conflicts among different interests. Thus, those responsible for creating, implementing, and enforcing policies must be able to reach decisions

accountable: being liable or answerable for decisions or actions; normally accomplished by specifying to whom a decisionmaker must report and is answerable

elite: as in "the elite," a group of people considered to be the best in a particular society or category, especially considered with respect to weath, power, and talent; in terms of wealth, refers to the population in the top 10 percent of a country's income distribution

transparent: describes a decision-making process that is open and accessible to stakeholders and other interested parties about ill-defined problem situations that usually are not well understood, have no one correct answer, and involve many competing interests.

Different Roles of Scientists and Policymakers

Given the above characteristics of the policy-making process, the needs of policymakers and scientists often are different. Policymakers usually focus on the short-term (commonly, the time until the next election), and on actions that will have tangible results and outcomes while minimizing risk. In contrast, scientists are interested in the long-term, in deferring action until understanding has been gained, and in recognizing the nature, extent, and magnitude of uncertainty. Thus, the policymaker normally is interested in the simple rather than the complex, the concrete rather than the abstract, and the immediate rather than a distant result. Policymakers also understand that sometimes conditions will be favorable for a decision or action, even if a technical understanding of the issue is incomplete. Waiting for more data, analysis, and interpretation may result in policymakers losing an opportune moment.

The personal characteristics of policymakers and scientists also are often different. The best policymakers are prepared and able to synthesize diverse information, move forward through acts of faith, make major leaps forward into the unknown, and effectively make prodigious bets.

In contrast, scientists are taught to be conservative and cautious, and to doubt results and conclusions until evidence and analysis support them. Scientists present their findings, conclusions, and recommendations qualified with many "if's" and "maybe's," because they recognize and appreciate the complexities and uncertainties associated with their knowledge. However, policymakers usually do not want qualified statements from scientific advisors. Instead, they want simple and clear answers.

The Disadvantage of Scientific Uncertainty.
Because various scientists may use different models and assumptions to guide their research, it is not unusual for different scientists to reach contradictory conclusions. For example, one scientist may determine that **groundwater** in an **aquifer** is being polluted; whereas another scientist may say it is not. If the aquifer is polluted, one scientist may conclude that the type and amount of **contaminants** in the aquifer is a threat to human health, but another scientist would disagree. One camp of scientists may say that climate warming is occurring, but others may say it is not.

As a result of these disagreements among experts, policymakers who do not like specific advice from a scientist usually can find another scientist who will provide a perspective that supports their preferred policy. The fact that scientists can disagree often confuses the public, who may be puzzled as to why scientists are not in agreement about a policy issue.

Despite the scientific uncertainty that may exist, policymakers are challenged to find optimal solutions that ideally have been identified through participatory processes that reflect the scientific consensus, and that balance the interests of various groups. SEE ALSO HYDROPOLITICS; INTEGRATED WATER RESOURCES MANAGEMENT; LAW, WATER; LEGISLATION, FEDERAL WATER.

Bruce Mitchell

---

groundwater: generally, all subsurface (underground) water, as distinct from surface water, that supplies natural springs, contributes to permanent streams, and can be tapped by wells; specifically, the water that is in the saturated zone of a defined aquifer

aquifer: a water-saturated, permeable, underground rock formation that can transmit significant quantities of water under ordinary hydraulic gradients to wells and springs

contaminant: as defined by the U.S. Environmental Protection Agency, any physical, chemical, biological, or radiological substance in water, including constituents that may not be harmful to the environment or human health

Bibliography

Cortner, Hanna J., and Margaret A. Moote. *The Politics of Ecosystem Management.* Washington, D.C.: Island Press, 1999.

Lasswell, Harold D. *The Decision Process: Seven Categories of Functional Analysis.* College Park, MD: University of Maryland, 1956.

Lindblom, Charles. *The Policy Making Process.* Englewood Cliffs, NJ: Prentice Hall, 1968.

MacKenzie, Susan Hill. *Integrated Resource Planning and Management: The Ecosystem Approach in the Great Lakes Basin.* Washington, D.C.: Island Press, 1996.

Simon, Herbert A. *Administrative Behavior: A Study of Decision Processes in Administrative Organization*, 2nd ed. New York: Free Press, 1965.

Pollution *See "Chemical" and "Pollution" entries; Acid Mine Drainage; Acid Rain; Erosion and Sedimentation; Land Use and Water Quality; Landfills: Impact on Groundwater; Nutrients in Lakes and Streams; Septic System Impacts.*

Pollution by Invasive Species

Plants and animals sometimes disperse naturally into new **habitats**, either by natural migrations or via floods, storms, and other events. Species also can be transported by humans, either deliberately or accidentally. When species are introduced into areas outside their natural ranges, they become known as nonindigenous (i.e., not originating from that geographic area), alien, or exotic.

Once introduced, exotic or nonindigenous species are separated from the predators, **parasites**, and diseases that kept them in balance in their native environments. With such controls lost or diminished, the species often become pests, some to the extent that they injure **ecosystems** and cause economic damage. In addition, some species have adaptations that allow them to overtake and possibly displace resident species. Exotic species that cause these negative impacts can be described as nuisance, injurious, or invasive, depending on the general degree of harm.

It should be remembered that nonindigenous species can be native to a larger geographic region: for example, nonindigenous to the Ohio River Valley, yet native to North America. Further, some native species can be invasive. However, this article focuses on invasive species that generally are nonindigenous, and that are introduced by human activities.

Introduction and Impact

A common method for accidental introduction of nonindigenous species in marine environments is through the **ballast** water of oceangoing vessels. To stabilize a ship, the crew fills the ballast tanks with water, along with any organisms in the water. When the ballast water is later discharged at a different port, possibly in another part of the world, these organisms are introduced into a new environment. Scientists estimate that up to 10,000 alien species per day are transported around the world in the ballast water of ships.

Aquaculture of fish, shellfish, and other aquatic organisms can accidentally or intentionally introduce cultivated species into natural habitats, where the nonnative species may harm indigenous populations. For example,

habitat: the environment in which a plant or animal grows or lives; the surroundings include physical factors such as temperature, moisture, and light, together with biological factors such as the presence of food and predators

parasite: an organism that lives within or on another organism, causing harm to the host organism

ecosystem: the community of plants and animals within a water or terrestrial habitat interacting together and with their physical and chemical environment

ballast: a heavy substance used to improve the stability and control the draft of a ship

aquaculture: the science, art, and business of cultivating marine or fresh-water animals or plants under controlled conditions

Pollution by Invasive Species

The northern snakehead is a predatory fish native to China's Yangtze River region. Here a fish-market vendor displays a snakehead, which is a popular food item in parts of Asia. Since the late 1990s, the snakehead has been detected in the United States, where it is highly invasive. Scientists fear a significant threat to aquatic ecosystems and native species, including crabs and clams.

tributary: a smaller stream that flows into a larger stream

various carp species imported from China, Russia, and Vietnam are used by some southern U.S. catfish farmers in their farm pond management. Some of these fish escaped into the Mississippi River Basin, probably when rivers overflowed their banks and flooded the ponds. Since their accidental introduction, they have spread rapidly along the Mississippi and its **tributaries**, reaching as far north as Illinois and Wisconsin, and ultimately threatening the Great Lakes.

Accidental or deliberate release of unwanted pets can be another introduction method. For example, aquatic reptiles such as the spectacled caiman (a small crocodile) and the red-eared slider (a turtle) both occur in Florida waters, and probably originated from pet releases. Exotic amphibians and tropical fish also are found throughout Florida.

Extent of U.S. Aquatic Invaders

Every state in the United States has been affected by aquatic nuisance species. For example, the nutria (a semi-aquatic rodent native to South America) has become particularly abundant in the southern United States, where it damages vegetation in waterways. Mosquito fish introduced in Arizona to control mosquito larvae now compete with and prey on native fish. Purple loosestrife grows wild in at least 42 of the 50 states. Eurasian watermilfoil has been recorded in 46 states.

In Lake Erie, a relatively new invader surfaced in the 1990s—the quagga mussel. This thumbnail-sized mollusk was introduced into the Great Lakes in the early 1990s, probably from ballast water. By 2002, quagga mussels were found in much of Lake Erie, Lake Ontario, the Erie Canal, the upper St. Lawrence River, and parts of Lake Huron. This mussel has the potential of outpacing the highly invasive zebra mussel (whose population exploded after 1988) and causing even more damage.

Two states—Hawaii and Florida—have been especially hard hit by nonindigenous species because both are geographically isolated and have a semi-

NUISANCE VERSUS INJURIOUS

Some exotic species do not seem to cause any particular harm. The Mediterranean gecko is now common across the southern United States. This small lizard prefers rock walls under lights, a niche that was not occupied by any native species. So the environmental impact of this exotic is minimal. Such exotic species are more a nuisance than an environmental threat. In contrast, exotic wildlife species such as the zebra mussel that pose an environmental threat are called "injurious."

tropical to tropical climate. For example, Florida has the largest total of established nonindigenous amphibian and reptilian species. Twenty-eight exotic fish species are reproducing in Florida's fresh-water lakes, rivers, and canals. Twenty-one of those are permanent residents. Overall, approximately 42 percent of Florida's reptile species, 22 percent of its amphibian species, 16 percent of its fish species, and 5 percent of its bird species are naturalized nonindigenous species.

Ecological and Economic Impacts. Numerous species of fish, oysters, clams, mussels, other shellfish, crustaceans, aquatic plants, and algae are native to the United States but have been moved to areas outside of their native range, often to improve sport fishing and to support aquaculture operations (see box on this page). Many of these species are highly valued, commercially cultured, and economically important.

However, many nonnative species can cause enormous ecological damage if released or transported to a different environment. The presence of bioinvaders can lead to the restructuring of established **food webs**, the importation of new diseases to the new surroundings, and competition with indigenous organisms for space and food. Other ecological changes occur when the nonnative species interbreed with native species, likely altering the gene pool. This may lead to hybridization and homogeneity, which reduces **biodiversity** and the ecosystem's ability to adapt to natural or human-induced changes. More than 40 percent of species on the federal list of **endangered** and **threatened** species are at risk primarily because of nonnative species.

Some invaders possess survival mechanisms that can give them an advantage over indigenous species. Although similar survival tools often can be found in indigenous species, the ability of bioinvaders to survive and adapt to harsh or unusual conditions compounds the problem already created by their introduction.

For example, the northern snakehead fish can survive in a moist, out-of-water environment for up to 3 days, and can even use its fins to push itself or "walk" short distances on land. The zebra mussel can survive out of the water for several days if harbored in a cool, moist environment, such as water in the bottom of sport fishing boats. The cane toad, if attacked by another animal, releases a milky bufotoxin that causes neurological impacts and even death in small animals. Thus the toad causes declines in predator populations because these predators have no natural immunity to the bufotoxin it secretes.

Economic consequences of nonnative species arise from impacts on commercial fisheries and water-based recreation, the degradation of wildlife habitat, the reduction of biodiversity, and the alteration of natural ecosystems. Natural resource, environmental, and fisheries managers must take action when invasive species threaten ecosystems, local and regional economies, and even the human environment.

The zebra mussel, for example, colonizes not only bottom **substrates**, but also a wide array of surfaces and structures, including pipes that supply water to power plants and public water supplies. Colonies can become so dense that the flow through the pipes is restricted. Zebra mussels also may attach themselves to other aquatic animals with hard shells, such as crustaceans. The zebra mussel's ecological and economic impacts factored

food web: a complex food chain, with several species at each level, so that there is more than one producer and more than one consumer of each type

biodiversity: a measure of the variety of the Earth's species, of the genetic differences within species, and of the ecosystems that support those species

endangered: describes a plant or animal species threatened with extinction by human-made or natural changes throughout all or a significant area of its range; designated in accordance with the 1973 Endangered Species Act

threatened: as defined by the 1973 Endangered Species Act (ESA), describes a plant or animal species that is likely to become endangered in the foreseeable future; an endangered species is in danger of becoming extinct throughout all or a significant portion of its range

substrate: the bottom sediments in lakes, rivers, and oceans that may contain living organisms

---

A LIST OF FISH

The U.S. Geological Survey maintains a comprehensive list of over 500 nonindigenous fish species in the United States. Of those, 317 are native to other parts of the United States, such as rainbow trout, brook trout, arctic char, channel catfish, striped bass, Pacific salmon, Atlantic salmon, and ornamental fish.

About 185 of the nonindigenous fish species were brought in from foreign countries. The species brought in from foreign countries include 71 that either have already established self-sustaining populations in open waters or are likely to do so.

heavily into passage of the U.S. Nonindigenous Aquatic Nuisance Prevention and Control Act of 1990 and its amendments via the National Invasive Species Act of 1996.

Water-Related Invaders

Thousands of invasive species worldwide are notorious for their distinctive habits, destructive potential, and ecological damage. Other invaders seem to have little environmental impact, but it can be many years after a species is introduced before adverse consequences appear. A few examples of bioinvaders threatening U.S. fresh-water ecosystems are listed below.

Northern Snakehead. The northern snakehead (*Channa micropeltes*) is a popular food item in parts of Asia, where some people believe it has medicinal benefits. Yet in the United States, this fish is a potentially devastating bioinvader that had been documented in more than eight states as of 2002. In Maryland, for example, if these aggressive and predacious northern snakeheads were to escape into the Little Patuxent River, just 75 meters (245 feet) away from a pond in which reproducing snakeheads were found, significant environmental impacts could result. Many states have banned the import or possession of northern snakeheads.

Asian Carp. Grass, bighead, silver, and black carp are imported by some U.S. catfish farmers. Some of these nonnative carp, which may reach up to 1.5 meters (5 feet) long and 68 kilograms (150 pounds) as adults, have escaped into natural waterways. Carp have been found dangerously close to the Chicago Sanitary Canal, which could serve as a gateway to the Great Lakes. In spring 2002, the U.S. Army Corps of Engineers installed an electric barrier in the canal to discourage the northward spread of carp and other undesirable fish, such as round gobies.

But carp have other ways of reaching the Great Lakes. Bighead carp have been found in Lake Erie, where they probably were directly introduced by an admirer of the nonindigenous species (e.g., a carp hobbyist or perhaps certain Asian immigrants for whom carp is a dietary staple or for whom it may have religious significance).

African Clawed Frog. The African clawed frog (*Xenopus laevis*) was widely used in human pregnancy testing in the 1940s and 1950s, and was shipped throughout the world. This completely aquatic frog is native to southern Africa, but is now found worldwide in suitable habitats. It is prolific and voracious, competing with native species for food such as insect larvae, small fish, and tadpoles. Although *Xenopus* remains a popular laboratory animal and is still available as a pet in many areas, some states require a permit for possessing it, and Washington state prohibits its importation.

Bullfrog. The common bullfrog (*Rana catesbeiana*) is native to the eastern half of the United States. It has been accidentally or deliberately introduced to several western states, where it has been implicated in declines of native frogs and toads. The bullfrog is large and voracious, and will eat a surprising array of plants and animals.

Cane Toad. The giant toad or cane toad (*Bufo marinus*) occurs naturally from the lower Rio Grande valley in Texas south to the Amazon basin. This fertile and highly productive toad is probably the most widely introduced amphibian in the world. It has been introduced to control insects in sugar

Pollution by Invasive Species

Opinions, hobbies, culture, and religion can hinder regulatory measures to control and eradicate some nonnative species. Mute swans, an aggressive bioinvader, ironically are admired by some waterside homeowners and other local interests. This large fiberglass replica greets visitors to a Michigan tourism office.

cane fields. However, there is no evidence that it has ever been a successful biological control, and it is now considered a nuisance species wherever it has been introduced. It successfully competes with native amphibians for food and habitat.

Mute Swans. In Maryland's Chesapeake Bay and along the northeast Atlantic Coast southward to Virginia, mute swans (*Cygnus olor*) are spreading rapidly. Mute swans are much larger and more aggressive than native waterfowl. They drive off and even kill native birds, and can eat up to 4.5 kilograms (10 pounds) of aquatic vegetation each day. Their copious droppings foul ponds and lawns.

Zebra Mussels. The zebra mussel (*Dreissena polymorpha*) originated in the Balkans, Poland, and areas within the former Soviet Union. The species was accidentally introduced into the Great Lakes in 1988 via the ballast water of ships. By 1990, it had spread to all the Great Lakes. In 1991, zebra mussels escaped the Great Lakes Basin and found their way into the Illinois River, giving them access to the entire Mississippi River Basin. As of 2002, they had expanded further, and were found from Virginia to New York, bringing the total number of states with documented occurrences to twenty-three.

Quagga Mussel. The quagga mussel (*Dreissena bugensis*) looks much like the now-infamous zebra mussel, and lives in many of the same habitats. The quagga mussel could potentially cause more economic and ecological damage than the zebra mussel. Scientists studying Lake Erie in 2002 found that the quagga outnumbered the zebra mussel 10-to-1 in some areas.

Sea Lamprey. The sea lamprey (*Petromyzon marinus*) is a primitive, jawless vertebrate with a slender, eel-like body. It attaches its circular mouth to its

Pollution by Invasive Species

anadromous: describes fish that move from the ocean up a river to spawn

wetland: an area that is periodically or permanently saturated or covered by surface water or groundwater, that displays hydric soils, and that typically supports or is capable of supporting hydrophytic vegetation

TROUT CRISIS IN YELLOWSTONE LAKE

Lake trout illegally introduced to Wyoming's Yellowstone Lake in the 1990s are imperiling the native cutthroat trout population. The cutthroat is a major food source for birds and mammals, including dozens that are either endangered or "of special concern." The region's cutthroat trout fishery is world-famous, historically significant, and an economic strength. Biologists and managers are developing an aggressive lake trout control program designed to protect the cutthroat population and thus the ecological character of the entire Yellowstone Lake basin.

prey (e.g., lake trout), tearing the flesh and blood vessels to obtain blood and body fluids. After the lamprey detaches, the host usually dies from direct predatory attack, or the host may succumb to bacterial and fungal parasites that enter through the open wound.

The sea lamprey originally was an **anadromous** fish. The anadromous form was found in much of eastern North America, including Lake Ontario. In 1833, the Welland Canal was completed, connecting Lake Ontario with Lake Erie. Sea lamprey migrated through the Welland Canal, eventually reaching the other four Great Lakes. Once trapped in the Great Lakes, sea lamprey adapted and became land-locked, fresh-water fish. Their predatory activity devastated the Great Lakes trout and whitefish fishing industry by the 1950s.

While it is impossible to completely rid the Great Lakes of sea lampreys, controlling them is vital to the health of the Great Lakes fisheries. The most effective treatment technique is applying a poison that kills sea lamprey larva but that is harmless to other fish and mammals. This has been an extremely successful yet expensive treatment. Less expensive alternatives include various kinds of barriers, trapping, and the release of sterile males. With effective control of the sea lamprey, the populations of commercially important fish in the Great Lakes have gradually recovered.

Purple Loosestrife Purple loosestrife (*Lythrum salicaria*) is a **wetland** plant from Europe and Asia. It was introduced into the East Coast of North America in the 1800s as an ornamental, and to provide nectar for honeybees. It escaped cultivation and spread along roadsides, canals, and drainage ditches. The plant is now found in 40 states and all Canadian border provinces.

Purple loosestrife forms dense, impenetrable stands that are unsuitable as cover, food, or nesting sites. It also has a dense mat of tangled roots that are difficult to pull up. By displacing native vegetation, purple loosestrife has adversely affected a wide range of wetlands as well as plants and animals that rely partly or totally on wetlands. Many rare and endangered wetland species are at risk because of purple loosestrife.

Addressing the Aquatic Threat

The best way to reduce ecological impacts of nonnative species is to prevent them from invading and becoming established. This often entails education campaigns to increase public awareness.

For example, although the zebra mussel's introduction to the Great Lakes was via ballast water from a commercial ship, the mussel is easily transported between lakes and rivers by recreational boaters and sport fishers. When a boat is pulled from the water, mussels may be trapped in the bilge water or inside the water intakes of the engine. Then, when the boat is launched into a different lake of river, the zebra mussel is flushed out. Eurasian watermilfoil (*Myriophyllum spicatum*) can be spread by similar means. Hence, many state natural resources agencies conduct public education programs that give practical guidance on how to help prevent the further spread of these bioinvaders.

Complete removal of an invasive species may be possible only early in an invasion, or if the invasion is limited to a restricted area. However, once an aquatic nuisance species becomes established in a new habitat, removal

A clump of purple loosestrife flourishes on a bank of the River Cherwell in central England. This member of the Lythrum family is among the most notorious of invasive plants that threaten fresh-water ecosystems.

is difficult, expensive, and generally impractical. The only practical approach is to try to limit the ecological and economic damage of the invading species by control or containment.

Control Methods. There are four basic techniques that have been used with some success: physical removal of the invading species; ecological control; chemical control; or biological control.

Physical removal of the invading species is usually possible only if the invasion is in its early stages or is limited to a small area. Physical removal is expensive and labor-intensive.

Ecological control consists of manipulating environmental factors such as presence of fire (burning off a marsh), or changing water levels or water flow to disrupt the life cycle of an invasive species. Obviously the environmental changes must be compatible with the life cycle of native species, but, if carefully chosen, ecological controls can provide native species an advantage in competing with invasive species. Unfortunately, nuisance species often are more tolerant of adverse conditions, such as drought and fire, than are native species.

Chemical control often is an effective method of controlling invasive species. However, chemical control has many adverse side effects. It is difficult to avoid harming other organisms. Chemical control also is expensive. However, chemical control is currently the most widely successful means of control and remains the technique of choice for most nuisance species.

Biological control may be the most environmentally sound way to control invasive species with minimal expense. However, selecting the appropriate biological control agent is a daunting task. On the one hand, biological control agents may not survive or thrive in the new environmental conditions; conversely, control agents can themselves become invasive. For example, the round goby (*Neogobius melanostomus*), a fish which was accidentally introduced into the Great Lakes, has potential to be an effective biological control for zebra mussels. Yet the goby takes over prime spawning sites traditionally used by native fish species, competes with them for habitat, and changes the balance of the ecosystem.

The Human Element. Control of nuisance species is often made more complicated by personal preference, values, opinions, and even religious beliefs. For example, there is an Asian cultural belief that the northern snakehead fish has curative powers. A related belief is that a live fish should be released for each fish consumed.

Control measures receive broad public support if the species causes a clearly harmful environmental or economic impact (e.g., zebra mussels and Asian carp), or if it has visibly "unpopular" habits (e.g., noisy and voracious bullfrogs that snatch ducklings from private ponds). But attractive species such as mute swans enjoy public popularity that can hinder the control efforts of resource managers.

In Maryland, for example, the Department of Natural Resources and some federal agencies have controlled mute swan populations to prevent their further spread. However, public opinion about the swans is mixed. Many people enjoy seeing the large, graceful birds, and do not want to see them harmed. Also, animal rights activists object to lethal control methods. Mute swan control activities are coupled with outreach activities to heighten public awareness to the problems these animals create.

Many scientists think the spread of exotic species is one of the most serious, yet largely unrecognized, threats to natural ecosystems and the overall environmental balance. Further, it is a permanent, nonreversible form of pollution. Safeguarding natural systems from bioinvaders involves preventing additional introductions; detecting new threats before major damage is done; controlling and managing existing invaders; and restoring native habitats. SEE ALSO AMPHIBIAN POPULATION DECLINES; BIRDS, AQUATIC; BIVALVES; CANALS; CHESAPEAKE BAY; CRUSTACEANS; ECOLOGY, FRESH-WATER; GLOBALIZATION AND WATER; GREAT LAKES; HUMAN HEALTH AND THE OCEAN; LAKE HEALTH, ASSESSING; MISSISSIPPI RIVER BASIN; POLLUTION OF LAKES AND STREAMS; PORTS AND HARBORS; REPTILES; STREAM HEALTH, ASSESSING; TRANSPORTATION; WETLANDS.

Cindy Clendenon and Elliot Richmond

Bibliography

Elton, Charles S. *The Ecology of Invasions by Animals and Plants.* Chicago, IL: University of Chicago Press, 2000.

Holt, Alan. "Hawaii's Reptilian Nightmare." *World Conservation* 28, no. 4 (1997): 31–32.

Kurdila, Julianne. "The Introduction of Exotic Species Into the United States: There Goes the Neighborhood." *Boston College Environmental Affairs Law Review* 16 (1995):95–118.

Lafferty, Kevin D., and Armand M. Kuris. "Biological Control of Marine Pests." *Ecology* 77, no. 7 (1996):1989–2000.

The Nature Conservancy. *America's Least Wanted: Alien Species Invasions of U.S. Ecosystems.* Arlington, VA: The Nature Conservancy, 1996.

Pimental, David et al. "Environmental and Economic Costs of Nonindigenous Species in the United States." *Bioscience* 50 (1999):53–65. Available online at http://www.news.cornell.edu/releases/Jan99/species_costs.html

Roberts, L. "Zebra Mussel Invasion Threatens U.S. Waters." *Science* 249 (1990): 1370–1372.

Simberlof, Daniel, Don C. Schmitz, and Tom C. Brown. *Strangers in Paradise: Impact and Management of Nonindigenous Species in Florida.* Washington, D.C.: Island Press, 1997.

Williamson, Mark H., and Bryan Griffiths. *Biological Invasions.* Norwell, MA: Kluwer Academic Publishers, 1997.

Internet Resources

"Exotic Species." *Fisheries Management.* Great Lakes Fishery Commission. <http://www.glfc.org/fishmgmt/exotic.asp>.

Global Invasive Species Programme. <http://jasper.stanford.edu/gisp/home.htm>.

"Invasive Species." *Oceans, Coasts, and Estuaries.* U.S. Environmental Protection Agency. <http://www.epa.gov/owow/invasive_species>.

"Invasive Species: A Threat to America's Biological Heritage." U.S. Geological Survey. <http://www.usgs.gov/invasive_species/plw>.

Reeves, Eric. "Exotic Policy: An IJC White Paper on Policies for the Prevention of the Invasion of the Great Lakes by Exotic Organisms." International Joint Commission. <http://www.ijc.org/milwaukee/wrkshps/ephist.html>.

Pollution of Groundwater

About half the population in the United States relies to some extent on **groundwater** as a source of drinking water, and still more use it to supply their factories with process water or their farms with **irrigation** water. However, if all water uses such as irrigation and power production are included, only about 25 percent of the water used nationally is derived from groundwater. Still, for those who rely on it, it is critical that their groundwater be unpolluted and relatively free of undesirable **contaminants**.

A groundwater pollutant is any substance that, when it reaches an **aquifer**, makes the water unclean or otherwise unsuitable for a particular purpose. Sometimes the substance is a manufactured chemical, but just as often it might be microbial contamination. Contamination also can occur from naturally occurring mineral and metallic deposits in rock and soil.

For many years, people believed that the soil and **sediment** layers deposited above an aquifer acted as a natural filter that kept many unnatural pollutants from the surface from infiltrating down to groundwater. By the 1970s, however, it became widely understood that those soil layers often did not adequately protect aquifers. Despite this realization, a significant amount of contamination already had been released to the nation's soil and groundwater. Scientists have since realized that once an aquifer becomes polluted, it may become unusable for decades, and is often impossible to clean up quickly and inexpensively.

Types of Groundwater Contamination

Groundwater pollution caused by human activities usually falls into one of two categories: point-source pollution and nonpoint-source pollution.

groundwater: generally, all subsurface (underground) water, as distinct from surface water, that supplies natural springs, contributes to permanent streams, and can be tapped by wells; specifically, the water that is in the saturated zone of a defined aquifer

irrigation: the controlled application of water for agricultural or other purposes through human-made systems; generally refers to water application to soil when rainfall is insufficient to maintain desirable soil moisture for plant growth

contaminant: as defined by the U.S. Environmental Protection Agency, any physical, chemical, biological, or radiological substance in water, including constituents that may not be harmful to the environment or human health

aquifer: a water-saturated, permeable, underground rock formation that can transmit significant quantities of water under ordinary hydraulic gradients to wells and springs

sediment: rock particles and other earth materials that are transported and deposited over time by geologic agents such as running water, wind, glaciers, and gravity; sediments may be exposed on dry land and are common on ocean and lake bottoms and river beds

Pollution of Groundwater

Fertilizers and pesticides applied to crops eventually may reach underlying aquifers, particularly if the aquifer is shallow and not "protected" by an overlying layer of low-permeability material, such as clay. Drinking-water wells located close to cropland sometimes are contaminated by these agricultural chemicals.

pesticides: a broad group of chemicals that kills or controls plants (herbicides), fungus (fungicides), insects and arachnids (insecticides), rodents (rodenticides), bacteria (bactericides), or other creatures that are considered pests

herbicides: a group of chemicals used to kill or reduce the growth of vegetation that is considered undesirable

Point-source pollution refers to contamination originating from a single tank, disposal site, or facility. Industrial waste disposal sites, accidental spills, leaking gasoline storage tanks, and dumps or landfills are examples of point sources. Chemicals used in agriculture, such as fertilizers, **pesticides**, and **herbicides** are examples of nonpoint-source pollution because they are spread out across wide areas. Similarly, runoff from urban areas is a nonpoint source of pollution.

Because nonpoint-source substances are used over large areas, they collectively can have a larger impact on the general quality of water in an aquifer than do point sources, particularly when these chemicals are used in areas that overlie aquifers that are vulnerable to pollution. If impacts from individual pollution sources such as septic system drain fields occur over large enough areas, they are often collectively treated as a nonpoint source of pollution.

Natural Substances. Some groundwater pollution occurs naturally. The toxic metal arsenic, for instance, is commonly found in the sediments or rock of the western United States, and can be present in groundwater at concentrations that exceed safe levels for drinking water.

Radon gas is a radioactive product of the decay of naturally occurring uranium in the Earth's crust. Groundwater entering a house through a home water-supply system might release radon indoors where it could be breathed.

Petroleum-based Fuels. One of the best known classes of groundwater contaminants includes petroleum-based fuels such as gasoline and diesel. Nationally, the U.S. Environmental Protection Agency (EPA) has recorded that there have been over 400,000 confirmed releases of petroleum-based fuels from leaking underground storage tanks.

Gasoline consists of a mixture of various hydrocarbons (chemicals made up of carbon and hydrogen atoms) that evaporate easily, dissolve to some extent in water, and often are toxic. Benzene, a common component of gasoline, is considered to cause cancer in humans, whereas other gasoline components, such as toluene, ethylbenzene, and xylene, are not believed to cause

Pollution of Groundwater

Aquifers in industrialized areas are at significant risk of being contaminated by chemicals and petroleum products. In most developed countries, various laws attempt to prevent land and water pollution, and to clean up contaminated areas when they occur. Developing countries and countries in economic distress are less likely than developed nations to assess the risk of groundwater contamination by land-use activities.

cancer in humans but may be toxic in other ways. One interesting property of gasoline is that it is less dense than water, and so it tends to float on top of the **water table**.

Chlorinated Solvents. Another common class of groundwater contaminants includes chemicals known as chlorinated solvents. One example of a chlorinated solvent is dry-cleaning fluid, also known as perchloroethylene. These chemicals are similar to petroleum hydrocarbons in that they are made up of carbon and hydrogen atoms, but the molecules also have chlorine atoms in their structure.

water table: the upper surface of the zone of saturation in an unconfined aquifer below which all voids in rock, sediment, and other geologic materials are saturated (completely filled) with water

As a general rule, the chlorine present in chlorinated solvents makes this class of compounds more toxic than fuels. Unlike petroleum-based fuels, solvents are usually heavier than water, and thus tend to sink to the bottoms of aquifers. This makes solvent-contaminated aquifers much more difficult to clean up than those contaminated by fuels.

Cleaning Up Contaminated Groundwater

Groundwater typically becomes polluted when rainfall soaks into the ground, comes in contact with buried waste or other sources of contamination, picks up chemicals, and carries them into groundwater. Sometimes the volume of a spill or leak is large enough that the chemical itself can reach groundwater without the help of infiltrating water.

Groundwater tends to move very slowly and with little turbulence, dilution, or mixing. Therefore, once contaminants reach groundwater, they tend to form a concentrated **plume** that flows along with groundwater. Despite the slow movement of contamination through an aquifer, groundwater pollution often goes undetected for years, and as a result can spread over a large area. One chlorinated solvent plume in Arizona, for instance, is 0.8 kilometers (0.5 miles) wide and several kilometers long!

plume: a concentrated area or mass of a substance that is emitted from a natural or human-made point source and that spreads in the environment; a plume can be thermal, chemical, or biological in nature

Cleanup Laws. Several federal laws focus on either preventing or remediating groundwater contamination, often caused by industrial, commercial, or petroleum pollutants. While these federal laws have provided an overall framework for these activities, the regulatory implementation of these laws is usually carried out by states in cooperation with local governments. Often, federal laws are adopted by the states largely unchanged.

The two major federal laws that focus on remediating groundwater contamination include the Resource Conservation and Recovery Act (RCRA) and the Comprehensive Environmental Response, Compensation, and Liability Act (CERCLA), also known as Superfund. RCRA regulates storage, transportation, treatment, and disposal of solid and hazardous wastes, and emphasizes prevention of releases through management standards in addition to other waste management activities. CERCLA regulates the cleanup of abandoned waste sites or operating facilities that have contaminated soil or groundwater. CERCLA was amended in 1986 to include provisions authorizing citizens to sue violators of the law.

The Cleanup Process

Several steps normally are taken to clean up a site once contamination has been discovered. Initially a remedial investigation is conducted to determine the nature and extent of the contamination. In the risk assessment phase, scientists evaluate if site contaminants might harm human health or the environment. If the risks are high, then all the various ways the site might be cleaned up are evaluated during the feasibility study. The record of decision is a public document that explains which of the alternatives presented in the feasibility study will be used to clean up a site.

Usually, the most protective, lowest cost, and most feasible cleanup alternative is chosen as the preferred cleanup method. The selected cleanup method is designed and constructed during the remedial design/remedial action phase. The operations and maintenance phase then follows. Periodically the remedial action is evaluated to see if it is meeting expectations outlined in the record of decision.

Methods of Cleanup

The various ways to respond to site contamination can be grouped into the following categories:

Containing the contaminants to prevent them from migrating from their source;

Removing the contaminants from the aquifer;

Remediating the aquifer by either immobilizing or detoxifying the contaminants while they are still in the aquifer;

Treating the groundwater at its point of use; and

Abandoning the use of the aquifer and finding an alternative source of water.

Containment. Several ways are available to contain groundwater contamination: physically, by using an underground barrier of clay, cement, or steel; hydraulically, by pumping wells to keep contaminants from moving past the wells; or chemically, by using a reactive substance to either immobilize or detoxify the contaminant. When buried in an aquifer, zero-valent iron (iron metal filings) can be used to turn chlorinated solvents into harmless carbon dioxide and water.

Removal. The most common way of removing a full range of contaminants (including metals, volatile organic chemicals, and pesticides) from an aquifer is by capturing the pollution with groundwater extraction wells. After it has been removed from the aquifer, the contaminated water is treated above ground, and the resulting clean water is discharged back into the ground or to a river. Pump-and-treat, as this cleanup technology is known,

---

MTBE: GASOLINE ADDITIVE

Methyl *tert*-butyl ether (MTBE) is used almost exclusively as a gasoline additive to help reduce harmful tailpipe emissions from motor vehicles. MTBE has been credited with improving air quality by significantly reducing carbon monoxide and ozone levels in areas where the additive has been used. Unfortunately, this is a case where the United States may have "robbed Peter to pay Paul": a growing number of studies have found that MTBE has contaminated groundwater and surface water in those same additive-use areas.

As a part of their National Water Quality Assessment, the U.S. Geological Survey (USGS) found MTBE in 21 percent of 480 wells located in specific areas of the United States that use MTBE in gasoline to abate air pollution. In the rest of the United States, MTBE detection frequency in groundwater was only about 2 percent. Furthermore, after controlling for factors such as population density, commercial and industrial land use, and the presence of gasoline stations, the USGS found that the use of MTBE in gasoline increases the probability of detecting MTBE in groundwater by a factor of about 4 to 6.

MTBE readily dissolves in water and can move rapidly through soils and aquifers. Because it is resistant to microbial degradation, it migrates faster and farther in the ground than other gasoline components, thus making it is more likely to contaminate public water-supply systems. According to the USGS, the vulnerability of aquifers to MTBE contamination appears to be most dependent on the chemical's use, the population density, and the presence of industry, commerce, and gasoline stations in the vicinity of sampled wells. Hydrogeologic factors such as well depth, groundwater level, and presence of roads seem to be less important.

There is widespread concern about MTBE in drinking-water sources because of potential human-health effects and its offensive taste and odor. The U.S. Environmental Protection Agency has tentatively classified MTBE as a possible human carcinogen, but has not yet established a drinking-water regulation. The agency, however, has issued a drinking-water advisory of 20 to 40 micrograms per liter (20 to 40 parts per billion) on the basis of taste and odor thresholds.

Although water can be treated using existing technologies such as air stripping or granular activated carbon (GAC), such treatment is difficult and time consuming because of MTBE's physical and chemical properties. Air stripping is a process in which contaminated water is passed through a large column filled with loose packing material while upward-flowing air evaporates volatile chemicals from the water. MTBE does not readily separate from water into the vapor phase, often requiring high air-to-water ratios.

The GAC treatment technique pumps contaminated water through a bed of activated carbon to remove organic compounds. Since MTBE does not adsorb well to organics such as carbon, high volumes of the contaminated water must pass repeatedly through a GAC system before MTBE is effectively removed.

Based on what is now known about MTBE, scientists and regulators have recommended significantly reducing or eliminating the use of MTBE in gasoline to protect drinking water. They are also recommending that safer alternatives to MTBE such as ethanol be used in gasoline to guarantee that clean air benefits are preserved.

can take a long time, but can be successful at removing the majority of contamination from an aquifer.

Another way of removing volatile chemicals from groundwater is by using a process known as air sparging. Small-diameter wells are used to pump air into the aquifer. As the air moves through the aquifer, it evaporates the volatile chemicals. The contaminated air that rises to the top of the aquifer is then collected using vapor extraction wells.

Remediation. **Bioremediation** is a treatment process that uses naturally occurring microorganisms to break down some forms of contamination into less toxic or non-toxic substances. By adding nutrients or oxygen, this process can be enhanced and used to effectively clean up a contaminated aquifer. Because bioremediation relies mostly on nature, involves minimal construction or disturbance, and is comparatively inexpensive, it is becoming an increasingly popular cleanup option.

Some of the newest cleanup technologies use surfactants (similar to dishwashing detergent), oxidizing solutions, steam, or hot water to remove contaminants from aquifers. These technologies have been researched for a number of years, and are just now coming into widespread use. These and other innovative technologies are most often used to increase the effectiveness of a pump-and-treat cleanup.

Treatment. Depending on the complexity of the aquifer and the types of contamination, some groundwater cannot be restored to a safe drinking quality. Under these circumstances, the only way to regain use of the aquifer is to treat the water at its point of use. For large water providers, this may mean installing costly treatment units consisting of special filters or evaporative towers called air strippers. Domestic well owners may need to install an expensive whole-house carbon filter or a **reverse osmosis** filter, depending on the type of contaminant. SEE ALSO ATTENUATION OF POLLUTANTS; CHEMICALS FROM AGRICULTURE; GROUNDWATER; LANDFILLS: IMPACT ON GROUNDWATER; LEGISLATION: FEDERAL WATER; MODELING GROUNDWATER FLOW AND TRANSPORT; POLLUTION OF GROUNDWATER: VULNERABILITY; SEPTIC SYSTEM IMPACTS.

William R. Mason

bioremediation: a method of waste cleanup using specialized, naturally occurring microorganisms with unique characteristics, and with metabolisms that allow them to break down organic pollutants

reverse osmosis: process in which dissolved substances are removed from water by forcing water, but not dissolved salts, through a semipermeable membrane under high pressure; commonly used to treat contaminated drinking water or process water; in desalinization, reverse osmosis is used to extract fresh water from salty water

Bibliography

Boulding, J. Russell. *Practical Handbook of Soil, Vadose Zone, and Ground-water Contamination: Assessment, Prevention, and Remediation.* Boca Raton, FL: Lewis Publishers, 1995.

Wiedemeier, Todd H. et al. *Natural Attenuation of Fuels and Chlorinated Solvents in the Subsurface.* New York: John Wiley & Sons, 1999.

Internet Resources

Johnson, Robert et al. "MTBE: To What Extent Will Past Releases Contaminate Community Supply Wells?" *Environmental Science & Technology* 34 no.9 (2000): 210A. <http://pubs.acs.org/hotartcl/est/2000/research/0666-00may_pankow.pdf>.

"Methyl Tertiary Butyl Ether (MTBE)." U.S. Environmental Protection Agency. <http://www.epa.gov/mtbe/>.

Swain, Walter. "Methyl Tertiary-Butyl Ether (MTBE)." <http://ca.water.usgs.gov/mtbe/>.

"Water Pollutants." *Recommended EPA Web pages.* U.S. Environmental Protection Agency. <http://www.epa.gov/ebtpages/watewaterpollutants.html>.

Pollution of Groundwater: Vulnerability

The water that individuals drink is the same water that falls in the form of rain on the fields that produce crops and graze livestock, the fertilized lawns in residential neighborhoods, and the oil-stained parking lots in major cities. Potential **contaminants** can be found in every rural area, in every suburban community, and on every city block.

Infiltrating precipitation moves through the **vadose zone,** (the unsaturated material above the **aquifer**), and can transport virtually any compound with which it comes into contact. Chemicals improperly stored at the surface, animal waste (e.g., livestock operations), septic tanks, and buried waste at landfills all have the potential to contaminate infiltrating precipitation, and, ultimately, groundwater.

Transport across the vadose zone depends on the chemical characteristics of the contaminant and the composition of the zone. Certain chemicals, such as TCE (trichloroethylene), will attach to organic matter in soil. This slows the rate at which TCE will move through the vadose zone.

Contaminated **recharge** water will contaminate the aquifer unless there are natural barriers that can slow or stop its downward migration. Surface activities have a dramatic effect on groundwater quality. It is easy to see the link between surface activities and groundwater quality by considering where groundwater comes from (i.e., precipitation) and where it has traveled (across the land surface and through the underlying soil and rock layers).

Aquifer Sensitivity

The sensitivity of an aquifer to contamination is based on the physical characteristics of the aquifer, the overlying geologic materials, and, for a specific contaminant, its chemical characteristics. "Sensitivity" is a relative term used to describe how well an aquifer is protected from infiltrating contamination. A highly sensitive aquifer would have little or no defense, whereas an aquifer with low sensitivity would be very well protected.

A shallow, unconsolidated sand-and-gravel aquifer is highly sensitive to contamination. The physical characteristics of the aquifer permit rapid infiltration of recharge. Rapid recharge leaves little time for contaminants to degrade naturally or be adsorbed before reaching the aquifer.

Conversely, a deep, confined, layered basalt aquifer has a very low sensitivity. Infiltrating recharge could take years to reach the aquifer, allowing time for contaminants to abate or degrade.

The sensitivity of an aquifer can vary greatly, depending on geologic conditions. Fractured or faulted terrain tends to conduct recharge much more quickly than unfractured rock because fractures act as conduits for fluid flow. Hence, faulted or fractured bedrock aquifers tend to be highly sensitive. Limestone terrain that has undergone dissolution (dissolving) by groundwater often forms karst topography, which is characterized by **sinkholes**, caves, and rapid underground drainage. With its many conduits connecting the surface and subsurface, karst terrain makes for a highly sensitive aquifer.

Highly impermeable strata, such as silt and clay, provide a physical barrier above an aquifer. Aquifers that are overlain by thick sequences of silt and clay or unfractured bedrock tend to be less sensitive to surface activities.

---

contaminant: as defined by the U.S. Environmental Protection Agency, any physical, chemical, biological, or radiological substance in water, including constituents that may not be harmful to the environment or human health

vadose zone: the subsurface zone between the water table (zone of saturation) and the ground surface where some of the spaces between the soil particles are filled with air; also referred to as the unsaturated zone or, less frequently, the zone of aeration

aquifer: a water-saturated, permeable, underground rock formation that can transmit significant quantities of water under ordinary hydraulic gradients to wells and springs

recharge: the process by which precipitation infiltrates below the surface and replenishes an aquifer

sinkhole: a depression in the Earth's surface caused by the collapse of underlying limestone, dolomite, salt, or gypsum

Pollution of Groundwater: Vulnerability

Although many problems stemming from industrial and other point-source pollution have been reduced in recent decades, new concerns revolve around the potential long-term effects of low levels of consumer chemicals (and their byproducts) on human and environmental health. Chemicals from household cleaning products are among those that may ultimately reach aquifers and hence appear in drinking-water wells.

Vulnerability Assessment

Two concepts have been introduced that can affect groundwater quality. The land-use activities that take place at the surface can affect groundwater quality, and the physical or geologic characteristics of the vadose zone and aquifer can provide protection from infiltrating contaminants.

Land-use activities and aquifer sensitivity are absolute terms that can be easily defined through observation and physical investigation. They are combined to define a relative term that is used to qualify the real risk to a given aquifer: vulnerability.

A vulnerability assessment defines the risk to an aquifer based on the physical characteristics of the vadose zone and aquifer and the presence of potential contaminant sources. This can be an important tool for communities and private well owners interested in protecting the long-term viability of their drinking-water source. The implementation of land-use planning or zoning overlays based on aquifer vulnerability can prevent aquifer contamination by carefully locating potential contaminant sources in areas of very low aquifer sensitivity.

For example, when a city planning board receives requests to locate an automobile service station, it poses a potentially great threat to groundwater because of the many wastes associated with vehicle maintenance and the storage and transfer of gasoline and other substances. The zoning decision-making process should be influenced by the relative risk of the potential source and the vulnerability assessment of the aquifer. The preferred result would place the gas station above the least-vulnerable regions of the aquifer.

How can the present generation ensure that future generations can depend on a safe, clean supply of groundwater? Assessing the vulnerability of aquifers is the first step toward careful management of groundwater resources. Implementation of groundwater protection strategies can enhance the long-term quality of an aquifer and raise public awareness of groundwater issues. SEE ALSO AQUIFER CHARACTERISTICS; "CHEMICAL" ENTRIES;

CLEAN WATER ACT; FRESH WATER, NATURAL CONTAMINANTS IN; GROUNDWATER; HYDROGEOLOGIC MAPPING; KARST HYDROLOGY; LAND USE AND WATER QUALITY; LAND-USE PLANNING; LANDFILLS: IMPACT ON GROUNDWATER; POLLUTION OF GROUNDWATER; SAFE DRINKING WATER ACT; SEPTIC SYSTEM IMPACTS; SUPPLIES, PROTECTING PUBLIC DRINKING-WATER.

Jeffrey Frederick

Bibliography

Driscoll, Fletcher G. *Groundwater and Wells.* St. Paul, MN: Johnson Division, 1987.

Stewart, Sheree, and Dennis Nelson. *Source Water Assessment Plan: Implementation of the Safe Drinking Water Act of 1996 Amendments.* Portland, OR: Oregon Department of Environmental Quality and Oregon Health Division, 1996.

Pollution of Lakes and Streams

Pollution is defined as "to make something impure"—in this case, the fresh water in lakes, streams, and **groundwater**. The pollution of water restricts its use for some human need or a natural function in the **ecosystem**.

Types of Pollutants

Physical. Physical pollutants to lakes and streams include materials such as particles of soil that are eroded from the landscape or washed from paved areas by flowing water. Once in a lake or stream, some particles settle out of the water to become bottom **sediments**. Chemical pollutants adsorbed (bound) to the particles are also incorporated into the sediments, where they may be permanently buried, or be carried by the water currents to other locations.

Another type of physical pollutant is heat that may be discharged from an industrial source, or runoff from hot surfaces in warm weather. The overclearing of shade trees along the shoreline of a lake or stream may also permit sunlight to warm waters above the normal temperature range.

Chemical. Fresh waters naturally contain chemicals dissolved from the soils and rocks over which they flow. The major **inorganic** elements include calcium, magnesium, sodium, potassium, carbon, chlorine, and sulfur as well as plant **nutrients**, such as nitrogen, silicon, and phosphorus. **Organic** compounds derived from decaying biological materials may also be present. In addition, nearly all fresh waters contain some human-made compounds, such as pesticides and other industrial and consumer products.

Chemicals resulting from human activities that increase the concentration of specific compounds above natural levels may cause pollution problems. Too much of a plant nutrient may lead to excessive plant growth, while synthetic organic compounds may cause physiological changes in aquatic organisms, or may become lethal at high concentrations. Pollutants can be taken up by plants and animals through contact with contaminated sediments, or directly from the water. Plants and organisms that become contaminated from these sources can pass the contamination up the **food chain** as predators consume them.

Biological. Although living organisms themselves are not generally thought of as pollutants, bacteria and plants that grow to nuisance propor-

---

groundwater: generally, all subsurface (underground) water, as distinct from surface water, that supplies natural springs, contributes to permanent streams, and can be tapped by wells; specifically, the water that is in the saturated zone of a defined aquifer

ecosystem: the community of plants and animals within a water or terrestrial habitat interacting together and with their physical and chemical environment

sediment: rock particles and other earth materials that are transported and deposited over time by geologic agents such as running water, wind, glaciers, and gravity; sediments may be exposed on dry land and are common on ocean and lake bottoms and river beds

inorganic: an element, molecule, or substance that did not form as the direct result of biologic activity

nutrients: a group of chemical elements or compounds needed for all plant and animal life; nitrogen and phosphorus are the primary nutrients; excessive or imbalanced nutrients in water may cause problems such as accelerated eutrophication

organic: pertaining to, or the product of, biological reactions or functions

food chain: the levels of nutrition in an ecosystem, beginning at the bottom with primary producers, which are principally plants, to a series of consumers—herbivores, carnivores, and decomposers

Pollution of Lakes and Streams

tions can impair the use of fresh waters. Such problems often arise when the plants die and decay, which is when bacterial decomposition consumes oxygen needed by aerobic aquatic organisms. An overabundance of algae or other plants provides more decaying material, and hence a greater reduction in oxygen as the material decomposes. Moreover, nonnative plants and animals that are introduced as a result of human activities can change the basic **ecology** of a lake or stream, often to great detriment.

Point Sources of Pollution

Point sources of water pollution are defined as those that originate from a known point, such as a pipe from which a pollutant may enter a lake or stream.

Nearly every city, town, and waterside settlement discharges some type of pollution to **surface waters**. Human wastes that are collected in sewers and piped to municipal sewage treatment plants ultimately are discharged to surface waters as treated wastewater. Older systems with combined sewer and stormwater systems discharge untreated sewage to rivers or lakes during heavy rainfall that overwhelms the drainage system. But in general, treatment processes remove solid material, many of the chemical pollutants, and then disinfect the treated sewage to kill disease-causing organisms before releasing the treated wastewater to the receiving waterbody.

Almost every industry uses water in its manufacturing process or in the production of raw materials and energy. Water can pick up pollutants when it is used to make a product or clean a manufacturing area. The pretreatment of wastes prior to discharge to sewers or directly to surface waters can recover metals and other valuable chemicals that save companies money while reducing pollution.

Large volumes of water are drawn from rivers and streams to remove excess heat from industrial processes. Cooling water is passed over heat exchange surfaces that transfer heat to the water, which increases the water temperature. The electric power industry is the largest user of cooling water in steam–electric power plants. Although the cooling water itself is not boiled to steam, its temperature may rise several degrees. If the temperature exceeds regulatory limits, the water must pass through cooling ponds or towers that lower the temperature of the water before it is discharged.

Nonpoint Sources of Pollution

Nonpoint sources of water pollution are those that cannot be traced to a specific point, such as an outfall pipe. Nonpoint pollution flows and seeps untreated into lakes, streams, and groundwater from urban lawns and gardens, paved surfaces, construction sites, hillsides, agricultural fields, forests, and other land areas.

Urban. In urban areas that have a high percentage of land covered by roofs, streets, and parking lots, rain and melting snow rapidly run off into lakes and rivers through drainageways and storm sewers. This urban runoff may contain nonpoint-source pollutants such as trash, pet wastes, lawn fertilizers, and herbicides, as well as oils, **heavy metals**, de-icing salts, and other pollutants from vehicles. In addition, the large volume of stormwater that rapidly enters streams from paved surfaces can produce flooding that erodes streambanks and destroys natural habitats.

The accidental release of cyanide from a precious metals recovery facility in Romania contaminated the Tisza River in 2000, killing aquatic and terrestrial animals, such as this horse who drank the poisoned waters 3 weeks after the spill. The pollution not only traveled downstream through Romania, Hungary, and Yugoslavia, but also entered the Danube River and ultimately the Black Sea. The release of an estimated 100 metric tons of cyanide caused what the United Nations called one of the worst pollution accidents in Europe.

ecology: the scientific study of the interrelationships of living things to one another and to the environment; also refers to the ecology of a given region

surface water: water found above ground and open to the atmosphere, such as the oceans, lakes, ponds, wetlands, rivers, and streams

heavy metals: a group of metals that have high density and are considered toxic at specified concentrations; with respect to soil management, such metals include copper, iron, manganese, molybdenum, cobalt, zinc, cadmium, mercury, nickel, and lead

Pollution of Lakes and Streams

Everyday activities such as doing the laundry, flushing the toilet, and using the in-sink garbage disposal add chemical and microbial pollutants to household (domestic) wastewater. If not treated and disposed properly via a septic system or a municipal sewage treatment plant, domestic wastewater can pose environmental and public health threats.

Agriculture. Modern agriculture depends on chemical fertilizers, pesticides, and irrigation to produce high-quality crops for animal and human consumption. To maximize the crop yield, nitrogen-based fertilizers are spread on the land. In addition, phosphorus and other essential minerals also may be applied where they are lacking or have been depleted in the soil. To improve production, herbicides to kill weeds and insecticides to kill insects are frequently applied to croplands. Not all of the fertilizers and pesticides stay where they are applied; consequently, some are released to the atmosphere, seep into groundwater, or are carried to lakes and streams by runoff, where they may create pollution problems.

Particles of soil that erode from tilled land can be carried by flowing water into lakes and streams. There, the fine particles fill the spaces between natural sand, gravel, and stones, thereby changing the surficial sediments comprising the benthic (bottom) **habitat** to a finer-grained silt and mud. Not only can bottom-dwellers be smothered by the fine particles, but sediment–water interactions can be changed.

Animal wastes create water pollution, such as when cattle or sheep are allowed to graze near streams. Wastes deposited by the animals can introduce nutrients and disease-causing organisms into the water, posing problems for aquatic organisms as well as human populations that use the water

habitat: the environment in which a plant or animal grows or lives; the surroundings include physical factors such as temperature, moisture, and light, together with biological factors such as the presence of food and predators

227

Confined disposal facilities, such as this one in Lake Erie, are areas where dredge spoil is disposed. Although the waste material is contained, contaminants still may seep into the receiving waterbody.

fossil fuel: substance such as coal, oil, or natural gas, found underground in deposits formed from the remains of organisms that lived millions of years ago

entrain: to draw in and transport (as solid particles or gas) by the flow of a fluid; for example, water droplets may become entrained in rising air currents

environment: all of the external factors, conditions, and influences that affect the growth, development, and survival of organisms or a community; commonly refers to Earth and its support systems

for domestic purposes. Feedlots often collect wastes from thousands of animals and store the wastes in central facilities from which they may be withdrawn and applied as fertilizer or soil conditioner. While this practice recycles nutrients to the soil, improperly maintained storage facilities or improperly applied fertilizer can create water pollution problems.

Airborne. Rain and snow are considered by some individuals to be relatively "pure," yet gases and particles introduced into the atmosphere by human activities and natural phenomena, such as volcanoes, can contaminate precipitation that falls back to Earth. The combustion of **fossil fuels** adds carbon dioxide to the atmosphere as well as nitrogen and sulfur compounds that tend to make precipitation more acidic. In areas of the world where the geology does not contain minerals that buffer the effects of the acids, the acidity of lakes and rivers may be increased as a result of "acid rain," derived, for example, from industrial areas. In addition, precipitation may also contain nutrients, heavy metals (e.g., mercury) and organic compounds that have entered the atmosphere and become **entrained** into precipitation.

Types of Impacts

Regardless of the type or source of a pollutant entering a lake or stream, the overall consequences to the **environment** may be the same—be it the degradation caused by soil erosion that eliminates the habitat of a stream organism, or the discharge of a chemical that interferes with a species' reproductive cycle. The use of water by humans can be compromised whether toxic chemical pollutants necessitate the treatment of drinking water, whether nutrients promote the growth of nuisance aquatic weeds that choke waterways, or whether bacteria close beaches.

The solution to water pollution is the elimination of pollutants at their source. This can be accomplished by reducing the use of polluting chemicals in the home and industry, by treating point sources to remove pollutants, and by each individual being more mindful of how human activities affect the landscape and the aquatic environment. SEE ALSO ACID RAIN;

ALGAL BLOOMS IN FRESH WATER; CHEMICALS FROM AGRICULTURE; ECOLOGY, FRESH-WATER; EROSION AND SEDIMENTATION; FISH AND WILDLIFE ISSUES; FRESH WATER, NATURAL COMPOSITION OF; HYDROLOGIC CYCLE; INSTREAM WATER ISSUES; LAKE HEALTH, ASSESSING; LAKE MANAGEMENT ISSUES; LAKES: BIOLOGICAL PROCESSES; LAKES: CHEMICAL PROCESSES; LAKES: PHYSICAL PROCESSES; LAND USE AND WATER QUALITY; LAND-USE PLANNING; MICROBES IN LAKES AND STREAMS; NUTRIENTS IN LAKES AND STREAMS; POLLUTION BY INVASIVE SPECIES; POLLUTION OF STREAMS BY GARBAGE AND TRASH; POLLUTION SOURCES: POINT AND NONPOINT; RUNOFF, FACTORS AFFECTING; SEPTIC SYSTEM IMPACTS; STREAM ECOLOGY: TEMPERATURE IMPACTS ON; STREAM HEALTH, ASSESSING.

Arthur S. Brooks

Bibliography

Ball, Philip. *Life's Matrix: A Biography of Water.* New York: Farrar Straus and Giroux, 2000.

Dodds, Walter K. *Freshwater Ecology, Concepts and Environmental Applications.* San Diego, CA: Academic Press, 2002.

Gleick, Peter et al. *The World's Water 2002–2003: The Biennial Report on Freshwater Resources.* Washington, D.C.: Island Press, 2002.

Internet Resources

Water. U.S. Environmental Protection Agency. <http://www.epa.gov/ebtpages/water.html>

Water Resources of the United States. U.S. Geological Survey. <http://water.usgs.gov>

Pollution of Streams by Garbage and Trash

Despite environmental regulations that protect the quality of streams, lakes, and wetlands, solid waste in the form of trash, litter, and garbage often ends up in these surface waters. Because surface waters collect in low-lying areas, anything that is dropped or blown into a **watershed** can eventually reach a drainageway. In urban areas, trash and litter (general terms for dry solid waste) often are transported by stormwater runoff. In both urban and rural areas, these items sometimes are illegally dumped directly into a waterbody or wetland, or deposited along riverbanks or lakeshores. Trash also comes from people who fish or participate in other forms of water-related recreation. Regardless of source or type, trash is a form of water **pollution**.

Ironically, in some circumstances, some discarded items (e.g., tires, plastic containers, and nonorganic construction debris) provide **habitat** for aquatic organisms. However, trash items are unsightly and are a sign of human neglect or disregard for aesthetic values and natural **ecosystems**. Despite increased environmental awareness, some people still use waterways as a repository for unwanted items, including couches and mattresses; cars and car parts; bicycles; shopping carts; bags of stolen property; fuel containers; and paint cans.

The most common litter in U.S. streams is household trash, including plastic cups, plastic bags and wrapping materials, fast-food wrappers, plastic bottles, and other plastic containers. Plastics can be especially hazardous to wildlife. Depending on their form they can either be ingested, causing internal organ failure, or they can cause a slow strangulation.✻

watershed: the land area drained by a river and its tributaries; also called river basin, drainage basin, catchment, and drainage area

pollution: any alteration in the character or quality of the environment, including water in waterbodies or geologic formations, which renders the environmental resource unfit or less suited for certain uses

habitat: the environment in which a plant or animal grows or lives; the surroundings include physical factors such as temperature, moisture, and light, together with biological factors such as the presence of food and predators

ecosystem: the community of plants and animals within a water or terrestrial habitat interacting together and with their physical and chemical environment

✻ See "Pollution of the Ocean by Plastic and Trash" for a photograph of a bird enshrouded by a piece of plastic.

Pollution of Streams by Garbage and Trash

A leader of the Mississippi River Beautification and Restoration Project sits on one of the barges used to collect household trash, tires, and construction debris. River cleanup committees and various "clean stream teams" are ways that citizens can directly participate in environmental quality initiatives.

toxic: describes a chemical substance that has the potential of causing acute or chronic adverse effects in plants, animals, or humans

nonpoint source: a pollutant release or discharge originating from a land use active over a wide land area (e.g., agriculture) rather than from one specific location (e.g., an outfall pipe from a factory)

Organic waste (e.g., wood wastes) can have chemical and biological impacts on rivers and streams. Among the many impacts are interfering with the establishment of aquatic plants, affecting the reproductive behavior of fish and other animals, and depleting the water of dissolved oxygen as the wastes decompose. Further, **toxic** materials can leak or leach out of certain kinds of trash (e.g., pressure-treated lumber, used oil filters, and lead-acid batteries).

Regulatory Context

In the United States, the Environmental Protection Agency (EPA) is charged with protecting the quality of surface waters, including streams, lakes, and coastal waters. The Clean Water Act as amended in 1977 provides the legal basis for the protection of the quality of surface water. The law uses a variety of tools to limit and control direct pollutant discharges into waterways, as well as the disposal of dredge or fill materials. The law also addresses pollutants coming from **nonpoint sources** (e.g., sediment-laden runoff from a farm field).

The control of solid waste falls under a different division within the EPA. The Office of Solid Waste regulates all solid waste under the Resource Conservation and Recovery Act (RCRA), which regulates the treatment and disposal of municipal solid waste and hazardous waste. Households create municipal solid waste, which consists mainly of paper, yard trimmings, glass, and other solid or semisolid materials. Industrial and manufacturing processes create mixtures of municipal solid wastes, hazardous waste, and other wastes such as construction-demolition debris. Some solid wastes, such as animal waste, radioactive materials, or medical waste, are managed by other government agencies and laws.

Although RCRA deals with waste once it reaches a regulated facility (such as a landfill), it does not directly address the problem of litter, even when the litter is in a watershed. Similarly, the Clean Water Act does not apply unless the materials are polluting a waterway; that is, it does not apply to trash or debris along a riverbank. Only when trash enters designated waterways and becomes "floatable debris," for example, does it become subject to regulations under the Clean Water Act, the Beaches Environmental Assessment and Coastal Health Act, or other applicable laws.

What is Pollution? The intended use of a waterbody is considered when deriving a working definition of "pollution," including human-derived litter. For example, if water is primarily designated to support wildlife habitat, oxygen is desirable but toxic compounds are not. But if water is to be converted to steam in a power plant, excess oxygen (that could corrode equipment) would be undesirable, and certain toxic chemicals may not be a concern within this very specific application.

Even with regulations that define pollution, sometimes no government or agency is willing to accept responsibility for litter and other debris in waterways. For example, the Rideau Canal and water in Ontario, Canada is a popular destination for boaters and other recreationists. A section of the canal in Ottawa forms the world's longest maintained skating rink. However, as the weather warms and the ice melts, low water levels reveal a variety of debris littering the bottom and floating on the water surface.

The National Capital Commission operates the rink and is responsible for keeping it clean during the winter. Once the ice melts, their responsibility ends. During the navigation season, Parks Canada takes over and raises the water level. Other government agencies that share responsibility for the canal say the water quality is good from an ecological perspective: that is, it meets criteria for waters that support ecosystems (e.g., sufficient oxygen and water clarity).

Who is Responsible? If no government agency has the responsibility or resources to clean up the banks of a stream or its littered streambed, then it is the responsibility of nongovernmental organizations and private citizens to do so. There are many opportunities for private citizens to participate in river and stream cleanups. For example, the U.S. Environmental Protection Agency sponsors an "Adopt-a-Watershed" program. Many states have "green team" or "stream team" opportunities, such as Vermont's Green-Up Day and Northern California's Riverwatch.

Volunteers who participate in stream cleanups often report a rewarding experience. In addition to providing an aesthetic and environmental benefit, cleanups reconnect citizens and the community to the waterways that have been a vital part of the nation's history and culture.

Research has shown than people are more likely to behave in ways that preserve our waterways if they are clean in the first place. If a stream bank or shoreline already has litter, people are more likely to continue littering. Individuals can take the initiative by cleaning up streamside trash and by disposing of trash properly.

The Challenge in Developing Nations

Most developed countries have environmental agencies, nongovernmental organizations, and special interest groups that support and participate in environmental protection activities. Environmental regulations have greatly reduced pollution of streams by sewage and by garbage or rubbish (i.e., wet wastes such as food byproducts). However, such efforts often are lacking in developing nations.

Pollution of Streams by Garbage and Trash

Waterways in developing countries often are used for dumping household rubbish and commercial wastes. While widespread dumping continues, some localities are beginning to address water-quality issues.

tributary: a smaller stream that flows into a larger stream

coliform: a group of bacteria predominantly inhabiting the intestines of humans or animals but also found in soil; while typically harmless, they are commonly used as indicators of the possible presence of pathogenic organisms

estuary: a tidally influenced coastal area in which fresh water from a river mixes with sea water, generally at the river mouth; the resulting water is brackish, which results in a unique ecosystem

In some developing countries, for example, navigable waterways are not only transportation and trade corridors, but also the site of floating villages. Canals often are lined with boats and floating shanties, and the canals are the repository of untreated sewage, rubbish, and trash. Elsewhere, impoverished villages in **tributary** watersheds dispose their wastes onto the ground or into small creeks, which eventually drain to larger waterbodies.

Despite the seemingly dismal outlook for waterways, international aid and governmental efforts are giving hope for local rehabilitation in some areas. For example, the Pasig River in the Philippines received industrial wastes, municipal solid wastes, and garbage. By the early 1990s, the river was considered biologically inactive and had dangerously high counts of fecal **coliform**. The river was a dark, murky color, and large rafts of floating garbage covered the surface of many river segments. Sunken boats and abandoned barges made navigation difficult and hazardous. Factories and makeshift shacks lined long stretches of the riverbank, as well as tributaries and **estuaries**. With help from assistance grants from the government of Denmark and the Asian Development Bank, the Pasig River Rehabilitation Commission was established in 1999 by an executive order from Philippines president Joseph Estrada. The government's goal is to upgrade the river's quality so it can sustain aquatic life and can be used for recreation by 2008.

Elliot Richmond (with Cindy Clendenon)

Bibliography

Chiras, Daniel D. *Environmental Science: A Systems Approach to Sustainable Development*, 5th ed. Belmont, CA: Wadsworth Publishing, 1998.

De Villiers, Marq. *Water: The Fate of Our Most Precious Resource*. Boston, MA: Houghton Mifflin, 2001.

Harms, Valerie. *The National Audubon Society Almanac of the Environment*. New York: Putnam Publishing, 1994.

Miller, G. Tyler, Jr. *Living in the Environment*, 6th ed. Belmont, CA: Wadsworth Publishing, 1990.

Nadakavukaren, Anne. *Our Global Environment: A Health Perspective*, 5th ed. Prospect Heights, IL: Waveland Press, 2000.

Newton, David. *Taking a Stand Against Environmental Pollution.* Danbury, CT: Franklin Watts, 1990.

Internet Resources

EPA Adopt Your Watershed. <http://www.epa.gov/adopt/resources/watersheds.html>.

Northern California River Watch. <http://www.northerncaliforniariverwatch.org/>.

Pollution of the Ocean by Plastic and Trash

Garbage has been discarded into the oceans for as long as humans have sailed the seven seas or lived on seashores or near waterways flowing into the sea. Since the 1940s, plastic use has increased dramatically, resulting in a huge quantity of nearly indestructible, lightweight material floating in the oceans and eventually deposited on beaches worldwide.

As the graph shows, trash items encompass a variety of materials. Sources of marine debris include:

- Items that are brought to the beach and left there by beachgoers;
- Garbage deliberately or accidentally discarded by ships at sea or from offshore oil platforms; and
- Material carried to sea by rivers and **estuaries**, especially from large coastal cities.

City storm sewers are a significant source of solid waste entering the sea from land sources.

estuary: a tidally influenced coastal area in which fresh water from a river mixes with sea water, generally at the river mouth; the resulting water is brackish, which results in a unique ecosystem

Effects on Wildlife

Aside from its unsightly appearance and potential impact to human health, marine debris has harmful effects on wildlife. Fish, birds, marine mammals, reptiles, and other animals can become entangled in discarded or lost nets that continue to do what they were designed to do—catch living animals—but now they catch them indiscriminately, a process called "ghost fishing." Items unintended for fishing become traps.

Woven plastic onion sacks floating in the sea have entrapped endangered hawksbill sea turtles. Plastic bags become invisible to birds diving for

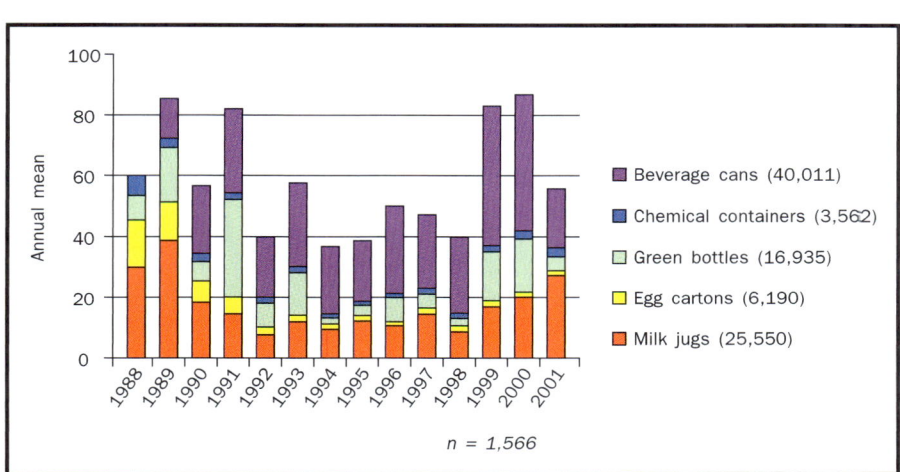

This graph shows the average number of trash items counted for several years along a popular 7-mile stretch of Mustang Island Gulf Beach, Texas. Beverage cans are single drink containers and include plastic bottles; chemical containers are 5-gallon pails and drums of chemicals; green bottles are bleach bottles from Mexico (common on Texas beaches); egg cartons and milk jugs are standard grocery items.

Pollution of the Ocean by Plastic and Trash

Almost all the marine debris collected at the North Jetty of the Aransas Pass (San Jose Island) has come from offshore. Even this remote Texas barrier island does not escape the floating plastics, Styrofoam® pieces, and other trash items.

fish and are skewered by the birds' sharp bills, usually resulting in their death. Plastic is also mistaken for food and is eaten at sea by birds, turtles, and even whales. This can choke them, poison them, or simply make them think they are full when this "food" has no nutritional value. Many marine animals get entangled in fishing lines used by recreational anglers.

Efforts to Reduce Debris

Unsightly littered beaches gained public attention in the early 1980s and efforts were made to find out how much garbage there was, where it came from, and what its effect was on ocean wildlife. Medical waste and other floatable debris washing ashore at public beaches of the eastern United States, primarily in New York and New Jersey, was a particularly disturbing and highly publicized aspect of ocean pollution by municipal solid waste. The incidents in New York and New Jersey spurred the 1988 passage of the federal Medical Waste Tracking Act (which has since transferred to individual states) as well as subsequent federal and state laws designed to better regulate the handling, treatment, transportation, and disposal of medical waste. Since the early 1990s, medical waste in U.S. coastal waters has faded from public view, and is not considered a public or environmental health threat associated with ocean dumping, primarily because it comprises only a small percent of municipal solid waste items found in the ocean. How-

Pollution of the Ocean by Plastic and Trash

Royal terns (*Sterna maxima*) are among several species of seabirds that dive from the air into the water to catch fish with their sharp beaks. A plastic bag floating at the surface would be invisible to the tern, and may even have attracted the fish in the first place. In this photograph, the tern's bill penetrated the plastic and left the bird wearing the bag around its neck like a shroud. The tern was treated at the University of Texas's Animal Rehabilitation Keep (ARK) and eventually was released back to the wild.

ever, plastic and trash remain major concerns due to their widespread presence.

Scientists and environmental groups since the 1980s have more closely addressed the problem of marine debris and its effect on public health and wildlife. The general public has helped in organized beach cleanups using data sheets to record the types and numbers of items found. Cleanups were at first limited to beaches bordering the United States ocean coastlines, but soon international efforts were organized, and cleanups were also done along riverbanks, lakeshores, **marinas**, and even by divers to recover submerged items along **jetties** and fishing piers.

MARPOL. Due in part to the public attention paid to marine debris, an international agreement (MARPOL, or The International Convention for the Prevention of Pollution by Garbage from Ships; Annex V) was reached by many of the world's governments to prohibit or limit the quantity of garbage that can be discharged at sea or in waterways that lead to the sea. (The United States ratified Annex V in 1987.) In some areas like the Gulf of Mexico, there is a total ban on discharging plastics into the sea. All vessels must carry signs informing crews of the laws and must provide containers for different types of materials that will be offloaded at the next port of call rather than dumped into the sea.

marina: a water-based facility used for storage, service, launching, operation, or maintenance of watercraft

jetty: a structure built out into the sea, a lake, or a river to counteract the effects of tides or currents

Persistent Problem. After the United States ratified MARPOL Annex V in 1987, the quantity of marine debris decreased, but has increased again in recent years. Solutions include:

- Improving the general public's awareness, concern, and attitude towards littering;

- Reducing the use of plastic and other materials for disposable packaging; and
- Enforcing existing laws, especially at sea, to punish habitual litterers.

As the twenty-first century opened, marine debris continued to wash ashore on beaches around the world, including the United States. There has been some reduction in the quantity of trash on America's ocean beaches, partly because of adherence to the law, partly because of self-imposed company rules, and partly owing to increased public awareness. Much of the credit for this must be given to the beach cleanups that have given the public a first-hand look at the problem. Even so, marine mammals, reptiles, birds, and other ocean life continue to sustain injuries from ingesting or becoming entangled in plastic debris and fishing gear. SEE ALSO BEACHES; COASTAL WATERS MANAGEMENT; HUMAN HEALTH AND THE OCEAN; OCEAN HEALTH, ASSESSING; POLLUTION OF STREAMS BY GARBAGE AND TRASH; POLLUTION OF THE OCEAN BY SEWAGE, NUTRIENTS, AND CHEMICALS; WASTEWATER TREATMENT AND MANAGEMENT.

Anthony F. Amos

Bibliography

Amos, Anthony F. *Solid Waste Pollution on Texas Beaches: A Post-MARPOL Annex V Study: OCS Study MMS 93-0013* New Orleans, LA: U.S. Department of the Interior, Minerals Management Service, Gulf of Mexico OCS Region, vol. 1, 1993.

Center for Marine Conservation. *A Citizen's Guide to Plastics in the Ocean: More than a Litter Problem*. Washington, D.C.: Center for Marine Conservation, 1994

Coe, James M., and Donald B. Rogers, eds. *Marine Debris: Sources, Impacts, and Solutions*. New York: Springer-Verlag, 1996.

Committee on Shipborne Wastes, Marine Board Commission on Engineering and Technical Systems, National Research Council. *Clean Ships, Clean Ports, Clean Oceans: Controlling Garbage and Plastic Wastes at Sea*. Washington, D.C.: National Academy Press, 1995.

Internet Resources

Assessing and Monitoring Floatable Debris. U.S. Environmental Protection Agency, Oceans and Coastal Protection Division. <http://www.epa.gov/owow/oceans/debris/floatingdebris/toc.html>.

Marine Debris. The Ocean Conservancy. <http://www.oceanconservancy.org/dynamic/issues/threats/debris/debris.htm>.

Marine Debris Abatement: Trash in Our Oceans—You Can Be Part of the Solution. U.S. Environmental Protection Agency, Ocean and Coastal Protection Division. <http://www.epa.gov/owow/oceans/debris/index.html>.

Pollution of the Ocean by Sewage, Nutrients, and Chemicals

Coastal waters receive a variety of land-based water pollutants, ranging from petroleum wastes to pesticides to excess sediments. Marine waters also receive wastes directly from offshore activities, such as ocean-based dumping (e.g., from ships and offshore oil and gas operations).

One pollutant in the ocean is sewage. Human sewage largely consists of excrement from toilet-flushing; wastewater from bathing, laundry, and dishwashing; and animal and vegetable matter from food preparation that is disposed through an in-sink garbage disposal. Because coasts are densely populated, the amount of sewage reaching seas and oceans is of particular concern because some substances it contains can harm **ecosystems** and pose a significant public health threat. In addition to the nutrients which can cause overenrichment of receiving waterbodies, sewage carries an array of potentially disease-causing microbes known as pathogens.

Animal wastes from feedlots and other agricultural operations (e.g., manure-spreading on cropland) pose concerns similar to those of human wastes by virtue of their microbial composition. Just as inland rivers, lakes, and **groundwater** can be contaminated by pathogenic microbes, so can coastal waters. Runoff from agricultural areas also contains nutrients such as phosphorus and nitrogen, which can cause overenrichment in coastal regions that ultimately receive the runoff.

The major types of ocean pollutants from industrial sources can be generally categorized as petroleum, hazardous, thermal, and radioactive. Petroleum products are oil and oil-derived chemicals used for fuel, manufacturing, plastics-making, and many other purposes. Hazardous wastes are chemicals that are toxic (poisonous at certain levels), reactive (capable of producing explosive gases), corrosive (able to corrode steel), or ignitable (flammable). Thermal wastes are heated wastewaters, typically from power plants and factories, where water is used for cooling purposes. Radioactive wastes contain chemical elements having an unstable nucleus that will spontaneously decay with the concurrent emission of ionizing radiation.

Sewage and Agricultural Wastes

Sewage originates primarily from domestic, commercial, and industrial sources. In many developed countries, these wastes typically are delivered either to on-site septic systems or to centralized sewage treatment facilities. In both methods, sewage is treated before being discharged, either underground (in the case of septic tanks) or to receiving surface-water bodies (in the case of sewage treatment plants), typically a stream, river, or coastal outlet.

Although sewage treatment facilities are designed to accommodate and treat sewage from their service area, partly treated or even untreated sewage sometimes is discharged. Causative factors include decayed **infrastructure**; facility malfunctions; or heavy rainfall events which overwhelm systems using combined sewers and stormwater drains (known as combined sewer overflows). In unsewered areas, improperly designed or malfunctioning septic tanks can contaminate groundwater and surface water, including coastal waters. In some developed regions (e.g., Halifax Harbor in Nova Scotia, Canada), raw sewage continues to pour into harbors, bays, and coastal waters. In developing countries with no on-site or centralized sanitation facilities, no opportunity exists for any type of treatment, and human wastes go directly into surface waters, including the coastal ocean.

Sewage Sludge.
Another source of ocean pollution by sewage-related waste is the disposal of biosolids, a semisolid byproduct of the sewage treatment process, often called sludge. Historically, sludge in developed nations was disposed in coastal waters: New York's twenty sewage treatment plants, for example, once disposed their sludge offshore in a region known as the New York Bight. Although today's environmental regulations in the United States prohibit this practice, sewage sludge is still disposed at sea in some countries.

---

ecosystem: the community of plants and animals within a water or terrestrial habitat interacting together and with their physical and chemical environment

groundwater: generally, all subsurface (underground) water, as distinct from surface water, that supplies natural springs, contributes to permanent streams, and can be tapped by wells; specifically, the water that is in the saturated zone of a defined aquifer

infrastructure: the permanent constructed system (e.g., pipes and other structures) that enables the treatment and delivery of water to support human habitation and activity, or that supports manufacturing activities and water projects (e.g., desalinization and hydropower plants)

Pollution of the Ocean by Sewage, Nutrients, and Chemicals

Disease-causing microbes are the primary human health risk in sewage-contaminated waters, and the main cause of recreational beach closures. Here a sign warns San Diego beachgoers of sewage in the waters.

estuary: a tidally influenced coastal area in which fresh water from a river mixes with sea water, generally at the river mouth; the resulting water is brackish, which results in a unique ecosystem

Agricultural Wastes. Animal wastes often reach waterbodies via runoff across the land surface, or by seepage through the surface soil layers. Hence, agricultural runoff containing animal wastes does not receive any "treatment" except what is naturally afforded by microbial activity during its transit to a waterbody. In coastal watersheds, these wastes can flow through river networks that eventually empty into the sea.

Coastal Eutrophication. Nutrients and organic materials from plants, animals, and humans that enter coastal waters, either directly or indirectly, can stimulate a biological, chemical, and physical progression known as eutrophication. Coastal eutrophication is commonly observed in **estuaries**, bays, and marginal seas. In a broad sense, coastal eutrophication mirrors the eutrophication of lakes. For example, as increased nutrients stimulate algal and other plant growth, light transmission decreases. The eventual bacterial decay of algae and other plants lowers the dissolved oxygen level in the water. In extreme cases, all of the oxygen can be removed.

Human-accelerated eutrophication (known as cultural eutrophication) can be triggered by inputs of sewage, sludge, fertilizers, or other wastes containing nutrients such as nitrogen and phosphorus. As recently as the 1980s, for example, the New York Bight was essentially lifeless due to oxygen depletion, caused largely by decades of sewage and sludge disposal. As of 2002, Halifax Harbor was still receiving a daily influx of raw sewage, creating serious ecological and public health concerns.

Nutrient-enriched runoff from agricultural land in the midwestern United States is the primary cause of the well-known Gulf of Mexico "Dead Zone." Half of the U.S. farms are located in the Mississippi River Basin, whose entire drainage basin empties into the gulf. Much of the nitrogen reaching the gulf is from agricultural fertilizers, with lesser amounts from residential fertilizers and other sources. The water of the 20,000-kilometer (7,728-square-mile) Dead Zone, extending from the mouth of the Mississippi River Basin to beyond the Texas border, has so little oxygen that essentially no marine life exists.

If human-accelerated eutrophication is not reversed, the entire coastal ecosystem ultimately may be changed. Sensitive species may be replaced by more tolerant and resilient species, and biologically diverse communities may be replaced by less diverse ones. Further, nutrient enrichment and the associated eutrophication in coastal waters is implicated in some harmful algal blooms, in which certain species of algae produce biotoxins (natural poisons) that can be transferred through the food web, potentially harming higher-order consumers such as marine mammals and humans.

Human Health. Sewage, particularly if partially treated or untreated, brings high microbe concentrations into the ocean. Human diseases can be caused by waterborne pathogens that contact the skin or eyes; waterborne pathogens that are accidentally ingested when water is swallowed; or foodborne pathogens found in the tissues of fish and shellfish consumed as seafood.*

* See "Human Health and Water" for a summary of common waterborne pathogens.

Beach pollution consequently is a persistent public health problem. Annually, thousands of swimming advisories and beach closings are experienced because high levels of disease-causing microbes are found in the water. Sewage often is responsible for the harmful microbial levels.

Seafood contaminated by sewage-related pathogens sickens untold numbers of people worldwide. Regulatory agencies will close a fishery when contamination is detected. However, many countries lack regulatory oversight or the resources to adequately monitor their fisheries.

Industrial Wastes

Industrial wastes primarily enter coastal waters from terrestrial (land-based) activities. Industries, like municipalities and other entities that generate wastes, dispose of many liquid wastes through wastewater systems (and ultimately to waterbodies), whereas they dispose of their solid wastes in landfills.

The quantity and characteristics of industrial wastewater depends on the type of industry, its water and wastewater management, and its type of waste pretreatment (if any) before delivery to a wastewater (sewage) treatment plant. Because industrial waste frequently goes down the same sewers as domestic and commercial nonindustrial waste, sewage often contains high levels of industrial chemicals and heavy metals (e.g., lead, mercury, cadmium, and arsenic).

Substances that are not removed by wastewater treatment processes are discharged via the treated effluent to a receiving stream, river, or coastal outlet. Inland waters ultimately reach the ocean, carrying with them some residual chemical that are not attenuated, stored, or degraded during their journey through the watershed. Other land-based sources of industrial pollutants in the ocean are pipeline discharges and transportation accidents, leaking underground storage tanks, and activities at ports and harbors. Intentional, illegal dumping in inland watersheds and in inland waterbodies also can deliver industrial wastes to drainageways, and ultimately to the ocean.

In coastal watersheds, some industries discharge their wastes directly to the ocean. Like industries located inland, these industries must first obtain a permit under the Clean Water Act. Industrial pollutants also can directly enter the ocean by accidental spills or intentional dumping at sea.

Wet and dry deposition of airborne pollutants is a sometimes overlooked, yet significant, source of chemical pollution of the oceans. For example,

SEWAGE FROM VESSELS

The Clean Water Act regulates the discharge of sewage from commercial and recreational vessels. The U.S. Environmental Protection Agency, Coast Guard, and individual states work jointly to protect human health and the aquatic environment from disease-causing microorganisms which may be present in sewage from boats. The act established standards for marine sanitation devices and no-discharge zone designations for vessels. As of 2002, seventeen coastal and Great Lakes states had designated part or all of their surface waters as no-discharge zones. Also that year, the Environmental Protection Agency and Coast Guard were assessing potential regulatory amendments that would more stringently regulate discharges from cruise ships in offshore waters.

sulfur dioxide from a factory smokestack begins as air pollution. The polluted air mixes with atmospheric moisture to produce airborne sulfuric acid that falls on water and land as acid rain. This deposition can change the chemistry and ecology of an aquatic ecosystem. The major transport of PCBs to the ocean, for example, occurs through airborne deposition.

Industrial chemicals can adversely affect the growth, reproduction, and development of many marine animals. Pollutants are appearing not only in the Pacific, Atlantic, and Indian Oceans and their marginal seas, but also in the more remote and once-pristine polar oceans. An array of contaminants have been found in the flesh of fish and marine mammals in polar regions. In addition to the environmental and ecological issues, there is growing concern over the potential human health impacts in aboriginal communities whose residents depend on fish and marine mammals for daily sustenance.

A major public health concern is the safety of seafood as it relates to the chemical pollution of waters used for commercial and recreational fishing and **mariculture**. Heavy metals (e.g., copper, lead, mercury, and arsenic) can reach high levels inside marine animals, and then be passed along as seafood for humans. A well-known case of human poisoning occurred in Japan, where one industry dumped mercury compounds into Minimata Bay from 1932 to 1968. Methyl mercury that accumulated in fish and other animals was passed along to humans who consumed them. Over 3,000 human victims and an unknown number of animals succumbed to what became known as "Minimata Disease", a devastating illness that affects the central nervous system.

mariculture: the science, art, and business of cultivating marine animals or plants under controlled conditions; a subcategory of aquaculture

Monitoring by fisheries, environmental, and public health agencies can prevent or minimize cases of human illness caused by chemical contaminants in seafood. Some shellfish-producing areas off the U.S. coasts have been either permanently closed or declared indefinitely off-limits by health officials as a result of this type of pollution. A large percentage of U.S. fish and shellfish consumption advisories are due to abnormally high concentrations of chemical contaminants in seafood.

Regulatory Controls

The 1890 River and Harbors Act prohibited any obstruction to the navigation of U.S. Waters, and hence regulated the discharge of dredged material into inland and coastal waters. By weight, dredged material comprises 95 percent of all ocean disposal on a global basis. Its regulation (administered by the U.S. Army Corps of Engineers) increasingly is being accomplished in concert with broader concerns, including ecological integrity and other public interests.

In 1972, the U.S. Congress passed the Marine Protection, Research, and Sanctuaries Act (Ocean Dumping Act) and the Federal Water Pollution Control Act Amendments (Clean Water Act) that, among other goals, prohibited the disposal of waste materials into the ocean, and regulated the discharge of wastes through pipelines into the ocean. The Ocean Dumping Act requires the federal review of all proposed operations involving the transportation of waste materials for the purpose of ocean dumping, and calls for an assessment of the potential environmental and human health impacts. The U.S. Army Corps of Engineers and U.S. Environmental Protection Agency implement the permit programs associated with these laws.

In the United States, ocean dumping of industrial wastes is prohibited. Yet the vastness of the open sea provides a haven for illegal dumping.

The Ocean Dumping Ban Act of 1988 significantly amended portions of the 1972 Ocean Dumping Act, and banned ocean dumping of municipal sewage sludge and industrial wastes (with limited exceptions) by phased target dates. The disposal of sewage sludge in waters off New York City was a major motivation for its enactment. Ocean disposal of sewage sludge and industrial waste was totally banned after 1991. Narrow exceptions were created for certain U.S. Army Corps of Engineers dredge materials that occasionally are deposited offshore. Dredging is necessary to maintain navigation routes for trade and national defense. Consequently, allowable ocean dumping in the United States since 1991 has essentially been limited to dredge material and fish wastes.

Two international conferences in 1972—the UN Conference on the Human Environment, and the Intergovernmental Conference on the Convention on the Dumping of Wastes at Sea—were the result of international recognition of the need to regulate ocean disposal from land-based sources on a global basis. These conferences resulted in an international treaty, the Convention on the Prevention of Marine Pollution by Dumping of Wastes and Other Matter (also known as the London Convention).

Another treaty addressing the issue of wastes disposed from vessels was adopted in 1973. The International Convention for the Prevention of Pollution from Ships (or MARPOL) calls for signatory nations to enforce bans on dumping oil and noxious liquids into the ocean from ships, but the disposal of hazardous substances, sewage, and plastics remains optional.

As per the U.S. regulations, the dumping of industrial wastes, radioactive wastes, warfare agents (chemical or biological), sewage, and incineration at sea are directly prohibited. Moreover, the ocean disposal of other waste materials containing greater than trace amounts of certain chemicals is strictly prohibited. Allowed under strictly regulated conditions are the ocean disposal of relatively uncontaminated dredged material (harbor sediments), geologic material, and some fish waste; burial at sea; and ship disposal.

In 2000, the U.S. Congress enacted the Beaches Environmental Assessment and Coastal Health Act (BEACH Act) to reduce the risk of disease to users of the nation's coastal and Great Lakes waters. Funds are being made available for states and tribes to establish monitoring programs for disease-causing microbes, and to notify the public when monitoring indicates and public health hazard. SEE ALSO ALGAL BLOOMS, HARMFUL; ALGAL BLOOMS IN THE OCEAN; CLEAN WATER ACT; COASTAL WATERS MANAGEMENT; ECOLOGY, MARINE; ESTUARIES; HUMAN HEALTH AND THE OCEAN; HUMAN HEALTH AND WATER; LAND USE AND WATER QUALITY; LAND-USE PLANNING; LEGISLATION, FEDERAL WATER; MICROBES IN LAKES AND STREAMS; MICROBES IN THE OCEAN; NUTRIENTS IN LAKES AND STREAMS; OCEAN HEALTH, ASSESSING; OIL SPILLS: IMPACT ON THE OCEAN; POLLUTION OF LAKES AND STREAMS; WASTEWATER TREATMENT AND MANAGEMENT.

Cindy Clendenon (with William Arthur Atkins)

Bibliography

Clark, Robert B. *Marine Pollution*, 4th ed. New York: Oxford Press, 1997.

Gorman, Martha. *Environmental Hazards: Marine Pollution*. Santa Barbara, CA: ABC-CLIO, 1993.

Internet Resources

Assessing and Monitoring Floatable Debris. U.S. Environmental Protection Agency, Oceans and Coastal Protection Division. <http://www.epa.gov/owow/oceans/debris/floatingdebris/toc.html>.

Beaches. U.S. Environmental Protection Agency. <http://www.epa.gov/waterscience/beaches/>.

Hypoxia in the Gulf of Mexico. Gulf of Mexico Hypoxia Assessment, National Oceanic Service, National Oceanic and Atmospheric Administration. <http://www.nos.noaa.gov/products/pubs_hypox.html#Intro>.

Marine Pollution Control Programs. U.S. Environmental Protection Agency, Oceans and Coastal Protection Division. <http://www.epa.gov/owow/oceans/regs/index.html>.

Marine Pollution in the United States. Pew Oceans Commission. Available online at <http://pewoceans.org/reports/022701report.pdf>.

"Oceans and Coastal Resources: A Briefing Book." Congressional Research Service Report 97-588 ENR, National Council for Science and the Environment. <http://www.crie.org/nle/crsreports/briefingbooks/oceans/a3.cfm>.

Pollution. The Ocean Conservancy. <http://www.oceanconservancy.org/dynamic/issues/threats/pollution/pollution.htm>.

Pollution Sources: Point and Nonpoint

All activities on Earth, both natural processes and human-made processes, produce some type of byproduct from that activity. Under normal conditions these byproducts, some known as **pollutants**, are returned back into the environment. In fact, natural environmental processes have the ability to assimilate some pollutants and correct most imbalances if given enough time. However, if a persistent overload of a pollutant is allowed to continue or the pollutant is not a substance that the environment can handle, then the environment has little chance to "self-clean."

Point and Nonpoint Pollution Sources

In the simplest of terms, a pollutant is a substance that enters the environment and elevates the "natural" background level of that substance. In many

pollutant: something that pollutes, especially a waste material that contaminates air, soil, or water

Pollution Sources: Point and Nonpoint

This drainage outlet delivering polluted runoff into the Ohio River is a point source of pollution because the pollution originates from a single, identifiable source.

cases, the natural system may not have any of the substance present until human activities add it to the environment.

Pollution originating from a single, identifiable source, such as a discharge pipe from a factory or sewage plant, is called point-source pollution. Pollution that does not originate from a single source, or point, is called nonpoint-source pollution. Liquid, solid, and airborne discharges from point sources as well as pollutants from nonpoint sources may go either into surface water or into the ground. (Airborne pollutants can be assimilated into rainwater and can affect water quality: acid rain is an example.) The ability for these pollutants to reach surface water or groundwater is enhanced by the amount of water available from precipitation (rain) or irrigation.

Point Sources

Point-source pollutants in surface water and groundwater are usually found in a **plume** that has the highest concentrations of the pollutant nearest the source (such as the end of a pipe or an underground injection system) and diminishing concentrations farther away from the source. The various types of point-source pollutants found in waters are as varied as the types of business, industry, agricultural, and urban sources that produce them.

Commercial and industrial businesses use hazardous materials in manufacturing or maintenance, and then discharge various wastes from their operations. The raw materials and wastes may include pollutants such as solvents, petroleum products (such as oil and gasoline), or **heavy metals**.

plume: a concentrated area or mass of a substance that is emitted from a natural or human-made point source and that spreads in the environment; a plume can be thermal, chemical, or biological in nature

heavy metals: a group of metals that have high density and are considered toxic at specified concentrations; with respect to soil management, such metals include copper, iron, manganese, molybdenum, cobalt, zinc, cadmium, mercury, nickel, and lead

243

Pollution Sources: Point and Nonpoint

Point sources of pollution from agriculture may include animal feeding operations, animal waste treatment lagoons, or storage, handling, mixing, and cleaning areas for pesticides, fertilizers, and petroleum. Municipal point sources might include wastewater treatment plants, landfills, utility stations, motor pools, and fleet maintenance facilities.

For all of these activities, hazardous materials may be included in the raw materials used in the process as well as in the **waste stream** for the facility. If the facility or operator does not handle, store, and dispose of the raw materials and wastes properly, these pollutants could end up in the water supply. This may occur through discharges at the end of a pipe to surface water, discharges on the ground that move through the ground with infiltrating rainwater, or direct discharges beneath the ground surface.

waste stream: the chemical composition and character of wastes produced at a facility; waste streams often differ from the composition of the initial products used; can refer to liquid or solid wastes

Groundwater. Some of the most persistent point-source pollutants in groundwater are **volatile organic compounds**, which include manufactured and refined toxic substances such as solvents, oils, paint, and fuel products. In general, it takes only a small amount of these chemicals to raise health concerns. For example, approximately 4 liters (about one gallon) of pure trichloroethylene, a common solvent, will contaminate over 1 billion liters (300 million gallons) of water. Once groundwater is contaminated, it is difficult, costly, and sometimes even impossible to clean up.

volatile organic compounds: organic compounds that can be isolated from the water phase of a sample by purging the water sample with inert gas, such as helium, and subsequently analyzed by gas chromatography

Surface Water. The most common point-source pollutants in surface water are:

- High-temperature discharges;
- Microorganisms (such as bacteria, viruses, and *Giardia*); and
- Nutrients (such as nitrogen and phosphorus).

Temperature increases and nutrients can result in excessive plant growth and subsequent decaying organic matter in water that depletes dissolved oxygen levels and consequently stressing or killing vulnerable aquatic life. Microorganisms can be hazardous to both human health and aquatic life. Pesticides and other toxic substances can also be hazardous to both human health and aquatic life, but are less commonly found in surface water because of high dilution rates.

Nonpoint Sources

Nonpoint-source pollution occurs as water moves across the land or through the ground and picks up natural and human-made pollutants, which can then be deposited in lakes, rivers, wetlands, coastal waters, and even groundwater. The water that carries nonpoint-source pollution may originate from natural processes such as rainfall or snowmelt, or from human activities such as crop irrigation or lawn maintenance.

Nonpoint-source pollution is usually found spread out throughout a large area. It is often difficult to trace the exact origin of these pollutants because they result from a wide variety of human activities on the land as well as natural characteristics of the soil, climate, and **topography**.

topography: the shape and contour of a surface, especially the land surface or ocean-floor surface

sediment: rock particles and other earth materials that are transported and deposited over time by geologic agents such as running water, wind, glaciers, and gravity; sediments may be exposed on dry land and are common on ocean and lake bottoms and river beds

The most common nonpoint-source pollutants are **sediment**, nutrients, microorganisms and toxics. Sediment can degrade water quality by contaminating drinking water supplies or silting in spawning grounds for fish and

This silt-laden runoff from a residential area contains not only soil and clay particles from nearby construction, but also is likely to contain small amounts of lawn chemicals, oil, grease, gasoline, and even residues from recent highway de-icing. These are all examples of pollutants released from nonpoint sources.

other aquatic species. Nutrients, microorganisms, and other toxic substances can be hazardous to human health and aquatic life.

People can contribute to nonpoint-source pollution without even realizing it. Nonpoint sources of pollution in urban areas may include parking lots, streets, and roads where stormwater picks up oils, grease, metals, dirt, salts, and other toxic materials. In areas where crops are grown or in areas with landscaping (including grassy areas of residential lawns and city parks), irrigation, and rainfall can carry soil, pesticides, fertilizers, herbicides, and insecticides to surface water and groundwater. Bacteria, microorganisms, and nutrients (nitrogen and phosphorus) are common nonpoint-source pollutants from agricultural livestock areas and residential pet wastes. These pollutants are also found in areas where there is a high density of septic systems or where the septic systems are faulty or not maintained properly. Other pollutants from nonpoint sources include salt from irrigation practices or road de-icing, and acid drainage from abandoned mines.

Preventing and Controlling Pollution

Over the years, federal laws and regulations have established a process for the U.S. Environmental Protection Agency (EPA) and the states to regulate point sources of pollution through issuing of permits that limit the types and amounts of pollutants a facility can discharge. In addition, there are many laws and regulations that mandate the ways that hazardous materials are handled, stored, and used. Those same laws and regulations often encourage voluntary pollution-prevention efforts to reduce and minimize the use of potential pollutants.

These laws, regulations, and voluntary efforts have helped clean up major water quality problems and reduced the amount of pollutants directly discharged to surface water and groundwater. However, EPA reports that more than one-third of the nation's waters are still not meeting water quality standards. Nonpoint sources of pollution have been identified as the primary reason for these continued problems.

watershed: the land area drained by a river and its tributaries; also called river basin, drainage basin, catchment, and drainage area

Addressing Nonpoint Sources. Preventing and controlling nonpoint-source pollution is primarily accomplished through regulation under the Clean Water Act and voluntary **watershed** protection efforts. Best management practices and pollution prevention can be applied at the local, state, and federal level to reduce and prevent nonpoint-source pollution. Some activities are federal and state responsibilities, such as ensuring that public lands are properly managed to reduce soil erosion, or developing legislation to govern chemical use. Many other regulatory approaches are best handled locally, such as by zoning or erosion-control ordinances. Each citizen can play an important role by being active in the community, learning more about the local watershed, practicing conservation, and by preventing pollution in homes, yards, and neighborhoods. SEE ALSO CHEMICAL ANALYSIS OF WATER; CHEMICALS FROM AGRICULTURE; CLEAN WATER ACT; LANDFILLS: IMPACT ON GROUNDWATER; POLLUTION OF GROUNDWATER; RUNOFF, FACTORS AFFECTING; SEPTIC SYSTEM IMPACTS.

Julie K. Harvey

Bibliography

Zoller, Uri, ed. *Groundwater Contamination and Control.* New York: Marcel Drekker, Inc., 1994.

Internet Resources

NonPoint Source Pointers (Factsheets). "NPS Home." U.S. Environmental Protection Agency (EPA), Office of Water. Nonpoint Source Pollution Pointer No's 1 through 7 EPA841-F-96-004(A through G) <http://www.epa.gov/OWOW/NPS/facts/index.html>.

Polluted Runoff: Nonpoint Source Pollution. U.S. Environmental Protection Agency, Office of Water. <http://www.epa.gov/OWOW/NPS/>.

Population and Water Resources

People use water for drinking, bathing, cooking, washing clothes, and maintaining lawns and gardens. Water also is used by the manufacturing sector to make products, by the agricultural industry to provide food, and by the energy industry to provide illumination, heat, and air conditioning.

The amount of water used directly by individuals is related to various human attributes such as age, education, cultural background, religious beliefs, and financial status. In general, more people use more water, even if the amount they use individually is reduced by education, the implementation of conservation practices, or technological improvements in water-supply systems.

Water sources in a specific region vary in the quantity and quality of water they contain at a given time, and in their rate and timing of replenishment. If projected withdrawals to meet population growth exceed the ability of the water sources that may be called upon to meet them, then new sources must be developed, if that is possible; otherwise, cutbacks in water use will be required. Yet demands can be decreased only so far until the decreases may endanger public health, damage the environment, or adversely influence the region's economy.

Population and Water Resources

Population Impacts on Future Water Sources

The impacts of population on the quantitative water needs of a locality are related to population density (that is, how the population is distributed geographically), and to the rate of increase or decrease in population growth. Because population changes affect such variables as the economy, the environment, natural resources, the labor force, energy requirements, infrastructure needs, and food supply, they also affect the availability and quality of the water sources that can be drawn upon for use.

Population is highly correlated with public water supply, about 56 percent of which is allocated for domestic (household) purposes. According to the U.S. Geological Survey, the average per capita public water use in the United States in 1995 was about 179 gallons per capita per day (gpcd) and that for domestic water use was about 101 gpcd. An average per capita figure for all water uses in the United States in 1995 (municipal, industrial, agricultural, etc.) was estimated to be about 1,280 gpcd.

The U.S. population in 2000 was about 275.6 million. Projections by the Population Research Bureau indicate a 2025 population of 373.8 million and a 2050 population of 403.7 million. The bureau also reported that the doubling time from the year 2000 for the population of the United States, at its current rate of growth, is about 120 years, for the world 51 years, and for the less developed countries, including China, 36 years. The importance

Sustainable growth levels in many localities around the world are already being exceeded because population growth outpaces water-supply capabilities. This single water tower amidst a sea of houses in Baguio, Luzon in the Philippines represents the population-related demands on limited water sources.

of these estimates can be seen when one notes that about 81 percent of the world's current population resides in less-developed countries.

Future Population Levels. The world's population reached just over 6 billion in 2000, and it is expected to peak at about 8 to 10 billion sometime this century. According to the Population Reference Bureau, about 7.8 billion people will inhabit the Earth by 2025, and by 2050, the total will reach about 9 billion. More than 90 percent of this growth is expected to take place in less developed nations (see graph), many of which are overpopulated and are stressing their water and other resources. If population projections prove to be reliable, many regions of planet Earth will be facing significant water shortages within the next 50 years.

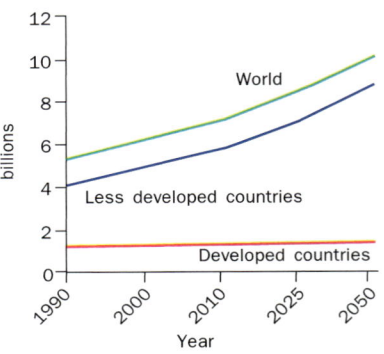

If the 1995 figure of 1,280 gpcd for all water uses in the United States is multiplied by the bureau's estimate of population in 2025, a rough estimate of water use in that year would be about 508 billion gallons per day. Although uncertainty exists in the population forecast, and technological changes and conservation practices could likely reduce the overall per capita water use in the future, significant increases are expected in population-related water use. However, in arid areas of the United States, large increases in water use may not be possible or **sustainable** unless water is imported or **brackish** or **saline** waters are desalinized.

Population Impacts on Future Water Quality

The impact of population on the ability of water sources to meet the demands placed on them by society is paralleled by the effects of population on the quality of water resources. People alter the properties of water as they use it, often degrading the quality with each successive use. Water used in households for drinking, bathing, and cooking becomes contaminated by various chemicals and other constituents introduced during its use. Drainage from water applied in agricultural irrigation carries away chemicals that have been applied to crops to enhance their growth and control weeds and pests. Industries introduce chemicals needed for the manufacture of their products.

As a result of human intervention, waters that have been used for a variety of purposes may contain harmful constituents, including sewage, that pose threats to the environment and to the public health. Their removal can be expensive and difficult.

Issues of water quantity and water quality are inseparable. If the quality of a water source is so degraded that restoring its quality for further use is not feasible, then the source is lost for all practical purposes. Remedial

---

sustainable: as in "sustainable development," describes efforts that guide economic growth in a manner that meets current needs without compromising the ability of future generations to meet their needs; in terms of natural resources, also encompasses development conducted in an environmentally sound manner, with an emphasis on natural resource conservation, including water and aquatic life

brackish: describes water having a salinity from 0.05 to 17 parts per thousand; typically a mixture of sea water and fresh water (e.g., as found in an estuary)

saline: describes water containing a high dissolved mineral content; in sea water, the dominant contributor to salinity is sodium chloride

actions are costly, and prevention rather than remediation should be the goal. To achieve it, the public, industries, governments, agencies, and a variety of organizations must all play a positive role.

Reducing Population Impacts

The impacts of future populations on the amount and quality of water resources available for use can be lessened by modifying the local rate of population increase, by modifying the per capita use of water, and by a combination of the two approaches.

A reduction in the per capita use rate for public water has already been demonstrated in the United States. Per capita use decreased from 184 gpcd in 1990 to 179 gpcd in 1995 even though the nation's population increased by 7 percent during that period. Education can play a major role in bringing about such changes.

Water-stressed regions should seek to slow their population growth and reduce their per capita water use to help alleviate their water supply problems. In general, developing nations are growing faster than industrialized nations. Between 2000 and 2050, most all of the world's population growth is projected to take place in developing nations. A reduction of population growth rate in these nations could significantly enhance the likelihood of achieving sustainability for their water supplies. SEE ALSO DEMAND MANAGEMENT; DEVELOPING COUNTRIES, ISSUES IN; FOOD SECURITY; SUPPLY DEVELOPMENT; SUSTAINABLE DEVELOPMENT; USES OF WATER.

Warren Viessman Jr.

Bibliography

Cunningham, William P., and Barbara Woodworth Saigo. *Environmental Science: A Global Concern*, 5th ed. New York: WCB/McGraw-Hill, 1999.

Gleick, Peter H. *The World's Water: The Biennial Report on Freshwater Resources 2002–2003*. Washington, D.C.: Island Press, 2002.

Tobin, Richard J. "Environment, Population, and the Developing World." In *Environmental Policy* (4th ed.). eds. Norman J. Vig and Michael E. Kraft. Washington, D.C.: CQ Press, 2000.

Viessman Jr., Warren, and Mark J. Hammer. *Water Supply and Pollution Control*, 6th ed. Menlo Park, CA: Addison-Wesley, 1998.

Ports and Harbors

Foreign trade is vitally important to the United States economy and the international movement of goods and materials has highlighted the significance of harbor access and development issues. Natural harbors, typically in bays, **estuaries**, and river mouths, occur where land and water converges in such a way as to protect ships from wind and waves as they enter and dock. Harbors can also be constructed using **jetties** and **breakwaters** to provide protection for ships.

Harbors include entrance channels, interior channels (to allow movement to anchoring areas or turning basins), and support facilities for refueling and repairing vessels. Harbors can be located either on the coast (such as the harbor at Long Beach, California, located on the coast of the Pacific Ocean) or on inland waterbodies (such as the harbor at Chicago, Illinois, located on Lake Michigan).

estuary: a tidally influenced coastal area in which fresh water from a river mixes with sea water, generally at the river mouth; the resulting water is brackish, which results in a unique ecosystem

jetty: a structure built out into the sea, a lake, or a river to counteract the effects of tides or currents

breakwater: a barrier that protects a harbor or shore from the full impact of waves

Ports and Harbors

About 95 percent of all U.S. trade passes through its nationwide system of ports. Terminal facilities like those visible here in the Port of Seattle handle container cargo moving between water-based and land-based transport systems.

Ports

Ports are transfer hubs for trade and are usually built near natural harbors, but they can also be located hundreds of miles up rivers or lakes. For example, in Texas, the Port of Corpus Christi is located along the coast of the Gulf of Mexico. About 320 kilometers (200 miles) north is the Port of Houston, a 40-kilometer (25-mile) long complex of public and private facilities located inland along the Houston Ship Channel.

Ports are land facilities constructed to transfer goods between water and land. They consist of major features such as:

- Docks or berths where vessels moor;
- Equipment and personnel to load and unload vessels;
- Connections to land transportation (such as highways, railways, and pipelines); and
- Cargo storage areas.

Ports are intermodal facilities, a place where rail, truck, barge, ship, and other transport methods converge. In this way, ports play a key role in moving products both to other countries and to the interior of the country. The port's terminal makes possible the docking and the handling, storage and transfer of cargo. Ports are designed to handle a wide variety of types of cargo: bulk or loose, breakbulk in packages (bundles, crates, barrels, pallets,

Ports and Harbors

Harbors vary in location, construction, and purpose. The inner harbor area at Erie, Pennsylvania exemplifies a lake harbor (Lake Erie, top). California's Crescent City harbor is a coastal harbor with breakwaters (Pacific Ocean, bottom).

etc.), liquid (e.g., petroleum), dry bulk (e.g., grain), and general cargo in steel containers.

There are over 260 coastal and inland ports throughout the United States. The states of Texas and Louisiana have the most commercial port facilities. Some of the largest U.S. ports are located on inland waterways, including Houston (Texas), Mobile (Alabama), New Orleans (Louisiana), and Portland (Oregon). The port city farthest from the ocean, Fairmont, West Virginia, is 3,355 kilometers (2,085 miles) via an inland waterway.

Issues in Port Planning

Navigation issues have long had a place in water resources planning. Building and maintaining harbors, ports, and waterways is necessary to the economic health of nearby communities and the nation as a whole. The federal

Ports and Harbors

Increasing public demand for recreational uses of harbors sometimes competes with plans for port development. This harbor in Maine is typical of U.S. coastal harbors used predominantly by recreational boaters and fishers.

government is charged with building and maintaining waterways when it is of substantial benefit to the country. The U.S. Army Corps of Engineers works with local areas to design, build, and maintain waterways; this can include dredging channels, building locks, and so on. However, it is the local port authority that is responsible for the economic planning to make the port a viable operation; this can include attracting clients, planning buildings and equipment, coordinating the distribution of imports and exports, and so on.

Well-maintained channels are vital to the success of any port. Typically the Army Corps of Engineers and various local sponsors work together as partners in the planning process and the costs of waterway projects are shared. Navigation planning must consider many factors, including channel width and depth, shipping and navigation technologies, terminal facilities, climate, seasonal variations, currents, tides, and physical limitations of a waterway (such as bottom conditions). Planners seek to determine the best locations for channels, harbors, and canals. Recent developments in hydraulic modeling have greatly aided these planning efforts. The overall goal of these planning projects is to maximize the benefits of the port to a community.

Available Land. Historically, ports were constructed in geographically favorable locations to expedite the transfer of goods. Active dockside communities emerged and thrived as waterborne trade flourished. The use of available land for landside facilities development was acceptable, meeting few obstacles because such expansion was in the name of economic prosperity.

However, as the twentieth century closed and the twenty-first century emerged, many port locations desired landside facility expansions but faced several obstacles, including land use. Changes in the shipping industry have changed the face of ports. For example, landside facilities capable of accommodating large containerships require long berth lengths, large cranes, and railway or highway access. In most cases, accommodating these increased demands for landside facilities requires the acquisition and development of adjacent land. Yet many port communities have encountered competing demands for waterside land for purposes other than those related to waterborne commerce: namely, commercial, residential, and recreational uses. The result has been disagreement over the use and development of available land.

Environmental Concerns. The increased awareness of environmental conditions has significantly influenced port planning and operations. The dredging or deepening of channels to maintain their navigability involves removing **sediment**, rock, and debris from the channel bottom. The bottom materials can vary from channel to channel, ranging from rock to sand to mud.

Because dredging disturbs the channel bottom material as well as the plants and animals living in the water, its environmental impact depends partly on the type of channel bottom. Where channel bottom material is contaminated, disturbing the sediment can further degrade the environment by distributing or releasing these **contaminants**. (An example is in Long Island Sound, where the effects of early American silver manufacturing and other industries are noted.) Although preventing contamination in the first place is the overarching goal, the key environmental issue when considering waterborne commerce usually is the disturbance of contaminated sediments when waterbodies are dredged, and then finding proper disposal sites for these contaminated sediments.

sediment: rock particles and other earth materials that are transported and deposited over time by geologic agents such as running water, wind, glaciers, and gravity; sediments may be exposed on dry land and are common on ocean and lake bottoms and river beds

contaminant: as defined by the U.S. Environmental Protection Agency, any physical, chemical, biological, or radiological substance in water, including constituents that may not be harmful to the environment or human health

Major U.S. Ports

The most common measures used to gauge or track trade are weight and value of goods handled. The value of goods in dollars unit of measure represents the economic impacts of trade.

Although the United States is serviced by several hundred deep-draft inland and coastal ports, about 76 percent of the value of goods can be attributed to eighteen ports (see box on page 254). According to the U.S. Army Corps of Engineers, the California ports of Los Angeles and Long Beach accounted for a combined 27 percent of value of trade (13.8 and 13.3 percent, respectively) in 2000, followed closely by the Port of New York City with 11.0 percent. The fourth-ranked port based on value of trade was the Port of Houston (Texas) with 5.9 percent. It is interesting to note the geographic distribution of the top four U.S. ports based on value of trade: two are on the Pacific coast, one on the Atlantic coast, and one in the Gulf of Mexico.

Another important aspect is the direction of the value of trade. Trade is based on the two-way exchange of goods: import and export. Of the total U.S. international waterborne trade in 2000, the United States imported

> **TOP U.S. PORTS**
>
> The following ports account for about three-fourths of U.S. trade, based on import–export total value for the year 2000. The ranking of the top ten ports remains fairly stable, but rankings thereafter may vary slightly from year to year.
>
> 1. Los Angeles, CA
> 2. Long Beach, CA
> 3. New York, NY and NJ
> 4. Houston, TX
> 5. Seattle, WA
> 6. Charleston, SC
> 7. Oakland, CA
> 8. Norfolk, VA
> 9. Baltimore, MD
> 10. Tacoma, WA
> 11. New Orleans, LA
> 12. Miami, FL
> 13. Savannah, GA
> 14. Port of South Louisiana, LA
> 15. Beaumont, TX
> 16. Portland, OR
> 17. Jacksonville, FL
> 18. Corpus Christie, TX
> 19. Port Everglades, FL
> 20. Philadelphia, PA

approximately 73 percent of value of the total trade and exported 27 percent. For the top eighteen U.S. ports (which accounted for 76 percent of value of trade in 2000) the balance of trade was also prevalent toward imports. With respect to the distribution of total value of trade between imports and exports, fourteen ports reported a ratio of imports to total trade greater than 60 percent; and four ports recorded a closer balance between imports and exports, with imports still favored between 50 and 60 percent.

SEE ALSO ECONOMIC DEVELOPMENT; NAVIGATION AT SEA, HISTORY OF; TRANSPORTATION.

Terri A. Thomas

Bibliography

Bryan, Leslie A. *Principles of Water Transportation.* New York, NY: The Ronald Press Company, 1939.

Hershman, Marc J. *Urban Ports and Harbors Management.* New York: Taylor & Francis, 1988.

Kendall, Lane C. *The Business of Shipping.* Centreville, MD: Cornell Maritime Press, 1986.

U.S. Department of Transportation. *An Assessment of the U.S. Marine Transportation System: A Report To Congress.* September, 1999.

Internet Resources

Marine Transportation System. U.S. Department of Transportation. <http://www.dot.gov/mts>.

Navigation Data Center. U.S. Army Corps of Engineers, Waterborne Commerce Statistics Center. <http://www.iwr.usace.army.mil/ndc/wcsc.htm>.

Ports: Your Connection to the World. American Association of Port Authorities Online. American Association of Port Authorities <http://www.aapa-ports.org/>.

U.S. Port Totals by Type Service. U.S. Foreign Waterborne Transportation Statistics Program, U.S. Army Corps of Engineers, U.S. Department of Transportation. <http://www.marad.dot.gov/statistics/usfwts/index.html>.

Powell, John Wesley

American Geologist
1834–1902

During America's Gilded Age, John Wesley Powell contributed significantly to a better understanding of the influence of running water on landscapes and the importance of a more rational use of water resources in the American West. His geomorphologic-based classification of rivers and drainage systems is still used today, more than 125 years after he first articulated the concepts.

Leadership and Exploration

John Wesley Powell left formal education, teaching, and natural history to command Union artillery units in the western theater during the American Civil War. Major Powell, who lost his right arm at the Battle of Shiloh (in Tennessee), returned in 1865 to teaching and museum curation. Summer excursions to the central Rocky Mountains generated productive exploration, when he led two daring reconnaissances by boat down the Colorado River in 1869 and 1871. Powell then managed four federal organizations funded by the U.S. Department of Interior: the U.S. Geographical and Geological Survey of the Rocky Mountain Region (from 1872 to 1879); the Bureau of Ethnology (from 1879 to 1902); the U.S. Geological Survey, or USGS (from 1881 to 1894); and the USGS Irrigation Survey (from 1888 to 1891).

John Wesley Powell began his river journeys by rowing up the Mississippi in 1855. He became a distinguished soldier, explorer, and manager of four federal agencies.

River Landscapes and Geologic Controls

Powell, reflecting on the dynamic relations between moving waters and the geologic structures they crossed or matched, presented in 1875 a three-fold classification of rivers and drainage systems based on their history and his concept of a base level of erosion below which streams could not cut. His "antecedent" streams predated geologic uplifts of plateaus and mountains and maintained their courses during elevation; "consequent" streams postdated these uplifts and were controlled by them; and "superimposed" streams exceeded uplift rates to expose older structural settings.

Using and Conserving the West's Water

In 1878, Powell formally proposed reforming land and water use in the West. A decade later Congress funded the Irrigation Survey to aid development in the region. When Powell refused to recommend promptly the dam and **reservoir** sites whose selection would have reopened the public lands to entry and released federal dowry lands to six new states, Congress terminated the Irrigation Survey.

Powell continued to promote reclaiming the West by wise communal irrigation and land use, including organizing drainage basins as counties in new states. In 1894, with the USGS under a new director, Congress restored funds for water-resource investigations by the agency and continued them thereafter. Powell lived to see the National Reclamation Act of 1902

reservoir: a pond, lake, basin, or tank for the storage, regulation, and control of water; more commonly refers to artificial impoundments rather than natural ones

lead to establishing the Reclamation Service within the USGS. SEE ALSO BUREAU OF RECLAMATION, U.S.; GEOLOGICAL SURVEY, U.S.; STREAM EROSION AND LANDSCAPE DEVELOPMENT.

Clifford M. Nelson

Bibliography

Darrah, William Culp. *Powell of the Colorado.* Princeton, NJ: Princeton University Press, 1951.

Powell, John Wesley. *Exploration of the Colorado River of the West and Its Tributaries.* Washington, D.C.: U.S. Government Printing Office, 1875.

——— *Report on the Lands of the Arid Region of the United States,* 2nd ed. Washington, D.C.: U.S. Government Printing Office, 1879.

Rabbitt, Mary C. *Minerals, Lands, and Geology for the Common Defence and General Welfare, Vol. 1: Before 1879* and *Minerals, Lands, and Geology for the Common Defence and General Welfare, Vol. 2: 1879–1904.* Washington, D.C.: U.S. Government Printing Office, 1979–1980.

Stegner, Wallace E. *Beyond the Hundredth Meridian: John Wesley Powell and the Second Opening of the West.* Boston, MA: Houghton Mifflin, 1954.

Worster, Donald. *A River Running West: The Life of John Wesley Powell.* New York: Oxford University Press, 2001.

Precipitation and Clouds, Formation of

Clouds are condensed droplets or ice crystals from atmospheric water vapor. Clouds form by the rising and cooling of air caused by convection, topography, convergence, and frontal lifting. Convection occurs when the Sun's radiation heats the ground surface, and warm air rises, cooling as it goes. Air also is cooled if an air mass is forced to move upward as a result of higher topography (e.g., a mountain range) in a process known as orographic lifting. Interestingly, when the air mass descends on the other side of the mountain, it warms and the clouds may disappear as the droplets transfer back to vapor.✱

✱ See "Climate Moderator, Water as a" for a diagram of orographic lifting and the rain shadow effect.

The counterclockwise motion of a low-pressure center draws air inward, and the convergence forces the air upward. Air also is lifted and cooled along either a cold front or a warm front. A cold front is the leading edge of an air mass that is colder than the air it is replacing. The front forms a wedge that pushes under the warmer air ahead, lifting it. A warm front is the leading edge of an air mass warmer than the air it is replacing. As the air mass pushes forward, the warm air slides up over the wedge of cold air ahead of it, as shown in the following figure.

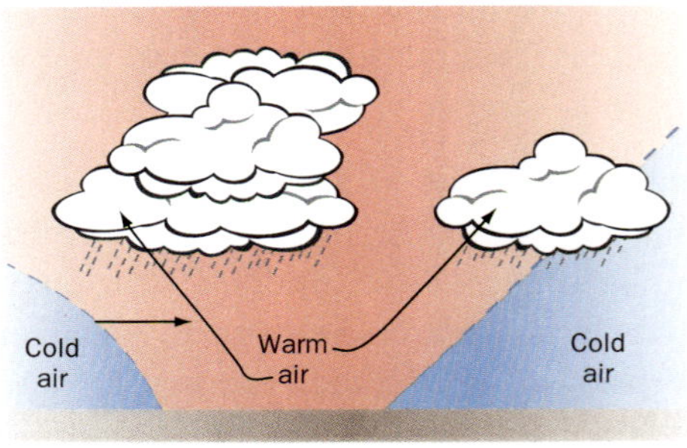

Classification of Clouds

Clouds are classified based on their shape and the height of the cloud's base above the ground. The most common shapes are cirriform, appearing feathery or fibrous; stratoform, appearing layered; and cumuloform, appearing as if piled up. Two additional words used to describe clouds are "nimbus," meaning rain, and "alto," meaning middle. Basic cloud types are based on height above the land surface and on the cloud's vertical development, as summarized below.

- High clouds (cloud base above 7 kilometers or 23,000 feet). Usually consisting of ice crystals, these include cirrus, cirrostratus, and cirrocumulus.
- Middle clouds (2 to 7 kilometers or 6,500 to 23,000 feet). Consisting of liquid droplets, these include altocumulus and altostratus.
- Low clouds (below 2 kilometers or 6,500 feet). Consisting of liquid droplets, these include stratus, stratocumulus, and nimbostratus
- Clouds of vertical development (cloud base generally is in the low cloud range, but the tops may reach great heights). These include cumulus clouds and the towering cumulonimbus.✻

✻ See "Precipitation, Global Distribution of" for a photograph of a cloud of vertical development.

Fog represents a special case of cloud-like formation. Although not truly a cloud, fog is essentially stratoform clouds on the ground.

Cumulus Clouds. Cumulus clouds are among the most interesting in terms of their shapes, which stir peoples' imagination and allow them to see a variety of "objects" or "scenes" in the sky. All cumulus clouds have two characteristics in common. They tend to be bulbous or popcorn-like on top, and have relatively flat bottoms. Why do they all share these features?

Cumulus clouds are classified as clouds of vertical extent. They form as air moves vertically, cooling until the water vapor in the air condenses. This vertical movement is the key to understanding the flat-bottomed character of these clouds. In a given area, for a given air mass, it is common to find that the cooling rate of ascending air is relatively constant. In other words, rising air will cool to a certain temperature at the same height above the ground throughout the area. Air moving vertically in that area will reach the condensation temperature at the same height above the ground and cloud formation will begin at that height. Consequently, the base of the cloud will be at the same height throughout and the cloud's base will appear flat.

As the air continues to rise and water vapor continues to condense, the cloud will extend vertically. The column of rising air actually consists of a number of currents with slightly different directions and may move as pulses. These currents move upward as they condense and give rise to the apparently independent bulbous lobes of the cloud.

Precipitation Types

Precipitation elements begin to form in the part of the cloud where ice crystals and cloud droplets coexist. Most precipitation starts out as snow, except for rain that comes out of very low clouds. Precipitation will remain snow unless it falls through a layer of warmer air, in which case it will melt and remain rain unless it falls through a colder layer of air, where it may freeze and become sleet or ice pellets (as shown in the following figure). When

Precipitation and Clouds, Formation of

the air at the ground level is below freezing, the raindrops can freeze when they hit the ground or other cold surfaces: it is then called freezing rain.

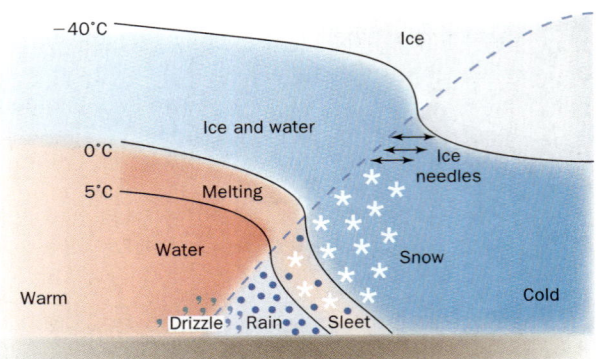

Hail is formed when a particle, like dust, attracts a drop of moisture to itself. The particle gets blown upward by strong updrafts in the cloud, and freezes as it goes through a colder layer of air. It is heavier and begins to fall, attracts more moisture and then gets forced upward again and again adding more frozen layers.

The varying intensities of rainfall have specific names. Liquid precipitation that is of a longer duration and larger drop size is called rain; when it falls in shorter spurts it is called a shower. Rain typically falls from low-level stratoform clouds with greater vertical extent. When the drops are very small, rain is called drizzle. A raindrop is about 1,000 times larger than drizzle.

Thunderstorms. Thunderstorms go through stages of development from the beginning cumulus stage, when the cloud starts to grow vertically; to the mature stage, when heights may reach from 12 to 18 kilometers (8 to 11 miles); to the dissipating stage (see figure below).

In the cumulus stage, there is an updraft of warm air throughout the cloud, as shown in part (a) of the figure. As the warm updraft increases, the

This schematic illustrates the three general stages of thunderstorm development. An elevation of 40,000 feet is equivalent to 12 kilometers or 8 miles.

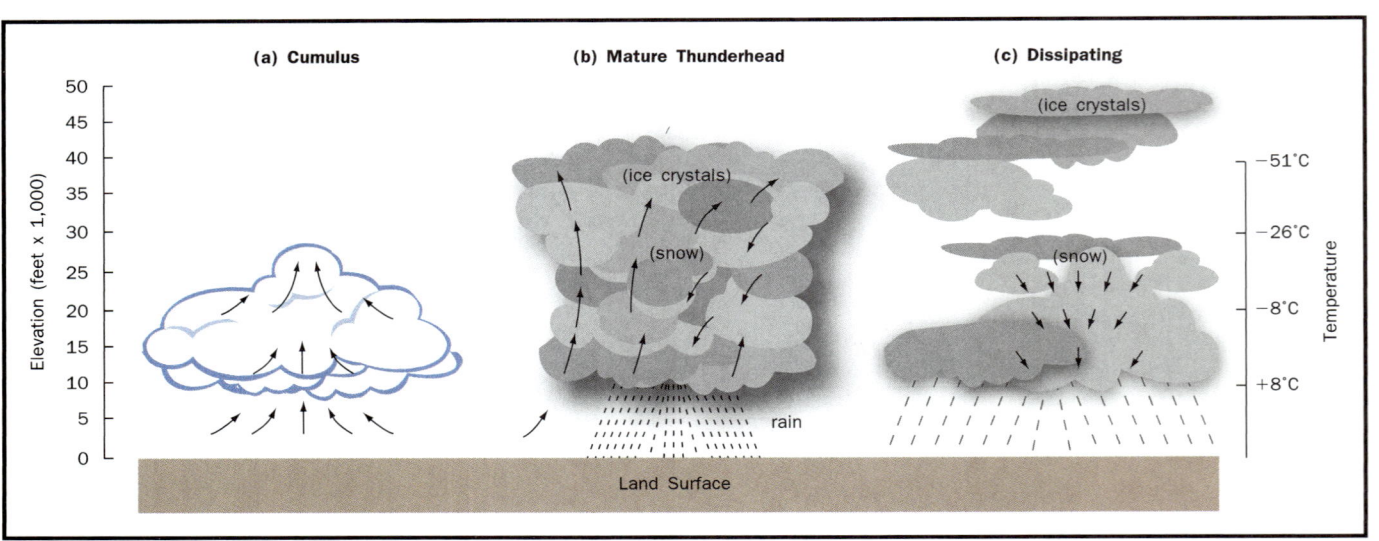

cloud builds well above the freezing level (the point at which the temperature of the rising air has cooled to water's freezing point). The precipitation particles grow larger and become heavier. The rising air soon cannot hold them up, and they begin to fall. The particles drag some of the air along with them, creating a downdraft, as shown in part (b) of the figure.

As the updraft pulls more dry air into the cloud, some of the raindrops evaporate, and the air cools, making it colder and heavier than the surrounding air. This usually strengthens the downdraft. The falling precipitation causes more downdrafts to form throughout the cloud, beginning the dissipating stage, as shown in part (c) of the figure. During the dissipating stage, precipitation occurs from the entire cloud base.

Some thunderstorms can develop into lines of severe thunderstorms, producing high winds, hail, frequent lightning, heavy rain, flash floods, and even tornadoes. SEE ALSO CLIMATE MODERATOR, WATER AS A; HYDROLOGIC CYCLE; PRECIPITATION, GLOBAL DISTRIBUTION OF; WEATHER AND THE OCEAN.

Timothy A. Chuey and Dennis O. Nelson

Bibliography

Ahrens, C. Donald. *Meteorology Today.* St. Paul, MN: West Publishing, 1985.

Internet Resource

"Clouds and Precipitation." WW2010 Weather World Project, Department of Atmospheric Sciences, University of Illinois at Urbana-Champaign. <http://ww2010.atmos.uiuc.edu/(Gh)/guides/mtr/cld/home.rxml>.

Precipitation, Global Distribution of

The global distribution of precipitation is influenced by the general circulation of the atmosphere, proximity to large bodies of water, and topography. Precipitation is most abundant where air rises, and least abundant where it sinks. It also tends to be greater near oceans and lakes, and in higher elevations.

General Circulation

The Earth's atmosphere is known to have regions characterized by large-scale rising air, and other regions with descending air; these vary by latitude and by season. Rising air is found primarily near the equator and in the mid-latitudes (40° to 60° North and South latitude), so these tend to be wet areas. Descending air dominates in the subtropics (20° to 30° North and South latitude) and the poles.✷ The global distribution of precipitation shows that the wettest areas on Earth are in the "rising air" zones, while the driest areas (subtropical deserts and the even drier polar areas) are in the "descending air" belts.

As the Earth revolves around the Sun during the year, the orientation of its axis relative to the Sun changes. This causes the apparent position of the Sun relative to the Earth to change, and creates distinct seasons. Between March and September, the axis of Earth is tilted toward the Sun, and hence the Sun shines more directly over the Northern Hemisphere, resulting in more sunlight, more heat, and the warmer temperatures of Northern summer. In the other 6 months, the Earth's axis is tilted away from the

✷ See "Climate and the Ocean" for illustrations of circulation zones and climate zones.

Sun, and the Sun shines more directly over the Southern Hemisphere, bringing summer to countries south of the Equator (and winter to the north).

The "rising" and "sinking" zones move northward and southward with the Sun's path. Thus, the wet area near the Equator moves northward into the Northern Hemisphere in its summer, and southward into the Southern Hemisphere during its summer. Similarly, the dry zones and wet zone at higher latitudes shift northward and southward throughout the year.

The result of these shifting zones are latitude bands with distinctive precipitation characteristics:

0–5° latitude: wet throughout the year (rising zone)

5–20° latitude: wet summer (rising zone), dry winter (sinking zone)

20–30° latitude: dry all year (sinking zone)

30–50° latitude: wet winter (rising zone), dry summer (sinking zone)

50–60° latitude: wet all year (rising zone)

60–70° latitude: wet summer (rising zone), dry winter (sinking zone)

70–90° latitude: dry all year (sinking zone)

If the Earth had no mountains, and oceans were homogeneous with respect to their heat content, the climate would occur in latitude bands like those listed above. However, mountains indeed exist, and they exert a strong influence on precipitation, as do warm and cold ocean currents.

Topography

When moving encounters a hill or mountain, it is forced to rise. Because rising air cools and condenses, precipitation is heaviest on the upwind side of a mountain, where the air is rising. This process is known as orographic lifting. On the downwind side, air descends, warms, and becomes drier.✷ In some parts of the world (such as the tropics), winds are steady throughout the year, and if these steady, moist winds encounter a mountain range, precipitation will occur frequently.

One example of a mountainous area receiving frequent rain is Mt. Waialeale on Kauai, Hawaii. It is a sharp peak directly in the path of steady trade winds which blow from the northeast most of the time. On the upwind (northeast) side of Waialeale, the air rises and condenses, resulting in almost constant clouds—nearly every day experiences rain. Waialeale is among the wettest places on Earth: over a 10-year period, annual rainfall averages more than 1,143 centimeters (450 inches). This is more than 8 times greater than rainfall experienced in many parts of the midwestern United States.

In contrast, on the downwind (southwest) side of Mt. Waialeale, the air descends, warms, and dries, in an area known as a "rain shadow." The result is a semiarid area with less than 51 centimeters (20 inches) of rain a year. This tremendous difference in precipitation—from 1,143 to 51 centimeters annually—occurs in a span of only 32 kilometers (20 miles).

Not every location receives such steady winds as Waialeale. But most mountainous areas have a wet side and a dry side, depending on the typical wind direction. Mountains can partially or completely override the rising

✷ See "Climate Moderator, Water as a" for a diagram of orographic lifting.

The formation of clouds and precipitation is a function of latitude, topography, and the presence of warm or cold oceans or large lakes, among other factors. This cumulus cloud is building into a cumulonimbus, which can bring heavy thunderstorms and even tornadoes.

and descending zones described above. Mountains can bring precipitation to normally dry areas, and downwind slopes of mountains can be dry, even in normally wet areas like the tropics.

Presence of Oceans or Lakes

According to the water cycle concept, water originates in the ocean, evaporates, rises into the air, condenses, falls as precipitation, and returns to the ocean.✳ This concept is generally true. However, the temperature of the water exerts a strong influence on how well the cycle operates.

✳ See "Hydrologic Cycle" for an illustration of the water cycle.

The warmer the surface of the ocean (or a large lake), the faster it evaporates, all other things being equal. Warm ocean waters evaporate tremendous amounts of water into the atmosphere, causing the dew point to be very high. (Dew point represents the actual amount of water vapor in the air.) Near warm waterbodies, the air usually has enough moisture to produce abundant precipitation—all it takes is for the air to rise, either by heating, or by encountering a mountain or other elevated terrain. On the other hand, cold ocean currents are relatively passive when it comes to evaporating water, so the dew points in such areas are lower (and the precipitation potential is lower) than near warm water.

As an example, consider two U.S. cities, both on the ocean at about the same latitude. Santa Monica, California, on the cool Pacific, is in the "wet winter, dry summer" zone between 30° and 50° North latitude, and receives only 36 centimeters (14 inches) of precipitation per year due to relatively low evaporation from the Pacific. Charleston, South Carolina, on the other hand, is on the very warm Atlantic. So much moisture is available to Charleston that it receives more rain in 2 months (July and August) than Santa Monica does all year, even though Charleston should also be a "wet winter, dry summer" location. For the year, nearly 127 centimeters (50 inches) are recorded in Charleston.

Moreover, the high dew points yield very high humidity at Charleston and other eastern U.S. cities, making them more uncomfortable during

CLOUD SEEDING

Cloud seeding is one form of human-induced weather modification, which also includes hail suppression and fog dispersal. All of these are attempts to change weather conditions at certain locales. The purpose of cloud seeding is mostly to augment precipitation (rain or snow) from storms. Activities involve the release into storm clouds of tiny condensation nuclei, which are dust or chemical particles that allow condensation to occur more quickly. If properly administered, cloud seeding can increase precipitation by about 10 percent.

summer. Meanwhile, in the "low dew point" city of Santa Monica, the residents are cooled by ocean breezes and enjoy steady sunshine during the typically very dry summers. SEE ALSO CLIMATE AND THE OCEAN; CLIMATE MODERATOR, WATER AS A; HYDROLOGIC CYCLE; PRECIPITATION AND CLOUDS, FORMATION OF; WEATHER AND THE OCEAN.

George Taylor

Bibliography

Rodgers, Alan, and Angella Streluk. *Precipitation*. Chicago, IL: Heinemann Library, 2003.

Internet Resources

Precipitation Trends in the 20th Century. Goddard Institute for Space Studies. <http://www.giss.nasa.gov/research/intro/delgenio_02/>.

Pricing, Water

An acquaintance once said that "every water faucet in New York City leaks." She was exaggerating, of course, but her point was that New Yorkers do not take the time or spend the money to repair leaks. Why? Most residents of the city pay a flat fee for their water. For a fixed monthly charge, residents can use as much water as they want. Marginal cost is zero. Hence, the wasted water costs them nothing, aside from the annoyance of listening to the drip.

But why should anyone worry about the cost of water? The figure shows recent prices of an additional 1,000 gallons per month in several U.S. cities in 2001, ranging from $1.25 to nearly $3.00. (Water rates in other countries often are considerably higher.) For most Americans, this is a very small part of their budgets. Even so, study after study has shown that most water users will indeed respond to a higher price by fixing leaks, using a broom instead of a hose to clean the driveway, and otherwise conserving water. Imposing a quantity charge, or raising it, forces users to rethink, even if only informally, their marginal benefit/marginal cost computations and adjust consumption accordingly.

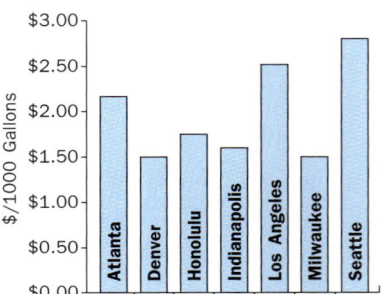

What is Water's True Cost?

Even when water utilities use a quantity charge instead of a fixed fee, they often set the quantity charge too low. Typical public water utilities design their rates to cover out-of-pocket costs, but such costs often fall short of the true economic value of extracting and distributing water.

Subsidies. First, governments often **subsidize** water **infrastructure**. Developers often must contribute ready-to-use water systems to the utility. These subsidies do not come directly from the utility company and hence do not show up in their accounting records.

Capital Equipment. Second, capital equipment—pumps, water mains, buildings, and so on—is a major element of total water cost and tends to last for several decades. Replacing a water main built, for example, 40 years ago would cost almost six times the original cost because of inflation alone. Yet few if any utilities update the value of aging capital equipment when they add up costs.

Scarcity Value. Third, water in the ground or in a stream is valuable because it is scarce. The right to divert water from a stream or to pump it from an **aquifer** is an asset of growing value to utility companies but again is often ignored in standard accounting practice. One study estimated **scarcity value** to be at least as large as all other conventionally reckoned costs together. Similarly, any environmental costs incurred in providing public water supplies should be added to water rates.

Pricing as a Conservation Incentive

For the reasons outlined above, water rate schedules based on the utility's out-of-pocket costs leave consumers paying less than they should. And since consumers pay too little, they use too much.

However, an increasing number of water utilities have recognized the potential of pricing to provide an incentive for their customers to conserve water. Some (Seattle, Washington and southern California, for example) have refined the notion, charging higher rates during droughts or in dry seasons or for unexpectedly large quantities.

But if water is priced at its full economic cost, what about the poor? Several major cities have taken at least tentative steps toward establishing what is called an "inclined-block" water rate schedule, as shown here:

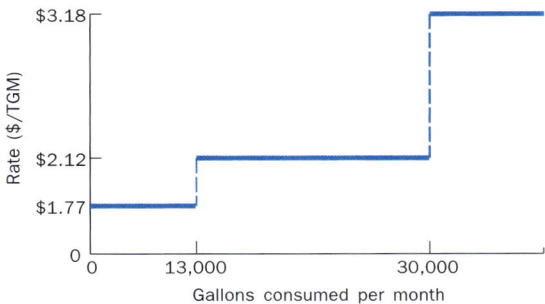

Ideally, customers under this system would pay a low rate for the first few thousand gallons used, but the rate would rise until they pay the full marginal cost for the last thousand gallons. The criterion of marginal benefits equals marginal costs would be met, and yet even the poorest could afford at least a basic amount.

Trending Toward Demand Management. On average, Americans use more than 1,000 gallons of water per day. The amount necessary to sustain life

---

subsidy: the sum of money granted by a government or other public body in order to assist an industry or business so that the price of the associated commodity or service will remain low or competitive; a common example is a farm subsidy

infrastructure: the permanent constructed system (e.g., pipes and other structures) that enables the treatment and delivery of water to support human habitation and activity, or that supports manufacturing activities and water projects (e.g., desalinization and hydropower plants)

aquifer: a water-saturated, permeable, underground rock formation that can transmit significant quantities of water under ordinary hydraulic gradients to wells and springs

scarcity value: the worth of something based on its limited, or lack of, availability; a resource is scarce when, at a zero price, more is wanted than is available; for water, the value of water above and beyond the cost of finding it, extracting it from the groundwater or surface-water body, treating it, and delivering it to users

EFFECTS OF SUBSIDIES

As anyone who has seen the 1939 film *Gone with the Wind* knows, cotton was once the staple crop of the southern United States. No more. Cotton, a very water-intensive crop, was well suited to the southern climate with its heavy rainfall. The center of cotton production has since shifted, however, to bone-dry parts of California. Why?

Several reasons exist for cotton's western migration, but clearly this shift could not have happened without federally funded irrigation projects. Generally, farmers who use water from these projects pay far less than its full economic value; they are thus subsidized. Subsidies are common among government-financed water resource projects. However, they are not free: taxpayers bear the cost in proportion to their tax payments.

processes is quite small, perhaps a few gallons per day. Of the difference, how much is really needed for bathing, laundry, housecleaning, car washing, lawn and garden care, filling the swimming pool, and so on? Clearly, if consumers are presented with the right incentive, they can conserve on water use.

This article has considered mainly residential water users, but similar considerations apply to businesses and farms that use water. Farmers, for example, have developed highly sophisticated means of conservation by determining exactly how much water each plant needs and applying just that amount, but adjusting for water costs. Hotels install low-flow showerheads and toilets. Car washes and many other businesses reuse water.

Most public utilities and other water-supply agencies try to accommodate growth in water demand by looking for additional water sources to develop; this is supply-side management. But growth can often be met by conservation in the use of existing sources; this is demand management. Some utilities go so far as to charge a premium during the dry season, for unanticipated high water demands, or during periods of drought. Pricing is a powerful tool of growing importance in the toolkits of water managers and environmentalists. SEE ALSO CONSERVATION, WATER; DEMAND MANAGEMENT; MARKETS, WATER; UTILITY MANAGEMENT.

James E. T. Moncur

Bibliography

Gardner, B. Delworth. "Water Pricing and Rent Seeking in California Agriculture." In *Water Rights: Scarce Resource Allocation, Bureaucracy, and the Environment.* Terry L. Anderson, ed. San Francisco, CA: Pacific Institute for Public Policy Research, 1983.

Howe, Charles W., and F. P. Linaweaver. "The Impact of Price on Residential Water Demand and Its Relation to System Design and Price Structure." *Water Resources Research* 3 (1967):12–32.

Moncur, James E. T., and Richard L. Pollock. "Accounting Induced Distortions in Public Enterprise Pricing." *Water Resources Research* 32, no. 11 (November 1996): 3355–3360.

———. "Scarcity Value for Water: A Valuation and Pricing Model." *Land Economics* 64, no.1 (February 1988):62–72.

Internet Resources

"Estimated Use of Water in the United States in 1995." U.S. Geological Survey. <http://water.usgs.gov/watuse/pdf1995/html>.

Photo and Illustration Credits

The illustrations and tables featured in Water: Science and Issues *were created by GGS Information Services. The photographs appearing in the text were reproduced by permission of the following sources:*

Volume 1

Photograph © Charles E. Rotkin/CORBIS: 2; Photograph © Joseph Sohm. ChromoSohm Inc./CORBIS: 6, 16, 173, 206; Photograph © Ted Spiegel/CORBIS: 8, 131; The Library of Congress: 11, 222; Photograph by Caroline Penn/CORBIS: 13, 248; Photograph © Galen Rowell/CORBIS: 17, 122; Photograph by Andrew Johnston. *Outdoor Indiana* Magazine, Indiana Department of Natural Resources: 23; Photograph by Mike Hutchings. © Reuters NewMedia Inc./CORBIS: 25; Photograph courtesy NASA/GSFC/LaRC/JPL, MISR Team: 27; Image by D. Grant Hokit: 30; Photograph by Craig Line. AP/Wide World Photos: 31; Photograph © David Samuel Robbins/CORBIS: 32; Photograph by Richard Fields. *Outdoor Indiana* Magazine, Indiana Department of Natural Resources: 33, 84; Photograph © Paul A. Souders/CORBIS: 35, 38, 130; Image © Lake County Museum/CORBIS: 37; OAR/National Undersea Research Program (NURP), National Oceanic and Atmospheric Administration: 43, 79 (top), 225; AP/Wide World Photos: 44, 128; Photograph © Jonathan Blair/CORBIS: 46; Photograph © Phil Schermeister/CORBIS: 48, 102; Corbis-Bettmann: 55, 211; Photograph by Cindy Clendenon: 56, 72, 73, 76, 79 (bottom), 180, 184, 187; The Kobal Collection: 57; Illustration by Don Davis. NASA: 59; Photograph by David Crisp and the WFPC2 Science Team (JPL/Caltech): 60; Photograph by Mark Wheeler. U.S. Army Corps of Engineers: 64; Photograph by Sherwin Crasto. AP/Wide World Photos: 66; Photograph by Robert J. Huffman. Field Mark Publications: 67, 81; © NASA/CORBIS: 68; Image provided by Orbimage. © Orbital Imaging Corporation; processing by NASA Goddard Space Center: 70; Photograph by Richard B. Mieremet. National Oceanic and Atmospheric Administration: 74; Florida Keys National Marine Sanctuary: 78; Photograph © Wolfgang Kaehler/CORBIS: 83; Photograph © Kelly A. Quin: 87, 147, 156, 157, 159, 209, 219, 259; Photograph © Robert Holmes/CORBIS: 92; Reuters/Archive Photos, Inc.: 93; Photograph © Jim Cummins/CORBIS: 95; Photograph by D. Walsh. U.S. Bureau of Reclamation: Pacific Northwest Region: 98; Photograph © Yann Arthus-Bertrand/CORBIS: 101; © The Mariners' Museum/CORBIS: 104; Photograph by Susan D. Rock: 105; Photograph © Chris Rainier/CORBIS: 108; Photograph by Christopher Sabine: 110; Photograph by John Maxwell. *Outdoor Indiana* Magazine, Indiana Department of Natural Resources: 113; Photograph by Ronald Crouse: 115; Photograph © John Zoiner/CORBIS: 116; Photograph © Roger Ressmeyer/CORBIS: 118; Photograph © Natalie Fobes/CORBIS: 119; Roger Ressmeyer/CORBIS: 120; Photograph © Caroline Penn/CORBIS: 124; Photograph by Animals Animals/© L. Gould—OSF: 126; Photograph © Robert Maass/CORBIS: 133; Photograph © Carl and Ann Purcell/CORBIS: 135; UPI/Bettmann Newsphotos:

Photo and Illustration Credits

137; Photograph by Jeffrey L. Clendenon: 139; Photograph © Stuart Westmorland/CORBIS: 142; Photograph by David M. Rohr: 143 (left); Photograph by Amos Nachoum/CORBIS: 143 (right); Photograph © George B. Diebold/CORBIS: 145; Photograph by Ed Young/CORBIS: 151; Photograph by Jack Dykinga. Courtesy of the Agricultural Research Service, USDA: 153; Illustration by Dr. Christian G. Daughton. U.S. Environmental Protection Agency (EPA): 161; National Oceanic and Atmospheric Administration, Department of Commerce: 165; Photograph by Mary Hollinger. National Oceanic and Atmospheric Administration, Department of Commerce: 167, 226; Photograph by John Wilkinson. © Ecoscene/CORBIS: 170; © ESA/PLI/CORBIS: 175; Photograph by Commander John Bortniak. NOAA Corps (ret.), National Oceanic and Atmospheric Administration: 185; Photograph by Annie Griffiths-Bell/CORBIS: 186; Photograph by Cydney Conger/CORBIS: 188; Photograph by Gregory Bull. AP/WideWorld Photos: 192; Photograph © Vince Streano/CORBIS: 197; CORBIS: 199; Photograph by Raveendran. © AFP/Corbis-Bettmann: 202; Photograph by Stephen Jaffe. © AFP/Corbis-Bettmann: 204; Photograph by Douglas Peebles/CORBIS: 212; Photo collection of Dr. James P. McVey. National Oceanic and Atmospheric Administration, Department of Commerce: 213; Photograph by William Harrigan. Florida Keys National Marine Sanctuary: 214 (top); Photograph by Anthony R. Picciolo. National Oceanic and Atmospheric Administration, Department of Commerce: 214 (bottom); Photograph by Dr. James P. McVey. National Oceanic and Atmospheric Administration, Department of Commerce: 216, 217; Sapelo Island National Estuarine Research Reserve: 224; Photograph by Peter Hulme. © Ecoscene/CORBIS: 228; Photograph © George D. Lepp/CORBIS: 230; Photograph by Martha Tabor/Working Images Photographs: 233, 235; Photograph by Jeff Christensen/Newsmakers. Getty Images: 237; Photograph © Kevin Fleming/CORBIS: 241; Photograph © Winifred Wisniewski. Frank Lane Picture Agency/CORBIS: 243; Photograph by Roger Ressmeyer/CORBIS: 245; Photograph by Rafiqur Rahman. © Reuters NewMedia Inc./CORBIS: 247; AFP/Corbis-Bettmann: 249; Photograph by Kathy Willens. AP/Wide World Photos: 251; Photograph by Michael Lewis/CORBIS: 254; Photograph © Caroline Penn/CORBIS: 255; © AFP/Corbis-Bettmann: 256; Photograph by Alan Towse. © Ecoscene/CORBIS: 258; National Archives and Records Administration: 261; Photograph by Tom Mihalek/Newsmakers. Getty Images: 264.

Volume 2

OAR/National Undersea Research Program (NURP), National Oceanic and Atmospheric Administration: 2, 50, 170; Photograph © Firefly Productions/CORBIS: 4; Photograph © James L. Amos/CORBIS: 6; Photograph. © Niall Benvie/CORBIS: 8; Photograph by Peter Johnson/CORBIS: 11; Image © Lake County Museum/CORBIS: 16; Photograph © Bates Littlehales/Corbis-Bettmann: 18; NASA JPL: 19 (left and right); Photograph by Richard Fields. *Outdoor Indiana* Magazine, Indiana Department of Natural Resources: 22, 57, 98; U.S. Fish and Wildlife Service: 23; Photograph © Robert Holmes/CORBIS: 25; Photograph © David Pu'u/CORBIS: 26; © AFP/Corbis-Bettmann: 28, 121, 184, 227; © Bettman/Corbis: 30, 161; Photograph © Mug Shots/CORBIS: 32; Photograph by Commander John Bortniak. NOAA Corps (ret.), National Oceanic and Atmospheric Administration: 34; Photograph by Cindy Clendenon: 35, 35, 79, 104, 144, 203, 268; Photograph by William Folsom. National Oceanic and Atmospheric Administration, Department of Commerce: 36; Photograph by April Bahen. National Oceanic and Atmospheric Administration, Department of Commerce: 38; Photograph by David Muench/CORBIS: 40; Photograph © Kelly A. Quin: 43; Photograph © Carl and Ann Purcell/CORBIS: 45; Photograph by Jeffrey L. Rotman/CORBIS: 49; Photograph by Jeffrey L. Clendenon: 51, 166; Photograph by Mary Hollinger. National Oceanic and Atmospheric Administration, Department of Commerce: 53; Photograph by Matthew Cavanaugh. AP/Wide World Photos: 55;

Photo and Illustration Credits

Photograph © Natalie Fobes/CORBIS: 59; Photograph © Gunter Marx Photography/CORBIS: 62; National Oceanic and Atmospheric Administration, Department of Commerce: 64; Alaska Fisheries Science Center, Marine Observer Program: 66; Photograph by Robyn Beck. © AFP/Corbis-Bettmann: 68; Photograph © Ivor Fulcher/CORBIS: 70; Photograph © AFP/Corbis-Bettmann: 71, 201; Photograph by Maury Tannen. © AFP/Corbis-Bettmann: 72; U.S. Army Corps of Engineers: 75; Photograph © Kevin Fleming/CORBIS: 77; Photograph © Yann Arthus-Bertrand/CORBIS: 80, 86, 264; Photograph by Savita Kirloskar. © Reuters NewMedia Inc./CORBIS: 82; Photograph by Grant A. Meyer: 88; Photograph © Robert Landau/CORBIS: 93; Photograph © Raymond Gehman/CORBIS: 95; Photograph © Layne Kennedy/CORBIS: 102; Photograph © Roger Ressmeyer/CORBIS: 106; Photograph created at University of Miami by Bob Evans, Peter Minnett, and coworkers. NASA: 108; AFP/Corbis-Bettmann: 110, 127; Photograph © Catherine Karnow/CORBIS: 113; Photograph © Joseph Sohm. ChromoSohm Inc./CORBIS: 114; Photograph by John Bortniak. National Oceanic and Atmospheric Administration, Department of Commerce: 115, 117; Photograph by Bernard Edmaier. Science Photo Library/Photo Researchers, Inc.: 119; Photograph by Giuseppe Zibordi. National Oceanic and Atmospheric Administration, Department of Commerce: 124; Photograph by Scott Bauer. Courtesy of the Agricultural Research Service, USDA: 128, 252; Photograph © Michael Weiss/CORBIS: 133; Photograph by Aizawa Toshiyuki. Hulton/Archive: 135; Photograph © Lester Lefkowitz/CORBIS: 138; Photograph © Mike Zens/CORBIS: 140; Photograph by Dan Newell. Illustration © 2002 Mark D. Heckman: 148; Photograph © Ted Spiegel/CORBIS: 159, 216; © Corbis-Bettmann: 164; National Undersea Research Program: 172; © Japack Company/Corbis: 176; Photograph © Stuart Westmorland/CORBIS: 177; Photograph by Commander Richard Behn. NOAA Corps, National Oceanic and Atmospheric Administration: 179, 209; Photograph © Liba Taylor/CORBIS: 182; Photograph by Susan D. Rock: 183; U.S. Geological Survey: 187; Photograph. © Bob Rowan, Progressive Image/CORBIS: 188; U.S. Army Corps of Engineers: 189; Photograph © Douglas Peebles/CORBIS 196; Illustration courtesy of Oregon State University, Department of Geosciences: 199; Photograph by Michael Van Woert. National Oceanic and Atmospheric Administration, Department of Commerce: 204, 207, 211; Photograph by Tas Van Ommen: 212; Photograph © Gary Braasch/CORBIS: 219; Photograph. © Bojan Breceli/CORBIS: 224; Photograph by Robert van der Hilst/CORBIS: 228; Photograph © Adam Woolfitt/CORBIS: 230; Hulton/Archive: 233; Photograph by George H. Huey/CORBIS: 236; Photograph by Jonathan Blair/CORBIS: 238; AP/Wide World Photos: 245; Photograph provided by the Earth Sciences and Image Analysis Laboratory, Johnson Space Center: 249; Photograph © Kevin Schafer/CORBIS: 257; Photograph © Peter Turnley/CORBIS: 266 (top and bottom).

Volume 3

Photograph © Joseph Sohm. ChromoSohm Inc./CORBIS: 4; Photograph by Richard Fields. *Outdoor Indiana* Magazine, Indiana Department of Natural Resources: 5, 245; AP/Wide World Photos: 8, 49, 64, 121; Photograph © James A. Sugar/CORBIS: 13, 118; Photograph by Daniel Aguilar. © Reuters NewMedia Inc./CORBIS: 15; Photograph © Robert Holmes/CORBIS: 20; AP/Wide World Photos/Yonhap: 25; Photograph © Alan Schein Photography/CORBIS: 27; Photograph by Marc Muench/CORBIS: 29; Photograph by Paul B. Southerland. AP/Wide World Photos: 33; Image © Seth Joel/CORBIS: 36, 36; UCB: 38; Alfred Russell. Corbis-Bettmann: 40; Photograph © Raymond Gehman/CORBIS: 45; Photograph by I. McDonald. OAR/National Undersea Research Program (NURP), National Oceanic and Atmospheric Administration: 47; OAR/National Undersea Research Program (NURP), National Oceanic and Atmospheric Administration: 50 (top and

Photo and Illustration Credits

bottom), 80, 130; Photograph by Anthony Bannister. CORBIS/Anthony Bannister; ABPL: 51; Image provided by Orbimage. © Orbital Imaging Corporation; processing by NASA Goddard Space Center: 55, 81, 147; Photograph by Cindy Clendenon: 59, 164, 213, 252; Photograph. © Bojan Breceli/CORBIS: 60; Photograph by Kevin Schafer/CORBIS: 61; Photograph by Commander Richard Behn. NOAA Corps, National Oceanic and Atmospheric Administration: 62; Photograph by Brandon D. Cole/CORBIS: 62; Photograph by Budd Christman. National Oceanic and Atmospheric Administration, Department of Commerce: 63; Image © AFP/Corbis-Bettmann: 68; AFP/Corbis-Bettmann: 70; Photograph © Ed Young/CORBIS: 72; Photograph © Custom Medical Stock Photo: 74; Photo Researchers, Inc.: 74; Photograph © Darrell Gulin/CORBIS: 76; Photograph © Layne Kennedy/CORBIS: 78, 230; Photograph © Roger Ressmeyer/CORBIS: 79; Photograph © Ralph White/CORBIS: 82; Photograph © Marie Tharp: 84; Photograph © Anthony Cooper. Ecoscene/CORBIS: 86; Photograph by Kevin Schafer. CORBIS Corporation (Bellevue): 89; Photograph. © Ed Kashi/CORBIS: 92; Image © Lake County Museum/CORBIS: 93; Photograph © Annie Griffiths Bell/CORBIS: 94; North Carolina Aquarium at Pine Knoll Shores: 95; Photograph © Macduff Everton/CORBIS: 96; Photograph by Cliff Schiappa. AP/Wide World Photos: 101; National Oceanic and Atmospheric Administration: 107, 112, 149; Photograph by Michael Van Woert. National Oceanic and Atmospheric Administration, Department of Commerce: 109, 168; UPI/Corbis-Bettmann: 111; Photograph by Robert J. Huffman. Field Mark Publications: 113; Photograph by Allen M. Shimada. NMFS: 115; Photograph by Richard T. Nowitz/CORBIS: 120; Photograph by Emmanuel Durand © AFP/Corbis-Bettmann: 124; Photograph by Jae-Hwan Kim. © AFP/Corbis-Bettmann: 125; Photograph © Tom Van Sant/CORBIS: 128; Engraving by James Poupard. National Oceanic and Atmospheric Administration: 137; Photograph by Mychele Daniau. © AFP/Corbis-Bettmann: 145; Photograph © Lowell Georgia/CORBIS: 153; NASA: 155; Photograph © Paul A. Souders/CORBIS: 158, 160, 232; © CORBIS: 162, 182; Photograph by William Van Woert. National Oceanic and Atmospheric Administration, Department of Commerce: 166; Photograph by Wolfgang Kaehler. Wolfgang Kaehler/Corbis-Bettmann: 171; Photograph © M. Dillon/CORBIS: 172; Photograph by Doug Wilson. Courtesy of the Agricultural Research Service, USDA: 175; Photograph by Robert Campbell. U.S. Army Corps of Engineers: 177, 251 (bottom); Photograph by Chinch Gryniewicz. Ecoscene/CORBIS: 178; © AFP/Corbis-Bettmann: 179, 183, 226; Photograph by Vince Streano/CORBIS: 180; Photograph by Susan D. Rock: 184; Photograph © Ted Spiegel/CORBIS: 187, 196; Photograph by Douglas P. Wilson. CORBIS/Douglas P. Wilson; Frank Lane Picture Agency: 188; National Archives: 192; Photograph © Charles E. Rotkin/CORBIS: 195, 243; Photograph © Bill Varie/CORBIS: 198; Photograph by Robert E. Wallace. U.S. Geological Survey: 202; Photograph by Lauren McFalls. AP/Wide World Photos: 207; © Reuters NewMedia Inc./CORBIS: 210; Photograph © Nick Hawkes. Ecoscene/CORBIS: 215; Photograph by Jack Dykinga. Courtesy of the Agricultural Research Service, USDA: 218; Photograph © David Turnley/CORBIS: 219; Photograph © Kelly A. Quin: 224, 227; Photograph by Ken Winters. U.S. Army Corps of Engineers: 228, 251 (top); Photograph by Anthony F. Amos: 234, 235; Photograph by Rick Doyle/CORBIS: 238; Greenpeace Photo: 241; Photograph © Michael S. Yamashita/CORBIS: 247; Photograph © Neil Rabinowitz/CORBIS: 250; The Library of Congress: 255; FMA Production: 261.

Volume 4

© Corbis-Bettmann: 1; Photograph by Alan Towse. © Ecoscene/CORBIS: 3; Photograph by Toby Talbot. AP/Wide World Photos: 6; Photograph by David Mercado. © Reuters NewMedia/CORBIS: 7; Photograph © Les Pickett/CORBIS: 9; © Reuters NewMedia Inc./CORBIS: 10, 18, 20; Photograph

Photo and Illustration Credits

© Horace Bristol/CORBIS: 12; © Underwood & Underwood/CORBIS: 14; Photograph © Robert Maass/CORBIS: 15; Photograph by Jeffrey L. Clendenon: 16; Photograph by Kamal Kishore. © Reuters NewMedia Inc./CORBIS: 26; AP/Wide World Photos: 27; Photograph © Kevin Fleming/CORBIS: 30, 208; Photograph by Gary Kazanjian. AP/WideWorld Photos: 31; Photograph © Bertil Ericson/CORBIS: 33; Photograph by Cindy Clendenon: 36, 37, 54, 118, 128, 132, 149, 169, 173, 185, 234, 235, 242; Photograph by Daniel Dal Zennaro. © AFP/Corbis-Bettmann: 38; Photograph by Dong Lin. California Academy of Sciences, Special Collections: 39; Photograph by Mufty Munir. © AFP/Corbis-Bettmann: 41; Illustration © CORBIS: 42; Photograph by Robert J. Huffman. Field Mark Publications: 43; Photograph © Jeffrey L. Rotman/CORBIS: 45; U.S. Fish and Wildlife Service: 46; Photograph © Marc Garanger/CORBIS: 48; Photograph © Annie Griffiths Bell/CORBIS: 50; Photograph. © Raymond Gehman/CORBIS: 53; © Bettman/Corbis: 55; Photograph © Tim Page/CORBIS: 57; Photograph © Buddy Mays/CORBIS: 61; Photograph by Grant A. Meyer: 63, 65; Photograph by Lawrence Migdale. Photo Researchers, Inc.: 67; Photograph © Gary Braasch/CORBIS: 70; Photograph © Natalie Fobes/CORBIS: 71; Photograph by Hope Alexander. EPA Documerica: 73; National Oceanic and Atmospheric Administration, Department of Commerce: 75; Photograph by Commander Richard Behn. NOAA Corps, National Oceanic and Atmospheric Administration: 76; Photograph © Layne Kennedy/CORBIS: 85; Photograph by Bob Heims: 87; Photograph © Bryan F. Peterson/CORBIS: 88; Image by Donald R. Edgar: 90, 91; Photograph © Tiziana and Gianni Baldizzone/CORBIS: 93; U.S. National Aeronautics and Space Administration (NASA): 99; NASA/JPL/Caltech: 101; Photograph © Ralph White/CORBIS: 102; Archive Photos: 105; Photograph © Paul A. Souders/CORBIS: 107, 251; Photograph © Michael S. Yamashita/CORBIS: 108; Photograph © Raymond Gehman/CORBIS: 110, 217; Photograph © Roger Ressmeyer/CORBIS: 111; Photograph © Carl and Ann Purcell/CORBIS: 112 (top right, [b]); Photograph © Macduff Everton/CORBIS: 112 (bottom right, [c]); Photograph © Yann Arthus-Bertrand/CORBIS: 112 (top left, [a]), 112 (bottom left, [d]), 116; Photograph by Richard Fields. *Outdoor Indiana* Magazine, Indiana Department of Natural Resources: 119, 123; Painting courtesy of OAR/National Undersea Research Program (NURP), National Oceanic and Atmospheric Administration: 138; Drawing courtesy of OAR/National Undersea Research Program (NURP), National Oceanic and Atmospheric Administration: 139; Corbis-Bettmann: 140, 191; OAR/National Undersea Research Program (NURP), National Oceanic and Atmospheric Administration: 142, 143, 204; Photograph © Kelly A. Quin: 92, 145; Photograph. © Bob Rowan, Progressive Image/CORBIS: 151, 198; Photograph © Peter Turnley/CORBIS: 154; Photograph by Akram Shahid. © Reuters NewMedia Inc./CORBIS: 155; Photograph © William West/CORBIS: 157; Photograph © David and Peter Turnley/CORBIS: 158; Photograph by Arthur Rothstein © CORBIS: 162; JLM Visuals: 167; Photograph by Robert Estel/CORBIS: 170; Photograph by Andy Newman. © Reuters NewMedia Inc./CORBIS: 174; Photograph © Galen Rowell/CORBIS: 175; Photograph © Tom Bean/CORBIS: 176; Photograph by Zoltan T. Asztalos. AP/Wide World Photos: 181; Photograph © Keren Su/CORBIS: 183; Photograph © Joel W. Rogers/CORBIS: 186; Photograph © Lloyd Cluff/CORBIS: 190; Photograph © James A. Sugar/CORBIS: 201; © CORBIS: 202; Photograph by Suhaib Salem. © Reuters NewMedia Inc./CORBIS: 207; Photograph © Lester Lefkowitz/CORBIS: 210, 211; Photograph by John Maxwell. *Outdoor Indiana* Magazine, Indiana Department of Natural Resources: 214, 246, 253; Photograph © Wolfgang Kaehler/CORBIS: 215; Photograph by Andrew Brown. © Ecoscene/CORBIS: 223; Photograph © David Pu'u/CORBIS: 226; Photograph by Sean Linehan. National Oceanic and Atmospheric Administration, Department of Commerce: 227; Photograph by Deborah Barr. National Oceanic and

Photo and Illustration Credits

Atmospheric Administration, Department of Commerce: 228; National Oceanic and Atmospheric Administration Historical Photo Collection: 230; Satellite photograph by National Hurricane Center, National Oceanic and Atmospheric Administration: 232; Satellite photograph by National Oceanic and Atmospheric Administration Historical Photo Collection: 233; Photograph © Khaled Zighari/CORBIS: 239; Photograph by Kevin Schafer/CORBIS: 244; Photograph by Larry Harwood: 248; © Ashanti Johnson Pyrtle: 250.

Glossary

100-year flood the flood from a river discharge that has a 1-percent chance of occurring in any given year

abiotic pertaining to any nonbiological factor or influence (not derived from living organisms), such as geological or meteorological characteristics

ablation a process that removes snow or ice from glaciers, including melting, evaporation, wind erosion, and sublimation

absolute describes the value of a parameter that is known precisely; for example, the temperature of the stream at a given point is 10.4°C (50.7°F); contrasts with a relative measure

absorption the incorporation of an atom, ion or molecule in another substance; the atom, ion, or molecule is not necessarily part of the substance's structure

accountable describes being liable or answerable for decisions or actions; normally accomplished by specifying to whom a decisionmaker must report and is answerable

accretionary prism wedge of sediment that accumulates in subduction zones; the sediment is scraped off the subducting plate and accreted to the overriding platearable

accuracy the exactness or degree to which a measurement or calculation approaches the actual quantity

acid mine drainage acidic water that flows into streams from abandoned mines, piles of mining waste or tailings, or sulfide-mineralized rock; water that becomes acidic through reactions that oxidize sulfide minerals in rocks, thereby producing sulfuric and other acids

acid rain rain that has become acidic due to the presence of dissolved substances such as sulfur oxides or nitrogen oxides; causation may be natural (e.g., volcanic eruptions) or human-induced (e.g., industrial pollution)

acid rock drainage see "acid mine drainage"

acute rapid in onset and of short duration (e.g., acute toxicity); contrasted with chronic, which is gradual or long-term (e.g., chronic toxicity)

acute toxicity the property of a chemical or microbe enabling it to cause symptoms of illness in a living organism only a short time after exposure; see "chronic toxicity"

Glossary

adsorb to collect a liquid, gas, or solid, in a condensed form, on a surface

aeration for water, any active or passive process by which close, sustained contact between air and water is assured, generally by spraying water in the air, bubbling air through water, or mechanical agitation of the water to promote surface absorption of air; for soil, the process of loosening or puncturing the soil by mechanical means in order to increase water and air permeability

aerobe an organism which requires oxygen for its life processes

aerobic describes organisms able to live only in the presence of air or free oxygen, and conditions that exist only in the presence of air or free oxygen

aerosol a suspension of colloidal (finely suspended) particles in a gas

aesthetic relating to the human senses, especially what is pleasurable or is deemed satisfactory or desirable

aestivation also spelled estivation, the sleeplike condition of partial or total inactivity during dry or hot weather conditions; seasonal opposite of hibernation

agrochemical a synthetic or naturally derived chemical used in agriculture, such as a fertilizer, pesticide, or hormone

albedo the fraction of incident solar radiation that is reflected by the surface

algae (singular, alga) simple photosynthetic organisms, usually aquatic, containing chlorophyll, and lacking roots, stems, and leaves

algal bloom also called algae bloom, a rapid increase in algae concentrations in fresh or marine waters that often is stimulated by nutrient enrichment; recurrent blooms may cause or accelerate eutrophication, and lead to a deterioration in water quality

algebra a branch of mathematics that deals with the use of letters and other symbols, as well as numbers, to represent quantities and to express generalizations about them

algorithm a formalized, step-by-step procedure or set of equations designed for the purpose of solving a particular type of problem

alkalinity the excess of strong bases over strong acids in water; a measure of a water's buffering capability, or ability to resist changes in pH; generally equivalent to the sum of bicarbonate (HCO_3^-) and carbonate (CO_3^{2-}); alkalinity generally increases with progressive water–rock chemical reactions

allocate to distribute resources for a particular purpose

alluvium a deposit of clay, silt, sand, gravel, or a mixture of these, that has been deposited by a stream or other body of running water in a streambed, on a floodplain, on a delta, or at the base of a mountain; alluvial refers to sediments deposited by rivers and streams

altimeter an instrument that determines height above ground surface, especially one mounted in an aircraft or satellite and incorporating a barometer or radar device

altitude in navigational usage, the angle measured from the horizon to a celestial body, such as the Sun, Moon, or stars; in general usage, the height above the ground surface, called elevation if referring to height above sea level

ambient describes an encompassing environment that is natural or nearly natural (e.g., ambient air temperature)

amphibian a cold-blooded, smooth-skinned vertebrate of the class *Amphibia*, such as a frog or salamander, that characteristically hatches as an aquatic larva with gills and then transforms into an adult having air-breathing lungs

anadromous describes fish that move from the ocean up a river to spawn

anaerobe an organism that does not require oxygen to maintain its life processes

anaerobic describes organisms able to live and grow only where there is no air or free oxygen, and conditions that exist only in the absence of air or free oxygen

anastomotic describes the character of individual stream paths within a braided stream; stream paths appear to interconnect with one another around sand bars

anion an ion that has a negative charge

anoxia the state in which water contains less than 0.1 milliliter of oxygen per liter, the threshold below which animal life diminishes significantly

anthropogenic of, relating to, or resulting from the influence of human beings on nature, as in water and air pollution

anthropomorphic having human-like or human-caused characteristics

antibiotic a substance produced by organisms, especially bacteria and fungi, which passes into the surrounding medium and is toxic to other organisms; for example, penicillin from the mold *Penicillin notatum* destroys many kinds of bacteria

appurtenances auxiliary components and facilities that accompany a water system, such as the valves, monitors, and other control devices in a distribution system

aquaculture the science, art, and business of cultivating marine or freshwater animals or plants under controlled conditions

aqualung equipment used by a person as an air supply while underwater; developed in 1943 by Jacques-Yves Cousteau and French engineer Émile Gagnan

aqueduct long, canal-like or pipe-like structure, either above or below ground, for transporting water some distance

aqueous relating to, similar to, containing, or dissolved in water; watery

aquifer a water-saturated, permeable, underground rock formation that can transmit significant quantities of water under ordinary hydraulic gradients to wells and springs

Glossary

arable used or suitable for growing crops

archaeology the historical study of humans through the excavation of sites and the analysis of artifacts and other physical remains

arid describes a climate or region where precipitation is exceeded by evaporation; in these regions, agricultural crop production is impractical or impossible without irrigation

artesian describes groundwater under pressure in a confined aquifer such that the water level rises in the well bore to a shallower level (i.e., closer to the ground surface) than the aquifer; if under sufficient pressure, artesian groundwater may naturally flow out of the top of the well, known as a flowing artesian well or simply a flowing well

arthropod a group of invertebrates (animals without backbones) that have segmented bodies and jointed limbs; includes insects and crustaceans

ash flow a hot, turbulent ash cloud that moves rapidly down a volcano's slope

assimilate to incorporate or take in; the ability of a body of water, air, or soil to reduce the impact of pollutants by mixing, dilution, or other processes

asteroid a rocky body, much smaller than a planet, that orbits the Sun; asteroidal means pertaining to or originating from an asteroid

asteroid belt the area between the orbits of Mars and Jupiter that contains most of the asteroids found orbiting the Sun

asthenosphere the zone inside the Earth beneath the lithosphere to a depth of less than 700 kilometers (approximately 400 miles), containing a low percentage of molten rock, and constituting the source of seafloor basalts

astrolabe an early instrument used to observe the position and determine the altitude of the Sun or other celestial body; used for navigation from the Middle Ages until the eighteenth century, when it was replaced by the sextant

astronomical unit one astronomical unit (AU) is the average distance between the Sun and the Earth: approximately 150 million kilometers (93 million miles)

atmosphere the gaseous layer surrounding the Earth, consisting of 78 percent nitrogen, 21 percent oxygen, and approximately 1 percent argon; atmospheric pressure at sea level is about 14.7 pounds per square inch, termed "one atmosphere"

atomic fission the action of dividing or splitting an atom or atoms into two or more parts

attenuation thinning, weakening, or lessening; for water velocity, the slowing, modification, or diversion of the flow of water as with detention and retention ponds; for water quality, the process of diminishing contaminant concentrations in water due to filtration, biodegradation, dilution, sorption, volatilization, and other processes

autonomous existing, reacting, or developing as an independent, self-regulating organism

autotroph an organism that is able to form nutritional organic substances from simple inorganic substances such as carbon dioxide

backshore the part of a shore between the foreshore and the landward edge that is above high water except in the most severe storms

backwash the seaward return of water following the uprush of the waves, also referred to as backrush; in water treatment, the reversal of flow through a rapid sand filter to wash clogging material out of the filtering medium and reduce conditions causing loss of head (pressure)

bait fish a fish used as bait in fishing to catch a larger fish

baleen a horned comb-like structure in the mouths of toothless whales, used for filtering small animals (e.g., small crustaceans) from the water

bank-full discharge a river discharge that raises the stream level to the top of the streambank; precipitation resulting in bank-full discharges occurs once every 1 to 2 years, on average

bar a ridge built up in a river or ocean by deposits of sand, rock particles, and other materials

basalt a dark, volcanic rock with abundant iron and magnesium and relatively low silica; common on all of the terrestrial planets

baseflow the portion of stream flow derived from groundwater seeping through the stream bottom and sides

base level the lowest level to which a river can erode; ultimate base level is sea level; local base level may be a larger river, a lake, or very resistant rock

basin (geologic) a low-lying geographic area that tends to collect sediments from higher land around it, thus building up rock sequences; a structural basin is a geologic feature where the sedimentary layers dip toward a central point

basin (lake) a topographic depression or low-lying area occupied by a lake

basin (ocean) the topographic low area occupied by oceans; the floor of ocean basins consists of basaltic crust that is more dense than typical continental rocks

basin (river) a geographic area drained by a river and its tributaries; consists of a drainage system that may be comprised of streams, wetlands, and often natural or artificial lakes; see "drainage basin," "catchment," and "watershed"

bathymetry the science of measuring the depths and underwater topography of seas, oceans, lakes, and reservoirs

beach nourishment also known as beach replenishment, the techniques used to temporarily mitigate erosion of beaches (e.g., bringing in sand to rebuild a beach)

beachdrift the net movement by longshore current of sand up or down and along the beach, depending on the direction of incoming waves; see "littoral drift"

bedload sediment particles, small and large, and including rocks, which bounce, slide, or roll along the bottom of a streambed

bedrock solid rock that lies beneath soil and other surficial material

beneficial use use of water for purposes considered worthwhile to society; includes domestic, municipal, industrial, agricultural and recreational uses

Benioff zone the seismically active boundary between the subducting plate and the overriding plate; the movement of the plates along the Benioff zone is the cause of high-magnitude, deep-seated earthquakes

benthos the plants and animals living on or closely associated with the bottom of a body of water, such as a river, lake, or sea; these organisms are described as benthic (bottom-dwelling), which means of or pertaining to the benthos

berm a narrow ledge or path at the top or bottom of a slope, streambank, or along a beach

bioaccumulative describes the increase in concentration of a chemical in organisms that reside in environments contaminated with low concentrations of various organic compounds

biochemical oxygen demand abbreviated as BOD, the amount of oxygen required for the biological decomposition of organic matter under atmospheric conditions

bioconcentration the increase in concentration of a chemical in an organism resulting from absorption levels exceeding the rate of metabolism and excretion

biodegradation breaking down of substances by microorganisms, which use the substances for food

biodiversity a measure of the variety of the Earth's species, of the genetic differences within species, and of the ecosystems that support those species

biofilm the layered growth of bacteria on surfaces, including the inner walls of fractured rock, pore spaces in sediments, and in pipes, wells, bathroom fixtures, etc.

biogeochemical cycle the cyclical system in which chemical elements are transferred between living and nonliving parts of the biosphere; includes the hydrologic, oxygen, carbon, and nitrogen cycles; biogeochemical cycles play a vital role in global ecology

biological contactors surfaces used in aerobic biological treatment of wastewater; they are surfaces upon which bacteria can grow and accumulate; when placed in the treatment process, biological contactors (by virtue of their bacteria) facilitate the breakdown of organic waste

biologically active capable of interacting with biological systems and producing a biochemical/physiological effect or response

biomass the total mass of living organic matter in a defined location; generally expressed as grams per unit volume or per unit area

biomass pyramid graphic model describing the distribution of biomass in an ecosystem or community at the trophic level

bioregenerative process the development of a part of a living organism to replace a similar structure that has been damaged or destroyed

bioremediation a method of waste cleanup using specialized, naturally occurring microorganisms with unique characteristics, and with metabolisms that allow them to break down organic pollutants

biosystem a biologically based system for the maintainable production and processing of biological materials (such as food) in parallel with the efficient utilization of natural and renewable resources

biota the plant and animal life of a region or ecosystem, as in a stream, lake, or ocean

biotic pertaining to life or living things, or caused by living organisms; also pertaining to biological factors or influences

biotransformation sum of processes by which a foreign chemical is subjected to chemical change by a living organism

black smoker a seafloor vent in which hot mineralized water from below the seafloor discharges into cold sea water; black color arises from the precipitation of dark sulfide minerals; compare with "white smoker"

bog a poorly drained wetland, usually found in a glacial depression, characterized by the presence of saturated organic soil (peat) and acidic water; plant decomposition is very slow in this environment

boiling point the temperature at which water becomes steam; the boiling point of water is 100°C (212°F) at sea level and decreases with decreasing pressure (i.e., with increasing elevation)

brackish describes water having a salinity from 0.05 to 17 parts per thousand; typically a mixture of sea water and fresh water (e.g., as found in an estuary)

breakwater a barrier that protects a harbor or shore from the full impact of waves

brine water containing a higher concentration of dissolved salts than normal sea water (which contains approximately 35 parts per thousand); produced in oceans through the evaporation or freezing of sea water, or in groundwater through extensive reaction with bedrock minerals

brine film a thin film of salty water surrounding grains of ice or soil that forms when salts are excluded during freezing of permafrost, ground ice, or sea ice; interstitial brines formed by freezing can remain liquid far below the freezing point of fresh water because dissolved salts depress water's freezing point

brine pockets the residual saline water within the ice after sea water has frozen; dissolved minerals are not readily incorporated into ice as it forms—therefore, as freezing takes place, the remaining water becomes increasingly saline; in most icebergs and in sea ice, trapped bubbles of saline water remain

calcite a mineral consisting of calcium carbonate

Glossary

calculus a branch of mathematics that involves computing or calculating quantities that change as functions of different variables

caldera a bowl-shaped crater at the top of a volcano caused by the collapse of the volcano's central part; collapse generally occurs coincident with massive eruptions

calibration the process of correlating the readings of an instrument with those of an established standard in order to check its accuracy of measurement

caliche a mineral deposit, generally containing calcite and gypsum, that forms in arid regions when rainwater that has infiltrated the ground is evaporated, causing the minerals to precipitate

canalization in general, the process of creating artificial canal-like waterways or enhancing natural waterways to drain or irrigate land or to allow navigation

canopy the network of limbs, leaves, and other vegetation high in a forest; the canopy offers habitat and influences ground temperature and precipitation through-fall

capillary action the action by which water is drawn around and into soil particles due to the forces of adhesion, cohesion, and surface tension acting in a liquid that is in contact with a solid

captive breeding the artificial propagation or maintenance of animals in captivity

capture fishery the removal of aquatic organisms from natural or enhanced waters

carbon-14 dating the technique of deriving an approximate age from organic material, such as bone, charcoal, or shell, by comparing the decay activity in the sample of radiocarbon (carbon-14), which begins to decrease at death and subsequent burial, with the decay activity of modern carbon-14; also called radioactive carbon dating, or radiocarbon dating

carbonaceous chondrites one of the three major groups of chondrites; little-altered, primitive meteorites containing significant amounts of carbon, water, and other volatile constituents

carbonate common minerals that are the principal constituents of sedimentary rocks such as limestone and dolostones; the most widespread carbonate minerals are calcite, aragonite, and dolomite

carcinogen a cancer-causing substance or agent

carnivore an animal that feeds mostly on other animals

carp a fresh-water fish, from the Family *Cyprinidae* (the minnow family), with a single back fin and barbels around the mouth; originally from Asia, but now found worldwide in lakes and slow-moving rivers, and farmed for food in large ponds; it prefers warm waters, feeding near the bottom of waters where it stirs up mud and uproots vegetation, often driving out more desirable fish

carrying capacity (animal) the maximum population that can be supported by the natural resources

catadromous describes fish that live in fresh water but move to salt water to spawn

catchment a river basin or a constructed basin or reservoir for the natural or artificial catching or collecting of water, especially rainfall

cation an ion that has a positive charge

centrifugal describes the tendency to move away from the center during rotation around a central point

chalk a pure form of limestone, formed from the deposits of microscopic, shelled animals

channel the bed of a stream, river, lake, impoundment, bay, or strait through which the main volume or current of water flows

channelization any excavation and construction activities intended to widen, deepen, straighten, or relocate a natural river channel; the term does not include maintenance activities on existing channels, such as the clearing of debris or dredging of accumulated sediments

chaotic terrain Martian surface having the appearance of jumbled and broken angular slabs or blocks; may be related to the melting of subsurface ice followed by collapse of the surface

chemical oxygen demand abbreviated as COD, the amount of oxygen required to degrade the organic compounds of wastewater; the larger the COD value of wastewater, the more oxygen the waste discharge demands from a waterbody

chemosynthesis the synthesis of organic compounds by bacteria or other such living organisms using energy from reactions involving inorganic chemicals, typically in the absence of sunlight

chitin a nitrogen-containing polysaccharide (carbohydrate whose molecules consist of a number of sugar molecules bonded together) forming a hard outer layer in many invertebrates, especially insects; also found in the cell walls of many fungi

chlorination the disinfection of drinking water through the addition of chlorine compounds, (e.g., chlorine gas, sodium hypochlorite); purpose is to inactivate disease-causing microbes

chlorofluorocarbon gases formed of chlorine, fluorine, and carbon whose molecules normally do not react with other substances; commonly used as spray-can propellants because of ability not to alter the material being sprayed

chloroplast the protoplasmic body or plastid in the cells of plants that contains chlorophyll and in which photosynthesis takes place

chondrites primitive meteorites with origins in the solar nebula

chronometer any instrument designed to measure time accurately, especially one designed to maintain accurate time in spite of motion or variations in temperature, air pressure, and humidity

Glossary

cistern an artificial reservoir or tank used for holding or storing water, often rainwater

civic pertaining to the rights and duties of the citizenship

clarifier a basin or tank of varied design in wastewater treatment that allows wastewater to stand and undergo progressive sedimentation; the upper portions of the wastewater in the basin are clarified as the solids slowly settle from it

climate the long-term average of weather conditions at a given location

climatology the science and study dealing with climate and climatic phenomena as exhibited by temperature, winds, and precipitation

coagulant flocculant; a substance that, when in a fluid, coagulates or flocculates (comes together in a coherent mass); contaminants such as microbes are attracted to the flocculant and are filtered out of the water

coagulation flocculation; in water treatment, the treating of raw water with chemicals in order to coagulate or flocculate (join together) many small particles

coccolith a minute, rounded calcareous platelet, numbers of which form the spherical shells of coccolithophores

coliform describes a group of bacteria predominantly inhabiting the intestines of humans or animals but also found in soil; while typically harmless, they commonly are used as indicators of the possible presence of pathogenic organisms

collaboration a process of working together or collectively to solve complex, interrelated problems; requires greater consultation, coordination, and public input into the process than largely independent efforts

collection basin a lake, reservoir, or other body of water fed by water drained from a watershed

colloid a particle size range of 10^{-3} to 10^{-6} millimeter (4×10^{-5} to 4×10^{-8} inches)

colluvium a general term for loose, unconsolidated material moved downslope under the influence of gravity

comet a celestial body, observed only in that part of its orbit that is relatively close to the Sun, having a head consisting of a solid nucleus, and comprising ice and rocky debris, termed a "dirty snowball," surrounded by a nebulous coma up to 2.4 million kilometers (1.5 million miles) in diameter and an elongated, curved vapor tail arising from the coma when sufficiently close to the Sun; cometary means pertaining to or originating from a comet

commission as in river basin commission, an independent regional body established to manage and coordinate federal, state, and local water management policies in a river basin, particularly regional basins and basins that cross state or international boundaries; river basin commissions almost always are created by treaty or legislation that outline the commission's mission, duties, and authority to carry out those duties

commissioning (dams) the process by which a dam is authorized and subsequently set into operation

common law a body of rules and principles based on court decisions, traditional usage, and precedent, rather than legislative enactments comprising codified written laws; compare with "statutory law"

common-property resource a resource used in common by many people, such as the air, oceans, and fisheries, wherein it is difficult to exclude anyone from appropriating the resource; sometimes called the "commons" or a "common-pool resource"

compact a formal contract or agreement between two or more parties, often governmental units

confined aquifer describes an aquifer in which groundwater is isolated from the atmosphere by impermeable formations; confined groundwater generally is subject to pressure greater than atmospheric pressure and often is artesian in character

conjunctive use the planned use of water from different sources, usually surface and groundwater sources, to optimize the benefit from available supplies

connate describes sea water that was trapped in the interstices of a sedimentary rock at the time of its deposition; see "evolved connate" and "fossil groundwater"

consequent system a stream drainage pattern that results solely from the influence of the slope of the land surface; pattern often is dendritic or treelike in appearance

conservation the organized management and planned use of living and nonliving natural resources; "water conservation" refers to strategies that increase the efficiency of water use, reuse, recycling, production, or distribution, or that decrease demand

Conservation Reserve Program a voluntary program created by the Food Security Act of 1985 ("Farm Bill") and administered by the U.S. Department of Agriculture; subsequent amendments have created new and innovative approaches to conservation, encompassing wetlands, forestlands, water quality, and sustainable land management

conservationist a person who believes in the regulated exploitation or wise use of natural resources so that irreparable damage is not incurred

conservative element an element or compound that does not readily participate in a chemical or biochemical reaction as it is transported through a system (e.g., an aquifer or river); it can be used to identify flow paths and velocity; see "conservative tracer"

conservative tracer an element or compound that does not readily react with the substance through which it is moving or that is carrying it; often used in hydrology to trace the movement of groundwater or to evaluate the flow characteristics of a stream; chloride and bromide are typical conservative tracers

Glossary

consolidated describes sedimentary rock in which the individual grains cling or are bound together, like mud or sandstone; as opposed to unconsolidated sediment, in which individual grains are clearly separated and can move freely, like silt or sand

consumptive use a use which lessens the amount of water available for another use; for example, water that is used for development and growth of plant tissue or consumed by humans or animals

contaminant as defined by the U.S. Environmental Protection Agency, any physical, chemical, biological, or radiological substance in water, including constituents that may not be harmful to the environment or human health

contamination impairment of the quality of water or the environment by natural or human-made substances to a degree that is considered undesirable for certain uses; this term usually implies a human or environmental health threat, but some types of contamination are merely nonaesthetic rather than harmful; see "pollution"

continental of or pertaining to the continents

continental ice sheet a relatively permanent layer of ice (a large ice cap) covering an extensive tract of land; also called continental glacier, as that of the Antarctic continent

continental margin region where continental crust meets oceanic crust; extending from the shoreline to the deep-ocean basin, this feature includes the continental shelf, continental slope, and continental rise

continental plate a large segment of Earth's crust and uppermost mantle (lithosphere) that supports a major landmass

continental shelf the relatively flat, submerged natural platform, about 1-degree slope, that extends seaward from the beach for about 70 kilometers (45 miles), with water depth up to 130 meters (425 feet) maximum, and ends where the slope and water depth increase

convection circulatory movement in an unevenly heated mass (liquid, solid, or gas); cooler material generally is denser and sinks in an area influenced by gravity, whereas warmer material usually is less dense and rises; convection takes place in the atmosphere, lakes, oceans, and Earth's mantle

cooperative agreement an agreement, typically voluntary, entered into by parties to achieve common goals

coral a marine organism that lives in colonies and excretes an external, calcium carbonate skeleton; groups of these anthozoan coelenterates often form large reefs in tropical seas

coral reef a resistant marine ridge or mound consisting chiefly of compacted coral together with algal material and biochemically deposited magnesium and calcium carbonates

coriparian landowners, either individual or states, sharing the area along a waterway

cost–benefit analysis an analytical technique that is used to guide policymakers by computing the present discounted value of benefits and costs for a set of policy alternatives

covalent bond a chemical bond between two atoms of the same or different elements, in which each atom contributes one or more electrons to be shared in pairs

critical habitat the minimum portion of the habitat that is essential for the survival of a species

cross-contamination phenomenon occurring when sewage is inadvertently transferred into potable (drinking) water pipes and distributed throughout the system; this can happen when sewer lines break, water pipes crack, or water pipes lose pressure

crude oil naturally occurring liquid composed of mixtures or organic chemicals called hydrocarbons; can be distilled to produce gasoline and many other products

crust the outermost portion of the Earth's lithosphere, above the mantle, and comprising rocks such as sandstone, limestone, granite, andesite, and basalt; the crust is enriched in SiO_2, Al_2O_3, K_2O, and Na_2O

crustacean arthropods with hard shells, jointed bodies, and appendages, and that primarily live in water; examples are shrimp, krill, crabs, and lobsters

cryptosporidiosis a condition in which a parasitic coccidian protozoan is found in the intestinal tract of humans

crystalline having an internal structure that is arranged in a repeating, orderly pattern

cyanobacteria also known as blue-green algae, primitive single-celled organisms structurally similar to bacteria, sometimes joined in colonies or filaments

dam failure the collapse of a dam or a portion of its components, often as a result of structural weaknesses that develop over time, of improper design or operation, or of larger-than-anticipated flood events

DDT abbreviation for dichlorodiphenyltrichloroethane, a colorless, odorless, water-insoluble, crystalline pesticide that acts as a nerve poison and is effective at killing insects; it tends to accumulate in ecosystems, and has toxic effects on many vertebrates; use as a pesticide is now prohibited in the United States

decommissioning (dams) the process by which a dam is permanently taken out of operation

decomposer any of various organisms, such as soil bacteria or fungi, that feed on and break down organic substances, such as dead plants and animals

deep well a relative term for a drilled well, generally greater than 50 meters (165 feet), with a motor-driven or engine-driven pump immersed below the water level in the well; the motor or engine is at ground level; the practical limit to well depth is about 200 meters (650 feet), although there is no theoretical limit

deep-sea of, relating to, or occurring in the deeper parts of the sea; often miles beneath the ocean's surface, where sunlight can no longer penetrate

Glossary

degradation breakdown of a chemical to yield usually simpler chemical products (e.g., molecules) by way of biocatalysis (e.g., metabolism), photolysis (e.g., sunlight), or physicochemical processes (e.g., hydrolysis)

delta an alluvial deposit of sediment at the mouth of a river where it enters quieter or deeper water, such as a lake or ocean

density mass per unit volume of a substance; with respect to sea water, the mass in kilograms of 1 liter of sea water; density is a complex function of its temperature and salt content, with warmer, fresher waters being less dense than colder, saltier waters

desalinization also spelled desalination, the process of removing salts and other dissolved solids from sea water or saline (salty) water, usually to make it drinkable

desertification the creation of deserts by climate change or by human-induced processes including overgrazing, the destruction of forest belts, exhaustion of the soil by intensive cultivation, and salinization of soils due to mismanaged irrigation

desorb detaching of a chemical constituent from a surface to which it had previously adsorbed; the relative rates in which a given constituent adsorbs versus desorbs is controlled by the water chemistry and the constituent's characteristics

detention pond a structure designed to temporarily store stormwater in order to reduce the potential for flooding

detritivore an animal that feeds on dead or decaying organic material (detritus)

detritus general term for dead and decaying organic material in an ecosystem, whether terrestrial or aquatic; examples include leaf litter on a forest floor, or fish carcasses in a bay

developing country any country whose Gross National Product (GNP) per capita is less than $3,000; in contrast, the GNP per capita of the richer developed countries including the United States, Canada, Japan, and most of Western Europe exceeds $25,000

development the process of improving the overall well-being of society, particularly through political, cultural, social, or economic means

diagenesis the process of chemical and physical changes that occur within sediments after their accumulation; includes the processes of compaction, the cementation of minerals to one another, recrystallization of minerals, and replacement of one mineral by another

diatom any of the microscopic unicellular or colonial algae constituting the class *Bacillarieae* that have a silicified cell wall, which persists as a skeleton after death and masses of which ultimately accumulate to form diatomite

dike an embankment to confine or control water, especially one built along the banks of a river to prevent overflow of lowlands

dinoflagellate a microscopic unicellular alga that moves by means of a flagellum, a threadlike or whiplike structure

discharge the volume of water or a watery solution flowing past a point per unit time; common units are cubic feet per second or cubic meters per second

discount rate a factor that translates future monetary values into today's values; in cost–benefit analysis, discount rate is represented as r in the discounting formula $1/(1 + r)^t$, when t is the number of years in a water project

discounted net present value for a water project, when all the costs and benefits for each year of the project are discounted using a common discount rate, the sum of the discounted cost minus discounted benefits gives the net present value (NPV); only when NPV is positive is a water project economically feasible

displaced population people who are forced out of their homes in search of safety during conflicts or natural disasters; during prolonged displacements, the international community often sets up camps where the people (refugees) may gain access to basic necessities, including water

dissect in landscape evolution, to cut by erosion into hills and valleys or into flat upland areas separated by valleys

dissociation product individual ions or neutral molecules formed through the chemical disassociation of a single molecule; for example, H_2CO_3 disassociates into H^+ and HCO_3^-

dissolution the action or process of dissolving or being dissolved

dissolved describes the chemical breakdown of a solid in a solution into individual atoms or molecules and their dispersement in the fluid medium; for example, describes the dissolved solids or dissolved gases in water

dissolved load all the material transported by a stream or river in solution, as contrasted with bedload and suspended load

dissolved oxygen concentration of oxygen, expressed in milligrams per liter, dissolved in water and readily available to fish and other aquatic organisms; strongly influenced by temperature, biologic activity, biochemical oxygen demand, and chemical oxygen demand

distillation process by which liquids may be purified or changed through evaporation and condensation (change from a gas to a liquid)

distribution system the system of pipes and storage tanks of a public water supply that deliver drinking water from the source or treatment plant to individual users

divergence a region where water flows outwards at the surface from a source, causing water to upwell from below the surface to replace it

dolomite $CaMg(CO_3)_2$; a mineral consisting of a carbonate of calcium and magnesium; often formed during diagenesis of limestone (calcite)

downdrift the direction of predominant movement of littoral (shore-related) materials in seas or lakes

drainage basin the land area drained by a river and its tributaries; also called catchment, drainage area, river basin, or watershed

Glossary

drainfield an arrangement, generally parallel, of buried perforated piping or tubing in which the fluid is discharged to the ground through seepage; most common use is with septic tanks, but can also be used for domestic or industrial wastewater disposal after other treatment methods

dredging the process of excavating sediments and other materials, usually from underwater locations, for the purpose of mining aggregate (sand and gravel), constructing new waterways, or maintaining existing waterway cross-sections

drip irrigation a system for slowly watering crops at points on or just below the soil surface so that a plant's root zone is thoroughly moistened, with little water being wasted via ponding or runoff

dripstone rock consisting of calcite deposited by precipitation from dripping water, as found in limestone caverns; dripstone may form stalactites and stalagmites

drought mitigation actions designed to lessen or reduce adverse impacts of drought events on individuals and communities

duplicate as in duplicate sample, a second sample collected in the same manner and within the same timeframe and analyzed separately from the primary sample in order to determine the precision of the analysis

dynamic equilibrium the state in which opposing reactions or processes are proceeding at the same rate such that no apparent change is occurring in the concentrations of components involved in the reaction; for surface water, a channel exhibits patterns of erosion and deposition but there is no net change in the input and materials; for groundwater, a condition in which the amount of recharge to an aquifer equals the amount of natural or artificial discharge

dynamic height the relative difference in depth of a pressure surface between two points, which depends on the density structure of the water column at each point

ecology the scientific study of the interrelationships of living things to one another and to the environment; also refers to the ecology of a given region

economic development quantitative and qualitative changes in an economy that enhance the well-being of a society or community

ecosystem the community of plants and animals within a water or terrestrial habitat interacting together and with their physical and chemical environment

ecotone the region of transition, generally gradational, between two ecological communities or ecosystems

efficiency the condition of minimal waste; in economics, a condition in which markets have optimally allocated resources

effluent a liquid that flows out of or away from an area of waste processing or containment; includes treated wastes from municipal sewage plants, brine wastewater from desalinization operations, and coolant waters from a nuclear power plant

ejecta blanket material ejected (thrown out) during impact crater formation and deposited around the crater, covering the surrounding terrain

El Niño an occasional warming of sea-surface temperatures in the equatorial Pacific off the coast of South America

elasticity a measure of the response to a given change, typically in the price of a product; for example, how much consumers increase their oil use when its price goes down, or how much less water they buy when its price goes up; elasticities are measured as percentage changes

elite as in "the elite," a group of people considered to be the best in a particular society or category, especially considered with respect to wealth, power, and talent; in terms of wealth, refers to the population in the top 10 percent of a country's income distribution

embayment a recess in a coastline that has formed a bay

emergent rising above a surrounding medium, especially a fluid; describes a plant with some part standing above the water surface and the rest submerged

empirical based on experience or observations, as opposed to reason or conjecture

emulsion a mixture of two liquids, in which one is in the form of fine droplets and is dispersed in the other

endangered describes a plant or animal species threatened with extinction by human-made or natural changes throughout all or a significant area of its range; designated in accordance with the 1973 Endangered Species Act

endocrine of, relating to, or denoting glands that secrete hormones or other such products directly into the bloodstream

endogenous originating from within, as opposed to coming from external sources

entomology the scientific study of insects

entrain to draw in and transport (as solid particles or gas) by the flow of a fluid; for example, water droplets may become entrained in rising air currents

entrenchment in landscape evolution, the process of a river rapidly downcutting as the result of climatic or base level changes such that the previous channel form or pattern is preserved (e.g., an entrenched meander)

environment all of the external factors, conditions, and influences that affect the growth, development, and survival of organisms or a community; commonly refers to Earth and its support systems

environmental impact statement often known as EIS, a detailed document that outlines potential impacts of projects being considered by federal agencies and that potentially have significant environmental implications; an EIS is required by the National Environmental Policy Act

environmental policy the set of laws, legislation, regulations, and political perspectives governing the environment and that provides a framework for achieving society's goals in managing environmental resources

Glossary

environmental science the interdisciplinary application of the physical, chemical, and biological sciences to the study of the environment, encompassing resources, population, and pollution

ephemeral stream a stream that flows only in direct response to precipitation, and thus discontinues its flow during dry seasons; in ephemeral wetlands, refers to holding water from weeks to months but not year-round

epicenter the location on Earth's surface directly above where an earthquake is generated (known as the earthquake focus)

epifauna aquatic animals that live on the surface of the seabed or a riverbed, or attached to submerged objects or other aquatic animals or plants

epilimnion the warm upper, well-mixed layer of a lake or sea that is thermally stratified, extending down from the water surface to the thermocline, which forms the boundary between the warmer upper layers of the epilimnion and the colder waters of the lower depths (hypolimnion)

epipelagic of, relating to, or denoting the organisms that live in the waters from the surface to depths generally not exceeding 200 meters

equilibrium constant for an equilibrium reaction, the ratio of the product of the concentrations of the individual reaction products divided by the product of the concentrations of the individual reactants; for a given temperature, this ratio (K) is constant; for example, for the equilibrium reaction A + B = C + D, the equilibrium constant (K) is expressed as K = ([C] $\times$ [D]) / ([A] $\times$ [B]), where the square brackets refer to molal concentrations; K is a function of temperature, and if K and the concentrations of any three of the components in the reaction are known, the fourth can be calculated

escarpment a steep slope or long cliff that results from erosion or faulting and separates two relatively level areas of differing elevations; also, the topographic expression of a fault

estuary a tidally influenced coastal area in which fresh water from a river mixes with sea water, generally at the river mouth; the resulting water is brackish, which results in a unique ecosystem

ethics a set of moral principles

eukaryote all living organisms other than the eubacteria and archaebacteria; all organisms that contain a cell or cells in which the genetic material is DNA in the form of chromosomes contained within a distinct nucleus

eustatic sea-level change a change in sea level that affects all the shorelines of the world simultaneously; for example, during ice ages, glaciers grow, resulting in the transfer of water from the ocean to snow and ice on land; compare with "isostatic sea-level change"

eutrophic describes a body of water that has become enriched with large nutrient concentrations, commonly phosphorus and nitrogen, resulting in high productivity; such waters often are shallow, and may experience periods of algal blooms and subsequent oxygen deficiency

eutrophication the process by which lakes and streams become enriched, to varying degrees, by concentrations of nutrients such as nitrogen and phosphorus; enrichment results in increased plant growth (principally algae) and decay, the latter of which reduces the dissolved oxygen content; highly eutrophic conditions may be considered undesirable, depending on the human use of the waterbody

evaporite sediment that forms as the result of the precipitation of minerals during the evaporation of water, primarily sea water, and that may form sedimentary rock; principal minerals are gypsum and halite

evapotranspiration water discharged to the atmosphere as a result of evaporation from the soil and surface-water bodies and by plant transpiration

evolved connate describes saline groundwater that owes its high mineral content to being in contact with, and slowly dissolving, rocks for long periods of time; may be found in sedimentary, igneous, and metamorphic rocks; see "connate" and "fossil groundwater"

exfoliation cracks fractures that parallel the surface of an object and tend to separate the object into thin sheets; may be caused by weathering or release of applied stress

exotic a general term that typically describes an organism or species that is not native to the area in which it is found (i.e., it is nonindigenous); exotic species may be invasive

externality the unintended or unwanted byproduct of production or consumption that must be borne by society in general; a negative externality arises from the detrimental effects of use or production (e.g., water pollution may represent a negative externality of watercraft operation); a positive externality arises from beneficial effects (e.g., decreased disease incidence arises from health vaccinations)

extrapolate the continuation, by means of simple estimation or sophisticated analysis, of a trend of time series data beyond its last observed value

extraterrestrial from beyond the Earth and its atmosphere

fault a fracture in a body of rock along which the mass of rock on one side of the fault moves against the mass on the other side; faults generate earthquakes

fault block a mass of rock bordered completely or partially by faults, and behaving as a unit during tectonic activity or faulting

federal–interstate compact in water resources, an agreement that forms an independent entity to coordinate federal, state, and local water management policies

feedback loop describes when natural processes respond to a disturbance in a manner that either reduces or increases the impact of the disturbance; for example, an increase in a stream's velocity will cause its channel to deepen and widen, thereby reducing the velocity (negative feedback); or, a reduction in vegetation on a slope will increase erosion, which will remove more plants, which will further increase erosion (positive feedback)

feldspar a silicate mineral that is the most common constituent mineral in Earth's crust

felsic describes igneous rocks that are high in silica and have high concentrations of sodium and potassium; granite and rhyolite are felsic igneous rocks

fen a saturated wetland characterized by the presence of basic or calcareous groundwater (as contrasted to a bog); often found as seepage areas on gentle slopes comprised of glacial deposits

fiduciary the law with respect to a trust relationship, especially between a trustee and a beneficiary

filter feeder an aquatic animal, such as a clam, barnacle, or sponge, that feeds by filtering particulate organic material from water

filtration a treatment process, under the control of qualified operators, for removing solid (particulate) matter from water by means of porous media such as sand or an artificially made filter; often used to remove particles that contain pathogens (disease-causing organisms)

finfish an aquatic animal with a backbone and fins, as opposed to a shellfish (an aquatic animal without a backbone and with a shell)

firmament the heavens or the sky, especially when regarded as tangible

fissure a surface of a fracture or crack in a rock along which there is a distinct separation of the rock on either side of the fissure

fissured sediment fractured (or cracked) sediments with distinct separation along the crack surfaces; fissures may be filled with mineral-bearing materials; in terms of water flow, water will infiltrate more rapidly along fissures than through porous sediment

flagellum any of various elongated, threadlike or whiplike appendages of plants or animals; plural is flagella

flash flood a sudden flood that crests in a short time (hours or minutes) and is often characterized by high-velocity flows; most common in deserts or areas of low vegetation

flocculation the agglomeration or clustering of colloidal and finely divided suspended matter after coagulation by gentle stirring by either mechanical or hydraulic means such that they can be separated from water or sewage; see "coagulation"

floe a contiguous piece of ice on the surface of water (e.g., rivers, lakes, or seas)

flood temporary inundation of normally dry land areas from the overflow of inland or tidal waters, or from the unusual and rapid accumulation or runoff of surface waters from any source; the rise in water may be caused by excessive rainfall, snowmelt, natural stream blockages, windstorms over a lake, storm surges on the ocean, or any combination of such conditions

flood hazard the degree of potential for inundation that presents risk to life, health, property, and natural floodplain values

floodplain the low-lying land adjoining a river that is sometimes flooded; generally covered by fine-grained sediments (silt and clay) deposited by the river at flood stage

floodplain management the societal process of decision-making to achieve the best use of floodplains, the low-lying land adjoining a river that is sometimes flooded

flow rate rate of flow of water expressed in units of length per time, such as meters per second or feet per day; compare with "flow velocity," "discharge," and "volumetric flow rate"

flow velocity the speed (rate) and direction of water flow, quantified as the rate of flow of water in a given direction; expressed in units of length per time (e.g., meters per second or feet per day) and direction (e.g., northeast); compare with "discharge" and "volumetric flow rate"

flowstone a layered deposit of calcium carbonate ($CaCO_3$) on rock over which water has flowed or dripped, as on the walls of a limestone cave; see "dripstone"

fluid inclusions small volumes of fluid (liquid and gas) trapped in imperfections as minerals grew or recrystallized around them; mostly water with minor CO_2 plus dissolved minerals; most are less than 0.1 millimeter (0.004 inches) in size

fluvial pertaining to the action of a river, stream, or flood flow, as in fluvial processes of erosion or the deposition of alluvium

food chain the levels of nutrition in an ecosystem, beginning at the bottom with primary producers, which are principally plants, to a series of consumers—herbivores, carnivores, and decomposers

food fish captured or farm-raised fish used for food

food web a complex food chain, with several species at each level, so that there is more than one producer and more than one consumer of each type

foreshore the part of a shore that lies between high and low watermarks; the part of a shore between the water and occupied or cultivated land

forest litter the accumulation of organic debris (e.g., limbs, leaves) on the forest floor

fossil a preserved plant or animal imprint or remains

fossil fuel substance such as coal, oil, or natural gas, found underground in deposits formed from the remains of organisms that lived millions of years ago

fossil groundwater also called connate water, a highly enriched brine that resides in ancient sediments often associated with oil-bearing formations

fracture in geology, a general term for any break in rock, which includes cracks, joints, and faults

frazil ice ice crystals in the water column, usually near the surface, having the appearance of slush; frazil ice is the first stage in the formation of sea ice; crystals start to form when the sea water cools to $-1.8°C$ ($28.8°F$)

front (atmospheric) a boundary between atmospheric air masses

front (ocean) a region in the ocean where a sudden change in temperature,

Glossary

velocity, or other parameter causes a sharp line of demarcation at the surface, often visible to the eye; a front may be small-scale and narrow, or larger-scale and extend across several kilometers; a front usually denotes a region of convergence, where water tends to sink

frost wedging an important mechanism of mechanical weathering of rocks wherein water freezes in cracks and, as a result of expansion during the formation of ice, forces the crack further apart, perhaps breaking the rock into smaller pieces

frustule the silicified cell wall of a diatom (a microscopic alga), consisting of two valves or overlapping halves

fugitive resource a natural resource such as water that moves from one location and one state (liquid, gas, or solid) to another

fumarole a hole or orifice in a volcanic region, usually in lava, from which high-temperature gases and vapors are expelled; mineral deposits frequently line the opening

gaining reach a length of stream that is receiving groundwater inflow along its length, resulting in a greater stream-water discharge downstream than upstream of the reach

game fish fish considered to possess sporting qualities on fishing tackle, such as salmon, trout, black bass, and striped bass

gamete a sex cell; in some of the simplest organisms, the gametes are not differentiated into egg and sperm

gangue the nonvaluable materials closely associated with the valuable minerals in ore deposits (also referred to as waste rock); common gangue minerals include quartz and calcite; typically must be removed and discarded to extract the valuable high-grade ore

gas hydrate a crystalline solid consisting of a gas molecule surrounded by water molecules

gastropod a large class (Gastropoda) of mollusks in which each animal has a head with eyes, a large flattened foot for movement, and often a single asymmetrical spiral shell; includes limpets, snails, and slugs

geochemistry the science that deals with the chemical composition of the Earth's materials and the chemical processes involved in their formation or modification; also refers to the chemical composition of Earth's materials in a given region

geodetic relating to the precise measurement of the Earth's surface or of points on its surface, such as in land surveying

geography the science of the Earth and life, especially the description of land, sea, air, and the distribution of plant and animal life, including humankind and its industries, with reference to the mutual relations among these diverse elements; also refers to the geographical features of an area—that is, the nature and relative arrangement of places and physical features

geoid the hypothetical shape of the Earth considered as a mean (average) sea-level surface extended continuously through the continents and the oceans

geologic unit a large volume of a certain kind of rock or sediment with recognizable and distinguishing characteristics and of a given age range

geology the scientific study of the Earth, its form and composition, and the physical, chemical, and biological processes that affect it; includes the study of ancient life on the planet; also refers to the geologic characteristics of a given region

geometry a branch of mathematics dealing with features of point configurations that are invariant under a specified group of mathematical transformations

geomorphology the scientific study of the physical characteristics of the land surface and landforms that are the result of specific geologic processes; also refers to the geomorphological characteristics of a given region

geophysical related to the physical characteristics and structure of the Earth, including geodesy, seismology, meteorology, oceanography, atmospheric electricity, terrestrial magnetism, and tidal phenomena

geostrophic current an ocean current that flows along a line of equal dynamic topography on the sea surface, oriented so that the high topographies are on the right in the Northern Hemisphere and on the left in the Southern Hemisphere

geothermal describes terrestrial heat, usually associated with water, as around hot springs

geyser a periodic thermal spring that results from the expansive force of superheated steam; also, a special type of thermal spring which intermittently ejects a column of water and steam into the air with considerable force

geyser reservoir the underground network of open volumes in the subsurface that serve to temporarily store water that will be heated and erupted from the geyser

gill net a net set upright in the water so that fish are caught in it when their gills become entangled in its meshes

glaciation the covering and modification of a landmass by a glacier or glaciers

glacier a huge mass of ice, formed on land by the compaction and recrystallization of snow, which moves slowly downslope or outward owing to its own weight

glacier cave a cave that formed at the base of a glacier along a stream of meltwater moving beneath the glacier

global warming an increase in the average temperature of the Earth's atmosphere, especially a sustained increase sufficient to cause climatic change

grab sample a water sample collected at a single location and at a single time as opposed to a sample composited over space or time

graded selected and arranged by size

gradient a measure of the change in magnitude of a parameter (e.g., temperature, elevation, chemical concentration) with distance; for example, a

Glossary

stream gradient would be 0.001 if the stream's elevation dropped 1 meter (3.3 feet) over a distance of 1,000 meters (3,300 feet); when a gradient exists, there is a tendency for a transfer to take place from the area of greater magnitude to the area of lesser magnitude

granite an igneous rock comprising quartz and potassium feldspar; high in silica and crystallized from a melt within Earth's crust

gravity flow the movement downslope of a mixture of unconsolidated material and water, solely in response to gravity; the resulting deposition shows delta-like patterns and structures

grease ice thin plates of organized ice crystals on the water surface; an early stage in the growth of sea-ice cover

greenhouse effect the phenomenon whereby a planetary body's atmosphere traps solar radiation; caused by the presence in the atmosphere of gases such as carbon dioxide, water vapor, and methane, that allow incoming sunlight to pass through but trap heat radiated back from the body's surface

greenhouse gas a gas in the atmosphere that traps heat and reflects it back to the planetary body

greywater wastewater from clothes-washing machines, showers, bathtubs, hand washing, lavatories, and sinks that is not used for disposal of chemicals or chemical–biological ingredients; less commonly spelled graywater or gray water

groin a wall placed perpendicular to the shoreline for the purpose of catching sediment to build up a beach

groundfish general term for more than 80 species that, with few exceptions, live on or near the ocean floor (e.g., rockfish, flounder, lingcod, ocean perch, and Pacific whiting)

groundwater generally, all subsurface (underground) water, as distinct from surface water, that supplies natural springs, contributes to permanent streams, and can be tapped by wells; specifically, the water that is in the saturated zone of a defined aquifer

groundwater mining the withdrawal of groundwater at a rate exceeding natural recharge, resulting in a permanent lowering of the water table

groundwater reservoir an aquifer or a previously unsaturated rock or sediment in which water is stored

gully general term for a defined channel, larger than a rill, produced by running-water erosion; compare with "rill"

guyot a flat-topped seamount on the ocean floor; formed as a volcano at a spreading center, the mountaintop is eroded by wave activity as the mountain sinks when the plate upon which it is riding moves away from the spreading center

gyre a circular pattern of currents in an ocean basin

habitat the environment in which a plant or animal grows or lives; the surroundings include physical factors such as temperature, moisture, and light, together with biological factors such as the presence of food and predators

half-life the time required for the initial concentration of a radioactive element to decrease, through radioactive decay, by 50 percent; in nonradiochemical usage, the time required for a pollutant to lose one-half of its original concentration (e.g., the biochemical half-life of DDT in the environment is 15 years)

hard water water that forms a precipitate with soap due to the abundance of calcium, magnesium, or ferrous ions in solution

harmful algal bloom also called harmful algae bloom, a rapid increase in algae concentrations in fresh or marine waters in which one or more algal species causes harm to animals or humans, particularly (but not exclusively) by virtue of their eventual mass decay (and the accompanying oxygen depletion) or by their generation of natural biotoxins; harmful marine blooms are a subcategory of "red tides," a popular term which encompasses both harmful and nonharmful blooms

hazardous waste any solid, liquid, or gas that, when disposed, exhibits the characteristics of ignitability, corrosivity, reactivity, or toxicity, as well as any industrial waste that has been specifically listed in the federal regulations as having hazardous properties

headland a point of land, usually high and with a sheer drop, extending out into a body of water; a promontory

headwaters the source or upper reaches of a stream; also the upper reaches of a reservoir

heat capacity the amount of heat required to raise the temperature of 1 gram of a substance 1 degree centigrade; water has a high heat capacity, and can absorb or evolve significant heat with minor temperature change

heat exchanger a mechanism whereby heat is exchanged from one medium to another

heavy metals a group of metals that have high density and are considered toxic at specified concentrations; with respect to soil management, such metals include copper, iron, manganese, molybdenum, cobalt, zinc, cadmium, mercury, nickel, and lead

hectare a metric unit of area equal to 10,000 square meters (107,639 square feet or 2.471 acres)

herbaceous with the characteristics of a herb; describes a plant with no persistent woody stem above ground

herbicides a group of chemicals used to kill or reduce the growth of vegetation that is considered undesirable

herbivore an animal that feeds mostly on plants

herpetology the scientific study of amphibians and reptiles

heterotroph an organism deriving its nutritional requirements from complex organic substances; animals or microorganisms that live on producers

high-grade ore rock or sediment that has high concentrations of valuable minerals

Glossary

holistic describes an approach to problem-solving that focuses not only on the individual components of the problem, but more importantly on its totality; with respect to water-related issues, typically describes an analytical and planning approach that considers the interrelated linkages and interdependencies of a socioeconomic system with resource use, pollution, environmental impacts, and preservation of an entire ecosystem

hormones biochemicals, usually proteins or steroids, released in very low concentrations by specialized cells (e.g., in glands) that travel to and are recognized by receptor cells, which in turn trigger a wide range of cascading biochemical events that regulate numerous aspects of metabolism or behavior

hot spot a location where a mantle plume rises toward the base of the Earth's lithosphere, producing magma, high heat flow, and volcanism at the surface; generally fixed relative to the moving lithospheric plate

hot spring a thermal spring that brings warm or hot water from the subsurface to the surface; water temperature usually is 8 Celsius degrees (15 Fahrenheit degrees) or more above the mean air temperature

hull the main body of a ship or other such vessel, including the bottom, sides, and deck but not the masts, rigging, superstructure, engines, and other fittings

hurricane a giant atmospheric circulation that forms over warm tropical ocean water, with a calm, clear eye at the center of the system, and winds of 65 knots or higher

husbandry in aquaculture, the rearing and careful management of captively held fish and other aquatic resources

hydrate to add water to a substance, system or compound

hydraulic gradient the change in hydraulic head between two points (e.g., the difference in water level between two points divided by the distance between the two points)

hydraulic head the potential energy of water as a result of its elevation and weight of overlying water; it is the driving force for natural water movement; for dams, the hydraulic head approximately equals the difference between upstream and downstream water depths

hydraulically connected a condition in which waterbodies and/or aquifers are in direct contact and water can move easily between them (e.g., water moving from a streambed into an aquifer, or vice versa)

hydraulics the scientific study of water in motion; modern hydraulics emphasizes the mechanical properties of water that describe the specific pattern and rate of movement in the natural environment or in artificial systems (for example, pipe systems)

hydric characterized by, relating to, or requiring an abundance of moisture; referring to a habitat characterized by wet or moist conditions rather than mesic (moderate moisture conditions) or xeric (dry conditions)

hydrocarbon a chemical compound that consists entirely of carbon and hydrogen, such as petroleum, natural gas, and coal

hydroelectric often used synonymously with "hydropower," describes electricity generated by utilizing the power of falling water, as with water flowing through and turning turbines at a dam

hydrogen bond in water, the type of chemical bond between two water molecules; caused by electromagnetic forces, and occurring when the positive (hydrogen) side of one water molecule is attracted to and forms a bond with the negative (oxygen) side of another water molecule

hydrogeology a branch of geology that deals with the occurrence and movement of groundwater in relation to Earth structures; also refers to the hydrogeologic characteristics of a given region

hydrograph a graphical representation or plot of changes in the flow of water or changes in the elevation of water level plotted against time

hydrologic cycle the solar-driven circulation of water on and in the Earth, characterized by the ongoing transfer of water among the oceans, atmosphere, surface waters (lakes, streams, and wetlands), and groundwaters

hydrology the science that deals with the occurrence, distribution, movement, and physical and chemical properties of water on Earth; also refers to the hydrologic characteristics of a given region

hydrolysis a chemical reaction with water, resulting in either the formation of carbonic acid or another weak acid, or the addition of a reaction product containing the hydroxyl ion (OH^-), which can act as a weak base

hydroperiod the seasonal and cyclical pattern of water in a wetland

hydrophyte plants typically found in wet habitats; any plant growing in water or on a wet substrate that is at least periodically deficient in oxygen as a result of excessive water content

hydroponics the cultivation of plants in nutrient solution rather than in soil

hydropower power, typically electrical energy, produced by utilizing falling water; see "hydroelectric"

hydrosphere liquid water and ice on the surface of the Earth and in underground reservoirs

hydrostatic referring to the pressure exerted by water at a point, related to the weight of the water above the point; for example, the hydrostatic pressure in pounds per square inch at the bottom of a tank is equal to the weight in pounds in a column of water one square inch in cross-section and having the height of the water in the tank

hydrothermal associated with hot water, especially with the action of hot water in dissolving, transporting, depositing, and otherwise changing the distribution of minerals in the Earth's crust

hydrothermal alteration zone a volume of rocks altered by the interaction of hydrothermal water with pre-existing rocks and minerals

hydrothermal vent an opening or other orifice on the seafloor through which hot watery solutions that have circulated through the underlying rock escape and mix with sea water; see "black smoker"

Glossary

hypersaline describes water with a salt concentration greater than 40 percent (parts per thousand); this extreme concentration is generally the result of evaporation of sea water

hypolimnion the lower layer of a thermally stratified lake, located below the thermocline, and in which the water is nearly uniformly cool and relatively quiescent

hyporheic exchange the movement of stream water into and out of the hyporheic zone

hyporheic zone the volume of sediment and porous space adjacent to a stream, and through which stream water exchanges

hypothesis a statement made about the condition or behavior of a variable or event that lends itself to rigorous testing for validity

hypoxia a condition in which natural waters have a low concentration of dissolved oxygen (about 2 milligrams per liter as compared with a normal level of 8 to 10 milligrams per liter); most game and commercial species of fish avoid such waters; compare with anoxia, which is less than 0.1 milliliter of oxygen per liter, and the threshold below which animal life diminishes significantly

ice age a cold period marked by episodes of extensive glaciation alternating with episodes of relative warmth; the formally designated "Ice Age" refers to the most recent glacial period, which occurred during the Pleistocene epoch

ice cap an extensive perennial accumulation of snow and ice that forms when glaciers completely fill their subglacial valleys and coalesce (join together); ice caps are smaller than ice sheets

ice pancake see "pancake ice"

ice sheet an extensive perennial accumulation of snow and ice completely covering the underlying topography; ice sheets are land-based or marine-based; present-day ice sheets cover Greenland and Antarctica; ice sheets are larger than ice caps

ice shelf a floating ice mass that is attached to the coast along at least one edge

iceberg a piece of floating ice that breaks off (calves) from an ice shelf, glacier, ice stream, or ice tongue

ichthyology the scientific study of fish

igneous describes rock that solidified from molten material (magma); the rock is extrusive (or volcanic) if it solidifies on the surface and intrusive (or plutonic) if it solidifies beneath the surface

individual quota the amount of a common-property resource assigned to a user, giving them rights to a fixed amount of access and harvesting; also called an individual transfer quota; in fisheries, the permit of each fisher to take a percentage of total allowable catch for a certain species during the fishing season—once an individual quota is attained, the fisher is restricted from fishing for that species until the next season

Industrial Revolution beginning in Great Britain around 1730, a period in the eighteenth and nineteenth centuries when nations in Europe, Asia, and the Americas moved from agrarian-based to industry-based economies

infauna aquatic animals that live in the substrate of a body of water, especially in a soft sea bottom

infiltration the process by which water enters the soil and that is controlled by the character of the soil and surface conditions, such as slope and amount of vegetation

infiltration rate the rate at which water from precipitation enters the soil; the maximum rate is known as infiltration capacity

infrastructure the permanent constructed system (e.g., pipes and other structures) that enables the treatment and delivery of water to support human habitation and activity, or that supports manufacturing activities and water projects (e.g., desalinization and hydropower plants)

inorganic an element, molecule, or substance that did not form as the direct result of biologic activity

insecticides a group of chemicals used to kill or otherwise control insects and arachnids that are considered undesirable

insolation the amount of solar radiation that reaches a given area

institution a custom, practice, relationship, or behavioral pattern of importance in the life of a community or society; also an established organization, especially one dedicated to public service

instream flow the amount of water remaining in a stream, without diversions, that is required to satisfy a particular aquatic environment or water use, such as the water required for fish and wildlife or for navigation

instream water use the use of water in place in a river or lake, without diversion or withdrawal; as opposed to offstream water use, in which water is diverted or withdrawn to be used elsewhere

integrated water management that blends coordinated viewpoints of social scientists with those of engineers and natural scientists or that coordinates in other ways different facets of management, such as competing purposes or different areas or interest groups

interests as in "business interests" or "environmental interests," the vested opinions, perspectives, and positions of stakeholders regarding gains and losses, real or perceived, stemming from decision-making outcomes

intergenerational equity ethical concept of fairness which holds that present generations should not degrade the environment to an extent that unreasonably constrains opportunities for future generations to meet their basic needs

intergovernmental existing or carried on between governmental bodies

internal phosphorus loading a process within a waterbody, typically a lake, whereby phosphorus is released from internal sources (for example, pore water in the bottom sediments, resuspension of sediments by wave action, or disruption of bottom sediment by burrowing organisms)

Glossary

interpolate to estimate intermediate values of a function between two known points; frequently used when certain periods of data are missing, but data surrounding these missing data values are available

interstate existing or carried on between states

interstate water according to law, interstate waters are defined as (1) rivers, lakes, and other waters that flow across or form a part of state or international boundaries; (2) waters of the Great Lakes; (3) coastal waters whose scope has been defined to include ocean waters seaward to the territorial limits and waters along the coastline (including inland streams) influenced by the tide

interstellar describes the region of space that occurs between individual stars, occupied by gas and dust as well as isolated molecules, including hydroxyl ions, water, sulfur oxide, as well as carbon-based molecules

intertidal coastal land that is covered by water at high tide and uncovered at low tide

intragenerational equity ethical concept of fairness which holds that people living on the Earth at the same time (i.e., within the current generation) should have similar opportunities to meet their basic needs and improve their basic standard of living, and that actions by wealthier people should not disadvantage poorer people

intragovernmental existing or carried on within a governmental body

intrastate existing or carried on within a state

invasive describes a plant or animal that moves in and takes over an ecosystem to the detriment of other species; often the result of environmental manipulation; see "exotic"

inverse estuary an estuary that receives no fresh-water input from a river in an area of high evaporation rates; circulation driven by density differences produced as evaporation and salinity increases inland; more saline water at the inner end sinks and moves seaward

invertebrate an animal without a backbone

ion an atom or molecule that carries a net charge (either positive or negative) because of an imbalance between the number of protons and the number of electrons

irrigable land arable land for which a water supply can be made available and which will respond well to irrigation

irrigation the controlled application of water for agricultural or other purposes through human-made systems; generally refers to water application to soil when rainfall is insufficient to maintain desirable soil moisture for plant growth

island arcs the arc-shaped chain of volcanoes that develop over the subducting plate inland from the trench as a result of melting processes associated with subduction; island arcs may develop on oceanic plates (ocean island arcs) or on continental plates (continental island arcs)

Glossary

isopach a line on a map that connects points of equal thickness with respect to a particular stratigraphic unit or group of units

isostasy describes the concept that the elevation of the Earth's surface (over tens of millions of years) seeks a balance between the weight of lithospheric rocks and the buoyancy of asthenospheric fluid (hot, plastic, partially molten rock); a mountain range where erosion has moved a significant amount of rock material may rise isostatically, whereas the basin that receives this eroded sediment may sink under the added weight; "isostatic" means pertaining to or related to isostasy

isostatic sea-level change a change in sea level that occurs owing to significant isostatic adjustments, such as large amounts of mass being loaded (deposition of sediments) or unloaded (erosion of the land) in a region, which causes the land to sink or rise; compare with "eustatic sea-level change"; see "isostasy"

isotope the one of two or more forms or varieties of a specific element that differ in their atomic mass; the proton number is the same for a given set of related isotopes, but the number of neutrons in the nucleus varies; for example, the common isotopes of oxygen, O-18, O-17, and O-16, all have 8 protons, but have 10, 9, and 8 neutrons, respectively

jetty a structure built out into the sea, a lake, or a river to counteract the effects of tides or currents

Jovian refers to the Jupiter-like (low-density) planets of Jupiter, Saturn, Neptune, and Uranus; the gas giants; compare with "terrestrial planets"

junior appropriator an individual whose right to appropiate water from a source is more recent in time than others with rights to the same source of water; a right with lower priority than all others

karst topography characterized by closed depressions or sinkholes, caves, and underground drainage formed by dissolution of limestone, dolomite, or gypsum

kelp forest a dense growth of seaweed (called giant kelp) that occurs in cool coastal waters where sunlight can reach the seafloor

keystone species a species on which the persistence of a large number of other species in the ecosystem depends

kilowatt a unit of power equal to 1,000 watts, wherein the watt corresponds to the rate of energy in an electric circuit

krill small, abundant, shrimp-like crustaceans that form an important part of the food chain in Antarctic waters

Kuiper Belt a disk-shaped region past the orbit of Neptune roughly 30 to 50 astronomical units (AU) from the Sun, containing many small icy bodies; considered the source of the short-period comets

La Niña an area of cooler-than-average ocean water in the tropical eastern Pacific off the coast of South America; the counterpart of El Niño

lacustrine pertaining to, produced by, or formed in a lake or lakes

Glossary

lag time the time period between a rainfall event in a watershed and the occurrence of peak discharge in the stream

land-use planning a generic term for a wide range of legislative and regulatory activities intended to limit or direct land development for the purpose of making its usage sustainable; large-scale land-use plans often are implemented by local zoning and land-use ordinances

landform a discernible natural landscape that exists as a result of wind, water, ice, or other geological activity, such as a plateau, plain, basin, or mountain

landscape development the progressive evolution of topography as a result of the actions of the geologic agents of wind, water, ice, and mass movements (landslides)

landslide a mass of material that has slipped downhill under the influence of gravity, frequently occurring when the material is saturated with water

latent heat of fusion the energy required to melt 1 gram of ice, or 80 calories per gram

latent heat of vaporization the energy required to vaporize 1 gram of water, or 540 to 640 calories per gram, depending on the water temperature ranging from 100°C to 0°C

latitude the angular distance north or south of the Earth's or another planet's equator, measured in degrees along a meridian

lattice the internal structure of a mineral, produced by the regular arrangement of the mineral's component elements or ions

lava a molten mass of rock material that is extruded at the surface by a volcano or through a fissure in the Earth

lava tube (cave) cave formed during solidification of a large lava flow; flow solidifies from the outside in; cooler solidified outer part of lava flow remains in place while inner molten part drains away, leaving a cavelike structure

leachate liquid that has moved through a substance, removing solids from the substance, generally by dissolution

lead any fracture or passageway through sea ice which is navigable by surface vessels

levee a natural or artificial earthen obstruction along the edge of a stream, lake, or river; also, a long, low embankment usually built to restrain the flow of water out of a riverbank and to protect land from flooding

Liebig's Law of the Minimum the recognition that crop yield or growth is proportional to the amount of the most limiting nutrient present

limnology the scientific study of fresh-water bodies such as lakes, ponds, streams, and rivers

lithosphere the rigid outer layer of Earth made up of the crust and the uppermost mantle

lithospheric plate a section of the lithosphere that acts as a single mass and interacts with other plates; lithospheric plates are created at spreading centers and destroyed at subduction zones; see "plate tectonics"

littoral the region along the shore of a nonflowing body of water; corresponds to riparian for a flowing body of water; more specifically, for marine waters, the zone of the sea flood lying between the tide levels

littoral transport the movement of sedimentary material in the zone extending seaward from the shoreline to just beyond the breaker zone by waves and currents; it includes movement parallel (long-shore drift) and sometimes also perpendicular (cross-shore transport) to the shore

load management steps taken to reduce water demand at peak load times, such as shifting some of it to off-peak times; may refer to peak hours or peak days

local base level see "base level"

lock one in a series of gates that allows vessels to pass through multiple water levels

logarithm the real number x satisfying the equation $b^x = a$, where the base b is a real number greater than 0 and not equal to 1

logarithmic scale a scale in which the distances that numbers are positioned from a reference point are proportional to their logarithms

longitude the angular distance measured east or west from the prime meridian (which runs through Greenwich, England), to the meridian passing through a position; expressed in degrees (or hours), minutes, and seconds

longshore transport the transport of sedimentary material parallel to the shore

losing reach a length of stream that is losing stream water along its length, resulting in a smaller stream-water discharge downstream than upstream of the reach

macroinvertebrates organisms visible with the naked eye; macroinvertebrates in water include insects, snails, bivalves, and sometimes amphipods or copepods

macrophyte a macroscopic form of aquatic vegetation; a plant, especially an aquatic plant, large enough to be seen by the naked eye

macropore an opening or zone of high permeability that provides a zone of rapid transport of water (and potentially waterborne contaminants) into the subsurface; examples include animal burrows and tree roots

mafic describes igneous rocks that are low in silica and have high concentrations of calcium, iron, and magnesium; basalt is the most common mafic igneous rock

magma molten rock found in the mantle and crust of the Earth (also found on planets, moons, and asteroids); when forced toward the surface, it cools and solidifies to become igneous rock; when it erupts at the surface, it is called lava

mangrove tropical evergreen trees and shrubs that have stilt-like roots and stems, and often form dense thickets along tidal shores

mantle the region of the Earth between the molten core and the outer crust, composed mainly of silicate rock, and roughly 2,900 kilometers (1,800 miles) thick; also the interior of another planet, moon, or large asteroid between the core and the crust

marginal sea a semi-closed sea associated with a continent and formed during rifting and early spreading

mariculture the science, art, and business of cultivating marine animals or plants under controlled conditions; a subcategory of aquaculture

marina a water-based facility used for storage, service, launching, operation, or maintenance of watercraft

marine snow small particulate matter that drifts down from the upper layers of the oceans; comprises debris from animals, plants, and nonliving matter, (e.g., fecal pellets, diatoms, dust); affects visibility and light transmission within the water; larger pieces may serve as food source for animals in deeper waters

market institution an arrangement that allows individuals to decide voluntarily how much of a good or service to produce, sell, buy, or consume based mainly on prices set by demand and supply conditions; in the alternative, the government decides who gets how much of a good or service, and at what price

marl in a lake environment, a sediment that accumulates on the lake floor and consists of a mixture of organic matter, clays, carbonates of calcium and magnesium, and remnants of shells; forms in the absence of significant land-derived sediments such as sand; useful as a fertilizer

mass media all of the widespread communications that reach a large audience, especially television, radio, newspapers, and the Internet

maximum contaminant level abbreviated as MCL, the allowable level of the specified contaminant in drinking water; established by the federal Environmental Protection Agency; state governments may set lower levels

mean the arithmetic average of a set of data

meander (noun) one of a series of somewhat regular bends in the course of a stream; (verb) to follow a winding course

media as in water treatment, the sand, gravel, small plastic beads, or other material designed for water filtration

mediator a person who is trained to help parties resolve their own conflicts by using established conflict resolution techniques

megawatt a unit of power equal to 1,000,000 watts, wherein the watt corresponds to the rate of energy in an electric circuit

melting point the temperature at which a solid becomes a liquid; the melting point at which ice becomes water: 0°C (32°F)

meltwater water resulting from the melting of snow, ice, or glacial ice

Mesoamerica the central region of America, from central Mexico to Nicaragua

metabolism the sum total of biochemical processes that occur within a living organism, or a portion of it, in order to maintain life; the biochemical changes by which energy is provided to living cells and new material is assimilated

metabolite any substance produced by metabolism or a metabolic process

metadata statistical information that describes the elements of a set of data

metamorphism the changes in the mineral assemblage and texture of a rock subjected to temperatures and pressures that are significantly different than the conditions under which the rock orginally formed; significantly enhanced by the presence of water; changes occur without melting the rock

metamorphosis the biological process of transformation from an immature form to an adult form in two or more separate stages

metasomatism a change in the composition of a rock due to the introduction or removal of chemical components

meteoric water atmospheric water that reaches the Earth's surface as rainfall or other form of precipitation; part of the hydrologic cycle

meteorite a solid body that has fallen from space to the surface of the Earth or another planet; meteoritic means pertaining to or originating from a meteorite

meteorology the science that deals with the atmosphere, especially with regard to climate and weather

methemoglobinemia a disease, primarily in infants, caused by the conversion of nitrates to nitrites in the intestines, and which limits the body's ability to receive oxygen; often referred to as "blue baby syndrome"

metric ton unit of weight equal to 1,000 kilograms; equivalent to 2,205 pounds or 1.1025 short tons

microbe a microscopic organism, or microorganism; the term encompasses viruses, bacteria, yeast, molds, protozoa, and small algae

microbial film see "biofilm"

microgravity the condition experienced in free fall as a spacecraft orbits Earth or another body; commonly called weightlessness; only very small forces are perceived in free fall, on the order of one-millionth the force of gravity on Earth's surface

microorganism a microscopic organism; see "microbe"

mitigation actions designed to lessen or reduce adverse impacts; frequently used in the context of environmental assessment

mitigation banking a mitigation bank with respect to wetlands is a wetland area that has been restored, created, enhanced, or (in exceptional circumstances) preserved, which is then set aside to compensate for later conversions of wetlands for development activities

model inversion working backwards with a model using observed values of what the model is supposed to predict to determine what the initial conditions were; often referred to as inverse modeling; commmonly used in calibrating a model (e.g., comparing observed water levels with values predicted by the model to determine optimal input parameters to the model)

Mohorovičić discontinuity abbreviated as Moho, the boundary between Earth's crust and the underlying mantle; recognizable by a sharp increase in seismic wave velocity

molality the number of moles of a dissolved solute per kilogram of solvent

molarity the number of moles of dissolved solute per liter of solution

mole a quantity of a given element or compound, defined as the formula weight (atomic or molecular weight) expressed in grams; for example, the formula weight for the water molecule (H_2O) is 18, so a mole of water is a quantity of water having a mass of 18 grams; a mole of a substance comprises 6.023×10^{23} atoms or molecules

molecular diffusion the movement of individual molecules through a solid, liquid, or gas in response to a concentration gradient; molecules will move from where their concentration is higher to where it is lower

mollusk an invertebrate animal with a soft, unsegmented body and usually a shell and a muscular foot; examples are clams, oysters, mussels, and octopuses

monsoon a wind system that influences large climatic regions and reverses direction seasonally; best known as a wet, warm-season wind carrying drenching rains; also can describe the wintertime wind shift that carries dry, cooler air

morphology the external shape, structure, form, and arrangement of landforms and waterbodies

mortality for a particular animal population, the number of deaths in a given area or period, or from a particular cause

mousse thick, foamy oil-and-water mixture formed when petroleum is subjected to mixing with water

mudpot a hot spring that reaches the surface through water-saturated fine-grained sediments; the hot water mixes with the sediments, producing a thick, pasty mud through which hot gases periodically escape, showering the immediate area with mud globules

multipurpose project a water project that is undertaken to meet a variety of objectives, ranging from water supply and hydropower to irrigation, recreation, or habitat maintenance

natural flow a doctrine developed by some riparian rights states that would require all water to be left in a watercourse

natural gas naturally occurring gas composed of methane and other light hydrocarbons

natural hazard a hazard event arising from geophysical processes or biological agents—such as those creating earthquakes, hurricanes, floods, or locust infestations—that affect the lives, livelihood, and property of people

navigable in general usage, describes a waterbody deep and wide enough to afford passage to small and large vessels; also can be used in the context of a specific statutory or regulatory designation

navigable waters surface-water bodies as specifically designated by statutes or regulations

nebula clouds of interstellar gas and dust

nekton assemblages of organisms that swim actively in water, such as fish, reptiles, and mammals; contrasts with organisms (plankton) that are simply carried along in water

neritic describes the area off the shallow regions of a lake or ocean that border the land; also used to identify the biota that inhabit the water along the shore of a lake or ocean

neritic zone the relatively shallow water zone that extends from the high tide mark to the edge of the continental shelf; also refers to such shallow water regions of lakes

net pen floating cages in coastal waterbodies (e.g., bays) that are used in mariculture operations

neurotransmitter a chemical substance released at the end of a nerve fiber by the arrival of a nerve impulse, enabling the transmission of the impulse between two nerve cells

nitrate the highly leachable form of soil nitrogen taken up by most plants through their roots; it is a common groundwater contaminant, especially in agricultural areas and locations with a high density of septic systems, that is regulated by the U.S. Environmental Protection Agency with a drinking water standard of 10 ppm (parts per million) of nitrogen in the nitrate form

nitrogen fixation the conversion of atmospheric nitrogen (N_2) into a form of nitrogen such as ammonia (NH_3) that can be used by plants and other biological agents

nitrogen fixer an organism capable of nitrogen fixation

noble gas a gas that is unreactive (inert) or reactive only to a limited extent with other elements; six noble gases make up a group on the periodic table: helium, neon, argon, krypton, xenon, and radon

nonmarket good a product or service that is not traded in a market and which does not have a market price that reveals how consumers value the good; examples include air and water quality, some parks and recreation amenities, endangered species, and some natural resources such as wetlands

nonpoint source a pollutant release or discharge originating from a land use active over a wide land area (e.g., agriculture) rather than from one specific location (e.g., an outfall pipe from a factory)

nonstructural measure an arrangement to manage, utilize, or control water and related lands that does not rely on constructed facilities; includes regulatory control and financial incentive

nonylphenol $C_9H_{19}OH$; a surface active agent used as a lube oil additive, and in stabilizers, fungicides, bacteriocides, dyes, drugs, adhesives, rubber chemicals, etc.

Glossary

nutrients a group of chemical elements or compounds needed for all plant and animal life; nitrogen and phosphorus are primary nutrients in aquatic systems; excessive or imbalanced nutrients in water may cause problems such as accelerated eutrophication

obligate describes organisms that require the specified condition (e.g., high pressure) for growth

oceanic plate the large segment of the lithosphere (i.e., crust of the Earth and uppermost mantle, the region just below the crust) that supports an ocean basin

oceanography the broad category of science that deals with oceans

offstream water use the diversion or withdrawal of water out of the source river or lake for use elsewhere; as opposed to instream water use, in which the water is used in place rather than being diverted or withdrawn

oligotrophic pertaining to a lake or other body of water characterized by low concentrations of nutrients (such as nitrogen and phosphorus) and having low to moderate productivity

omnivore an animal that will feed on many different kinds of food, including both plants and animals

ooid a small (0.5 to 2 millimeter), spherical object composed of concentric layers of calcium carbonate that has grown in concentric rings, often around some kind of nucleus such as a quartz grain

Oort Cloud a spherical cloud of comets around the Sun at a distance far from Neptune's orbit, extending from approximately 50,000 and 100,000 astronomical units (AU) from the Sun

open-access resource a resource for which there is no private-property-right system governing access and withdrawals

ordinance a law or rule enacted by an authority, such as a city government

ordnance military weapons such as cannon and artillery and their ammunition

ore deposit usually refers to a vein (or veins) of ore (or massive mineralized or otherwise economically valuable rock or sediment) that can be mined as a unit

organic pertaining to, or the product of, biological reactions or functions

organochlorine any chemical compound that contains carbon and chlorine

ornithology the scientific study of birds

orogeny the tectonic processes that lead to mountain-building; orogenies occur at convergent plate boundaries (i.e., the boundary between two lithospheric plates that are moving directly against one another)

orographic lift the process whereby air is forced to rise as it blows across upward-sloping terrain or against mountain ranges

overallocation the action or process of too widely allocating or distributing something, such as water resources

overcapitalization the existence of more capital applied in an industry than is necessary for its efficient operation, as has been the case in the fishing industry

overfishing the removal of such a large number of certain fish from a body of water that breeding stocks are reduced to levels that will not support the continued presence of the fish in desirable quantities for sport or commercial harvest

oxbow the crescent-shaped body of shallow, standing water formed by a stream meander cut-off; sometimes called an oxbow lake

oxidant the oxidizing agent (oxygen) that is used in water treatment processes to break down organic waste or chemicals such as cyanides, phenols, and organic sulfur compounds in sewage by bacterial and chemical means

oxidation a chemical reaction involving the loss of one or more electrons from a specific element; results in an increase in the charge of the element; for example, iron (II) (Fe^{2+}) is oxidized to iron (III) (Fe^{3+}) through the loss of an electron; such reactions often take place in the presence of free oxygen

oxidation–reduction reaction a chemical reaction involving both oxidation and reduction; the electron(s) lost by the oxidized element is (are) gained by the reduced element

oxide a compound consisting of a metal and oxygen (e.g., SiO_2, Al_2O_3, FeO)

ozone a chemical compound composed of three oxygen atoms (triatomic oxygen), used in water treatment; a blue, gaseous allotrope of oxygen O_3, formed naturally from diatomic oxygen by electric discharge or exposure to ultraviolet radiation

ozone layer a layer within the stratosphere enriched in ozone (O_3) produced through the interaction of cosmic radiation and atmospheric oxygen; the ozone layer effectively screens out approximately 99 percent of harmful ultraviolet radiation from the Sun

pack ice blocks of floating ice compacted together to form a solid surface on the sea; generally speaking, any area of sea ice other than fast ice

paleothermometry a method using the isotopic composition of certain elements within a substance (e.g., oxygen in ice or fossil shells) to determine the temperature of the water or atmosphere when the substance formed

palustrine describes fresh-water habitats, especially wetlands, other than those that are lake-related (lacustrine) or river-related (riverine)

pancake ice collectively refers to plates of floating ice, each resembling a pancake or lily pad when viewed from above; formed from a slushy mixture of thickened grease ice, the ice plates have rounded outer boundaries and upturned edges from jostling against one another

parameterization the use of simplified or approximate forms of the physical processes involved; simple equations are used as a substitute for complex physics-based models to reduce computer simulation time to reasonable values

parasite an organism that lives within or on another organism, causing harm to the host organism

parent–daughter compounds term usually used in radiochemistry to describe the radioactive element (parent atom) and the decay product (daughter atom); the analogy for environmental organic chemistry is to describe chemical transformation processes (e.g., degradation) acting on a starting molecule (parent) and producing products (daughters)

pathogen a disease-producing agent, usually a living organism, and commonly a microbe (microorganism)

PCBs abbreviation for polychlorinated biphenyls, a group of chemicals once commonly used as insulator fluid for electric condensers and as an additive for high-pressure lubricants

pelagic referring to open waters at all depths, excluding the benthic zone

peninsula a piece of land that projects into a body of water and is connected with the mainland by an isthmus

percolation the migration of water through the active soil profile into greater depths where it may become groundwater

percolation rate the rate, usually expressed as a velocity, at which water moves through saturated granular material; also applies to quantity per unit of time of such movement

permafrost permanently frozen ground in the Artic and sub-Arctic regions that may extend to several thousand feet below the surface

permeability the capacity of a porous medium to transmit a fluid; highly depends on the size and shape of the pores and their interconnections

pesticides a broad group of chemicals that kills or controls plants (herbicides), fungus (fungicides), insects and arachnids (insecticides), rodents (rodenticides), bacteria (bactericides), or other creatures that are considered pests

petroleum naturally occurring hydrocarbon compounds, derived from organic matter (e.g., plankton) that has been buried and broken down into simpler organic molecules over geologic time

petroleum reservoir a porous and permeable rock in which petroleum accumulates; primarily marine sedimentary rocks such as sandstone and limestone

pH a measure of the acidity of water; a pH of 7 indicates neutral water, with values between 0 and 7 indicating acidic water (0 is very acidic), and values between 7 and 14 indicating alkaline (basic) water (14 is very alkaline); specifically defined as $-\log_{10}(H^+)$, where (H^+) is the hydrogen ion concentration, more appropriately the hydronium ion concentration (H_3O^+)

phosphate the general term for phosphorus-containing derivatives of phosphoric acid (H_3PO_4); phosphates can be environmentally harmful when phosphate-rich wastewaters reach waterbodies; in surface waters, phosphates can act as a primary nutrient source for algae, whose accelerated growth and subsequent death and decay can deplete the oxygen needed for aquatic organisms

photic zone the upper water layers from the water surface and extending down to the depth of effective light penetration where photosynthesis balances respiration; this level (the compensation level) usually occurs at the depth of 1 percent light penetration (for example, 1 percent of surface light intensity) and forms the lower boundary of the zone of net metabolic production

photolysis the breakdown of a material by sunlight; for example, nitrogen dioxide (NO_2) is broken into nitric oxide (NO) and atomic oxygen (O) by the ultraviolet energy in sunlight; photolysis is an important degradation mechanism for contaminants in surface water and in the terrestrial environment

photosynthesis the process by which plants manufacture food from sunlight; specifically, the conversion of water and carbon dioxide to complex sugars in plant tissues by the action of chlorophyll driven by solar energy

phthalate a derivative of phthalic acid, produced through a reaction of the acid and an alcohol; commonly used as a plasticiser to provide flexibility in plastics; some varieties also are used in synthetic lubricants in the automobile industry

physical chemistry the branch of chemistry concerned with the physical properties of materials, such as their physical, electrical, or magnetic behavior

phytoplankton microscopic floating plants, mainly algae, that live suspended in bodies of water and that drift about because they cannot move by themselves or because they are too small or too weak to swim effectively against a current

piscivore a species that feeds preferably on fish

planktivore a species that eats plankton, the tiny, often microscopic plants and animals floating or drifting in water

plankton an assemblage of small, often microscopic aquatic organisms encompassing aquatic plants (phytoplankton) and aquatic animals (zooplankton) that float or drift passively with water current or drifting organism, including large plants and animals

plate see "lithospheric plate"

plate tectonics the theory that the Earth's lithosphere can be divided into a few large plates that are slowly moving relative to one another; plate sizes change and intense geologic activities occur at plate boundaries (e.g., earthquakes, volcanism, mountain-building); continents drift on the plates and therefore their position with respect to latitude and longitude and with respect to one another have changed over geologic time

Pleistocene epoch of, belonging to, or designating the geologic time, rock series, and sedimentary deposits of the earlier of the two epochs of the Quaternary Period; this epoch (commonly referred to as the Ice Age) was characterized by the alternate appearance and recession of northern glaciation and the appearance of the progenitors of human beings

plucking a process of glacial erosion by which fairly large fragments of bedrock that have been weakened along joints or fissure planes by the ac-

Glossary

tion of freezing water are loosened, pried off, and carried away as the glacier advances

plume a concentrated area or mass of a substance that is emitted from a natural or human-made point source and that spreads in the environment; a plume can be thermal, chemical, or biological in nature

point bar a landform that represents a sequence of deposition in which coarser materials are at the bottom and finer materials at the top; in streams, a bank on the inside of a stream's meander bend that has built up due to sediment deposition where the stream velocity is lowest; in lakes, point bars are spatially related to tributaries and sediment inputs

point source a pollutant release or discharge originating from one specific location (e.g., an outfall pipe from a factory) rather than over a wide land area (e.g., water runoff from a farm field)

polar molecule an asymmetric molecule with respect to the distribution of electrons associated with constituent elements; electrons are more strongly attracted to one element over the other(s), and a separation of charges thus occurs resulting in the molecule having a positive side (pole) and a negative side (pole); examples include water (H_2O), sulfuric acid (HCl), and ammonia (NH_3)

polarity a relative measure of the distribution of electron density for a molecule; its significance for environmental chemistry is that it determines whether a chemical prefers to associate with water (polar, or hydrophilic) or fat/lipid (nonpolar, or hydrophobic or lipophilic)

policy a pattern of goal-oriented choice and action; a plan of action

pollutant something that pollutes, especially a waste material that contaminates air, soil, or water; see "contaminant"

pollution any alteration in the character or quality of the environment, including water in waterbodies or geologic formations, which renders the environmental resource unfit or less suited for certain uses; see "contamination"

polyculture the simultaneous development or exploitation of several crops or kinds of animals

polygonal terrain a major morphologic component of the Utopia Planitia region of the northern plains on the planet of Mars; thought to be similar to outflow channels and a possible ancient ocean; its formation process is a subject of debate

polymers a group of chemical compounds composed of small molecules linked together to form larger molecules with repeating structures

polynya any nonlinear open area of sea water enclosed in sea ice

ponding the natural formation of a pond in a stream whose normal streamflow has been interrupted (e.g., ponding behind a landslide or temporary debris dam); also can refer to the development of standing water on natural and human-made surfaces

population density the number per unit area of individuals of any given species, including humans, at a given time

Glossary

population growth rate the percentage increase in population over a defined time period

pore an open space within an otherwise solid material; common pore spaces include the openings between constituent grains in a sediment such as sand

pore network a pattern of interconnected pore spaces, often leading to a higher permeability in an aquifer

pore water water that occurs within the open spaces (pores) of soil, sediment, or rock below the ground surface

position a point of view; positions often are voiced with passion, "this is my stand"

post-audit the systematic study of decision-making after plans have been implemented to evaluate how effectively the project's goals were met

post-glacial rebound the recovery of land following a heavy period of glacial ice in which the land surface gradually rises in response to the melting of snow and ice

potable drinkable; specifically, fresh water that generally meets the standards in quality as established in the U.S. Environmental Protection Agency

potential evapotranspiration the maximum quantity of water capable of being evaporated from the soil and transpired from the vegetation of a specified region in a given time interval under existing climatic conditions; expressed as a depth of water (e.g., centimeters)

ppb abbreviation for parts per billion; also expressed as micrograms per liter (μg/L)

ppm abbreviation for parts per million; also expressed as milligrams per liter (mg/L)

precipitate (verb) in a solution, to separate into a relatively clear liquid and a solid substance by a chemical or physical change; (noun) the solid substance resulting from this process

precipitation in the atmosphere, the downward movement of water in liquid or solid form (rain, sleet, hail, snow) from the atmosphere following condensation in the atmosphere due to cooling of the air below the dew point; in chemistry, the separation of a solid phase (precipitate) from solution (dissolved state)

precision the reproducibility or repeatability of the results of a test, measurement, or experiment; in a series of tests, refers to the ability to arrive at the same answer each time under the same set of circumstances or sampling criteria

predation the preying of one species on other species

predator an animal that hunts and kills other animals for food

preservation an approach to natural resource management based on the idea that natural resources—and nature—should be valued and protected for its own sake and not merely for its utility to humans

Glossary

preservationist a person who believes that nature should be protected for its own sake, not just for the uses it provides for humans

pressure gradient the change in pressure over a given distance; driving force for atmospheric circulation; in an ocean context, when winds blow across the sea surface, they tend to "pile up" water at one side of an ocean basin; the water then tries to flow horizontally to make the surface level again; pressure gradients also can occur when water of different densities is found on opposite sides of an ocean basin, because pressure equals depth times density

prey (verb) to hunt and kill another animal for food; (noun) an animal that is hunted and killed by another for food

primary consumer an organism that eats a primary producer

primary producer an organism capable of using the energy from light or a chemical substance to manufacture organic compounds

primary productivity the rate at which biomass is produced by photosynthetic and chemosynthetic organisms in the form of organic substances

primary treatment the removal of particulate materials from domestic wastewater, usually done by allowing the solid materials to settle as a result of gravity; typically, the first major stage of treatment encountered by domestic wastewater as it enters a treatment facility, and generally removes 25 to 35 percent of the Biochemical Oxygen Demand (BOD) and 45 to 65 percent of the total suspended matter

prior appropriation a concept or doctrine in water law under which the first person to take a quantity of water and put it to beneficial use has a higher priority of right than a subsequent user; that is, "first in time is first in line"; contrast with riparian water rights

private right in terms of property rights, a right held by the owner of a resource (e.g., water) that entitles them to access, withdraw, manage, exclude others (from using the resource), and sell their ownership to someone else

process a series of experiences, actions, or functions that brings about a particular result; the steps of a prescribed procedure

profundal describes the body of deep water below the depth of effective light penetration

prokaryote a microscopic, single-celled organism that has neither a distinct nucleus with a membrane nor other specialized organelles; includes organisms such as bacteria and cyanobacteria

property right a generic term that refers to any type of right to specific property, whether it is personal or real property, tangible or intangible; as an example, a landowner has a property right to use water attached to the land

public investment large-scale social expenditures made with public monies to create value for society

public right a right given to the public's common need, such as public rights to water (e.g., using surface waters for navigation); contrast with private (property) rights

public trust an historical and presently evolving concept relating to the ownership, protection, and use of essential natural and cultural resources, the purpose of the trust is to preserve resources in a manner that makes them available to the public for certain public uses

public water system a system for provision to the public of piped, or otherwise conveyed, water for human consumption, if such system has at least 15 service connections (such as households) or regularly serves at least 25 individuals daily for at least 60 days out of the year (such as businesses or schools)

pumice a highly vesicular, glassy, volcanic rock, compositionally similar to rhyolite and often light enough to float on water

pyroclastic flow an ash flow that takes place at any temperature

quahog a rounded, edible clam found in temperate and boreal waters on both sides of the North Atlantic Ocean

qualitative describes the assessment of the quality of something rather than its quantity; also a term applied to an approximate analysis, such as determining the presence, but not the concentration, of a dissolved constituent

quantitation limit also spelled quantification, the detection limit, or lowest concentration of a given constituent that can be determined by the analytical procedure in a quantitative manner

quantitative describes the measurement of the quantity of something rather than its quality

radioactive describes a substance such as uranium or plutonium that emits energy in the form of streams of particles because of the decay of its unstable atoms

radioactivity the emission of ionizing radiation or particles caused by the spontaneous disintegration of atomic nuclei

radionuclide a type of atom that exhibits radioactivity; radioactive chemicals may be artificial or naturally occurring and may be found in drinking water

rain shadow an area of little rainfall that lies downwind of mountain ranges

rainbow trout a type of trout highly prized as a game fish; native to cold coastal streams and lakes on both sides of the Pacific Ocean and commonly found around the world

rank (verb) to place a series of numbers in order, commonly from the lowest value to the highest value; for example, in a series of 10 numbers, the lowest number would have a rank of 1 while the highest number would have a rank of 10; in some rankings the highest number would be ranked as 1 and the lowest number as 10; (noun) the position in a ranking process

recharge the process by which precipitation infiltrates below the surface and replenishes an aquifer

recharge area the geographic region over which recharge takes place for a given aquifer

Glossary

reclamation in terms of conservation, the process of restoring land to its prior state, such as converting old mineland back to forestland; in historical use, the process of converting land to a more desired use, such as draining a marsh for human development; also refers to treating wastewater in a way it can be reused

recruitment the increase in a natural population as offspring grow and immigrants arrive

red tide a visible coloration of the sea caused by the excessive growth (bloom) of microscopic algae, commonly dinoflagellates; the red, brown, green, purple, or yellow tint in the water is a result of the high concentration of algal pigments; some red tide events are harmful to marine animals and/or humans; see "harmful algal bloom"

redox potential a characterization of an environment with respect to whether oxidizing or reducing reactions are favored; a high redox potential generally will lead to the occurrence of the oxidized form of an element as opposed to the reduced form (e.g., nitrogen in nitrate [NO_3^{2-}] as opposed to ammonia [NH_3])

reduction a chemical reaction in which an element gains one or more electrons; results in a reduction of the charge of the element; for example, iron (III) (Fe^{3+}) will be reduced to iron (II) (Fe^{2+}) through the gaining of an electron; typically takes place in oxygen-poor or anoxic environments

reef a strip or ridge of rocks, sand, or coral that rises to, or near the surface of a body of water

refraction the change in direction of propagation that occurs when a wave passes from one medium to another; for ocean waves, when a wave approaches the shoreline at an angle, part of the wave enters a shallower water and slows relative to the rest of the wave, causing the wave to "bend" toward the shoreline

regional cooperative agreement a multi-agency agreement to coordinate policies to achieve common goals

relative describes the value of a parameter in a system that is known only in comparison to the value of the same parameter in another system; for example, the lake water's temperature is warmer than the stream water's temperature, or the groundwater in the deep aquifer is older than the groundwater in the shallow aquifer; contrasts with an absolute measure

relative humidity the ratio of the amount of water vapor in the atmosphere to the amount necessary for saturation at the same temperature; expressed in terms of percent and measures the percentage of saturation

remediation the cleanup, through a variety of methods, to remove or contain a toxic spill or hazardous materials from a contaminated site

remote sensing the collection and interpretation of information about an object without being in physical contact with the object; most often, it refers to satellite-based collection of data to map and monitor the environment and resources on Earth

reserves referring to petroleum, the amount of petroleum that can be extracted, depending on economics and technology

reservoir a pond, lake, basin, or tank for the storage, regulation, and control of water; more commonly refers to artificial impoundments rather than natural ones

residence time the average time an element spends in a given environment between the time it arrived and the time it is removed by some process; in the ocean, residence time is defined as the concentration in sea water relative to the amount delivered to the ocean per year; in groundwater, it is the time elapsed between water being recharged to the aquifer; in lakes and reservoirs, it is the time elapsed between a parcel of water entering the waterbody and leaving it

resin in water treatment, a manufactured chemical substance that is designed to attract certain contaminants

respiration the oxidative process occurring within living cells by which the chemical energy of organic molecules (for example, substances containing carbon, hydrogen, and oxygen) is released in a series of metabolic steps involving the consumption of oxygen (O_2) and the liberation of carbon dioxide (CO_2) and water (H_2O)

restoration the act or process of bringing something back to a previous condition or position; for example, the establishment of natural land contours and vegetative cover following extensive degradation of the environment caused by activities such as surface mining

retention pond a permanent drainage area (such as a pond and lake) where stormwater runoff accumulates but does not escape during a given period

reverse osmosis process in which dissolved substances are removed from water by forcing water, but not dissolved salts, through a semipermeable membrane under high pressure; commonly used to treat contaminated drinking water or process water; in desalinization, reverse osmosis is used to extract fresh water from salty water

rhyolite a pale, fine-grained volcanic rock, similar to granite in composition and commonly exhibiting flow characteristics

rill a small channel, typically less than a few inches deep, formed by uneven removal of soil by running-water erosion, and that can be obscured by normal tillage operations

rip tide a strong, narrow surface current that flows rapidly away from the shore, returning the water carried landward by waves; generally the result of wave convergence in embayments; also referred to as rip current

riparian pertaining to the banks of a river, stream, waterway, or other (typically flowing) body of water, as well as to plant and animal communities along these waterbodies

riparian water right a doctrine governing the legal rights of an owner whose land abuts water; specifically, the person who owns land adjacent to a stream has the right to make reasonable use of water from the stream; as contrasted with prior appropriation

riprap irregular blocks of rock too large to be easily moved by streamflow, and placed along the streambank for stabilization

Glossary

risk management process of evaluating and selecting regulatory and non-regulatory responses to environmental risks

river basin see "drainage basin"

river basin commission see "commission"

riverine relating to, formed by, or resembling a river including tributaries, streams, and brooks

runoff the part of precipitation that does not infiltrate, evaporate, or transpire and that subsequently collects in surface-water bodies; also refers to the movement of water across the land surface

saline describes water containing a high dissolved mineral content; in sea water, the dominant contributor to salinity is sodium chloride

salinity the concentration of dissolved materials carried in an aqueous (watery) solution; typically expressed in grams per liter (parts per thousand) or milligrams per liter (parts per million)

salt a compound consisting of a metal and a base, as is formed when an acid has its hydrogen replaced by a metal; common salt is the sodium salt of hydrochloric acid

salt deposit a sedimentary deposit, typically of gypsum ($CaSO_4[H_2O]_2$) and halite (NaCl), generally formed through the evaporation of sea water

salt dome a geologic structure produced when low-density salt deposits, buried beneath other sediments, rise buoyantly, deforming the overlying rocks into dome-like structures; common in the Middle East, Texas, and Louisiana

salt-water intrusion the invasion of sea water into coastal aquifers, generally caused by overpumping fresh water from those aquifers; the sea water occupies a portion of the aquifer formerly occupied by fresh water and prevents the fresh water from returning, thereby permanently reducing the long-term capacity of the aquifer

sand a sediment wherein the individual particles range in size from 0.625 to 2 millimeters

saprolite soft, decomposing igneous rock that remains where it was located when solid; formed by heavy weathering in a humid environment

saprophyte an organism, especially a fungus or bacterium, that grows on and derives its nourishment from dead or decaying organic matter, and that enhances natural decomposition of organic matter in water

saturated with respect to hydrogeology, refers to the condition of having all open spaces within soil, sediment, and rock filled with water

saturated overland flow water from precipitation or snowmelt that flows over the land surface after the soil becomes saturated with water because the rate of precipitation or snowmelt exceeds the rate at which water can percolate down through the saturated soil

saturated thickness the thickness of the portion of the aquifer in which all pores, or voids, are filled with water; specifically, in a confined aquifer, this is generally the aquifer thickness, whereas in an unconfined aquifer, this is the distance between the water table and the base of the aquifer

saturated zone an area where pore spaces within the soil are entirely filled with water

scarcity value the worth of something based on its limited, or lack of, availability; a resource is scarce when, at a zero price, more is wanted than is available; for water, the value of water above and beyond the cost of finding it, extracting it from the groundwater or surface-water body, treating it, and delivering it to users

scavenger an animal that eats animal wastes, dead plant material, and dead bodies of animals not killed by itself

scientific method a systematic method of inquiry regarding a specific question or problem that includes the objective collection of data relating to that question, the development of tentative hypotheses or solutions to the problem, collecting more data to test a proposed solution to the problem, and the rational determination of the hypothesis most successful in explaining the problem

scoria a cindery, vesicular crust formed on the surface of basaltic or andesitic lava as a result of the escape and expansion of gases before solidification

scour to clear, dig, or remove by or as if by a powerful current of water, such as when waves undercut material on the coastal shore; the erosive action of running water in streams, which excavate and carry away material from the bed and banks

scree weathered and broken rock fragments that have fallen downslope, often forming or covering a slope on a mountain

scrimshaw any of various carved or engraved articles made by whalers, usually from either baleen or whale ivory

scuba an apparatus for breathing underwater consisting of a portable canister of compressed air and a mouthpiece; the acronym for self-contained underwater breathing apparatus

scute a thickened, external horny or bony plate or scale on some animals, especially snakes, turtles, and other reptiles

sea cave cave formed from wave action where the waves repeatedly force water into the cracks in rock, breaking apart the rock

sea ice a general term for any form of ice found at sea which has originated from the freezing of sea water; includes types of ice such as grease ice, frazil ice, pancake ice, and pack ice

seamount an isolated conical submarine mountain rising 1,000 meters (3,280 feet) or more above the seafloor; most form as submarine volcanoes at spreading centers and are transported to the deep ocean by plate movement

seawall a structure built along a portion of a coast, lake, or river, primarily to prevent erosion and other damage by wave action; to perform its function a seawall retains earth against its shoreward face

Glossary

secondary treatment the treatment that follows primary wastewater treatment that involves the biological process of reducing suspended, colloidal, and dissolved organic matter in effluent from primary treatment systems and which generally removes 80 to 95 percent of the biochemical oxygen demand (BOD) and suspended matter

sediment rock particles and other earth materials that are transported and deposited over time by geologic agents such as running water, wind, glaciers, and gravity; sediments may be exposed on dry land and are common on ocean and lake bottoms and river beds

sediment load the combination of bed load, suspended load, and dissolved load carried by a stream

sedimentation in geology and geomorphology, a process in which sediment is transported and deposited in a new location; in water treatment, the settling of solids or of flocculated or coagulated particles

seiche an oscillation of the water surface of a lake or other body of water due to variations of atmospheric pressure, wind, or minor earthquakes; the oscillation may be a foot or more in amplitude and may last several hours

semiarid a climate or region where moisture is normally greater than under arid conditions but still limits the growth of most crops; either dryland farming methods or irrigation generally are required for crop production

senior appropriator in water rights, the holder with the highest priority for water use; the oldest water right

septic system also called an on-site system, a common method of sewage disposal in which sewage enters a holding tank from the home or business; in the tank, solids settle; then liquid effluent flows from the tank into a system of perforated pipes buried beneath the ground, where the effluent percolates into the soil

sequester to remove or render inactive a specific chemical or chemical group from a solution

sextant a navigational instrument incorporating a telescope and an angular scale that is used to measure latitude and longitude

shadow zone pertains to the zone at Earth's surface between 105 and 140 degrees from an earthquake's epicenter where direct seismic waves do not occur because of refraction of seismic waves during their passage through Earth's interior

shale a fine-grained, laminated sedimentary rock; produced by the compaction of clay, silt, or mud; typically composed of equal proportions of quartz, clay, and miscellaneous materials and organic matter

shallow well a relative term for a drilled or dug well with depth below ground surface not exceeding 10 meters (33 feet); water is lifted out of the well using manual or animal-driven devices; a motorized pump, if provided, is at the ground surface with the end of its suction pipe immersed below the water surface in the well

silica silicon dioxide (SiO_2); occurs in crystalline form (quartz) or amorphous form (opal), or as a component in rocks such as granite and sandstone

silicates the mineral group in which the basic structure is a molecule of silica, consisting of a silicon atom surrounded by four oxygen atoms; in most silicates the silica molecules are linked by other ions such as calcium, magnesium, sodium, iron, and potassium; common silicate minerals include quartz and feldspar

sill the shallow area that separates coastal bays or marginal seas from the adjacent oceans or that separates two basins from one another

silt a sediment wherein the individual particles range in size from 0.004 to 0.625 millimeters; smaller than a sand particle but larger than a clay particle

sink a substance or process that removes a component of concern from the active environment; for example, the adsorption of metals on the surfaces of organic matter serves as a sink for these elements as it removes them from a solution

sinkhole a depression in the Earth's surface caused by the collapse of underlying limestone, dolomite, salt, or gypsum

slough a backwater area or remnant of a former river channel that contains standing water and serves as the main river channel only during high water

sludge the accumulated solids that remain after the treatment of wastewater

smectite a type of clay mineral that expands when exposed to water

snow line the general altitude to which the continuous snow cover of high mountains retreats in summer, such as the snowcap of a mountain, chiefly controlled by the depth of the winter snowfall and by the summer temperature

snowpack a field of naturally packed snow that ordinarily melts slowly during the early summer months

soft water water that contains low concentrations of metal ions such as calcium and magnesium; does not precipitate soaps and detergents

soluble that which can be dissolved; able to pass into solution

solute the dissolved solids in a solution

solution cave a type of cave formed by slightly acidic groundwater circulating through fractures in carbonates (such as limestone), dissolving the rock and leaving behind an opening; less commonly called solutional cave

solvent a substance capable of dissolving other substances; in a solution, it is the liquid that has dissolved the solids (solutes)

sorb a term that encompasses a number of processes by which a given compound is removed from solution by surface or near-surface reactions; for example, adsorption, absorption, etc.

sorbent describes a substance that has the property of collecting molecules of another substance by the process of sorption

sorption processes that remove solutes from the fluid phase and concentrate them on the solid phase of a medium; used to encompass absorption and adsorption

sovereign possessing independent authority or power; (of a group) fully independent and able to determine its own affairs; (of affairs) subject to a specified control but without outside interference

spa as in water-related usage, a resort having mineral springs or hot springs, and often providing therapeutic baths or mud baths

spatial describes the characteristics of a given area; for example, the spatial distribution of whales in the ocean or the spatial distribution of aquifer thickness

species the narrowest classification or grouping of organisms according to their characteristics; members of a species can reproduce only with others of that group

specific heat the amount of heat (energy), measured in calories, required to raise the temperature of 1 gram of a substance by 1 degree Celsius (1°C); for water, the specific heat is 1 calorie

specific yield the volume of water released by gravity from a unit area (square meter or square foot) of an unconfined aquifer when the water table drops a unit length (meter or foot); expressed as a fraction; for example, if 0.2 cubic meter of water will drain from 1 cubic meter of aquifer sand, the specific yield is 0.2 (20 percent)

sphagnum any moss of the genus *Sphagnum*, in the family Sphagnaceae

spike in chemical analysis, a prepared solution with components or isotopes in known proportion that is added to the sample containing a constituent or an isotope of interest, and used to facilitate the accurate determination of the constituent's or isotope's concentration

spreading center a plate boundary where lithosphere is created by igneous activity; plates move away from the spreading center in either direction

spreadsheet computer-based program to facilitate computations and manipulations involving numerical and alphanumeric values

spring location where a concentrated, natural discharge of groundwater emerges from the Earth's subsurface as a definite flow onto the surface of the land or into a body of surface water, such as a lake, river, or ocean

stakeholder an individual or group impacted by a potential decision or action; term is usually associated with a limited number of individuals representing the interests of other like-minded individuals or groups

static water level the level of water in a well that is not being affected by withdrawal of groundwater

statutory law law enacted by Congress, or a state legislature, as opposed to common law

steady state a state of a system in which reactions are occurring or processes are happening, but the system has reached a state of balance such that all components remain at a constant concentration

stenohaline pertaining to an aquatic organism unable to withstand wide variation in salinity of the surrounding water

steroid any of a class of lipid proteins, such as sterols, bile acids, sex hormones, or adrenocortical hormones, containing a cyclopentanoperhydrophenanthrene nucleus; most have specific physiological action

stock (noun) a distinct population of fish or aquatic resource, defined on the basis of population biology and commonly used in the context of conservation biology and the particular fishery; (verb) to add fish or aquatic animals to a waterbody for the purpose of either conservation or sport

stormwater runoff from precipitation events in which precipitation rate exceeds infiltration rate or falls directly on an impermeable surface; stormwater often is discharged directly to streams and may carry pollutants such as bacteria, petroleum products, and metals

strata distinct horizontal layers in geological deposits; each layer may differ from adjacent layers in terms of texture, grain size, chemical composition, or other geological criteria; also applied to layering of other material such as the atmosphere

stratification the arrangement of a body of water, such as a lake, into two or more horizontal layers of differing characteristics, such as temperature and density; also applies to other substances such as sediments, soil and snow

stratigraphy the geologic study of the formation, composition, sequence, and correlation of unconsolidated rock layers

stratosphere layer of the atmosphere extending from 11.3 to 48.3 kilometers (7 to 30 miles) above the Earth's surface, lying between the troposphere and the mesosphere

stream channel the bed where a natural stream of water runs or may run; the long, narrow depression shaped by the concentrated flow of a stream and covered continuously or periodically by water

stream order the extent of tributary development above a given stream segment; a first-order stream has no tributaries above it; a second-order stream is formed when two first-order streams join; a third-order stream is formed below the confluence to two second-order streams, etc.

stromatolite a laminated structure formed in quiet water when a layer of filamentous algae traps sedimentary particles, chiefly carbonate; another layer of algae grows on this sedimentary surface, trapping another layer, thus building up a dome shape or a column

structural measure a facility or component that has been constructed to manage water, such as a dam, canal, or levee

structure contour a contour line drawn on a map, connecting points of equal elevation of a particular geologic structure to represent the shape or configuration of that structural unit in the subsurface, such as a dipping sedimentary layer or a folded rock layer, to enhance the interpretation of subsurface geology; useful in oil exploration and hydrogeology

subducting plate a lithospheric plate that is undergoing subduction

Glossary

subduction the process by which one lithospheric plate is forced to move under another plate, moving in the opposing direction; site of volcanoes and deep earthquakes (e.g., the Andes, Cascades, Japan, Aleutian Islands, and islands of the Southwest Pacific Ocean)

submergent describes a plant anchored to the bottom by roots or rhizomes; its foliage is either entirely submersed or some floating leaves may also be present; some common examples include pondweed, watermilfoil, and waterweed

subsequent stream a stream whose drainage pattern is controlled by the relative resistance of rock upon which the drainage developed; streams develop preferentially in easily eroded rock material, such as along fault lines or in shale as opposed to sandstone or igneous rock; drainage typically appears to consist of straight line segments or orthogonal patterns

subsidence the sinking of the land surface due to a number of factors, including the accumulation of sediments in a basin, loading of glacial ice, and groundwater extraction

subsidy the sum of money granted by a government or other public body in order to assist an industry or business so that the price of the associated commodity or service will remain low or competitive; a common example is a farm subsidy

substrate the bottom or underlying materials; in ecology, the bottom sediments in lakes, rivers, and oceans that may contain living organisms; in microbiology, the foodstuff for microorganisms, supplying energy and carbon; in wastewater treatment, the organic matter in wastewater that serves as the foodstuff for microorganisms involved in breaking down sewage

subsurface of, relating to, or situated in an area beneath a surface, especially the surface of the Earth or of a body of water

sulfate a combination of sulfur in the oxidized state (S^{6+}) and oxygen, and a part of naturally occurring minerals in some soil and rock formations; a common constituent in groundwater and surface water; sulfate minerals tend to be highly soluble

sulfide any compound of sulfur in the reduced state (S^-) and another element; common in igneous rock and some sedimentary rocks such as shale; heavy metal sulfides are generally insoluble

supercritical fluid a type of thermal treatment using moderate temperatures and high pressures to enhance the ability of fluid (such as water) to break down large organic molecules into smaller, less toxic ones; oxygen injected during this process combines with simple organic compounds to form carbon dioxide and water

supernatant the clear liquid that can be poured off of a mixture of liquid and particles after the particles have been allowed to settle to the bottom of the vessel

surface tension a phenomenon caused by a strong attraction towards the interior of the liquid action on liquid molecules in or near the surface in such a way as to reduce the surface area; the tension that results usually is expressed in dynes per centimeter or ergs per square centimeter

surface water water found above ground and open to the atmosphere, such as the oceans, lakes, ponds, wetlands, rivers, and streams

surficial relating to, being at, or covering the surface of the Earth

suspended describes a particulate remaining in a fluid for a long period of time because of its slow settling velocity in water or air; for example, a fine-grained sediment remaining suspended in water, or a fine-grained volcanic ash remaining suspended in the upper atmosphere

suspended load sediment carried in suspension in the water column; generally silt and clay which, as a result of their small size, have a very low settling velocity; all the material transported by a stream or river, neither in contact with the river bottom (bedload) nor in solution (dissolved load)

sustainable as in "sustainable development," describes efforts that guide economic growth in a manner that meets current needs without compromising the ability of future generations to meet their needs; in terms of natural resources, also encompasses development conducted in an environmentally sound manner, with an emphasis on natural resource conservation, including water and aquatic life

Sverdrup abbreviated as Sv, 1 Sverdrup is 1 million cubic meters of water per second

symbol something that represents something else

synthetic produced artificially rather than naturally; often refers to a product made by chemical synthesis

taking a scenario wherein a governmental body appropriates the private property from an owner for public purposes based on the Fifth Amendment of the U.S. Constitution

talus cave cave formed from huge rocks that have fallen from cliffs, leaving spacious chambers within the boulder piles

task force an *ad hoc* organization with a focused mission, such as to solve water quality problems in a certain watershed

taxonomy classification of organisms that reflects their natural relationships

tectonic formed by tectonism, the shaping by deformation of the crust of a planet or moon

tectonic cave a naturally hollowed-out place in the ground formed by any geological force that causes rocks to move apart

tectonic forces the types of stresses (i.e., compression, tension, and shear) that develop within segments of Earth's crust during deformation

tectonic plate see "lithospheric plate"

tectonism process of deformation in the Earth's crust as a result of geological forces acting within or below the crust; includes faulting, folding, uplift, and down-warping of the crust

telemetry the remote measurement or the remote collection of physical, environmental, or biological data

Glossary

terrace (agricultural) an embankment or combination of an embankment and channel constructed across a slope to control erosion by diverting and temporarily storing surface runoff instead of permitting it to flow uninterrupted down the slope

terrace (marine) an ancient beach area perched above the current beach level; often flat and gently sloping toward the sea; has been elevated relative to current beach level by a lowering of sea level or uplift of the coastal area

terrace (river) an old alluvial floodplain, ordinarily flat or undulating, bordering a river, but at a higher level than the current floodplain; results from a river's downcutting ability being accelerated, leaving remnants of the former floodplain perched above the new stream level as terraces; stream terraces are frequently called second bottoms (as contrasted with floodplains) and are seldom subject to overflow

terrestrial living or growing on land rather than in water or air

tertiary treatment selected biological, physical, and chemical separation processes to remove organic and inorganic substances, particularly nutrients, that resist conventional treatment practices

thermocline in a thermally stratified waterbody, the water layer of rapid temperature change over a short vertical interval; it serves as a barrier to water-column mixing

thermohaline pertaining to large-scale circulation in the ocean driven by density differences, where density is controlled by temperature ("thermo") and salinity ("haline")

threatened as defined by the 1973 Endangered Species Act, describes a plant or animal species that is likely to become endangered in the foreseeable future; an endangered species is in danger of becoming extinct throughout all or a significant portion of its range

tideline an artificial indicator marking the high-water or low-water limit of the tides

tilapia an African fresh-water perch-like fish that has been introduced to many areas for the purpose of food

time step in a model that describes the progressive change in some condition, the length of time modeled between solutions (e.g., modeling the change in river discharge over a 24-hour period in 1-hour increments or time steps)

topography the shape and contour of a surface, especially the land surface or ocean-floor surface

torpedo a self-propelled underwater missile designed to be fired from a ship or submarine or dropped into the water from an aircraft and to explode upon reaching its target

total allowable catch a fishery management regime in which a person can buy in advance an individual portion (quota) of the total allowed catch of a certain species for a fishing season, and then is allowed to catch that amount or to trade it to someone else

total dissolved solids a measure of the amount of dissolved minerals in water (e.g., calcium, sodium, chloride, and sulfate)

Total Maximum Daily Load the maximum quantity of a particular water pollutant that can be discharged into a body of water without violating a water quality standard; the amount of pollutant is set by the U.S. Environmental Protection Agency

toxic describes chemical substances that are or may become harmful to plants, animals, or humans when the toxicant is present in sufficient concentrations

toxicant a chemical substance that has the potential of causing acute or chronic adverse effects in plants, animals, or humans

toxicity the ability of a chemical substance to cause acute or chronic adverse health effects in plants, animals, or humans when swallowed, inhaled, or absorbed

trace element in the ocean, elements that occur at concentrations of less than 1 part per million (ppm), or 1 milligram per kilogram of water; in solid substances, elements that can substitute for a major component (e.g., strontium can substitute for calcium in calcite [$CaCO_3$], a common constituent in the shells of marine organisms)

tracer a substance introduced into a physical system or biological organism so that its later distribution can be easily followed from a distinctive feature such as fluorescence, color, and radioactivity

transboundary describes rivers, streams, lakes, and other waterbodies that cross political or administrative boundaries

transformation product an intermediate breakdown product that occurs during the stepwise breakdown of a chemical in the environment

transgression the gradual rise in sea level resulting in the progressive onshore submergence of land, as when sea level rises or land subsides

transparent describes a decision-making process that is open and accessible to stakeholders and other interested parties

transpiration the process by which water is evaporated from plants, primarily through microscopic air spaces in their leaves

transport modeling a type of computer model that predicts the movement of a specific contaminant in groundwater or surface water; considers the characterics of the aquifer or streamflow and the chemical characteristics of the contaminant

travertine a mineral consisting of a massive, usually layered, calcium carbonate (e.g., calcite) formed by deposition from springs waters or especially hot springs; also forms dripstone in limestone caverns; compare "tufa"

trench an elongated surface feature on the sea floor that marks the location where the lithosphere bends downward at a subduction zone

tributary a smaller stream that flows into a larger stream

Glossary

trigonometry a branch of mathematics dealing with trigonometric functions, triangles, and solutions of plane or spherical triangles

trophic level a group of organisms related by their place on the chain of energy transfer, and the number of hierarchical steps relative to primary producers; primary producers are first trophic level, herbivores second, and successive levels of predators follow

troposphere lowest layer of the atmosphere, extending from the Earth's surface to an altitude of approximately 11.3 kilometers (7 miles)

trunk stream the largest or principal stream of a given area or drainage system; also called main stream and master stream

tsunami a long-wavelength sea wave caused by a great disturbance under an ocean such as a strong earthquake, volcanic eruption, submarine landslide, or some other major movement of the Earth; wave is less than 1 meter (3 feet) high in the open ocean but may grow to heights of 20 meters (65 feet) or more as it moves onshore; also known as a seismic sea wave; incorrectly referred to as a tidal wave

tufa a highly porous calcium carbonate deposit generally associated with emergence of CO_2-rich groundwater at the surface or into a lake or stream; the CO_2 (carbon dioxide) escapes, causing calcite to precipitate; one example is the "tufa towers" of Mono Lake, California; compare "travertine"

turbidity a measure of the cloudiness (reduced transparency) of water, determined by the amount of light reflected by particulate matter in the water

turbidity current a gravity current resulting from a density increase brought about by increased water turbidity; possibly initiated by some sudden force, such as an earthquake, the turbid mass continues under the force of gravity down a submarine slope

turbulent a flow condition characterized by rapidly changing flow direction and velocity, as in a turbulent stream flow

turnover the physical process in which a thermally stratified waterbody (e.g., lake or reservoir) loses its stratification and mixes from top to bottom, yielding a uniform temperature throughout the water column

typhoon a tropical storm occurring in the region of the Indian or western Pacific oceans

unconfined describes an aquifer whose upper surface is the water table that is free to fluctuate under atmospheric pressure

unconsolidated with reference to sediments, consisting of loose, separate, and unattached grains or particles (e.g., sand, gravel, silt, and clay); contrast with consolidated sediments, in which individual grains cling or are bound together

unsaturated zone the zone between the ground surface and the water table that contains both air and water; see "vadose zone"

updrift the direction to which the predominant long-shore movement of beach material approaches

upland in general, the elevated lands above a floodplain or other low-lying areas

uplift through a geological process, upward movement and the resulting pressure on the base of a structure

uprush a sudden upward surge or flow

upwelling in marine environments, the movement of nutrient-rich water from great depths to the ocean surface; in hydrogeology, the upward movement of groundwater in areas of discharge (i.e., streams and springs); upward movement of water in a spring-fed pond or pool

vadose of, relating to, or being water that is located in the subsurface unsaturated zone (zone of aeration) between the ground surface and the saturated zone

vadose zone the subsurface zone between the water table (zone of saturation) and the ground surface where some of the spaces between the soil particles are filled with air; also referred to as the unsaturated zone or, less frequently, the zone of aeration

values abstract concepts of what is right and wrong, and what is desirable and undesirable

vaporization the change of a substance from a liquid or solid state to the gaseous state

vaporize to convert into vapor or gas, usually by heating liquid to its boiling point

vector an organism such as a biting insect or tick that transmits a parasite or disease from one plant or animal to another

vesicle a small cavity in volcanic rock that is produced by gas bubbles in the rock-forming lava before it hardens

virulence the ability of a pathogenic organism to overcome the natural defenses of the infected organism

viscosity a measure of the resistance of a fluid to flow; for liquids, viscosity increases with decreasing temperature; expressed as mass per length–time (for example, kilograms per meter–second)

volatile easily vaporized at moderate temperatures and pressures

volatile organic compounds organic compounds that can be isolated from the water phase of a sample by purging the water sample with inert gas, such as helium, and subsequently analyzed by gas chromatography

volcanism the activity and phenomena of volcanoes on Earth

volumetric flow rate the rate, expressed as volume per time, of water moving through a river, canal, or aquifer; often expressed as cubic meters per second, cubic feet per second, or gallons per second

vulnerability assessment an analysis of the potential for a particular negative impact to occur as a result of some event or action, consisting of the identification of the parts of the system that are sensitive to the event and

Glossary

the probability that the event will occur; for example, vulnerability assessments are conducted to evaluate the vulnerability of an aquifer to pollution, a computer system to hacking, or a facility to terrorist attack

waste stream the chemical composition and character of wastes produced at a facility; waste streams often differ from the composition of the initial products used; can refer to liquid or solid wastes

wastewater pond an earthen pond created for the treatment of wastewater generated by an agricultural or industrial facility

water column a vertical section through the sea or lake, highlighting the differences in properties of the water at different levels

water institution see "institution"

water management the application of practices to obtain added benefits from precipitation, water, or water flow in any of a number of areas, such as irrigation, drainage, wildlife and recreation, water supply, watershed management, and water storage in soil for crop production

water right a legal right to the use of water; a legally protected right, granted by law, to take possession of water occurring in a water supply and to divert the water and put it to beneficial use

water table the upper surface of the zone of saturation in an unconfined aquifer below which all voids in rock, sediment, and other geologic materials are saturated (completely filled) with water

water treatment processes undertaken to purify water that is acceptable to some specific use (for example, drinking); most water treatment processes include some form, or combination of forms, of sedimentation, filtration, and disinfection, commonly chlorination

water use the use of water for any purpose, including drinking, irrigation, processing of goods, power generation, and so on

watershed the land area drained by a river and its tributaries; also called river basin, drainage basin, catchment, and drainage area

watershed response function the characterization of a watershed in terms of factors, such as slope, vegetative cover, and percent impermeable surface that control the amount of runoff relative to precipitation and the time to peak flow (discharge)

weather the condition of the atmosphere at any given time and location, including the temperature, pressure, and humidity of the air; wind direction and speed; and phenomena such as clouds, rain, and snow

weathering the decay or breakdown of rocks and minerals through a complex interaction of physical, chemical, and biological processes; water is the most important agent of weathering; soil is formed through weathering processes

weir a dam or other structure placed across a river or canal to raise, direct, or divert the water, as for a millrace, or to regulate or measure the flow

wetland an area that is periodically or permanently saturated or covered by surface water or groundwater, that displays hydric soils, and that typically supports or is capable of supporting hydrophytic vegetation

white smoker a seafloor vent in which hot mineralized water from below the seafloor discharges into cool sea water; white color arises from the precipitation of minerals rich in barium, calcium, and silica; compare with "black smoker"

win-win solution the collaborative outcome where the conflicting parties each feel like they gain something from the decision-making outcome (and do not "lose" due to the decision)

zeolite a naturally occuring mineral used in some water treatment processes

zoning usually a legislative process by which a county or city is divided into separate zones or districts, each with its own unique requirements; this process can serve many purposes, including preservation of open spaces and prioritization of land uses (e.g., agricultural, residential, commercial, industrial)

zoning overlay a land-use practice in which a mapped area of special concern (such as a drinking-water protection area, wetlands, or a specific habitat) is placed over the city's map of existing zoning requirements; the city may apply additional restrictions for industry or development within the area

zoology the branch of biology that studies animals, including their structure, function, growth, origin, evolution, and distribution

zooplankton microscopic animals that live suspended in bodies of water and that drift about because they cannot move by themselves or because they are too small or too weak to swim effectively against a current; composed primarily of protozoans, microcrustacea (copepods, cladocera, rotifers) and larval stages of certain invertebrates

zooxanthellae a yellowish-brown, microscopic, symbiotic alga; the alga lives within the tissue of the reef-building coral, contributing to the calcification capability of corals by extracting carbon dioxide from the animal's body fluids

Cumulative Index

Page numbers in **boldface type** indicate article titles; those in *italic* type indicate illustrations. The number preceding the colon indicates the volume number; the number after a colon indicates the page number. The letter *f* after a page number indicates a figure; the letter *t* indicates a table.

100-year flood, defined, **2:**72, and Glossary

A

Ablation, defined, **1:**199, and Glossary
Absolute integrity of the river, **3:**19, 22
Absorption, chemical, **1:**147
Abyss, as marine ecosystem, **2:**14
Accountable, defined, **3:**207, and Glossary
Accuracy, defined, **1:**146, and Glossary
Acid mine drainage, **1:1–5**
    and acid rock drainage, **1:**2
    mineral deposits, **1:**2–4
    problem and cleanup, **1:**4
Acid rain, **1:5–11**
    aquatic ecosystems, **1:**7, *8*
    chemical controls of water composition, **2:**90
    Clean Air Act Amendments, **1:**9–10
    forests and soils, **1:**8
    human health and human environments, **1:**8–9
    lakes and streams, **3:**126
    sources and forms, **1:**5–6
Acid rock drainage, **1:**2
Acidification
    and amphibians, **1:**29
    lake chemical processes, **2:**264
    See also pH

Acoustic Thermometry of Ocean Climate (ATOC), **4:**83, 104
Activated-sludge method of wastewater treatment, **4:**211
Acute toxicity, defined, **1:**156, and Glossary
Additive effect, **1:**148
Aeration, defined, **1:**34, and Glossary
Aerobes, in lakes and streams, **3:**76–77
Aerobic, defined, **1:**63, and Glossary
Aerosol
    defined, **1:**19, **2:**127, and Glossary
    ocean chemical processes, **3:**136–137
Aesthetic, defined, **2:**251, **4:**89, and Glossary
Aesthetic quality, of fresh water, **2:**94–97
Aestivation, defined, **2:**260, and Glossary
Agassiz, Louis, **1:11–12**
Agates, **2:**153
Agricultural chemicals. *See* Chemicals from agriculture
Agricultural pollutants, **3:**227–228
Agricultural reuse, **4:**30, *31*, 32–33
Agricultural wastes, **3:**238
Agriculture and water, **1:12–16**
    agriculture and water quality, **1:**15
    hydroponic production, **1:**15–*16*
    irrigation, **1:**13–14
    land-use planning, **3:**7–*8*
    rain-fed agriculture, **1:**12–13
    water-supply constraints, **1:**14–15
Agrochemical, defined, **1:**159, and Glossary
Air sparging, **3:**222
Air stripping, **3:**221
Airborne pollutants

    lakes and streams, **3:**228
    ocean, **3:**239–240
Albatrosses, **1:**82
Albedo, defined, **2:**207, and Glossary
Aldicarb, **3:**3
Alexander, the Great, **4:***138*
Alexandrium, 176
Algae
    defined, **1:**4, **2:**1, **3:**49, **4:**78, and Glossary
    food chain, **1:**18, 21, 26, 28
    golden, **2:**9*t*
    green, **2:**9*t*
    snow, **3:**45
    and ultraviolet radiation, **1:**27, 28
Algae, blue-green. *See* Cyanobacteria
Algae, fresh-water. *See* Phytoplankton
Algal bloom, defined, **2:**96, and Glossary
Algal blooms, harmful, **1:16–20**
    aquaculture, **1:**19–20
    human health and the ocean, **2:**175
    human impacts and intervention, **1:**19–20
    mechanisms of harm, **1:**17–18
    types and examples, **1:**18–19
Algal blooms in fresh water, **1:21–24**
    algae in aquatic ecosystems, **1:**21
    color, **4:**91
    control, **1:**23–24
    occurrences and impact, **1:**21–22
    toxic blooms, **1:**22–23
Algal blooms in the ocean, **1:24–28**
    algae and photosynthesis, **1:**26
    benefits, **1:**26–28
    Black Sea, **3:***55*
    human health and the ocean, **2:**175
    marine ecology, **2:**13

Algal blooms in the ocean (continued)
  radiometer data, **3**:162
  requirements for bloom, **1**:25–26
  types, **1**:26–28
Algorithms, defined, **3**:104, and Glossary
Allocate, defined, **1**:14, **2**:200, **3**:9, **4**:181, and Glossary
Allocation, defined, **4**:5, and Glossary
Alluvium, **1**:40
  defined, **1**:100, **2**:233, and Glossary
Alternative dispute resolution (ADR), **1**:204
Altimeter, defined, **3**:161, **4**:75, and Glossary
Altitude, defined, **1**:176, **3**:120, and Glossary
Alvin (submarine), **4**:141–142, *143*
Amazon River, **2**:56–57, **4**:*61*, 62
Ambient, defined, **3**:18, **4**:115, and Glossary
Amphibian, defined, **1**:80, and Glossary
Amphibian population declines, **1**:28–31
  malformations and deformities, **1**:30–*31*
  toxic substances, **1**:29
  ultraviolet radiation effects, **1**:29–30
  water quality factors, **1**:28–29
Anacapa Island (CA), **2**:*236*
Anaconda, **4**:*45*, 46
Anadromous, defined, **2**:22, **3**:214, **4**:48, and Glossary
Anaerobes, in lakes and streams, **3**:76–77
Anaerobic, defined, **1**:63, **3**:56, and Glossary
Anastomotic, defined, **3**:100, and Glossary
Ancient civilizations. *See* Irrigation systems, ancient; Waterworks, ancient
Ancient waterworks. *See* Waterworks, ancient
Animal kingdom, **1**:22
Animal wastes, **3**:125, 227–228, 237
Anion, defined, **2**:263, **3**:12, and Glossary
Annelida, **3**:189
Anoxia, defined, **1**:22, **2**:268, and Glossary
Anoxic environment, **3**:46–47
Antagonistic effect, **1**:148
Antarctic, in literature, **1**:54

Antarctic Bottom Water (AABW), **3**:168, 169
Antarctic Circumpolar Current (ACC), **3**:141, 167, 168
Antarctic Deep Water (AADW), **3**:168
Antarctic Intermediate Water (AAIW), **3**:168–169
Antarctic Ocean, **1**:77, **3**:129, 141, 167–169, 170–171
Antarctic sea ice, **2**:206–207
Antibiotic, defined, **1**:162, and Glossary
Apalachicola River, **4**:133
Aphotic zone
  lakes, **2**:259
  life in extreme water environments, **3**:46–47
  marine environment, **3**:49
  ocean, **3**:52–53
Appurtenances, defined, **2**:221, and Glossary
Aquaculture, **1**:31–36
  algal bloom effects, **1**:19–20
  ancient and modern, **1**:31–32
  commercial success criteria, **1**:32–33
  defined, **1**:19, **2**:27, **3**:7, and Glossary
  pollution by invasive species, **3**:209–210
  potential adverse effects, **1**:35–36
  purposes, **1**:34
  species diversity, **1**:33–34
  types, **1**:34–35
  See also Mariculture
Aqualung, defined, **1**:222, and Glossary
Aquariums, **1**:36–39
  foundations, **1**:36–37
  modern, **1**:37–39
  public interest renewal, **1**:37
Aquatic biodiversity, **1**:77–78, **2**:87
Aquatic birds. *See* Birds, aquatic
Aquatic ecosystems
  acid rain effects, **1**:7, *8*
  algae, **1**:21
  assessing, **2**:53, **4**:122–123
  lakes, **2**:259–260, 260f
Aqueduct, **4**:*223*
  ancient, **4**:221, 222
  defined, **1**:100, **2**:213, **3**:67, **4**:13, and Glossary
  systems in U.S., **1**:100, *101*
Aquifer
  artificial recharge, **1**:51–52, 52f
  defined, **1**:49, **2**:76, **3**:1, **4**:28, and Glossary
  sensitivity, **3**:223

Aquifer characteristics, **1**:39–43
  types, **1**:39–40, **2**:151–152
  well reports and pump tests, **1**:40–*42*
Arabian Peninsula, **1**:*70*
Arable, defined, **2**:36, **3**:20, and Glossary
Aral Sea, **2**:218, 250, **4**:*158*
Arbitration, binding, **1**:204
Archaea, **3**:80–81
Archaeology, defined, **1**:138, and Glossary
Archaeology, underwater, **1**:43–47
  Mediterranean, **1**:43–45
  methods, **1**:43, *44*–46
Archimedes Screw, **4**:13–*14*
Archipelagos, **2**:237
Arctic Deep Water, **3**:169
Arctic Ocean, **3**:169
  basin, **3**:129–130
  biodiversity, **3**:170–171
  sea ice, **2**:206
Arctic Surface Water, **3**:169
Arid, defined, **1**:6, **2**:87, **3**:18, **4**:30, and Glossary
Army Corps of Engineers, U.S., **1**:47–49
  Columbia River Basin, **1**:197
  dams, **1**:229–230
  development projects, **3**:31
  Mississippi River Basin, **1**:99–100
  nineteenth century, **1**:47
  ports and harbors, **3**:252
  today, **1**:49
  twentieth century, **1**:47–49
  waste dumping in navigable rivers, **3**:29
  water resources planning and management history, **3**:191, 193–194
  wetlands protection, **1**:172
Arno River canals, **3**:36–37
Arrow worm, **3**:189
Arsenic, **2**:97–98, **3**:218
Artesian, defined, **4**:238, and Glossary
Arthropoda, **3**:190
Artificial recharge, **1**:49–52
  aquifer storage and recovery, **1**:51–52
  benefit to streams, **1**:51
Arts, water in the, **1**:52–58
  environmental art, **1**:56–57
  film, **1**:56, *57*
  fine art and popular art, **1**:54–*55*, **1**:*56*
  literature, **1**:53–54
  music and song, **1**:55–56

Asteroid
   defined, **1**:198, and Glossary
   water on, **4**:100
Asteroid belt, defined, **1**:200, **4**:98, and Glossary
Asthenosphere, **3**:203
   defined, **3**:84, and Glossary
Astrobiology, **1**:58–62
   Mars, **1**:*60–62*
   water and planetary habitability, **1**:58–60
Astrolabe, **3**:120–121
Astronomical unit, defined, **4**:99, and Glossary
Atlantic Ocean basin, **3**:127–128
Atlantic Water, **3**:169
Atlantis, **1**:44
Atmosphere, defined, **3**:69, **4**:141, and Glossary
ATOC (Acoustic Thermometry of Ocean Climate), **4**:83, 104
Atomic fission, defined, **4**:140, and Glossary
Atomic number, **2**:239–240
Attenuation of pollutants, **1**:62–64
   applications, **1**:64
   chemical attenuation, **1**:63–64
   contaminant behavior and attenuation, **1**:62–63
   perchlorethylene (PCE), **1**:62, 63
   physical attenuation, **1**:63
   trichlorethylene (TCE), **1**:62, 63
Autonomous, defined, **2**:200, and Glossary
Autonomous underwater vehicles (AUVs), **4**:142
Autotroph, defined, **3**:188, and Glossary

B

Backshore, **1**:72
Bacteria, **3**:188
   fecal, **3**:77
   in groundwater, **3**:72
   human health and water, **2**:180, 181*t*
   land use and water quality, **3**:3
   microbes in the ocean, **3**:79–80, *82*
   and stream health, **4**:122
   sulfate-reducing, **4**:*92*
Baffin Bay (North Atlantic Ocean), **2**:208
Bail bail, **4**:12
Bait fish, defined, **1**:34, and Glossary
Balancing diverse interests, **1**:65–67
   California, **1**:100–102
   legislation, **1**:65–66
   water management, **1**:66–67
Ballard, Robert, **1**:44
Ballast water
   human health and the ocean, **2**:176–177
   microbes in the ocean, **3**:82–83
   pollution by invasive species, **3**:209, 213
Baltic Sea, **3**:56
Bar, defined, **1**:73, and Glossary
Barnacles, **1**:226
Barrages, tidal, **2**:24–25
Barrier islands, **1**:185, *186*
   impacts of rising sea level, **4**:*73*
   migration, **1**:73
   National Park Service, **3**:119
Basalt, defined, **2**:167, **3**:137, **4**:214, and Glossary
Base flow, **2**:87
   defined, **4**:117, and Glossary
Base level, defined, **4**:120, and Glossary
Basin (ocean), defined, **1**:59, **2**:3, **3**:89, **4**:165, and Glossary
   See also Ocean basins
Basin (river), defined, **2**:223, and Glossary
Baths, **4**:222
Bathymetric chart, **3**:149*f*
Bathymetry, defined, **3**:56, and Glossary
Bathymetry, ocean-floor. *See* Ocean-floor bathymetry
Bays, gulfs, and straits, **1**:67–71
   bays, **1**:*68–69*, **4**:170–171
   gulfs, **1**:69–*70*
   straits, **1**:70–71
Beach nourishment, **1**:75–76, **4**:74
   defined, **4**:74, and Glossary
Beach pollution. *See* Pollution of the ocean by plastic and trash; Pollution of the ocean by sewage, nutrients, and chemicals
Beachdrift, defined, **1**:73, and Glossary
Beaches, **1**:71–77
   cleanups, **3**:235, 236
   currents, **1**:73, **2**:*238*
   erosion and sea level, **4**:74
   erosion-control nonstructural alternatives, **1**:75–76
   erosion-control structures, **1**:74–75
   lake *vs.* ocean, **1**:74
   Mustang Island Gulf Beach (TX), **3**:233*f*
   sands, **1**:183–184, *185*
   St. Petersburg Beach (FL), **2**:*238*
   and storms, **1**:73–74
   zones, **1**:72–73, 184
Beaches Environmental Assessment and Coastal Health Act (BEACH Act), **3**:242
Bears, polar, **3**:63
Bedload, defined, **4**:113, and Glossary
Bedrock, defined, **2**:164, **4**:100, and Glossary
Beneficial use, defined, **4**:1, and Glossary
Benthic, defined, **2**:71, and Glossary
Benthic species, **3**:50
Benthic zone, **2**:259–260, 260*f*
Bergy bits, **2**:208
Best management practices, **2**:256
Bioaccumulative, defined, **1**:35, and Glossary
Bioconcentration, **2**:14–15
Biodegradation, of pollutants, **1**:63
Biodiversity, **1**:77–80
   aquatic, **1**:77–78, **2**:87
   defined, **1**:7, **2**:87, **3**:54, **4**:56, and Glossary
   and lake health, **2**:252–253
   polar oceans, **3**:170–171
   threats to, **1**:80
   types, **1**:77
   value, **1**:78
Bioengineering, in watershed restoration, **4**:218
Biofilm, defined, **4**:131, and Glossary
Biogeochemistry, ocean. *See* Ocean biogeochemistry
Biological diversity. *See* Biodiversity
Biological oceanography. *See* Oceanography, biological
Biological opinions, **2**:22–23
Biological processes of lakes. *See* Lakes: biological processes
Biological pump, **1**:109–110
Bioluminescence, **3**:*79*
Biomass
   defined, **2**:91, **3**:50, **4**:105, and Glossary
   marginal seas, **3**:55
Biomass pyramid, **2**:12
"Biopackaging," plankton, **3**:155, 156
Bioregenerative processes, and space travel, **4**:105–106
Bioremediation, defined, **3**:222, and Glossary
Biosystem, defined, **1**:242, and Glossary

Biota
- defined, **3:**152, **4:**133, and Glossary
- polar oceans, **3:**170–171

Biotechnology, and extremophiles, **3:**48

Bioterrorism, **4:**87, 88
See also Terrorism

Biotic index, **2:**53

Biotoxins, marine, **1:**18*t*, **3:**82

Biotransformation, **1:**147

Birds, aquatic, **1:**80–82
- American widgeon, **1:***81*
- Canada goose, **1:**82
- conservation and endangered species, **1:**81–82
- diving, **1:**80–*81*
- Hawaiian common gallinule, **2:***23*
- mute swans, **1:**167–168, **3:***213*, 216
- nuisance waterbirds, **1:**82
- royal terns, **3:***235*
- wading, **1:**80
- whooping cranes, **1:**81–82

Bivalves, **1:**82–86
- characteristics, **1:**83–84
- Isthmus of Panama, **2:**238–239
- species, **1:**84–85

Black Sea, **3:***55*, 56

Black smoker, **2:***170*, **3:**85, 91
- defined, **3:**137, and Glossary

"Blue baby syndrome." *See* Methemoglobinemia, defined

Blue holes, **1:**140

"Blue water," **2:**200, 201

Blue-green algae. *See* Cyanobacteria

Bogs, **4:**243

Boron, **2:**99

Bottled water, **1:**86–88
- *vs.* public and domestic water supplies, **4:**148
- quality, **1:**87–88
- for travelers, **2:**185
- types, **1:**86–87

Boulder Dam. *See* Hoover Dam

Boundaries (plate tectonics), **3:**204–205

Brackish
- defined, **1:**51, **3:**86, **4:**152, and Glossary

Brackish water, and coastal management needs, **1:**189

Brahmaputra River (Bangladesh), **4:**41, 60

Breakwater, **1:**75
- defined, **3:**249, and Glossary

Bretz, J Harlen, **1:**88–91

Bridges, causeways, and underwater tunnels, **1:**91–94
- bridges, **1:**91–92
- causeways, **1:**92
- underwater tunnels, **1:***93*–94

Brine
- defined, **1:**61, **2:**113, **3:**142, **4:**77, and Glossary

Brine film, defined, **3:**44, and Glossary

Brines, natural, **1:**94–97, 95*f*
- geothermal energy, **2:**113
- ocean brine formation, **1:**94–96
- oil-field brines, **1:**96–97
- sea ice, **4:**76–77

Brooklyn Bridge, **1:**91

Buffer capacity (ocean), **1:**111

Buffering capacity (soil), **1:**7, 8

Bureau of Reclamation, U.S., **1:**97–99
- dams and reservoirs, **1:**230, **3:**29
- water reclamation, **4:**31
- water resources planning and management history, **3:**191, 193, 194

Bushnell, David, **4:***139*

C

Cadillac Desert (Reisner), **4:**39

Caffeine, in water and sewage, **1:**155, 156

CAFOs (Concentrated Animal Feeding Operations), **3:**4

Calcium carbonate, **1:**183, 215, 216

Calculus, defined, **1:**132, and Glossary

Caldera, defined, **2:**101, and Glossary

Calibration, defined, **1:**146, and Glossary

California, water management in, **1:**99–103
- balancing diverse interests, **1:**100–102
- prior appropriation, **4:**1
- surface water movement, **1:**100

Calvin cycle, **1:**26

Calypso, **1:**223

Campylobacter jejuni, **2:**181*t*, 183

Canals, **1:**103–107
- barge, **1:**106
- construction, **1:**106–107
- Industrial Revolution era, **1:**103–105
- for irrigation, **2:**232–235
- locks era, **1:**103
- Mars, **3:**69
- navigation, **1:**105–106
- pre-industrial canal era, **1:**103
- ship, **1:**106
- types and area served, **2:**229

Canoeing, **4:**106–107, 174–175

Cap-and-trade, **1:**9–10

Cape Cod (MA), **2:**236–237

Cape Hatteras Lighthouse (NC), **1:**75

Capes. *See* Islands, capes, and peninsulas

Capillary action, **2:**101, **4:**224
- defined, **3:**175, and Glossary

Capitol Building (U.S.), acid rain effects, **1:**9

Captive breeding, defined, **1:**38, and Glossary

Capture, rule of, **2:**66–67, **3:**28

Capture fishery, defined, **1:**31, **2:**61, and Glossary

Carbon
- lake chemical processes, **2:**264
- ocean biogeochemistry, **3:**132
- removal of, **1:**111
- taxes, **2:**136

Carbon cycle, **1:**109–110, **3:**131–132, 131*f*, 160

Carbon dioxide in the ocean and atmosphere, **1:**107–112
- anthropogenic carbon dioxide uptake, **1:**110–112
- balances in carbon dioxide levels, **1:**107–109
- emission regulations, **1:**109
- global warming, **1:**178–179, **2:**130–131, 130*f*, 131*f*, 132, 132*f*
- natural ocean carbon cycle, **1:**109–110

Carbon isotopes, in groundwater, **2:**242

Carbon-14 dating, **2:**157–158
- defined, **2:**157, and Glossary

Carbonaceous chondrites, defined, **1:**198, and Glossary

Carbonate, defined, **1:**72, **2:**243, and Glossary

Carbonation, **4:**92

Carbonic acid, **2:**90, **4:**235

Carcinogen, **1:**147
- defined, **2:**99, and Glossary

Carcinogenic, defined, **3:**3, and Glossary

Careers in environmental education, **1:**112–116
- academic preparation, **1:**114–115
- employment opportunities, **1:**114
- fresh-water education, **1:***113*

from interpretation to education, 1:112–113

marine education, 1:*115*

work types, 1:114

Careers in environmental science and engineering, **1:116–117**

Careers in fresh-water chemistry, **1:117–118**

Careers in fresh-water ecology, **1:118–119**

Careers in geospatial technologies, **1:120–121**

Careers in hydrology, **1:121–122**

Careers in international water resources, **1:123–125**

career opportunities, 1:123

education and skills, 1:124–125

international water management profession, 1:123–124

Careers in oceanography, **1:125–131**

employment opportunities, 1:129–130

fields of study, 1:127–128

key career aspects, 1:130

oceanographers and limnologists, 1:126–127

Careers in soil science, **1:131–132**

Careers in water resources engineering, **1:132–134**

beyond graduate degree, 1:133–134

engineering and society, 1:132

preparation, 1:132–133

related fields, 1:133

Careers in water resources planning and management, **1:134–136**

job opportunities, 1:134–135

other disciplines and experience, 1:135–136

qualifications, 1:135

rewards, 1:136

Carlsbad Caverns (NM), 1:140–141

Carnivore, 3:63

defined, 2:48, 4:47, and Glossary

Carp, 1:*32*, 2:80, 3:212

defined, 1:32, and Glossary

Carrying capacity (animal)

defined, 1:263, and Glossary

Carson, Rachel, **1:136–138**, 2:29

Carter, Jimmy, 1:106

Cartilaginous fishes. *See* Fishes, cartilaginous

Cascades. *See* Waterfalls

Catadromous, defined, 4:48, and Glossary

Catastrophism, 1:89

Catchment, 4:26–27

defined, 4:27, and Glossary

Catchment-cistern conveyances, 4:*27*, 28

Cation, defined, 1:63, 3:12, and Glossary

Causeways, 1:92

See also Bridges, causeways, and underwater tunnels

Cavern development, **1:138–141**

karst hydrology, 2:244

large U.S. caves, 1:140–141

sinkholes, 1:140

solutional caves, 1:138, *139*, 140

subterranean caves, 2:244–245

Centrifugal, defined, 4:165, and Glossary

Cephalopods, **1:141–144**

characteristics, 1:141–142

Nautilus, 1:141, 142, *143*

octopuses, cuttlefish, and squid, 1:142–143

CERCLA (Comprehensive Environmental Response, Compensation, and Liability Act), 3:220

Cetaceans, 3:61–62

CFC. *See* Chlorofluorocarbon

Chaco irrigation system, 2:234, 235

Chang Jiang (China), 2:57, 58, 59, 4:59–60

Channel control, 2:52

Channel Islands National Park (CA), 2:*236*

Channel Tunnel, 1:*93*–94

Channelization, defined, 1:196, 2:10, 3:191, 4:136, and Glossary

Chaotic terrain, defined, 3:69, and Glossary

Chemical analysis of water, **1:144–146**

Quality Assurance Project Plan, 1:144–146

results, 1:146

Chemical concentration units, 1:146

Chemical oceanography. *See* Oceanography, chemical

Chemical oxygen demand, defined, 3:12, and Glossary

Chemical pollutants

pollution of lakes and streams, 3:225

Chemical processes of lakes. *See* Lakes: chemical processes

Chemical processes of oceans. *See* Ocean chemical processes

Chemicals: combined effect on public health, **1:147–150**

regulatory considerations, 1:149

risk assessments, 1:148–149

Chemicals, pollution by. *See* Pollution of the ocean by sewage, nutrients, and chemicals

Chemicals from agriculture, **1:150–154**

history, 1:153–154

hooded sprayers, 1:*153*

nationwide sampling studies, 1:150–153

Chemicals from consumers, **1:155–158**

groundwater pollution, 3:*224*

impacts on humans and wildlife, 1:156–158

new research, 1:158

septic system impacts, 4:97–98

in water and sewage, 1:155–156

Chemicals from pharmaceuticals and personal care products, **1:158–164**

biochemical targets and nontargets, 1:162

conventional and nonconventional pollutants, 1:159–160

disposal of pharmaceuticals, 1:160

in environment, 1:160–162, 161*f*

pharmaceuticals in water and sewage, 1:156, 157

role of individuals, 1:162–163

Chemistry careers. *See* Careers in fresh-water chemistry

Chemosynthesis

bacteria, 2:14, 3:49

defined, 2:12, and Glossary

hot springs on the ocean floor, 2:170–171

Chernobyl reactor accident, 4:*20*, 25

Chesapeake Bay, **1:164–169**

basin and bay, 1:164–166

environmental concerns, 1:166–168

saving, 1:168

toxic contamination, 1:166

Chesapeake Bay Bridge–Tunnel, 1:93

Chesapeake Bay Program, 3:33

China, grain production, 2:83, 83*f*

Chip log, 3:121

Chitin, defined, 1:224, 3:188, and Glossary

Chloride, 3:4–5

Chlorinated solvents, 3:219

Chlorination, defined, 4:210, and Glossary

Chlorofluorocarbon

defined, 2:157, and Glossary

and groundwater age, 2:157

Chlorofluorocarbon (continued)
  tracers of ocean-water masses, **4**:178–179, 179f
Chloroplast, defined, **1**:21, and Glossary
Choke point, **1**:71
Cholera, **2**:177, 178, 181t, 182–183, 185, **3**:77, 82
Chondrites, defined, **1**:199, and Glossary
Chordata, **3**:190
Chronometers, **3**:121–122
Chunnel (English Channel Tunnel), **1**:93–94
Circulation. See Water circulation
Circuses, **1**:36–37
Cistern, **4**:27–28
  defined, **4**:27, and Glossary
Civic, defined, **4**:6, and Glossary
Clams, **1**:83, 84
Clark, William. See Lewis, Meriwether and William Clark
Clean Air Act. See Acid rain
Clean Water Act, **1**:65, **169–174**, **2**:32
  amendments, **1**:171–172, **3**:30
  aspirations and deficiencies, **1**:173–174
  Columbia River Basin, **1**:197
  Concentrated Animal Feeding Operations, **3**:4
  federal responsibility, **1**:169–170
  Great Lakes, **2**:145
  pollution of streams by garbage and trash, **3**:230–231
  pollution of the ocean, **3**:240
  reassessment needs, **1**:172
  regulatory objectives, **1**:170–171
  sewage from vessels, **3**:239
  sewage treatment plants, **3**:124
  stormwater management, **2**:255
  thermal pollution of streams, **4**:116
  wastewater treatment, **4**:210
  water quality, **1**:169–170, **3**:29–30
  wetlands, **4**:246
Cleanups
  beach, **3**:235, 236
  groundwater pollution, **3**:220–222
  radioactive chemicals, **4**:20–21
  river and stream, **3**:231
Climate, defined, **3**:116, and Glossary
Climate and the ocean, **1**:174–179
  climate change, **1**:177–179
  climate zones, **1**:176–177, 177f

Earth's orbit, **1**:175, 178
factors affecting Earth's climate, **1**:174–176, 176f
solar energy, **1**:174–175, 178
Climate change
  climate and the ocean, **1**:177–179
  human health and the ocean, **2**:178–179
  ice at sea, **2**:207–208, **4**:74
  Mars, **3**:71
  See also Glaciers, ice sheets, and climate change; Global warming
Climate mechanisms, in ice ages, **2**:204–205
Climate moderator, water as a, **1**:179–183
  air temperature moderation, **1**:182
  water content in air, **1**:180–182
Clouds. See Precipitation and clouds, formation of
Cnidaria, **2**:178, **3**:49, 189
Coagulant, defined, **1**:258, and Glossary
Coagulation
  defined, **4**:137, and Glossary
  drinking-water treatment, **1**:258–259
Coastal ocean, **1**:183–186
  barrier islands, **1**:185
  beaches, **1**:183–185
  tides, **1**:185–186
Coastal waters management, **1**:187–190
  brackish waters, **1**:189
  Coastal Zone Management Act, **1**:189–190
  fresh waters, **1**:189
  National Park Service, **3**:119
  need for, **1**:187–189
  ocean, **1**:188
Coastal Zone Management Act, **1**:189–190
Coccolithophores, **1**:27–28, **3**:187
Codes of ethics, **2**:44
Codex Leicester (Leonardo da Vinci), **3**:35, 36
Coelenterata, **2**:178, **3**:49, 189
Coliform, **3**:77, **4**:122, 240
  defined, **3**:232, **4**:32, and Glossary
  Total Coliform Rule, **3**:74–75
Colluvium, defined, **3**:17, and Glossary
Colonization, of hot springs on the ocean floor, **2**:172–173
Colorado River, **4**:61–62
  color, **4**:90
  water flows, **2**:218, **4**:135–136

Colorado River Basin, **1**:190–194
  the basin, **1**:191–193
  management efforts, **1**:193–194
  map, **1**:191f
  tectonic processes and erosion, **4**:59
Columbia River Basin, **1**:194–198
  Corps of Discovery, **3**:42
  dams, **1**:196, 197, **2**:219
  Environmental Protection Agency, U.S., **1**:197
  institutional complexities, **1**:196–198
  map, **1**:195f
  Native and non-Native settlement, **1**:195–196, 197
  planning, **4**:56
  river, **4**:61
  salmon, **1**:195, 196, 197, **2**:190–191, **4**:48, 69
Comb jellies, **3**:189
Comet, defined, **1**:198, and Glossary
Comets and meteorites, water in, **1**:198–200
  comets, **1**:199, **4**:100
  Halley's Comet, **1**:199
  meteorites, **1**:200
  solar components of water, **1**:199
Commerce, waterborne. See Canals; Ports and harbors; Transportation
Commercial water use category, **4**:193
Commission, defined, **3**:33, and Glossary
Common law, defined, **3**:26, **4**:3, and Glossary
Common-property resource, defined, **2**:67, and Glossary
Commons, tragedy of the, **2**:67
Community development, **2**:16
Compact, defined, **3**:32, and Glossary
Compass, mariner's, **3**:120
Compass rose, defined, **3**:120, and Glossary
Comprehensive Environmental Response, Compensation, and Liability Act (CERCLA), **3**:220
Comprehensive Everglades Restoration Plan, **2**:47, **3**:31
Computers and technology. See Data, databases, and decision-support systems; Geospatial technologies; Modeling groundwater flow and transport; Modeling streamflow
Concentrated Animal Feeding Operations (CAFOs), **3**:4

Conductivity test, **4:**220
"Cone of depression," **1:**41, 41f, **2:**154, 154f
Confederation Bridge (Canada), **1:**92
Confined aquifer, defined, **1:**86, and Glossary
Confined disposal facilities, **3:**228
Conflict and water, **1:201–206**
   conflict resolution, **1:**204–205
   dams, **1:**202, 203, *204*
   disputes over water, **2:**198–199
   international water law, **3:**18
   Law of the Sea, **3:**24
   nature of conflict, **1:**201
   resolving water-use conflicts, **3:**26–27
   sources of conflict, **1:**201–204
   water law, **3:**26–27
   See also War and water
Congo River, **4:**60
Conjunctive use, **2:**222
   defined, **1:**49, and Glossary
Connate water, **2:**94, **4:**91–92
Conservation, defined, **2:**28, **3:**117, **4:**31, and Glossary
Conservation, water, **1:206–210**
   conservation planning, **1:**206–207
   conserving water resources, **1:**206
   demand management, **1:**207–209
   food security, **2:**84
   social acceptability, **1:**209–210
   water conservation-based rate structure, **1:**208
Conservative tracer, defined, **4:**130, and Glossary
Conshelf living stations, **1:**223
Constant-rate pump tests, **1:**41, 42f
Constitution (U.S.) and public water rights, **4:**49, 51–52
Consumer chemicals. *See* Chemicals from consumers
Consumptive use, defined, **2:**77, **3:**26, **4:**27, and Glossary
Containment
   floodwaters, **4:**113
   groundwater pollution, **3:**220–221
Contaminant
   defined, **1:**189, **2:**94, **3:**30, **4:**66, and Glossary
   emerging, **3:**6
   in public and domestic water supplies, **4:**148
Contaminated site risk assessments, **1:**148–149
Contamination
   defined, **1:**257, **3:**1, **4:**28, and Glossary
   security and water, **4:**87–88
   See also "Chemicals" and "Pollution" entries
Continental, defined, **1:**12, and Glossary
Continental ice sheet, defined, **3:**99, and Glossary
Continental margin, defined, **3:**126, and Glossary
Continental plate, defined, **4:**200, and Glossary
Continental shelf
   Arctic Ocean, **2:**206
   defined, **2:**63, **3:**24, and Glossary
Continental-continental convergence, **3:**205, 205f
Convection, defined, **4:**105, and Glossary
Convention on the Prevention of Marine Pollution by Dumping of Wastes and Other Matter, **3:**241
Convergent boundaries, **3:**204–205, 205f
Cook, James, **1:210–211**, **1:**242, **3:**122
Cooperative agreement, defined, **3:**33, and Glossary
Copepods, **3:**190
Coral, defined, **1:**45, **2:**237, and Glossary
Coral reef, defined, **1:**37, **2:**1, and Glossary
Corals and coral reefs, **1:212–220**
   atoll reefs, **1:***212*, 215
   barrier reefs, **1:**215
   brain coral, **1:***79*
   coral bleaching, **2:**20
   coral types, **1:**213–215
   distribution and roles, **1:**212–213
   and dust, **1:**218
   fire coral, **1:**213
   fringing reefs, **1:**215
   hazards and global stresses, **1:**218–219
   as historic recordkeepers, **1:**216
   as marine ecosystem, **2:**13–14
   physical and biological controls, **1:**216–218
   preservation, **1:**219
   reef types, **1:**215–216
   scleractinian corals, **1:**214–215
   symbiosis, **1:**215
   tropical oceans, **3:**173
Cordilleran Ice Sheet, **2:**202
Core dating, **2:**212–213
Core (Earth), **3:**202–203
Coriolis effect, **4:**231
   circulation cells, **1:**175–176, 176f
   hurricanes, **4:**232
   ocean currents, **3:**139, 140, 141–142
   tides, **4:**169, 170
Cost–benefit analysis, **1:220–221**
   Army Corps of Engineers, U.S., **3:**193
   defined, **1:**48, and Glossary
   evaluating dollars over time, **1:**221
   guidelines, **3:**30
   practical economic analysis, **1:**220–221
   value of nonmarket goods, **1:**220
Cotton, as a crop, **3:**263
Council on Environmental Quality, **3:**113
Cousteau, Jacques, **1:**37, **1:221–223**
Covalent bond, defined, **2:**100, and Glossary
Crabs, **1:**224–225
   blue, **1:**166–*167*
   fiddler, **1:**224–225
   ghost, **1:***224*
   hermit, **1:**224
   horseshoe, **1:***226*
Creep (landslides), **3:**17
Crocodilians, **4:***46*–47
   See also Reptiles
Cross-contamination, defined, **4:**208, and Glossary
Crown-of-thorns starfish, **1:**217
Crude oil, defined, **3:**181, and Glossary
Cruise ships, **4:**172, *174*
Crust (Earth), **3:**201–202
Crustaceans, **1:224–227**
   barnacles, **1:**226
   crabs, **1:**224–225
   defined, **1:**33, **2:**79, and Glossary
   krill, **1:**226
   lobsters, **1:***225*
   other aquatic arthropods, **1:**226–227
   shrimp, **1:**225–226
Cryptosporidiosis, **3:**82
Cryptosporidium, **2:**181t, 183, **3:**74, **4:**87
Crystalline, defined, **2:**5, and Glossary
Ctenophora, **3:**189
Currents, ocean. *See* Ocean currents
Cuttlefish, **1:**142
Cyanide, **3:***226*
Cyanobacteria
   characteristics, **2:**9t
   defined, **3:**76, and Glossary

Cyanobacteria (continued)
  in fresh water, **1:**21, 22–23
  in the ocean, **1:**26, **3:**79–*80*
Cyclones, **3:**148, **4:**230, 232, *232f*, 233

D

Dam failure, defined, **1:**228, and Glossary
Dams, **1:227–231**
  arch, **1:**227–228, *230*
  Bonneville Dam, **4:***87*
  building, **1:**229–230
  Bureau of Reclamation, U.S., **1:***97–98*, 99
  buttress, **1:**228
  classes, **1:**227–228
  Colorado River Basin, **1:**192–193
  Columbia River Basin, **1:**196, 197
  and conflict, **1:**202, 203, *204*
  Dalles Dam (OR), **2:***219*
  developing countries, **1:**249, 250, **4:**57–58
  embankment, **1:**227
  Glen Canyon Dam (AZ), **1:***230*
  Grand Coulee Dam (WA), **1:***98*, 196, **2:**189, 190
  gravity, **1:**227
  history, **3:**193
  instream water issues, **2:**218, *219*
  Lower Granite Dam on the Snake River (WA), **2:***189*
  multipurpose, **4:**55
  Pen-y-Garreg Reservoir Dam (Wales), **1:***228*
  planning, **1:**228–229
  regulations, **1:**47–48
  salmon decline and recovery, **4:***70*, 71
  Tellico Dam (TN), **2:**23–24, 218
  Tennessee Valley Authority, **4:**162–163
  terrorist threats, **4:**86–87
  Three Gorges Dam (China), **1:**229, 249, **4:**60
  tidal barrages, **2:**24–25
  traditional supply planning, **3:***195*
  See also Hoover Dam; Hydroelectric power
Danube River, **4:**61
Danube River Case (Hungary v. Slovakia), **3:**21, **4:***181*
Darcy, Henry, **1:231–232**
Data, databases, and decision-support systems, **1:232–236**
  data collection, **1:**233–234
  data types, **1:**232–233
  databases, **1:**234–235
  decision-support systems, **1:**235
  large-scale data, **1:**234
  long-term data, **1:**234
  water resources data, **2:**105
Davis, William Morris, **1:236–237**
DDT, **1:**137, 151, **3:**5
  defined, **2:**29, **3:**5, and Glossary
Dead Zone
  Gulf of Mexico, **1:**69–70, 188, **3:***99f,* 101, 238
  Lake Erie, **2:**146
Decay rate, **3:**103
Decision-support systems, **1:**235
  See also Data, databases, and decision-support systems
Decommissioning (dams), defined, **1:**230, and Glossary
Decomposer
  defined, **2:**12, and Glossary
  in fresh-water ecosystems, **3:**76–77
Decomposition, and lake health, **2:**252
Deep wells, **2:**229
  defined, **2:**150, and Glossary
Deep-sea trenches, **3:**150–151, 204, 206, **4:**205–206, *205f*
Deforestation, **1:**249
Deformities, in amphibians, **1:**30–*31*
Degradation, defined, **1:**160, and Glossary
Delaware River Basin Commission, **3:**33
Delta, **3:***162,* **4:**59
  defined, **2:**228, **3:**98, and Glossary
Demand management, **1:237–239**
  demand curve, **1:***238f*
  drought, **1:**265
  integrated resource planning, **3:**199
  least-cost planning, **3:**197–198
  supply development, **4:**152–153
  tools and techniques, **1:**238
  utility programs, **1:**238–239
  water conservation, **1:**207–209
  water demand, **1:**238
  water pricing, **3:**263–264
Dengue, **2:**184
Density
  fresh water, **2:**102–103
  ocean, **4:**81–82, *82f*
Department of Agriculture, U.S., **1:**15, 235
Department of Energy, U.S., **4:**20–21

Deposition, cloud, **1:**6
Deposition, of rivers, **2:**249–250
Derby effects, **2:**66
Desalinization, **1:239–242**
  defined, **1:**245, **2:**27, **4:**207, and Glossary
  desert hydrology, **1:**245
  freezing, **1:**240–241
  heat distillation, **1:**239–240
  iceberg towing, **1:**242
  ion extraction, **1:**240
  reverse osmosis, **1:***241*, 242
  solar humidification, **1:**241
Desert hydrology, **1:242–246**
  coastal deserts, **1:**243
  desalinization, **1:**245
  desert crust, **1:**244
  hydrologic aspects of deserts, **1:**243–245
  polar regions, **1:**243
  salt accumulation, **1:**244–245
  subtropical deserts, **1:**243
  vadose zone studies, **1:**244
  vegetation control, **1:**244
  water management issues, **1:**245
  winter deserts, **1:**243
Desiccation, **3:**45–46
Detergents, household, **3:**124, *125*
Detritivore, defined, **2:**9, and Glossary
Developed countries
  groundwater pollution, **3:**219
  irrigation management, **2:**228
  population projections, **3:***248f*
  sustainable development, **4:**158
Developing countries
  human health and water, **1:**246–247, 254–255
Developing countries, issues in, **1:246–251**
  dams, **1:**249, 250, **4:**57–58
  deforestation, **1:**249
  drinking water and society, **1:**254–255
  floods, **1:**247–248
  groundwater pollution, **3:**219
  inadequate supply and sanitation, **1:**246–247, *256*
  irrigation, **1:**248, **2:**229–230
  pollution of streams by garbage and trash, **3:**231–232
  population projections, **3:***248f*
  problem-solving aspects, **1:**250
  river basin planning, **4:**57–58
  sustainable development, **4:**158–159
  transportation, **4:***183*

water pollution, **1**:247
water-supply development, **4**:153
Developing country, defined, **2**:82, and Glossary
Development, economic. *See* Economic development
Dew point, **3**:261–262
Diagenesis, defined, **3**:90, and Glossary
Diamonds, **3**:91
Diarrhea, **2**:181*t*, 185, **3**:77
Diatom, **1**:26–27, **3**:187
   characteristics, **2**:9*t*
   defined, **1**:18, **3**:132, and Glossary
   marine, **3**:80
Diatomaceous earth, **3**:86–87
Dichlorodiphenyltricholoroethane. *See* DDT
Diffraction, wave, **4**:226, *227*
Diffusers (pumps), **4**:9
Dike, defined, **2**:41, **4**:85, and Glossary
Dimethyl sulfide, **1**:27–28
Dinoflagellates, **1**:27, **3**:187
   characteristics, **2**:9*t*
   defined, **1**:16, and Glossary
   marine, **3**:82
Dip slope (landslides), **3**:16
Discharge, defined, **3**:98, **4**:58, and Glossary
Discount rate, defined, **1**:221, and Glossary
Disease, waterborne. *See* Drinking water and society; Human health and the ocean; Human health and water; Microbes in groundwater; Microbes in lakes and streams; Microbes in the ocean; War and water
Disinfection, **1**:259, **4**:149
Dispersivity, **3**:103
Disphotic zone, **3**:52
Displaced population, defined, **4**:208, and Glossary
Dissect, defined, **3**:70, and Glossary
Dissociation product, defined, **1**:109, **2**:263, and Glossary
Dissolution (water composition), **2**:90, 101, 103
Dissolved, defined, **2**:95, **3**:56, and Glossary
Dissolved inorganic carbon (DIC), **2**:242
Dissolved load, defined, **4**:113, and Glossary
Dissolved organic carbon (DOC), **2**:242
Dissolved oxygen

defined, **1**:144, **4**:122, and Glossary
   nutrients in lakes and streams, **3**:123
   in sea water, **4**:78–79*f*
   and watershed water quality, **4**:220
Divergent boundaries, **3**:204
Diversion of water, **2**:51–52
Diversity, biological. *See* Biodiversity
Diving, **4**:106, 173–174
Diving bells, **4**:*138*
Doldrums, **3**:142
Dolomite, defined, **4**:215, and Glossary
Dolostones, **2**:243
Dolphins, **3**:62
   bottlenose, **3**:62, *64*
   human connection to, **3**:65
   killer whales, **3**:*61*
Domestic water supplies. *See* Supplies, public and domestic water
Domestic water use category, **4**:193, 195
Domoic acid, **1**:19
Douglas, Marjory Stoneman, **1**:251–253
Downdrift, defined, **1**:75, and Glossary
Downwellings, in tropical oceans, **3**:172–173
Dowsing, **2**:160, *161*
Drainage basin, defined, **3**:99, **4**:117, and Glossary
Drainfield, septic system, **4**:95–96, 96*f*
Drawdown
   pump tests, **1**:41, 41*f*, 42*f*, **4**:239–240
   pumping and overpumping, **2**:154–155, 154*f*, 155*f*
Drawers of Water (White, Bradley, and White), **4**:248–249
Dredging
   defined, **1**:75, and Glossary
   estuaries, **2**:41
   ports and harbors, **3**:253
Drillers' logs, **1**:40, 41
Drinking water
   arsenic in, **2**:98
   chemicals from pharmaceuticals and personal care products in, **1**:160, 162
   municipal reuse, **4**:33
   pollution sources, **4**:144–145
   radon in, **4**:19
   See also see Groundwater supplies, exploration for; Pollution of groundwater; Safe Drinking

Water Act; Supplies, protecting public drinking-water; Supplies, public and domestic water
Drinking water and society, **1**:253–257
   developing countries, **1**:254–255
   inadequate water supplies, **1**:255–256
   infrastructure *vs.* search for water, **1**:253–255
Drinking-water treatment, **1**:257–260
   disinfection, **1**:259
   filtration, **1**:*259*–260
   flocculation, **1**:258–259
   groundwater, **1**:257–258
   other treatment technologies, **1**:259–260
   surface water, **1**:258–259
Drip irrigation, defined, **1**:15, and Glossary
Drought
   agricultural, **1**:262
   demand management, **1**:265
   economic losses, **1**:263
   environmental impacts, **1**:263
   hydrological, **1**:262
   meteorological, **1**:262
   social impacts, **1**:263
   socioeconomic, **1**:262–263
   supply augmentation, **1**:265
Drought management, **1**:260–265
   drought as natural hazard, **1**:260–261
   drought characteristics and definition, **1**:261
   drought impacts, **1**:263
   drought preparedness and mitigation, **1**:263–265
   drought types, **1**:262–263
   United States drought, **1**:262*f*
Drought mitigation, defined, **1**:265, and Glossary
Drug benefits from the sea, **2**:175
Dry deposition, **1**:6
Dublin Statement, **2**:226
Duplicate (sample), defined, **1**:146, and Glossary
Dust Bowl, **1**:*261*

E

E. coli
   in groundwater, **3**:*74*
   and human health, **2**:181*t*, 183, 185
   in lakes and streams, **3**:77
   in the ocean, **3**:82
   and watershed water quality, **4**:219

Earle, Sylvia, **2:1–3**
Earth, **2:3–5**
   early Earth, **2:**3
   structure, **3:**201–203
   water in solar system, **4:***99,* 100
   water movement, **2:**3–5
Earthquakes
   and estuary formation, **2:**39
   water levels in wells as earthquake detectors, **2:**7
Earthquakes, underwater. *See* Tsunamis
Earth's interior, water in the, **2:5–7**
   influence of water, **2:**6–7
   water states, **2:**6
East Pacific Rise, **3:**83–84, 84*f*, 127
Echinodermata, **3:**189
Ecology, defined, **1:**65, **3:**52, and Glossary
Ecology, fresh-water, **2:7–11**
   human influences, **2:**10
   lakes, **2:**9–10
   role of microbes, **3:**76–77
   streams, **2:**7–9
   wetlands, **2:**10
Ecology, marine, **2:11–15**
   abyss, **2:**14
   bioconcentration, **2:**14–15
   coral reefs, **2:**13–14
   examples of marine ecosystems, **2:**13–14
   hydrothermal vents, **2:**14
   kelp forests, **2:**14
   phytoplankton and zooplankton, **2:**12–13
   polar oceans, **2:**14
   tidal pools, **2:**14
   trophic levels and biomass pyramids, **2:**11–12
Ecology careers. *See* Careers in fresh-water ecology
Economic development, **2:15–17**
   community development, **2:**16
   national development, **2:**15–16
Economic losses, from drought, **1:**263
Economics. *See* Cost–benefit analysis; Markets, water; Pricing, water
Ecosystem, defined, **1:**5, **2:**12, **3:**7, **4:**52, and Glossary
Ecotone, defined, **4:**131, and Glossary
Ecotourism, **4:**172–173
   See also Tourism
Eddies, **3:***147–*148, 162
Edmund Fitzgerald, **1:**46

EEZ (Exclusive Economic Zone), **3:**25
Effluent, defined, **3:**124, **4:**210, and Glossary
Ejecta blanket, defined, **3:**70, and Glossary
Ekman, Valfrid, **3:**139–140
El Niño, defined, **3:**167, and Glossary
El Niño and La Niña, **2:17–21**
   climatology and effects of El Niño, **2:**20, 20*f*
   El Niño ocean conditions, **2:**18–19
   glaciers as record of, **2:**122
   and human health, **2:**178, **3:**82
   La Niña, **2:**19
   marine fisheries, **2:**63
   studying El Niño, **2:**20–21
   understanding El Niño, **2:**17*–*19
   weather and the ocean, **4:**232–233
Elasticity, defined, **2:**136, and Glossary
Electrical conductivity, of fresh water, **2:**100, 101, **4:**220
Electrodialysis, **1:**240
Elite, defined, **3:**207, and Glossary
Embayment, defined, **2:**250, and Glossary
Emissions trading, **1:**9–10
Empirical, defined, **2:**126, **3:**35, and Glossary
Empress of Ireland, **1:**45
Endangered, defined, **1:**81, **2:**55, **3:**27, **4:**44, and Glossary
Endangered species, **1:**81–82, **2:**22
Endangered Species Act, **1:**65–66, **2:21–24**, 55
   aquatic examples, **2:**23–24
   Columbia River Basin, **1:**196, 197
   prior appropriation, **4:**2
   purpose and administration, **2:**21–23
Endogenous, defined, **1:**162, and Glossary
Energy, fresh water physics and chemistry, **2:**102, 103
Energy, geothermal. *See* Geothermal energy
Energy from the ocean, **2:24–27**
   energy from temperature, **2:**26–27
   energy from tides, **2:**24–25
   energy from waves, **2:**25–26
Engineering careers. *See* Careers in environmental science and engineering; Careers in water resources engineering

English Channel Tunnel, **1:**93–94
Entomology, defined, **1:**118, and Glossary
Entrain, defined, **2:**97, **3:**137, and Glossary
Entrenchment, defined, **4:**120, and Glossary
Environment, defined, **1:**65, **2:**15, **3:**112, and Glossary
Environmental art, water in, **1:**56–57
Environmental education careers. *See* Careers in environmental education
Environmental engineering careers. *See* Careers in environmental science and engineering
Environmental impact statement, **3:**113–114
   defined, **3:**39, and Glossary
Environmental Information Management System, **1:**235
Environmental literature, **1:**54, 137–138
Environmental movement, role of water in the, **2:27–31**
   emergence, **2:**28–29
   historical background, **2:**27–28
   modern movement, **2:**29–30
   water, people, and environment, **2:**30
Environmental Protection Agency, U.S., **2:31–32**
   Clean Water Act, **1:**170–171, 172, 173, 197, **2:**32, **3:**30
   Columbia River Basin, **1:**197
   counterterrorism, **4:**88
   databases, **1:**235
   landfill standards, **3:**11, 13–14
   methyl *tert*-butyl ether (MTBE), **3:**221
   Office of Solid Waste, **3:**230
   Office of Water, **2:**32
   public drinking water regulation, **1:**87, 88, 254, **2:**98, **3:**30
   public water system regulations, **3:**74–75
   radon standard for drinking water, **4:**19
   Safe Drinking Water Act, **4:**66, 67, 68
   sewage from vessels, **3:**239
   water infrastructure security, **2:**215, **4:**199
   water-quality standards, **1:**150, 151–152, **2:**214
Environmental science careers. *See* Careers in environmental science and engineering

Environmental terrorism, **4**:85–86, 207
  See also Terrorism
EPA. *See* Environmental Protection Agency, U.S.
Ephemeral stream, defined, **1**:244, and Glossary
Epilimnion
  defined, **2**:262, and Glossary
  lakes, **2**:267, **3**:50
Epipelagic zone, **3**:49
Equatorial region, **1**:176, 177*f*, **3**:141–142
Equilibrium constant, defined, **2**:90, and Glossary
Equitable utilization, rule of, **3**:19–20, 21–22
Erie, Lake, **2**:146
Erosion and sedimentation, **2**:33–37
  accelerated sedimentation, **2**:34–35
  differential valley, **4**:215
  ecological and economic impacts, **2**:35–36
  erosive forces, **2**:33–34, 35
  gully erosion, **2**:33, 88
  lake formation, **2**:249–250
  prevention and control, **1**:74–76, **2**:36–37
  rill erosion, **2**:33, 88
  sheet erosion, **2**:33
  stream sedimentation, **2**:34–35
  tunnel erosion, **2**:33
  waterfalls, **4**:213, 215
  See also Stream erosion and landscape development
Escarpment, defined, **4**:215, and Glossary
Escherichia coli. *See* E. coli
Estuaries, **2**:37–42
  common features, **2**:40–42
  formation, **2**:38–39
  ocean chemical processes, **3**:135–136
  tides, **4**:170–171
  water circulation, **2**:39–40
Estuary, defined, **1**:18, **2**:24, **3**:6, **4**:70, and Glossary
Ethics, defined, **2**:42, and Glossary
Ethics and professionalism, **2**:42–44
  codes of ethics, **2**:44
  ethical decision-making, **2**:42–44
Eukaryotes, **1**:22
  defined, **1**:26, and Glossary
Euphotic zone
  lakes, **2**:259, 262
  ocean, **3**:52
Euphrates River, **2**:233

Euryhaline species, **3**:51
Eurythermal species, **3**:52
Eutrophic, defined, **2**:265, and Glossary
Eutrophication
  Chesapeake Bay, **1**:166
  coastal, **3**:238–239
  cultural, **2**:251, 255, 255*f*, **3**:77, 238–239
  defined, **1**:31, **3**:6, **4**:137, and Glossary
  estuaries, **2**:41–42
  lake management issues, **2**:254–256, 255*f*
  in lakes and streams, **3**:123–124
  natural, **2**:255*f*
  See also Estuaries; Lake management issues; Nutrients in lakes and streams
Evaporation
  ocean chemical processes, **3**:137
Evaporation, and ocean brine formation, **1**:94
Evaporite
  defined, **1**:59, **2**:263, **3**:88, and Glossary
  and karst hydrology, **2**:243
Evapotranspiration, defined, **1**:208, **2**:46, **3**:16, **4**:135, and Glossary
Everglades, **2**:44–47
  Comprehensive Everglades Restoration Plan, **2**:47, **3**:31
  Douglas, Marjory Stoneman, **1**:251, 252–253
  drainage, **2**:84
  fresh and salt waters, **2**:45
  geology, **2**:46
  human development, **2**:46–47
  Leopold, Luna, **3**:39
  water pollution, **2**:85
  weather and rainfall, **2**:45–46
The Everglades: River of Grass (Douglas), **1**:251
Exceedance probability, **3**:105
Exclusive Economic Zone (EEZ), **3**:25
Excretion, chemical, **1**:147
Exotic
  defined, **1**:28, **2**:219, **4**:116, and Glossary
Exotic species, **2**:256–258
Externality, defined, **2**:139, **3**:196, and Glossary
Extrapolate, defined, **2**:191, and Glossary
Extraterrestrial, defined, **1**:58, and Glossary
Extraterrestrial life. *See* Astrobiology

Extreme water environments. *See* Life in extreme water environments
Extremophiles research. *See* Life in extreme water environments
Exxon Valdez, **1**:82, **3**:178, 180

F

Fall overturn, **2**:267, *268*, 269
Fault, defined, **1**:61, **2**:39, **4**:109, and Glossary
Fault block, defined, **2**:247, and Glossary
FDA (Food and Drug Administration), **1**:87–88
Federal Emergency Management Agency, U.S., **2**:75–76
Federal Interagency River Basin Committee, U.S., **4**:56
Federal water legislation. *See* Legislation, federal water
Federal Water Pollution Control Act, **3**:29
Federal Water Pollution Control Act Amendments of 1972. *See* Clean Water Act
Federal Water Power Act, **3**:29
Feedback loop, defined, **2**:128, and Glossary
FEMA (Federal Emergency Management Agency), **2**:75–76
Ferrell cells, **1**:176, 176*f*, **4**:231
Ferries, **4**:*185*
Fertilizers
  nitrate contamination, **2**:242
  pollution of lakes and streams, **3**:125–126, 227
  residential use, **1**:154
  See also Chemicals from agriculture
FIARBC (Federal Interagency River Basin Committee), **4**:56
Fiduciary, defined, **4**:50, and Glossary
Film, water in, **1**:56, 57
Filter feeder, defined, **1**:226, **2**:175, and Glossary
Filtration, **4**:148–149
Fine art, water in, **1**:54–55
Finfish, defined, **1**:31, and Glossary
Fires, effect on forest hydrology, **2**:87, 88, 89
Firmament, defined, **4**:41, and Glossary
Firn, **2**:211
Fish, **2**:48–50
  cutthroat trout, **3**:214
  flotation, **3**:51
  lake trout, **3**:214

Fish (continued)
  in lakes, **2:**10
  locomotion, protection, and feeding, **2:**49, *50*
  nonindigenous fish species list, **3:**211
  physical characteristics, **2:**48–49
  pufferfish, **2:**70, 175
  reproduction, **2:**49–50
  round goby, **3:**216
  salinity and habitat, **2:**48
  snakehead fish, **2:**258, **3:***210*, 211, 212, 216
  tropical porcupine fish, **2:***49*
  wolf eel, **2:***177*
Fish and wildlife issues, **2:**50–54
  impacts and assessments, **2:**52–53
  oil spill impact, **3:**178, *179*
  pollution of ocean by plastic and trash, **3:**233–234, *235*
  prior appropriation, **4:**2–3
  water management practices, **2:**50–52
Fish and Wildlife Service, U.S., **2:54–55**
  Columbia River Basin, **1:**197
  conserving fish and wildlife, **2:**55
  Endangered Species Act, **2:**22–23, 55
  national wildlife refuges, **2:**54–55, **3:**100
Fish culture, **1:**31
Fish ladders, **4:***70*
Fisheries, fresh-water, **2:56–61**
  economic importance, **2:**56–57
  future, **2:**60
  Great Lakes, **2:**144–145
  management, **2:**60
  threats, **2:**58–59
Fisheries, marine, **2:61–65**
  assessment, **2:**63–64
  cod fishery, **2:**63, 65, 80
  management, **2:**63–64
  oil spill impact, **3:**178
  production, **2:**61–63
Fisheries, marine: management and policy, **2:65–69**
  alternative management regimes, **2:**66–69
  capture fishery decline, **2:**65
  management, **2:**65–66
Fishery councils, **2:**63
Fishes, cartilaginous, **2:69–71**
  rays and skates, **2:**70
  sharks, **2:**71
Fishing, **2:**59, **4:**107, 173
  See also Overfishing

Fissure, defined, **4:**204, and Glossary
Fissured sediment, defined, **1:**244, and Glossary
Flagellum, defined, **1:**27, **3:**187, and Glossary
Flash flood, defined, **2:**72, **4:**64, and Glossary
Flocculation
  defined, **4:**210, and Glossary
  drinking-water treatment, **1:**258–259
Floe, defined, **3:**169, and Glossary
Flood Control Act, **1:**48, 49, 220, **3:**100, 193
Flood hazard, defined, **4:**247, and Glossary
Floodplain, defined, **1:**135, **2:**8, **3:**85, **4:**59, and Glossary
Floodplain management, **2:71–76**
  benefits and costs, **2:**72–73
  defined, **3:**191, **4:**248, and Glossary
  floods, **2:**71–72
  National Park Service, **3:**119
  public policy, **2:**75–76
  strategies, **2:**73–75
Floods
  control, **1:**48–49, **3:**100, **4:**248
  defined, **2:**71, and Glossary
  developing countries, **1:**247–248
  effects on stream channels, **4:**113
  floodplain management, **2:**71–72
  insurance, **2:**74–75, **3:**33
  Mississippi River Basin, **3:**100–*101*
  Missoula Floods, **1:**89–91, *90f*
  runoff, **3:**105–106, **4:**64–65
  stream hydrology, **4:**127
Florida, water management in, **2:76–79**
  water districts, **2:**77, *78f*
  water management issues and tools, **2:**78–79
Flotation, in aquatic life, **3:**51
Flows (landslides), **3:**16
Flowstone, **1:***139*, 140
Fluid inclusions
  defined, **2:**6, and Glossary
  volcanoes and water, **4:**200
Fluoride, as natural contaminant in fresh water, **2:**99
Fluvial, defined, **3:**173, and Glossary
Fog harvesting, **4:**28
Folk art, water in, **1:**55
Food and Drug Administration, U.S., **1:**87–88
Food chain

  algae, **1:**18, 21, 26, 28
  defined, **1:**18, **2:**11, **3:**49, and Glossary
  marine, **2:**11–15, 62, **3:**49
  planktivores, **1:**33
  polar oceans, **2:***11*, 14
Food fish, defined, **1:**34, **2:**54, and Glossary
Food from the sea, **2:79–81**
  human health and the ocean, **2:**175, 176
  ocean pollution by sewage, nutrients, and chemicals, **3:**239, 240
  seafood and society, **2:**81
  See also Mariculture
Food security, **2:82–85**
  food shortage prevention, **2:**83–84
  global warming, **2:**84
  grain production, **2:**83, *83f*
  Green Revolution, **2:**82–83
  irrigated agriculture, **2:**84–85, *84f*
  land and wetland drainage, **2:**84
  land management, **2:**84
  water conservation, **2:**84
  water management, **2:**83
  water pollution, **2:**85
Food web
  defined, **1:**78, **2:**8, **3:**123, and Glossary
  hydrothermal vents, **2:**14
  lakes, **2:**252–253, 260–261, *261f*
  marine, **2:**12, *13f*
  phytoplankton, **3:**187
  stream, **2:**8–9
  tidal pools, **2:**14
Foraminifera, **2:**203, 204
Foreshore, **1:**72–73
Forest hydrology, **2:85–89**
  aquatic biodiversity, **2:**87
  effects of acid rain, **1:**8
  effects of fires, **2:**87, 88, 89
  effects of roads, **2:**88–89
  forest management and watershed quality, **2:**87–89
  forests and hydrologic cycle, **2:**86–87
  riparian areas, **2:**87
  water and forested ecosystems, **2:**85–87
Formosa Mine (OR), **1:**1–2
Fossil, defined, **1:**61, **3:**35, and Glossary
Fossil fuel, defined, **1:**5, **2:**113, **3:**64, and Glossary
Fractionation, **2:**203–204

Fracture, defined, **2**:5, **4**:119, and Glossary
Fracture zones, mid-ocean ridges, **3**:84–85
Fram, **2**:208, **3**:111
Frazil ice, **2**:206, **3**:169–170
Freedom of the Seas doctrine, **3**:24
Freezing of sea water. *See* Sea water, freezing of
Frequency analysis (flood flow model), **3**:105–106
Fresh water, natural composition of, **2:89–94**
    chemical controls, **2**:90–91, 91*t*
    dissolution, **2**:90
    groundwater, **2**:92–94
    natural acidity, **2**:90
    oxidizing-reducing reactions, **2**:91
    saline groundwater, **2**:92–94
    streams and lakes, **2**:91–92
Fresh water, natural contaminants in, **2:94–99**
    aesthetic quality, **2**:94–97
    appearance, **2**:96–97
    arsenic, **2**:97–98
    color, **2**:95, **4**:89–*91*, 93
    hardness, **2**:95
    natural chemical constituents of health concern, **2**:97–99
    odor, **2**:96
    radon, **2**:98–99
    selenium, fluoride, and boron, **2**:99
    taste, **2**:96
Fresh water, physics and chemistry of, **2:100–103**
    density, **2**:102–103
    dissolved materials, **2**:103
    dissolving capacity, **2**:101
    distinct states, **2**:101–*102*, 103
    electrical conductivity, **2**:100, 101
    energy, **2**:102, 103
    pH, **2**:101
    surface tension, **2**:100–101
Fresh water, tracers in. *See* Tracers in fresh water
Fresh water and the senses. *See* Senses, fresh water and the
Fresh water minerals. *See* Mineral resources from fresh water
Fresh-water algae. *See* Phytoplankton
Fresh-water careers. *See* Careers in fresh-water chemistry; Careers in fresh-water ecology
Fresh-water ecology. *See* Ecology, fresh-water
Fresh-water fisheries. *See* Fisheries, fresh-water

Frogs
    African clawed, **3**:212
    bullfrog, **3**:212
    Cascades, **1**:29, *30*
Front (atmospheric), defined, **2**:127, and Glossary
Front (ocean), defined, **3**:147, and Glossary
Fugitive resource, defined, **1**:201, and Glossary
Fujiwhara effect, **4**:*233*
Fulton, Robert, **4**:139
Fumarole, defined, **2**:165, and Glossary
Fundy, Bay of, **4**:*169*, 170–171
Fungi, **1**:22

G

Gagnan, Émile, **1**:222
Gallatin Report, **1**:47
Game fish, defined, **1**:144, and Glossary
Gamete, defined, **1**:19, **3**:189, and Glossary
Ganges River, **4**:42, 60
Gangue, defined, **1**:1, and Glossary
Garbage. *See* Pollution of streams by garbage and trash; Pollution of the ocean by plastic and trash
Garrels, Robert, **2:103–104**
Gas hydrate, **3**:181
Gases in sea water. *See* Sea water, gases in
Gastroenteritis, **2**:175, 181*t*, 183, 185
Gastropod, defined, **2**:171, and Glossary
GCP (Global Carbon Program), **3**:132
Genetic diversity, **1**:77
Geodes, **2**:153
Geodetic, defined, **4**:75, and Glossary
Geographic Information Systems (GIS), **1**:45, 120, **2**:109–111, **3**:10
Geography, defined, **1**:236, and Glossary
Geoid, **4**:72
Geologic cross-sections, **2**:192, 193, 193*f*
Geologic mapping, **2**:159–160, 192
Geologic unit, defined, **2**:191, and Glossary
Geological oceanography. *See* Oceanography, geological
Geological Survey, U.S., **2:105–106**
    historical overview, **2**:105–106
    nonindigenous fish species list, **3**:211

    Water Resources Division, **3**:37–38
Geology, defined, **1**:236, **4**:111, and Glossary
Geomorphology, defined, **1**:233, **2**:7, **3**:38, **4**:129, and Glossary
Geophysical, defined, **2**:173, and Glossary
Geophysical data, in hydrogeologic mapping, **2**:192
Geospatial technologies, **2:106–111**
    Geographic Information Systems (GIS), **1**:45, 120, **2**:109–111, **3**:10
    Global Positioning System (GPS), **1**:45, 120, **2**:109, *110*, **3**:122, 204
    remote sensing, **2**:106–109
Geospatial technology careers. *See* Careers in geospatial technologies
Geostrophic current, **3**:140
    defined, **3**:167, and Glossary
Geothermal, defined, **3**:92, **4**:114, and Glossary
Geothermal energy, **2:111–114**
    direct use, **2**:112
    electricity generation, **2**:112–113
    environmental considerations, **2**:113–114
    first plants, **2**:112
    heat pumps, **2**:112
    using, **2**:112–114
Geyser
    defined, **2**:111, **3**:43, and Glossary
    See also Hot springs and geysers
Giardia, **2**:181*t*, 183, **3**:77
Giardiasis, **2**:181*t*, 185, **3**:82
Gill net, defined, **3**:62, and Glossary
GIS (Geographic Information Systems), **1**:45, 120, **2**:109–111, **3**:10
Glacial periods, **1**:178
Glaciation, **1**:12
    defined, **2**:191, and Glossary
Glacier, defined, **1**:12, **2**:3, **3**:22, **4**:144, and Glossary
Glaciers, ice sheets, and climate change, **2:118–123**
    Antarctic ice shelf breakups, **2**:120–122
    changes in glaciers, ice sheets, and ice shelves, **2**:119–122
    modern glacier retreat, **2**:120
    records of climate change, **2**:122
    See also Global warming and glaciers
Glaciers and ice sheets, **2:115–118**
    accumulation and ablation, **2**:116

Glaciers and ice sheets (continued)
 calving glacier, **2:**115, 208
 characteristics, **2:**116–117
 Columbia Glacier (AK), **2:**119
 deposits, **2:**116–117
 as erosive force, **2:**33
 and estuary formation, **2:**38
 evidence of ice flow, **2:**117–118
 and lake formation, **2:**247, 249
 Mendenhall Glacier (AK), **2:**117
 movement, **2:**116
 temperature, **2:**116
Global Carbon Program (GCP), **3:**132
Global cycles, **1:**58–59
Global Ocean Conveyor Belt, **2:**5
Global Ocean Ecosystems Dynamics (GLOBEC), **3:**108, 132
Global Ocean Observing System (GOOS), **3:**108
Global Positioning System (GPS), **1:**45, 120, **2:**109, 110, **3:**122, 204
Global warming
 defined, **1:**188, and Glossary
 food security, **2:**84
 high latitudes, **2:**207
 human health, **2:**178–179
 long-term streamflow changes, **4:**136
Global warming: policy-making, **2:134–137**
 national-level efforts, **2:**136
 obstacles, **2:**134–135
 policy instruments, **2:**135–136
Global warming and glaciers, **2:123–125**
 See also Glaciers, ice sheets, and climate change
Global warming and the hydrologic cycle, **2:126–129**
 consequences of global warming, **2:**127–129
 evidence of hydrologic changes, **2:**126–127
Global warming and the ocean, **1:**178–179, 187, **2:129–134**
 carbon dioxide, **2:**130–131, 130f, 131f, 132, 132f
 and coral reefs, **1:**218
 effects of global warming, **2:**132–133
 factors in global warming, **2:**129–131
 ocean water and temperature, **2:**131
 paleoclimatology, **2:**131–132, 132f
Globalization and water, **2:137–141**
 globalization challenge, **2:**140–141

globalization trends, **2:**137–138
 impacts on environment and water resources, **2:**138–140
 public policy, **2:**138
 water issues, **2:**139–140
GLOBEC (Global Ocean Ecosystems Dynamics), **3:**108, 132
Gold, **3:**87, 88, 91, **4:**1
Golden Gate Bridge (CA), **1:**91–92
GOOS (Global Ocean Observing System), **3:**108
Gossan, **1:**3
GPS (Global Positioning System), **1:**45, 120, **2:**109, 110, **3:**122, 204
Grab sample, defined, **1:**146, and Glossary
Gradient
 defined, **2:**24, **3:**142, **4:**131, and Glossary
 horizontal pressure, **3:**140, 142
Grain production, and food security, **2:**83, 83f
Grand Coulee Dam (WA), **1:**98, 196, **2:**189, 190
Granular activated carbon, **3:**221
Gravel, **3:**86, 90
Gravity flow, defined, **2:**269, and Glossary
Grease ice, **2:**206
Great Lakes, **2:141–149**
 diversion of waters, **2:**146–147, 148
 environmental degradation, **2:**145–146
 fisheries, **2:**56, 58, 59, 60, 144–145
 future, **2:**147–148
 International Joint Commission, **4:**181
 Lake Erie, **2:**146
 Lake Superior, **2:**143, 146, 147
 lakes, **2:**142–143
 map, **2:**142
 pollution by invasive species, **3:**212, 213, 214, 216
 recreation, **2:**143–144
 transportation, **2:**143
 water levels, **2:**146
 watershed, **2:**142–143
Great Salt Lake (UT), **2:**247
Green Book, **3:**193
Green Revolution, **2:**82–83
Green River (UT), **4:**90
"Green water," **2:**200, 201
Greenhouse effect, defined, **1:**182, **2:**129, **3:**131, **4:**99, and Glossary
Greenhouse gas, **2:**129–131, 130f, 130t, 131f

defined, **1:**108, **2:**24, **3:**47, **4:**136, and Glossary
Greenland ice cap, **3:**110–111
Greywater, **4:**34
 defined, **4:**28, and Glossary
Groins, for erosion control, **1:**75
Groundfish, defined, **2:**65, and Glossary
Groundwater, **2:149–156**
 agates, geodes, and petrified wood, **2:**153
 aquifers, **2:**150, 151–152
 baseflow, **4:**126, 126f
 carbon isotopes, **2:**242
 contamination, **3:**1–3, 2f, 175
 defined, **1:**14, **2:**5, **3:**1, **4:**17, and Glossary
 drinking-water treatment, **1:**257–258
 fecal contamination, **3:**73–75
 flow rate, **2:**153
 fresh water composition, **2:**92–94
 geologic controls on flow, **2:**191
 hydrologic cycle, **2:**195–196
 isotopes, **2:**241
 land use, **2:**152
 movement, **1:**40, **2:**152–154, 152f
 origin and occurrence, **2:**149–152, 149f, 151f
 point sources of pollution, **3:**244
 public and domestic water supplies, **4:**148
 pumping and overpumping, **2:**154–156
 recharge, **2:**151f, 153–154
 saline, **2:**92–94
 septic system impacts, **4:**96
 subsidence, **2:**156
 as supplementary supply in California, **1:**100
 well interference, **2:**154–156, 154f, 155f
 zones, **2:**150
Groundwater, age of, **2:157–158**
 in millennia, **2:**157–158
 in years, **2:**157
Groundwater, microbes in. *See* Microbes in groundwater
Groundwater flow and transport, modeling. *See* Modeling groundwater flow and transport
Groundwater hydraulic head, **3:**103
Groundwater mining, **2:**156
 defined, **3:**174, and Glossary
Groundwater pollution. *See* Pollution of groundwater
Groundwater Rule, **3:**75

Groundwater supplies, exploration for, **2:158–162**
 geophysical techniques, 2:160–162
 remote sensing, 2:160
 test wells, existing wells, and geologic mapping, 2:158–160
Growlers, **2:**208
Guantánamo Bay (Cuba), **1:**69
Gulf of Mexico
 Dead Zone, **1:**69–70, 188, **3:**99f, 101, 238
 salt domes, **1:**96
 water circulation, **3:**57
Gulf Stream
 flow rate, **4:**160
 and ice ages, **2:**205
 map, **3:**138, 139f
 satellite image, **2:**108
Gulfs, **1:**69–70
 See also Bays, gulfs, and straits
Guyot, **3:**150
Gymnodinium breve, **1:**19
Gypsum, **3:**87–88, 90
Gyre, **3:**140–141, **4:**169
 defined, **2:**5, **3:**147, and Glossary
Gyroscopic compass, **3:**122

H

Habitat
 critical, **2:**22
 defined, **1:**2, **2:**11, **3:**5, **4:**40, and Glossary
 fish, **2:**48
 flowing-water, **3:**51
 microbial, **3:**72–73
 specialized, **2:**10
 still-water, **2:**7, **3:**50
Habitat alteration
 fresh-water fisheries, **2:**58
 salmon decline and recovery, **4:**70
 as threat to ocean's health, **3:**145
Hadley cells, **1:**176, 176f, **4:**231
Half-life, defined, **1:**162, **2:**158, and Glossary
Halite, **3:**87
Halley's Comet, **1:***199*
Halocline, **4:**76
Harbors. *See* Ports and harbors
Hard water, defined, **1:**257, and Glossary
Hardness, of fresh water, **2:**95, **4:**92–93
Harmful Algal Bloom and Hypoxia Research and Control Act, **1:**20

Harmful algal blooms. *See* Algal blooms, harmful
Hawaii, water markets in, **3:**67
Hazardous waste, defined, **1:**62, **3:**11, and Glossary
Headwaters, defined, **1:**190, **2:**8, **4:**64, and Glossary
Heat distillation, and desalinization, **1:**239–240
Heat pumps, **2:**112
Heat transfer, **1:**175–176, **4:**231–232
Heavy metals, defined, **3:**12, **4:**32, and Glossary
Hectare, defined, **1:**34, and Glossary
Helminths, **2:**181–182, 181t
Hem, John D., **2:162–163**
Hepatitis A, **2:**176, 181t, 183, **3:**73, 82
Herbaceous, defined, **4:**242, and Glossary
Herbicides, **1:**150–151
 defined, **1:**151, **3:**218, and Glossary
 See also Chemicals from agriculture; Pesticides
Herbivore, defined, **1:**218, **2:**9, **3:**62, and Glossary
Herpetology, defined, **1:**119, and Glossary
Heterosigma, **1:**19–20
Heterotroph, defined, **3:**188, and Glossary
High plains, **4:**244
High Plains Aquifer. *See* Ogallala Aquifer
Hohokam irrigation system, **2:**234–235
Holistic, defined, **4:**248, and Glossary
Holland, John, **4:***140*
Holoplankton, **3:**49
Hoover Dam, **2:163–165**
 base surface, **2:**163–164
 benefits and impacts, **2:**165
 photograph, **2:***164*
 pouring concrete, **2:**164
 reasons for building, **2:**164
 river basin planning, **4:**55
 See also Colorado River; Lake Mead reservoir
Hormones, in water and sewage, **1:**157, 162
Hot spots, **2:**167, **4:**204
Hot spring, **2:***93*, **3:***45*
 defined, **2:**111, **4:**114, and Glossary
Hot springs and geysers, **2:165–169**
 automobile "geysers," **2:**168
 geyser eruptions, **2:**168

 heat source, **2:**167
 plate tectonics role, **2:**167
 underground water circulation system, **2:**167
 water supply, **2:**166–167
Hot springs on the ocean floor, **2:169–173**
 chemosynthesis, **2:**170–171
 chimneys, black smokers, and mineral deposits, **2:**170, **3:**85, 91
 colonization, **2:**172–173
 as marine ecosystem, **2:**14
 microbes, **2:**171
 multicellular organisms, **3:**47
 tubeworms, **2:**171, *172*
 vent biocommunities, **2:**171–173
 vent circulation and chemistry, **2:**169–170
Hubbert, Marion King, **2:173–174**
Human Adjustments to Floods (White), **4:**248
Human health and the ocean, **2:174–180**
 animals, **2:**178
 changes in ocean currents and climate, **2:**178–179
 drug benefits from sea, **2:**175
 harmful algal blooms, **2:**175
 inorganic and organic wastes, **2:**175–176
 microbes, **2:**176–177, **3:**82–83
 pollution by sewage, nutrients, and chemicals, **2:**176, **3:**239
 seafood, **2:**70, 175, 176
 storms and rough seas, **2:**178, *179*
Human health and water, **2:180–186**
 acid rain effects, **1:**8–9
 bacteria, **2:**180
 developing countries, **1:**246–247, 254–255
 global availability of safe water, **2:**185
 helminths, **2:**181–182
 microbes, **2:**180–183, 181t, **3:**77
 mosquito-borne diseases, **2:**184
 protozoa, **2:**181
 radioactivity found in typical adult human body, **4:**21–22
 viruses, **2:**181
Humidity, **1:**180–182, **4:**105
Hunley, **4:**139
Hurricanes, **3:**148, **4:**230, 232, 232f, 233
Husbandry, defined, **1:**31, and Glossary
Hutton, James, **2:**104, **186–187**

"Hydraulic civilizations," 2:228
Hydraulic gradient, defined, 1:39, 3:12, and Glossary
Hydraulic head, defined, 1:227, and Glossary
Hydraulic ram pump, 4:15–16
Hydraulics
   defined, 3:34, and Glossary
   Leonardo da Vinci, 3:35, *36*
Hydrocarbon, defined, 3:176, and Glossary
Hydrochloric acid, 4:91
Hydrocorals, 1:213
Hydroelectric, defined, 1:65, 2:24, 3:29, 4:197, and Glossary
Hydroelectric power, **2:187–191**
   early twentieth century, 2:188
   function, 2:190
   impacts and trends, 2:190–191
   Tennessee Valley Authority, 4:163
   today, 2:189
   See also Dams
Hydrogen bond, defined, 1:58, 2:100, and Glossary
Hydrogen isotopes, 2:241
Hydrogen sulfide gas, in groundwater, 4:92
Hydrogeologic mapping, **2:191–194**
   data sources, 2:192–193
   depicting hydrogeologic data, 2:193–194
   geologic controls on groundwater flow, 2:191
   geologic cross-sections, 2:192, 193, 193*f*
   geologic maps, 2:192
   geophysical data, 2:192
   isopach mapping, 2:193–194
   structure contour mapping, 2:194
   well logs, 2:192–193
Hydrogeologist, defined, 4:164, and Glossary
Hydrogeology, defined, 1:51, 2:191, and Glossary
Hydrograph, 4:126*f*
Hydrolab, 4:141, *142*
Hydrologic address, 2:32
Hydrologic budget, 3:106
Hydrologic cycle, 1:58–59, **2:194–198**
   defined, 1:201, 2:5, 3:7, 4:29, and Glossary
   focus of water managers, 2:197
   forest hydrology, 2:86–87
   irrigation impacts, 2:218
   isotopes, 2:241
   movement of water, 2:194–197, 195*f*

precipitation, 2:194–195
   surface water and groundwater, 2:195–196
   uses of water, 4:192–193
   water movement on Earth, 2:5
   water temperature effects, 3:261–262
   water vapor, 2:195
   watershed concept, 2:196–197
   See also Global warming and the hydrologic cycle
Hydrologic Unit Codes, 2:32
Hydrology, defined, 1:120, 2:162, 3:1, 4:19, and Glossary
Hydrology careers. *See* Careers in hydrology
Hydrolysis, defined, 4:235, and Glossary
Hydropolitics, **2:198–199**, 225
Hydroponics, 1:15–*16*
   defined, 4:106, and Glossary
Hydropower
   defined, 1:47, 2:50, 4:162, and Glossary
   See also Hydroelectric power
Hydrosolidarity, **2:200–202**
   interconnecting factors, 2:200–201
   role of institutions, 2:201–202
Hydrosovereignty, 2:200
Hydrosphere, defined, 1:200, and Glossary
Hydrostatic, defined, 2:265, 3:43, 4:78, and Glossary
Hydrothermal alteration zone, defined, 1:2, and Glossary
Hydrothermal processes, 3:133, 137
Hydrothermal solutions, 2:7
Hydrothermal vents. *See* Hot springs on the ocean floor
Hypersaline, defined, 2:247, and Glossary
Hypolimnion
   defined, 2:263, 4:94, and Glossary
   lakes, 2:262, 267, 268, 3:50
Hyporheic zone of a stream. *See* Stream, hyporheic zone of a
Hypothesis, defined, 3:35, and Glossary
Hypoxia, defined, 2:268, 3:101, and Glossary

I

Ice, pure, 2:211
Ice Age, 1:12, 2:*236*
Ice age, defined, 1:89, 2:237, and Glossary

Ice ages, **2:202–206**
   climate mechanisms, 2:204–205
   island formation, 2:237
   isotope applications, 2:203–204
   Pleistocene glacial cycles, 2:202–204
Ice at sea, **2:206–210**, 4:76
   Arctic and Antarctic sea ice, 2:206–207
   and climate change, 2:207–208, 4:74
   as erosive force, 2:33
   formation, 4:76–77
   iceberg harvesting, 2:208–210
   movement of sea ice and icebergs, 2:208
   and ocean brine formation, 1:95
   polar oceans, 3:*168*, 169–170
Ice cap
   defined, 2:118, 3:111, and Glossary
   See also Glaciers, ice sheets, and climate change; Glaciers and ice sheets
Ice cores and ancient climatic conditions, **2:210–213**
   core dating, 2:212–213
   ice cores, 2:*212*–213
   Lake Vostok, 2:248
   pure ice, 2:211
   snow and ice signals, 2:210–211
   snow impurities, 2:210–211
   spaces between snow grains, 2:211
Ice sculpting, 1:54
Ice sheets
   defined, 2:118, and Glossary
   See also Glaciers, ice sheets, and climate change; Glaciers and ice sheets
Ice shelf, 2:*204*, 209, 210
   defined, 2:208, and Glossary
Icebergs
   density, 2:102
   harvesting, 2:208–210
   movement, 2:208
   tourism, 2:*209*
   towing, 1:242
Igneous, defined, 1:40, 2:91, and Glossary
Iguanas, marine, 4:44–45
Iguaza Falls (South America), 4:*215*
Impellers (pumps), 4:9
"Incidental take permits," 2:23
"Inclined-block" water rate schedule, 3:263, 263*f*
Indian Ocean basin, 3:128–129
Individual quota, 2:64, 67–69

defined, **2**:67, and Glossary
Indus River (Asia), **4**:60
Industrial Revolution
    canals, **1**:103–105
    dams, **2**:187–188
    defined, **2**:28, **4**:136, and Glossary
    environmental movement, **2**:28
Industrial wastes, **3**:239–240
Industrial water use, **4**:34, 193, 195, 196
Infiltration, defined, **1**:51, and Glossary
Infiltration rate, defined, **1**:244, and Glossary
Infrastructure, defined, **1**:207, **3**:237, **4**:4, and Glossary
Infrastructure, water-supply, **2**:213–217
    components, **2**:213–214, 214f, 216
    leaks and breaks, **2**:214
    maintenance and safety, **2**:214–215, **4**:199
    needs, **2**:215
    as pollution source, **4**:144–145
    security, **2**:215–216
Injection wells, **1**:50–51
Injurious species, **3**:210
In-lake management, **2**:258
Inorganic, defined, **1**:26, **2**:12, **3**:132, **4**:66, and Glossary
Inorganic wastes, **2**:175–176
Insecticides, **1**:29, 151
    See also Chemicals from agriculture; Pesticides
Insolation, **2**:204–205
    defined, **3**:71, and Glossary
Institution, defined, **3**:199, and Glossary
Instream flow, defined, **3**:27, **4**:3, and Glossary
Instream water issues, **2**:217–220
    dam operations, **2**:218, *219*
    importance of instream flows, **2**:217–218
    integrity of instream flows, **2**:219–220
    mining, **3**:86
    water rights and protection, **2**:218–220
    water use, **4**:151, 192f
    withdrawals and diversions, **2**:218
Integrated, defined, **2**:220, and Glossary
Integrated resource planning, **3**:198–200
Integrated water resources management, **2**:220–223

    challenges, **2**:222
    components and viewpoints, **2**:220–221, 221f
    coordination and cooperation, **2**:222–223
    geographic regions, **2**:222
    governmental and interest groups, **2**:222
    interdisciplinary perspectives, **2**:222
    multiple purposes, **2**:221
    Total Water Management, **2**:222–223
Interests
    defined, **1**:201, **4**:5, and Glossary
    See also Balancing diverse interests
Intergenerational equity, defined, **4**:158, and Glossary
Interglacial periods, **1**:178, **2**:205
Intergovernmental, defined, **3**:33, and Glossary
Intergovernmental Conference on the Convention on the Dumping of Wastes at Sea, **3**:241
International Convention for the Prevention of Pollution from Ships (MARPOL), **3**:235, 241
International cooperation, **2**:223–227
    1997 United Nations Convention, **2**:226
    hydropolitics, **2**:225
    international agreements, **2**:226
    irrigation management, **2**:231–232, 231t
    shared watercourses, **2**:224–225
    treaties, **2**:225
    White, Gilbert, **4**:248–249
    World Bank, **2**:226
International Court of Justice, **3**:21, **4**:180
International Joint Commission, **2**:145, 148, **4**:181
International maritime fleets, **4**:186–187
International river basins, **2**:198, 199f, **3**:18
International Seabed Authority, **3**:25
International Tribunal for the Law of the Sea, **3**:25
International water law. *See* Law, international water
International water resources careers. *See* Careers in international water resources
International Whaling Commission (IWC), **3**:65
Interpolate, defined, **2**:193, and Glossary

Interstate, defined, **3**:32, and Glossary
Interstellar, defined, **3**:152, and Glossary
Intertidal, defined, **1**:72, **4**:45, and Glossary
Intertidal zone, **3**:50
Intragenerational equity, defined, **4**:158, and Glossary
Intrastate, defined, **3**:32, and Glossary
Invasive species, pollution by. *See* Pollution by invasive species
Ion, defined, **1**:29, **2**:7, **4**:80, and Glossary
Ion exchange, **1**:240
Ion extraction, **1**:240
Iron bacteria, in groundwater, **4**:92
"Iron hypothesis," and carbon removal, **1**:111
Irrigation
    agriculture and water, **1**:13–14
    Bureau of Reclamation, U.S., **1**:97–99
    center-pivot sprinklers, **3**:*175*
    defined, **1**:47, **2**:27, **3**:3, **4**:9, and Glossary
    developing countries, **1**:248
    food security, **2**:84–85, 84f
    hours ordinances, **1**:208
    impact on hydrologic cycle, **2**:218
    leading countries, **1**:14t
    Nile River, **2**:*224*
    Ogallala Aquifer, **3**:174–175
    supplemental, **2**:84
    as water use category, **4**:193–194, 195, 196
Irrigation management, **2**:227–232
    canals, **2**:*227*
    developed countries, **2**:228
    developing countries, **2**:229–230
    help by international agencies, **2**:231–232, 231t
    poverty, **2**:230
    small land holdings, **2**:229
    Water User Associations, **2**:230–231
Irrigation Survey, U.S., **3**:255
Irrigation systems, ancient, **2**:232–235
    Egypt and Mesopotamia, **2**:232–234
    North America: Chaco and Hohokam systems, **2**:234–235
    prehistoric Mexico, **2**:234
Islands, capes, and peninsulas, **2**:235–239

Islands, capes, and peninsulas
(continued)
capes, **2:**236–237
islands, **2:**237, 237t
peninsulas, **2:**237–239
Isopach, defined, **2:**193, and Glossary
Isopach mapping, **2:**193–194
Isostasy, defined, **3:**111, and Glossary
Isotope, defined, **1:**63, **2:**99, **4:**17, and Glossary
Isotopes: applications in natural waters, **2:239–243**
applications to hydrology, **2:**241–243
carbon isotopes in groundwater, **2:**242
delta values, **2:**240–241
hydrogen isotopes, **2:**241
hydrologic cycle, **2:**241
ice ages, **2:**203–204
isotopes, **2:**211, 239–241
mass spectrometry, **2:**242
nitrogen isotopes, **2:**242–243
origin of groundwater, **2:**241
oxygen isotopes, **2:**240, 241–242, **4:**176–177
paleotemperature, **2:**241–242
IWC (International Whaling Commission), **3:**65

J

Jefferson, Thomas, **3:**39, 40–41
Jellyfish, **2:**178, **3:**49, 189
Jetty
defined, **3:**235, and Glossary
for erosion control, **1:**75
Jökulhlaup, **4:**203
Jordan River, **2:**198, 199
Jovian, defined, **4:**100, and Glossary
Juan de Fuca Ridge, **4:**205
Junior appropriator, defined, **4:**2, and Glossary

K

Kankakee River (IN), **4:***128*
Karst, defined, **1:**140, and Glossary
Karst hydrology, **2:243–246**
disappearing streams, **2:**244–245
hydrologic considerations, **2:**245–246
karst features, **2:**243–245, 244f
karst springs, **2:**245
sinkholes and caverns, **2:**243–244, *245*
vs. surface streams, **2:**246
unknown flow paths, **2:**246

Kayaking, **4:**106–107, 174–175, *186*
Keiko (killer whale), **3:***61*
Kelimutu Volcano (Indonesia), **4:***93*
Kelp forests, **2:**14
See also Seaweed
Kelvin waves, **4:**229
See also Waves
Keystone species, defined, **1:**217, and Glossary
Kilimanjaro, Mount, **2:**120
Killer whales, **3:***61*, 62
Kilowatt, defined, **2:**188, and Glossary
Kingdoms, **1:**22
Krill, **1:**226, **2:***11*, 14
Kudzu, **4:**218
Kuiper Belt, defined, **1:**199, and Glossary
Kyoto Protocol, **1:**109, **2:**134, *135*

L

La Niña. See El Niño and La Niña
Labrador Sea, **3:**143
Lagoons, **1:**185
Lake formation, **2:247–251**
changing lakes, **2:**250
glacial activity, **2:**247, 249
natural processes leading to, **2:**247, 249–250
tectonic activity, **2:**247
volcanic activity, **2:**249
Lake health, assessing, **2:251–254**
assessment factors, **2:**251–253
decomposition, **2:**252
ecosystem functions, **2:**251–252
ecosystem structure, **2:**252–253
external inputs, **2:**253
internal processes, **2:**253
management goals and societal values, **2:**253
production, **2:**251
Lake management issues, **2:254–258**
eutrophication, **2:**254–256, 255f
exotic species, **2:**256–258
in-lake management, **2:**258
overuse, **2:**256, *257*
Lake Mead reservoir, **4:**47, *48*
See also Hoover Dam
Lake pollution. See Pollution of lakes and streams
Lake Ponchartrain Causeway, **1:***92*
Lakes
eutrophic, **2:**251, **3:**50
fresh-water, **2:**9–10, 91–92
humic, **2:**264
hypereutrophic, **2:**251

mesotrophic, **2:**255f
oligotrophic, **2:**255f, **3:**50
oxidation-reduction chemistry, **2:**262
photosynthesis-respiration process, **2:**262
precipitation, global distribution of, **3:**261–262
volcanic, **2:**264
world's deepest, **2:**248t
world's largest, **2:**248t
See also specific lakes
Lakes: biological processes, **2:259–262**
adaptations, **2:**260
aquatic communities, **2:**259–260, 260f
food web, **2:**260–261, 261f
light, **2:**259
Lakes: chemical processes, **2:262–267**
acidification, **2:**264
biotic influences, **2:**262
exploding lakes, **2:**265–266
limiting nutrient concept, **2:**264–265
pH and salinity, **2:**263–264
sediment-water interactions, **2:**265
Lakes, microbes in. See Microbes in lakes and streams
Lakes, nutrients in. See Nutrients in lakes and streams
Lakes: physical processes, **2:267–270**
oxygen levels, **2:**268–269
sedimentation, **2:**269
thermal processes, **2:**267–268, 269
Land breeze, **1:**182
Land use and water quality, **3:1–7**
bacteria, **3:**3
Concentrated Animal Feeding Operations, **3:**4
emerging contaminants, **3:**6
groundwater contamination and land use, **2:**152, **3:**1–3, *2f*
nitrate, **3:**2–3, 4
pesticides, **3:**3
pollution of groundwater, **3:**224
pollution sources and contaminant pathways, **3:**4–5
road salt, **3:**3
sediment, **3:**5
surface-water contamination and land use, **3:**4–6
volatile organic compounds, **3:**3
wastewater discharges, **3:**5–6

Landfills: impact on groundwater, **3:11–14**
   ash landfills, **3:**12
   design standards, **3:**11, 13–14
   dry-tomb landfills, **3:**14
   hazardous waste landfills, **3:**11
   leachate generation and composition, **3:**11–12, 12*t*
   leachate release and migration, **3:**12–13
   municipal solid waste landfills, **3:**11, 12
   solid waste landfills, **3:**11
   wood-waste landfills, **3:**12
Landform
   defined, **1:**89, **3:**38, and Glossary
   and tides, **4:**169–170
Landsats, **2:**107
Landscape development and stream erosion. *See* Stream erosion and landscape development
Landslides, **3:14–17**
   classification, **3:**16–17, 17*f*
   creep, **3:**17
   debris avalanches, **3:**16
   dip slope, **3:**16
   driving forces, **3:**15–16
   El Salvador, **3:***15*
   factors, **3:**15–16
   flows, **3:**16
   recognizing landslide terrains, **3:**17
   resisting forces, **3:**15–16
   vs. sinkholes, **2:**244
   slope and materials, **3:**16
   vegetation as resisting force, **3:**16
   water as driving force, **3:**15–16
Land-use planning, **3:7–11**
   coastal mountain watershed, **3:***9f*
   defined, **3:**32, and Glossary
   described, **3:**8–10
   future, **3:**10
   Geographic Information Systems (GIS) and natural resource management, **3:**10
   impacts and benefits of land uses, **3:**7–8
   levels and process, **3:**10
   watershed perspective, **3:**7
Larsen B Ice Shelf, **2:**120–*121*
Latent heat vaporization, defined, **1:**175, and Glossary
Latitude
   defined, **3:**120, and Glossary
   and water circulation, **3:**259–260
Latrine, **4:**222
Laurentide Ice Sheet, **2:**202–203

Lava
   defined, **2:**167, and Glossary
   pillow, **4:***204*
Law, international water, **3:18–23**
   customary, **3:**18–20, 22–23
   equitable sharing of waters, **3:**21–22, 23
   future challenges, **3:**22–23
   maritime fleets, **4:**186–187
   Nile River as case study, **3:**22–23
   no-harm rule, **3:**20–21, 22
   priority of use, **3:**22–23
   riparian nations, **3:**19–20
   United Nations Convention, **3:**20–22
   water and potential for conflict, **3:**18
Law, water, **3:26–28**
   groundwater and surface-water rights, **3:**27–28
   from private rights to public values, **3:**26–27
   relationship between rights, **3:**27
   resolving water-use conflicts, **3:**26–27
   water rights and hydrologic cycle, **3:**27–28
Law of the Minimum, **2:**264–265
Law of the Sea, **3:24–26**
   Exclusive Economic Zone (EEZ), **3:**25
   growing conflict over ocean uses, **3:**24
   ocean's future, **3:**26
   United Nations Conference on the Law of the Sea, **3:**24
   United Nations Convention on the Law of the Sea, **3:**25–26
Law of the vendetta, **3:**22
Leachate
   defined, **3:**11, and Glossary
   generation and composition, **3:**11–12, 12*t*
   release and migration, **3:**12–13
Leaks, **1:**209, **2:**214
Least-cost planning, **3:**196–198
Legislation, **1:**65–66
Legislation, federal water, **2:**31–32, **3:28–31**
   development and restoration, **3:**31
   drinking-water standards, **3:**30
   history, **2:**29–30
   navigation and water supply, **3:**29
   water quality, **3:**29–30
   water resources planning, **3:**30
Legislation, state and local water, **3:31–34**

   cooperation of water authorities, **3:**33–34
   regional and local water authority, **3:**32–33
   riparian doctrine *vs.* prior appropriation, **3:**32
Lentic ecosystem, **2:**7
Leonardo da Vinci, **3:34–37**
   Arno canals, **3:**36–37
   brief chronology, **3:**34–35
   hydraulics, **3:**35, *36*
   hypotheses, **3:**35–36
   legacy, **3:**37
   Notebooks, **3:**35–36
Leopold, Luna, **3:37–39**
Levee, defined, **1:**47, **2:**76, **3:**100, **4:**84, and Glossary
Levoli (tanker), **3:***145*
Lewis, Meriwether and William Clark, **3:39–43**
   Corps of Discovery, **3:**40–42
   explorers, **3:**39–40
   homeward bound, **3:**42
   illustration, **3:**40*f*
   route map, **3:**42*f*
Lewis and Clark: The Journey of the Corps of Discovery, 3:41
Lidar (light detection and ranging), **4:***228*
Liebig's Law of the Minimum, **2:**264–265
Life in extreme water environments, **3:43–48**, **3:**44*t*
   aphotic and anoxic environments, **3:**46–47
   chemical extremes, **3:**45–46
   desiccation, **3:**45–46
   implications and benefits, **3:**47–48
   pH, **3:**45
   physical extremes, **3:**43–45
   pressure, **3:**44–45
   radiation, **3:**45
   salinity, **3:**45
   temperature, **3:**43–44
Life in water, **3:48–52**
   benthic species, **3:**50
   challenges of aquatic life, **3:**51–52
   distance from shore, **3:**50
   flotation, **3:**51
   flowing-water habitats, **3:**51
   fresh-water environment, **3:**50–51
   marine environment, **3:**49–50
   pelagic organisms, **3:**49
   salinity, **3:**51
   still-water habitats, **3:**50
   temperature, **3:**52

Light, and lake biological processes, **2**:259
Light transmission in the ocean, **3**:52–54, **4**:82
  light attenuation, **3**:53
  light penetration, **3**:53
  light spectrum, **3**:53
  measuring, **3**:53
  reflection, refraction, and color, **3**:52–53
Limestone, **2**:242, 243, **3**:90
Limiting nutrient concept, **2**:264–265
Literature, water in, **1**:*53*–54
Lithosphere, **3**:203
  defined, **3**:44, and Glossary
Lithospheric plate, defined, **2**:167, **3**:83, and Glossary
Little Ice Age, **2**:119, 205
Littoral transport, defined, **1**:73, and Glossary
Littoral zone, **2**:259, 260f
Livestock water use category, **4**:194, 195
Lizards, **4**:44–45
  See also Reptiles
Load management, defined, **1**:237, **3**:197, and Glossary
Lobsters, **1**:*225*
Local water legislation. *See* Legislation, state and local water
Lock, defined, **1**:103, and Glossary
Logarithm, defined, **2**:101, **4**:126, and Glossary
London Convention, **3**:241
Long-black-spine sea urchin, **1**:217–218
Longitude, **3**:121–122
  defined, **3**:120, and Glossary
Longshore currents, **1**:73
Longshore drift, **1**:184–185
Longshore transport, defined, **1**:72, and Glossary
Loop Current, **3**:57
Loosestrife, purple, **3**:210, 214, *215*
Loran (Long Range Navigation), **3**:122
Lotic ecosystem, **2**:7
Lourdes water, **4**:42
Low Velocity Zone, **2**:6

M

Machiavelli, Niccolo, **3**:36–37
Macroinvertebrates, defined, **2**:8, and Glossary
Macroplankton, **3**:186
Macropore, defined, **1**:244, and Glossary

Magma
  chambers, **2**:169
  characteristics, **4**:200–201
  defined, **2**:169, **3**:203, **4**:200, and Glossary
  viscosity, **4**:201
Magnesium, **3**:90
Magnetic field reversals, **3**:203
Magnetites, **1**:61
Malformations, in amphibians, **1**:30–*31*
Mammals, marine. *See* Marine mammals
Mammoth Cave (KY), **1**:140–141
Management. *See* Planning and management, history of water resources; Planning and management, water resources
Manatees, **3**:*62*–63
Manganese nodules, **3**:90, 154
Mangrove, **4**:*244*
  defined, **2**:45, and Glossary
Mantle
  defined, **1**:200, **2**:6, **3**:83, and Glossary
  plate tectonics, **3**:202, 206
Mapping, geologic, **2**:159–160, 192
Mapping, hydrogeologic. *See* Hydrogeologic mapping
Mapping, isopach, **2**:193–194
Mapping of the ocean floor, **3**:163
Marginal sea, defined, **3**:170, and Glossary
Marginal seas, **3**:54–57
  biomass production and primary productivity, **3**:55–56
  human impacts, **3**:54–55
  water circulation, **3**:56–57
Mariculture, **1**:31, **3**:58–60
  considerations, **3**:59–60
  defined, **2**:79, **3**:240, and Glossary
  types of operations, **3**:58–59
  See also Aquaculture
Marina, defined, **3**:235, and Glossary
Marine ecology. *See* Ecology, marine
Marine fisheries. *See* Fisheries, marine; Fisheries, marine: management and policy
Marine mammals, **3**:60–66
  carnivores, **3**:63
  cetaceans, **3**:61–62
  human connection to, **3**:64–65
  marine otters, **3**:63
  mysticetes, **3**:61
  odontocetes, **3**:61–62
  sirenians, **3**:62–63

Marine protected areas (MPAs), **1**:219
Marine Protection, Research, and Sanctuaries Act, **3**:240–241
Marine snow, **3**:*130*, 131
Marine Transportation System, **4**:185–*186*, 188
Market institution, defined, **3**:67, and Glossary
Markets, water, **3**:66–68
  balancing cost and benefit, **3**:66–67, 66f
  and efficiency, **3**:67
  Hawaii, **3**:67
Marl, defined, **2**:46, and Glossary
MARPOL (International Convention for the Prevention of Pollution from Ships), **3**:235, 241
Mars, exploration for life, **1**:60–62
Mars, water on, **3**:68–71, **4**:100
  early evidence: cold and dry, **3**:69
  evidence in minerals and sedimentary features, **3**:71
  later evidence: geologic and fluvial activity, **3**:69
  ocean on Mars, **3**:70
  other geologic evidence, **3**:70
  reconciling evidence, **3**:71
Marshland, **4**:218
Mass spectrometry, **2**:242
Mauna Kea (HI), **4**:204
Maximum contaminant level, **4**:67, 68
  defined, **1**:152, and Glossary
Maximum sustainable yield, **2**:63–64
Meander, defined, **4**:113, and Glossary
Media, defined, **4**:211, and Glossary
Mediation, **1**:204
Mediator, defined, **1**:205, and Glossary
Mediterranean Sea, **3**:*55*, 56–57
Megaplankton, **3**:186
Megawatt, defined, **1**:196, **2**:189, and Glossary
Mekong River, **2**:57, 60, **4**:60
Mekong River Basin Commission, **4**:*57*, 60
Meltwater, defined, **3**:99, and Glossary
Mercury, water on, **4**:99
Mercury poisoning, **2**:15
Meroplankton, **3**:49
Mesoamerica, defined, **4**:26, and Glossary
Metabolism, **4**:114
  defined, **1**:155, **3**:188, **4**:210, and Glossary

Metabolite, defined, **1**:155, and Glossary
Metamorphic, defined, **1**:40, **2**:4, and Glossary
Metamorphism, **2**:6–7
  defined, **1**:200, **2**:6, and Glossary
Metamorphosis, defined, **1**:29, and Glossary
Metasomatism, **2**:7
Meteoric water, defined, **2**:241, and Glossary
Meteorite
  defined, **1**:198, and Glossary
  impact of, **1**:59f, **2**:250
  Mars, **1**:61, **3**:71
  water in, **1**:200
Meteorology, defined, **1**:236, **2**:212, **3**:106, **4**:160, and Glossary
Methanogens, **3**:47
Methemoglobinemia, defined, **1**:150, **2**:242, **3**:2, and Glossary
Methyl *tert*-butyl ether (MTBE), **3**:3, 221
Metolachlor, **3**:3
Microbe
  defined, **1**:50, **2**:97, **4**:131, and Glossary
  hot springs on the ocean floor, **2**:171
  human health and the ocean, **2**:176–177
  human health and water, **2**:180–183, 181t
  mat-forming, **2**:171
  plume, **2**:171
  symbiotic, **2**:171
Microbes and health. *See* Algal blooms, harmful; Drinking water and society; Human health and water; Microbes in groundwater; Microbes in lakes and streams; Microbes in the ocean; Safe Drinking Water Act; Supplies, protecting public drinking-water
Microbes in groundwater, **3**:72–75
  commonly occurring microbes, **3**:72–73
  disease-causing pathways, **3**:73–74
  microbial habitats, **3**:72–73
  pathogen movement and persistence, **3**:73–74
  protecting public health, **3**:74–75
Microbes in lakes and streams, **3**:75–78
  aerobes and anaerobes, **3**:76–77
  ecological roles of fresh-water microbes, **3**:76–77
  microbes and human health, **3**:77

treating and preventing microbial pollution, **3**:77–78
Microbes in the ocean, **3**:78–83
  animal and human health impacts, **3**:81–83
  Archaea, **3**:80–81
  bacteria, **3**:79–80, *82*
  human pathogens, **3**:82–83
  occurrence and role in the ocean, **3**:79–81
  plankton, **3**:79
  viruses, **3**:79
Microbial film, defined, **4**:211, and Glossary
Microcystins, **1**:23
Microgravity, defined, **4**:105, and Glossary
Microorganism
  defined, **1**:63, **2**:180, **3**:53, **4**:94, and Glossary
  See also "Microbe" entries
Microplankton, **3**:186, 189
Mid-Atlantic Ridge, **3**:84, 84f, 85, 128, 204
Mid-ocean ridges, **3**:83–85, 84f
  fracture zones, **3**:84–85
  hydrothermal processes, **3**:133, 137
  plate tectonics, **3**:204
  ridge types, **3**:83–84
  submarine volcanoes, **4**:204–205
  water and minerals, **3**:85
Mill privilege, **4**:53–*54*
  See also Prior appropriation
Mineral resources from fresh water, **3**:85–88
  acid mine drainage, **1**:2–4
  aggregate minerals, **3**:86
  diatomaceous earth, **3**:86–87
  gold, **3**:87, 88
  heavy minerals, **3**:88
  peat, **3**:87
  salt deposits, **3**:87–88
Mineral resources from the ocean, **3**:88–91
  hot springs on the ocean floor, **2**:170, **3**:85
  limestone and gypsum, **3**:90
  magnesium, **3**:90
  manganese nodules, **3**:90
  metal deposits associated with volcanism and seafloor vents, **3**:91
  phosphorites, **3**:90–91
  placer gold, tin, titanium, and diamonds, **3**:91
  potassium, **3**:89–90
  salt, **3**:*89*

sand and gravel, **3**:90
water, **3**:91
Mineral waters and spas, **3:92–93**
  health benefits, **3**:92–93
  mineral water baths, **3**:92–93
  types of mineral waters, **3**:92
"Minimata Disease," **3**:240
Mining
  gold, **4**:*1*
  gravel, **3**:86
  groundwater, **2**:156
  instream, **3**:86
  off-channel, **3**:86
  sand, **3**:86
  as water use category, **4**:194
Minorities in water sciences, **3:93–98**
  academic focus, **3**:94–95
  institutions, **3**:95
  off-campus programs, **3**:95–98
  organizations of interest, **3**:97, 98
Mississippi River Basin, **3:98–102**
  contemporary management issues, **3**:100–101
  Dead Zone, **1**:69–70, 188, **3**:*99f*, 101, 238
  flood control, **3**:100
  history of river management, **3**:99–100
  major floods, **3**:100–*101*
  map, **3**:99f
  navigation, **3**:99–100
  preventing natural channel change, **3**:100
  river, **3**:99, **4**:61
  runoff, **4**:63
Missoula Floods, **1**:89–91, 90f
Mitigation, defined, **1**:263, **2**:74, **4**:246, and Glossary
Mitigation banking, defined, **3**:193, and Glossary
Mixing, ocean. *See* Ocean mixing
Modeling groundwater flow and transport, **3:102–104**
  input data: flow and transport, **3**:103
  model calibration, **3**:104
  overview of modeling process, **3**:102
Modeling streamflow, **3:104–106**
  flood runoff, **3**:105–106
  methods of analysis, **3**:105–106
  water supply, **3**:106
Models
  solute transport computer model, **4**:130–131
  Stommel model, **3**:142

Mole, defined, **2**:90, **3**:135, and Glossary
Molecular diffusion, defined, **1**:109, and Glossary
Mollusca, **3**:189
Mollusk, defined, **1**:33, **2**:79, **3**:49, and Glossary
Mollusks. *See* Bivalves; Cephalopods
Monera, **1**:22
Mono Lake (CA), **1**:100–101, *102*, **4**:50
Monsoon, defined, **2**:19, **3**:140, **4**:28, and Glossary
Monterosi, Lake (Italy), **2**:255
Moon, water on, **4**:100
Moorings and platforms, **3**:106–110
    equipment and instrumentation, **3**:*107*, 108, *109*
    platforms, **3**:108–110
Moraines, **2**:116–117
Morphology, defined, **3**:70, and Glossary
Mortality, defined, **1**:29, **3**:178, and Glossary
Mosquito-borne diseases, **2**:184
Mountains
    coastal watershed, **3**:9*f*
    fossils and shells on, **3**:35–36
    tallest, **4**:204
    wetlands, **4**:*242*, 243–244
    See also specific mountains
Mousse, defined, **3**:177, and Glossary
MPAs (marine protected areas), **1**:219
MTBE (methyl *tert*-butyl ether), **3**:3, 221
Mudpot, **4**:*111*
    defined, **2**:165, **3**:43, and Glossary
Mudstone slab, **4**:*118*
Muir, John, **2**:29
Multiple-step pump tests, **1**:41–42, 42*f*
Municipal reuse, **4**:33
Museums, natural history, **1**:36–37
Music and song, **1**:55–56
Mussel Watch program, **3**:145
Mussels, **1**:85
    contaminants in, **3**:145
    quagga, **3**:210, 213
    zebra, **3**:100, 210, 211–212, 213, 214, 216
Mustang Island Gulf Beach (TX), **3**:233*f*
Mute swans, **1**:167–168, **3**:*213*, 216
Mysticetes, **3**:61

N

Nanoflagellates, **3**:188
Nanoplankton, **3**:186, 189
Nansen, Fridtjof, **3**:110–112
National Aeronautics and Space Administration (NASA)
    Lake Vostok, **2**:248
    Mars exploration, **1**:61–62
    radar studies, **2**:108–109
National Environmental Policy Act, **3**:112–114
    components and requirements, **3**:113–114
    visionary scope, **3**:114
    water-supply development, **4**:150
National Environmental Satellite, Data, and Information Service (NESDIS), **3**:116
National Flood Insurance Program, **2**:74–75, **3**:33
National Marine Fisheries Service (NMFS), **2**:22–23, **3**:116
National Marine Sanctuary Program, **3**:119
National Ocean Service (NOS), **3**:116
National Oceanic and Atmospheric Administration (NOAA), **3**:114–117
    major divisions, **3**:115–116
    Mussel Watch program, **3**:145
    National Environmental Satellite, Data, and Information Service (NESDIS), **3**:116
    National Marine Fisheries Service (NMFS), **2**:22–23, **3**:116
    National Ocean Service (NOS), **3**:116
    National Weather Service (NWS), **3**:116
    Office of Oceanic and Atmospheric Research (OAR), **3**:116
    Pacific Ocean conditions, **2**:20–21
National Park Service, **3**:117–119
    activities, services, and administration, **3**:117
    floodplains and wetlands, **3**:119
    large and small properties, **3**:117
    shorelines and barrier islands, **3**:119
    water resources, **3**:118–119
    watershed and stream processes, **3**:119
National Pollutant Discharge Elimination System (NPDES), **1**:171, **3**:4, 30, **4**:210
National Primary Drinking Water Regulations (NPDWR), **4**:67
National Resources Inventory, **1**:235
National Secondary Drinking Water Regulations (NSDWR), **4**:68
National Weather Service (NWS), **3**:116
National wildlife refuges, **2**:54–55, **3**:100
Nationwide Survey on Recreation and the Environment (NSRE), **4**:36–37
Native Americans, **1**:195–196, *197*, **2**:142, **4**:41, 69, *197*
Natural flow, **4**:53
    defined, **3**:28, and Glossary
Natural gas, defined, **3**:181, and Glossary
Natural hazard, defined, **1**:260, and Glossary
Natural history museums, **1**:36–37
Naturally occurring radioactive material (NORM), **4**:18
Nautical chart, **3**:120
Nautilus (mollusk), **1**:141, 142, *143*
Nautilus (submarine), **4**:139, 141
Navigable, defined, **1**:47, **2**:85, **3**:28, **4**:3, and Glossary
Navigable waters, defined, **4**:49, and Glossary
Navigation
    federal water legislation, **3**:29
    public water rights, **4**:49–50
Navigation at sea, history of, **3**:119–122
    early navigational tools, **3**:120–122
    modern navigation, **3**:122
    tall ships, **3**:*121*
Nekton, **3**:49
Neritic zone, **3**:50
NESDIS (National Environmental Satellite, Data, and Information Service), **3**:116
Net pens, **1**:*35*, **3**:59
    defined, **1**:20, and Glossary
Neurotransmitter, defined, **1**:157, and Glossary
Neutrons, **2**:239, 240
NGOs (nongovernmental organizations), **4**:6
Niagara Falls, **2**:*144*, **4**:215, 216
Nile River, **4**:60
    international water law, **3**:22–23
    irrigation, **2**:224, 232–233
Nitrate
    and amphibians, **1**:29
    defined, **1**:6, **3**:103, **4**:96, and Glossary
    land use and water quality, **3**:2–3, 4
    ocean biogeochemistry, **3**:133*f*
    in private wells, **4**:240
    septic system impacts, **4**:96

and watershed water quality, **4**:220
Nitrogen
   atmospheric, **3**:126
   isotopes, **2**:242–243
   lake chemical processes, **2**:264
   nutrients in lakes and streams, **3**:123–126
   in sea water, **4**:79
Nitrogen oxides (NO_x) emission reduction program, **1**:9, 10
NMFS (National Marine Fisheries Service), **2**:22–23, **3**:116
NOAA. *See* National Oceanic and Atmospheric Administration (NOAA)
Noble, defined, **4**:18, and Glossary
Noble gas
   defined, **2**:99, and Glossary
   in sea water, **4**:78
No-harm rule, **3**:20–21, 22
Nongovernmental organizations (NGOs), **4**:6
Nonmarket good, defined, **1**:220, and Glossary
Nonpoint source
   defined, **1**:15, **3**:30, and Glossary
   groundwater pollution, **3**:218
   lake management, **2**:254, 255–256
   pollution of lakes and streams, **3**:226–228
   pollution sources, **3**:244–245, 246
   surface-water contamination, **3**:4
   watersheds, **3**:8
   See also Pollution sources: point and nonpoint
Nonpotable reuse, direct, **4**:29–30
Nonylphenol
   defined, **1**:155, and Glossary
   salmon decline and recovery, **4**:70–71
Noria, **4**:13
NORM (naturally occurring radioactive material), **4**:18
North Atlantic bathymetric chart, **3**:149*f*
North Atlantic Deep Water (NADW), **3**:142–143, 168
Norwalk agent, **2**:176, 181*t*, 183
NOS (National Ocean Service), **3**:116
Notebooks (Leonardo da Vinci), **3**:35–36
NPDES (National Pollutant Discharge Elimination System), **1**:171, **3**:4, 30, **4**:210
NPDWR (National Primary Drinking Water Regulations), **4**:67

NSDWR (National Secondary Drinking Water Regulations), **4**:68
NSRE (Nationwide Survey on Recreation and the Environment), **4**:36–37
Nutrient cycling, **2**:252
Nutrients
   defined, **1**:8, **2**:8, **3**:46, **4**:32, and Glossary
   See also Pollution of the ocean by sewage, nutrients, and chemicals
Nutrients in lakes and streams, **3**:123–126
   eutrophication and its impacts, **3**:123–*124*
   sources, **3**:124–125
   stream health assessment, **4**:122
NWS (National Weather Service), **3**:116
Nyos, Lake (Cameroon), **2**:265–266

O

Obligate, defined, **3**:44, and Glossary
Ocean, microbes in the. *See* Microbes in the ocean
Ocean, radionuclides in the. *See* Radionuclides in the ocean
Ocean and global warming. *See* Global warming and the ocean
Ocean and weather. *See* Weather and the ocean
Ocean basins, **3**:126–130, 128*f*
   Arctic Ocean basin, **3**:129–130
   Atlantic Ocean basin, **3**:127–128
   Indian Ocean basin, **3**:128–129
   Pacific Ocean basin, **3**:127
   Southern Ocean, **3**:129
Ocean biogeochemistry, **3**:130–134
   carbon, **3**:132
   cycles, **3**:131–133
   hydrothermal processes, **3**:133
   nitrate, **3**:133*f*
   oxygen, **3**:133, 133*f*
   phosphate, **3**:133*f*
   silicon, **3**:132–133
   silicon dioxide, **3**:133*f*
Ocean chemical processes, **3**:134–138
   addition-removal processes and considerations, **3**:134–138, 135*f*
   aerosols, **3**:136–137
   biological processes, **3**:137
   estuaries, **3**:135–136
   evaporation, **3**:137
   hydrothermal processes, **3**:137

   pore-water interactions, **3**:137–138
   residence time, **3**:135
Ocean currents, **3**:138–144
   Antarctic Circumpolar Current, **3**:141, 167, 168
   beach currents, **1**:73, **2**:*238*
   causes, **3**:138–140
   Coriolis effect, **3**:139, 140, 141–142
   deep currents, **3**:142–144
   density-dependent currents, **3**:139, 143
   Ekman, Valfrid, **3**:139–140
   geostrophic currents, **3**:140
   human health and the ocean, **2**:178–179
   longshore currents, **1**:73
   ocean mixing, **3**:147
   rip currents, **1**:73
   role of water masses, **3**:143–144, 144*f*
   Stommel model, **3**:142
   surface current patterns, **3**:140–142, 141*f*
Ocean Drilling Program (ODP), **3**:164, 165
Ocean Dumping Act, **3**:240–241
Ocean floor mapping, **3**:163
Ocean health, assessing, **3**:145–146
   evaluation methods, **3**:145
   looking to future, **3**:146
   ocean threats, **3**:145–146
Ocean mixing, **3**:146–148
   currents and eddies, **3**:*147*–148
Ocean pollution. *See* Pollution of the ocean by plastic and trash
Ocean temperature. *See* Temperature, ocean
Ocean Thermal Energy Conversion (OTEC), **2**:26–27
Ocean water. *See* "Sea water" entries
Ocean-floor bathymetry, **3**:148–152
   abyssal plain, **3**:150
   bathymetric techniques, **3**:148–149
   continental shelf, **3**:149
   continental slope and rise, **3**:150
   hypsographic curve, **3**:151–152, 151*f*
   mid-ocean ridge, **3**:150
   ocean floor in cross-section, **3**:149–151
   rift valley, **3**:150
   subduction zones, **3**:150–151
Ocean-floor sediments, **3**:152–157
   manganese nodules, **3**:154
   preservation of sediment, **3**:156

Ocean-floor sediments (continued)
   production of sediment, 3:153–154
   study of, 3:164
   transport of sediment, 3:154–155
Oceanic plate, defined, 2:6, 4:200, and Glossary
Oceanic zone, 3:50
Oceanic-continental convergence, 3:204, 205f
Oceanic-oceanic convergence, 3:204, 205f
Oceanography, biological, **3:157–159**
   areas of research, 3:158–159
   funding sources, 3:159
   medical research, 1:129–130
   tools and technology, 3:157–158
Oceanography, chemical, **3:159–161**
Oceanography, defined, 3:110, and Glossary
Oceanography, geological, **3:163–165**
   programs, 3:165
   tools and techniques, 3:163–165
Oceanography, physical, **3:165–167**
Oceanography careers. *See* Careers in oceanography
Oceanography from space, **3:161–163**
   altimeter data, 3:161
   ocean color, 3:162
   radiometer data, 3:161–162
Oceans, polar, **3:167–171**
   Antarctic Ocean, 3:167–169
   Arctic Ocean, 3:169
   biological diversity and biota, 3:170–171
   ice, 3:169–170
   as marine ecosystem, 2:14
   polynyas, 3:170
   primary productivity, 3:170
Oceans, tropical, **3:172–173**
The Oceans (Sverdrup, Johnson, and Fleming), 4:161
Ocean-water masses, tracers of. *See* Tracers of ocean-water masses
Octocorals, 1:213–214
Octopuses, 1:142
Odontocetes, 3:61–62
Odor, of fresh water, 2:96, 4:91–92
ODP (Ocean Drilling Program), 3:164, 165
Offshore drilling and exploration, 3:182–183
Offstream water use, 4:151, 192f
Ogallala Aquifer, **3:173–176**
   characteristics, 3:173–174
   managing for future, 3:176

   map, 3:174f
   reducing contamination, 3:175
   slowing depletion rate, 3:174–175
   using and protecting, 3:174–176
Oil spills, **3:176–181**
   cleanup and recovery, 3:179–180
   costs and prevention, 3:180–181
   damage to fisheries, wildlife, and recreation, 1:82, 3:178
   environmental recovery rates, 3:179–180
   examples of large spills, 3:178
   long-term fate of oil on shore, 3:178–179
   oil spill behavior, 3:177–178
   oil spill interaction with shoreline, 3:177–178
   penguins, 3:179
   prevalence during drilling *vs.* transportation, 3:176
   See also Petroleum from the ocean
Old Faithful geyser (WY), 2:165, 166, 167
Olivine, in beach sand, 1:184
Omnivore, defined, 2:48, and Glossary
Ooid, defined, 1:183, and Glossary
Oort Cloud, defined, 1:199, and Glossary
Open-access resource, defined, 2:67, and Glossary
Orcas, 3:61, 62
Ordnance, defined, 4:207, and Glossary
Oregon Plan for Salmon and Watersheds, 4:71
Organic, defined, 1:4, 2:11, 3:3, 4:67, and Glossary
Organic wastes, 2:175–176, 242
Organizations, water science, 3:97, 98
Organochlorine, defined, 1:151, and Glossary
Ornithology, defined, 1:119, and Glossary
Orographic lifting, 1:180–181, 181f, 3:256, 260
Oscillating Water Column, 2:25
OTEC (Ocean Thermal Energy Conversion), 2:26–27
"Our Common Future" (World Commission on Environment and Development), 4:156
Outgassing, 1:59–60
Outwash, defined, 1:89, and Glossary
Overcapitalization, defined, 2:66, and Glossary

Overfishing
   defined, 2:65, and Glossary
   fresh-water fisheries, 2:59
   marine fisheries, 2:65, 3:145–146
Oxbow, defined, 4:113, and Glossary
Oxidation, defined, 1:63, 2:171, and Glossary
Oxidation-reduction chemistry, 2:91, 262
Oxide, defined, 4:235, and Glossary
Oxygen
   in sea water, 3:133, 133f, 4:78–79f
   in streams, 4:114
Oxygen isotopes, 2:240, 241–242, 4:176–177
Oysters, 1:85
Ozone, defined, 1:9, 2:157, 4:149, and Glossary

P

Pacific Ocean basin, 3:127
Pack ice, 2:206
   defined, 4:81, and Glossary
PAHs (polycyclic aromatic hydrocarbons), 1:30
Paleoclimatology, 2:131–132, 132f
Paleothermometry, 2:241–242
   defined, 2:241, and Glossary
Panama, Isthmus of, 2:238–239
Panama Canal, 1:105, 106
Pancake ice, 2:206, 207
Paradigms, 3:200
Parasite
   defined, 1:31, 2:14, 3:209, and Glossary
   fresh-water fisheries, 2:58–59
Pasig River (Philippines), 3:232
Pathogen
   defined, 1:28, 2:176, 3:3, 4:93, and Glossary
   movement and persistence, 3:73–74
PCBs (polychlorinated biphenyls) defined, 3:5, and Glossary
PCE (perchlorethylene), 1:62, 63
Peat, 3:87
Pelagic, defined, 2:71, and Glossary
Pelagic organisms, 3:49
Pelagic zone, 2:259, 260f
Penguins, 1:81, 82
   adelie, 3:171
   oil spill impact, 3:179
Peninsulas. *See* Islands, capes, and peninsulas
Perchlorethylene (PCE), 1:62, 63

Percolation rate, defined, **3**:175, and Glossary
Perge, **4**:222
Permafrost, defined, **4**:64, and Glossary
Permeability
　aquifers, **1**:40
　defined, **1**:51, and Glossary
　hydrogeologic mapping, **2**:191
Permits
　"incidental take," **2**:23
　pollution, **1**:149
　trading, **2**:136
　water, **1**:171
Perspiration, **2**:102
Peru, **2**:17–20, 178
Pesticides
　defined, **1**:15, **2**:36, **3**:2, and Glossary
　environmental effects, **1**:137–138
　land use and water quality, **3**:3
　pollution of lakes and streams, **3**:227
　residential use, **1**:154
　See also Chemicals from agriculture
Petrified wood, **2**:153
Petroleum, defined, **1**:69, and Glossary
Petroleum from the ocean, **3**:181–186
　coastal waters management, **1**:188
　discovery and use, **3**:181–182
　economic value, **3**:185–186
　gas hydrate, **3**:181
　locations of resources, **3**:181–182
　marine transportation, **3**:*184*, 185
　offshore drilling, **3**:183
　offshore exploration, **3**:182–183
　platforms, **3**:110, *182*
　production, **3**:184–185
　tanker carriers, **3**:*184*, 185
　technological advances, **3**:182–185
Petroleum reservoir, defined, **1**:141, **3**:107, and Glossary
Pfiesteria, **1**:18–19
pH
　defined, **1**:3, **2**:90, **3**:12, and Glossary
　fresh water composition, **2**:90
　fresh water physics and chemistry, **2**:101
　lake chemical processes, **2**:263–264
　life in extreme water environments, **3**:45
　and watershed water quality, **4**:219–220
Pharmaceuticals and personal care products (PPCPs). *See* Chemicals from pharmaceuticals and personal care products
Phosphate
　defined, **3**:124, and Glossary
　ocean biogeochemistry, **3**:133*f*
Phosphorites, **3**:90–91
Phosphorus
　lake chemical processes, **2**:264–265
　nutrients in lakes and streams, **3**:123–126
　and watershed water quality, **4**:220
Photic zone, **3**:49
Photosynthesis
　and algae, **1**:26
　and carbon dioxide levels, **1**:108
　dark reactions, **1**:26
　defined, **1**:21, **2**:3, **3**:46, **4**:78, and Glossary
　in lakes, **2**:259, 262
　light reactions, **1**:26
　process, **1**:215
　temperature impacts on, **4**:114
Photosynthesis-respiration process, **2**:262
Phthalate, defined, **1**:155, and Glossary
Physical oceanography. *See* Oceanography, physical
Physical processes of lakes. *See* Lakes: physical processes
Physics and chemistry of sea water. *See* Sea water, physics and chemistry of
Phytoplankton, **3**:186–189
　carbon dioxide in the ocean and atmosphere, **1**:109–110, 111
　characteristics, **2**:9*t*
　defined, **1**:17, **2**:9, **3**:58, **4**:78, and Glossary
　flotation, **3**:51
　in food web, **3**:187
　in lakes, **2**:261, 261*f*
　marine ecology, **2**:12–13, **3**:49, 79
　microscopic, **3**:*188*
Picoplankton, **3**:186
Pipelines, underwater, **3**:185
Pisciculture, **1**:31
Piscivores, in lakes, **2**:261, 261*f*
Planktivore
　defined, **1**:33, and Glossary
　in lakes, **2**:261, 261*f*
Plankton, **3**:186–190
　carbonate, **3**:156
　defined, **1**:226, **2**:8, **3**:79, and Glossary
　microbes in the ocean, **3**:49, 79
　siliceous, **3**:156
　See also Phytoplankton; Zooplankton; *specific kinds of plankton*
Planning and management, history of water resources, **3**:190–194
　Army Corps of Engineers, U.S., **3**:191, 193–194
　Bureau of Reclamation, U.S., **3**:191, 193, 194
　changing goals and values, **3**:192–194
　environmental era, **3**:193–194
　Green Book, **3**:193
　major water management agencies, **3**:191–192
　Tennessee Valley Authority, **3**:191–*192*
Planning and management, water resources, **3**:194–201
　balancing diverse interests, **1**:66–67
　channel control, **2**:52
　desert hydrology, **1**:245
　diversion, **2**:51–52
　federal water legislation, **3**:30
　fish and wildlife issues, **2**:50–52
　food security, **2**:83
　hydrologic cycle focus, **2**:197
　integrated resource planning, **3**:198–200
　least-cost planning, **3**:196–198
　planning in practice, **3**:200–201
　storage, **2**:50–51
　total water management, **3**:200
　traditional supply planning, **3**:194–196
Plant kingdom, **1**:22
Plants
　and carbon dioxide, **2**:130–131, 131*f*
　pitcher plants, **2**:10
　purple loosestrife, **3**:210, 214, *215*
　riparian, **2**:52, **4**:115, *116*, 123, 218
　and runoff, **4**:*63*, 64
　wetlands, **4**:242–243
Plastic. *See* Pollution of the ocean by plastic and trash
Plate tectonics, **3**:201–206
　boundaries, **3**:204–205
　convergent boundaries, **3**:204–205, 205*f*

Plate tectonics (continued)
  defined, **1**:59, **2**:3, **3**:83, **4**:216, and Glossary
  divergent boundaries, **3**:204
  Earth structure, **3**:201–203
  hot springs and geysers, **2**:167
  lake formation, **2**:247
  plate movements, **3**:206
  theory development, **3**:203–204
  transform boundaries, **3**:205
  water introduction, **4**:200
  and water movement on Earth, **2**:3–5
Plateaus, **4**:244
Platforms, **3**:108–110
  See also Moorings and platforms
Playas, **4**:244
Pleistocene epoch
  defined, **2**:247, and Glossary
  glacial cycles, **2**:202–204
Plucking, defined, **4**:119, and Glossary
Plumbing fixtures, ultra-low-volume, **1**:208, *209*
Plume, defined, **1**:63, **2**:107, **3**:12, **4**:25, and Glossary
Pluto Water, **3**:93
Point bar, defined, **2**:269, **4**:113, and Glossary
Point source
  defined, **2**:176, **3**:30, and Glossary
  groundwater pollution, **3**:218
  lake management, **2**:254, 255
  pollution of lakes and streams, **3**:226
  pollution sources, **3**:243–244
  surface-water contamination, **3**:4
  See also Pollution sources: point and nonpoint
Polar bears, **3**:63
Polar cells, **1**:176, *176f*
Polar molecules, **2**:100
Polar oceans. *See* Oceans, polar
Polar regions, **1**:177, *177f*
Polarity, defined, **1**:160, and Glossary
Policy, defined, **3**:206, and Glossary
Policy-making process, **3:206–209**
  aspects, **3**:206–208
  complex issues, **3**:207–208
  guidance for policymakers, **3**:207
  public demands, **3**:207
  roles of scientists and policymakers, **3**:208
  and scientific uncertainty, **3**:208
  special interest groups, **3**:207
Politics. *See* Hydropolitics

Pollutant
  agricultural, **3**:227–228
  airborne, **3**:228, 239–240
  biological, **3**:225–226
  chemical oceanography, **3**:160
  defined, **1**:62, **2**:14, **3**:242, **4**:32, and Glossary
  traps for, **2**:41
  types, **3**:225–226
Pollutant attenuation. *See* Attenuation of pollutants
Pollution
  defined, **1**:80, **2**:28, **3**:29, **4**:219, and Glossary
  developing countries, **1**:247
  food security, **2**:85
  fresh-water fisheries, **2**:58
  marginal seas, **3**:54–55
  sources, **1**:171–172
  as threat to ocean's health, **3**:145
  See also Acid mine drainage; Acid rain; "Chemical" and "Pollution" entries; Erosion and sedimentation; Land use and water quality; Landfills: impact on groundwater; Nutrients in lakes and streams; Septic system impacts
Pollution by invasive species, **3:209–217**
  addressing aquatic threat, **3**:214–216
  control methods, **3**:215–216
  ecological and economic impacts, **3**:211–212
  extent of U.S. aquatic invaders, **3**:210–211
  fresh-water fisheries, **2**:58–59
  human element, **3**:216
  introduction and impact, **3**:209–212, **4**:218
  water-related invaders, **3**:212–214
Pollution of groundwater, **3:217–222**
  chlorinated solvents, **3**:219
  cleanup laws, **3**:220
  cleanup process, **3**:220
  containment, **3**:220–221
  natural substances, **3**:218
  petroleum-based fuels, **3**:218–219
  remediation, **3**:222
  removal, **3**:221–222
  treatment, **3**:222
  types of groundwater contamination, **3**:217–219
Pollution of groundwater: vulnerability, **3:223–225**
  aquifer sensitivity, **3**:223

  vulnerability assessment, **3**:224
Pollution of lakes and streams, **3:225–229**
  agricultural pollutants, **3**:227–228
  airborne pollutants, **3**:228
  biological pollutants, **3**:225–226
  chemical pollutants, **3**:225
  nonpoint sources of pollution, **3**:226–228
  physical pollutants, **3**:225
  point sources of pollution, **3**:226
  types of impacts, **3**:228
  types of pollutants, **3**:225–226
  urban pollutants, **3**:226
Pollution of streams by garbage and trash, **3:229–233**
  developing nations, **3**:231–*232*
  regulatory context, **3**:230–231
  responsibility for, **3**:231
  working definition of pollution, **3**:231
Pollution of the ocean by plastic and trash, **3:233–236**
  effects on wildlife, **3**:233–234, *235*
  efforts to reduce debris, **3**:234–236
  medical waste, **3**:234
Pollution of the ocean by sewage, nutrients, and chemicals, **3:236–242**
  agricultural wastes, **3**:238
  airborne pollutants, **3**:239–240
  coastal eutrophication, **3**:238–239
  human health, **3**:239
  human health and the ocean, **2**:176
  industrial wastes, **3**:239–240
  regulatory controls, **3**:240–242
  sewage and agricultural wastes, **3**:237–239
  sewage from vessels, **3**:239
  sewage sludge, **3**:237
Pollution permits, **1**:149
Pollution sources: point and nonpoint, **3:242–246**
  addressing nonpoint sources, **3**:246
  groundwater, **3**:244
  land use and water quality, **3**:4–5
  nonpoint sources, **3**:244–245
  point and nonpoint pollution sources, **3**:242–243
  point sources, **3**:243–244
  preventing and controlling pollution, **3**:245–246
  public drinking-water protection, **4**:144–145

surface water, **3**:244
Polychlorinated biphenyls. *See* PCBs (polychlorinated biphenyls)
Polyculture, defined, **1**:34, and Glossary
Polycyclic aromatic hydrocarbons (PAHs), **1**:30
Polygonal terrain, defined, **3**:70, and Glossary
Polynyas, **2**:207, **3**:170
Ponds, in mariculture, **3**:59
Popular art, water in, **1**:54–55, **1**:*56*
Population and water resources, **3:246–249**
  future population levels, **3**:248
  impacts on future water quality, **3**:248–249
  impacts on future water sources, **3**:247–248
  population projections, **1**:254, **3**:247–248, 248*f*
  reducing population impacts, **3**:249
Pore spaces, **2**:*149*, 150, 156, 211, **4**:63–64
Pork barrel projects, **1**:221
Porpoises, **3**:62, *63*
Ports and harbors, **3:249–254**
  available land, **3**:252–253
  coastal harbors, **3**:*251*, 252
  environmental concerns, **3**:253
  lake harbors, **3**:*251*
  major U.S. ports, **3**:253–254
  planning issues, **3**:251–253
  ports, **3**:250–251, **4**:185
Position, defined, **4**:5, and Glossary
Post-audit, defined, **4**:249, and Glossary
Post-glacial rebound, defined, **4**:74, and Glossary
Potable, defined, **1**:51, and Glossary
Potable reuse, **4**:29–30
Potassium, **3**:89–90
Potential evapotranspiration, defined, **1**:262, and Glossary
Potentiating effect, **1**:148
Potholes, **4**:*119*
Powell, John Wesley, **3**:191, **255–256**, **4**:54–55
PPCPs (pharmaceuticals and personal care products). *See* Chemicals from pharmaceuticals and personal care products
Precipitate, defined, **1**:3, **2**:92, **3**:45, and Glossary
Precipitation
  defined, **1**:241, **3**:90, and Glossary
  hydrologic cycle, **2**:194–195
  as origin of groundwater, **2**:241
  types, **3**:257–259, 258*f*
Precipitation, global distribution of, **3:259–262**
  general circulation, **3**:259–260
  presence of oceans or lakes, **3**:261–262
  topography, **3**:260–261
Precipitation and clouds, formation of, **3:256–259**
  classification of clouds, **3**:257
  cloud deposition, **1**:6
  cloud formation, **1**:27–28
  cloud seeding, **3**:261
  cumulus clouds, **3**:257, *261*
  precipitation types, **3**:257–259, 258*f*
  thunderstorms, **3**:258–259, 258*f*
Precision, defined, **1**:146, and Glossary
Predation, defined, **4**:44, and Glossary
Predator, defined, **1**:28, and Glossary
Preservationist, defined, **2**:28, and Glossary
Prey, defined, **1**:28, **4**:44, and Glossary
Pricing, water, **3:262–264**
  capital equipment, **3**:263
  as conservation incentive, **1**:208, **2**:214, **3**:263–264
  demand management trend, **3**:263–264
  developing countries, **1**:250
  scarcity value, **3**:263
  subsidies, **3**:263
  true cost of water, **3**:262–263
  U.S. cities, **3**:262*f*
Primary producer, defined, **2**:11, **3**:187, and Glossary
Primary productivity
  defined, **2**:41, **3**:55, and Glossary
  marginal seas, **3**:56
  polar oceans, **3**:170
"Principles and Standards for Planning Water and Related Land Resources," **2**:15
Prior appropriation, **4:1–3**
  defined, **3**:28, **4**:54, and Glossary
  evolution of doctrine, **4**:2–3
  fish and wildlife protection, **4**:2–3
  history, **4**:1
  legal aspects, **3**:28, 32, **4**:2
  mill privilege, **4**:53–54
  public trust doctrine, **4**:2–3
  water markets, **3**:67
  water quality protection, **4**:3
Priority of use (law), **3**:22–23
Private right, defined, **3**:27, and Glossary
Private water supplies. *See* Supplies, public and domestic water
Privatization of water management, **2**:139, **4:3–5**, 198
  considerations, **4**:4–5
  effective design, **4**:4
  operating revenues of private U.S. water companies, **4**:4*t*
  potential barriers, **4**:4–5
  public *vs.* private, **4**:3–4
  safeguards for cities, **4**:4
Process, defined, **4**:5, and Glossary
Professionalism. *See* Ethics and professionalism
Profundal zone, **2**:259, 260*f*
Progressive Era, **2**:27–28, **3**:192
Prokaryote, **1**:22
  defined, **3**:188, and Glossary
Property right, defined, **2**:66, **4**:2, and Glossary
Protists, **1**:22
Protochlorophytes, **3**:188
Protons, **2**:239–240
Protozoa, **3**:189
  in groundwater, **3**:72
  human health and water, **2**:181, 181*t*
Prozac, effect on animals, **1**:157–158
Pseudo-nitzschia, **1**:19
Public drinking-water supplies, protecting. *See* Supplies, protecting public drinking-water; *See* Algal blooms, harmful; Drinking water and society; Human health and water; Microbes in groundwater; Microbes in lakes and streams; Microbes in the ocean; Safe Drinking Water Act; Supplies, protecting public drinking-water
Public Health Security and Bioterrorism Preparedness and Response Act, **4**:68
Public opinion polls, on water issues, **2**:29
Public participation, **4:5–8**
  importance, **4**:6
  nongovernmental organizations (NGOs), **4**:6
  stakeholders, **4**:5–6
  successful, **4**:7
Public right, defined, **3**:26, **4**:3, and Glossary
Public supply water use category, **4**:194, 195
Public trust
  defined, **1**:101, **4**:53, and Glossary

Public trust (continued)
    prior appropriation, **4**:2–3
    public water rights, **4**:50
Public water rights. *See* Rights, public water
Public water supplies. *See* Supplies, public and domestic water
Public water system, defined, **4**:145, and Glossary
Pump and treat method, **3**:221–222
Pump tests, **1**:40–42, **2**:158–*159*
Pumps, modern, **4:8–11**, **4**:*151*
    centrifugal pumps, **4**:9
    jet pump, **4**:10–11
    reciprocating pumps, **4**:8
    rotary pump, **4**:8–10
    shallow-well and deep-well pumps, **4**:10–11
    turbine pumps, **4**:9–10
    for wastewater treatment, **4**:*210*
Pumps, traditional, **4:11–16**
    Archimedes Screw, **4**:13–*14*
    bail bail, **4**:12
    buckets, wheels, and paddles, **4**:11–13
    hydraulic ram pumps, **4**:15–*16*
    lift and hand pumps, **1**:*249*, **3**:*78*, **4**:14–*15*
    noria, **4**:13
    shaduf, **2**:*232–233*, **4**:12
    wheels and loops, **2**:*228*, **4**:*12–13*
Pycnocline, **4**:82, 82*f*

Q

Quahog, defined, **2**:67, and Glossary
Quantitative, defined, **3**:37, and Glossary
Quotas
    individual, **2**:64, 67–69
    trading permits, **2**:136

R

Radar, **2**:108–109, **3**:122, 149, 161
Radiation, **3**:45, **4**:17
Radioactive, defined, **1**:244, **2**:98, **4**:109, and Glossary
Radioactive chemicals, **4:17–21**
    anthropogenic radioactive waste, **4**:18–19
    classifying, **4**:17–20
    disposal and cleanup, **4**:20–21
    in drinking water, **4**:19
    NORM (naturally occurring radioactive material) and radon, **4**:18
    nuclear waste cleanup candidates, **4**:20–21
    quantifying radioactivity and radiation, **4**:17
    radioactivity and types of radiation, **4**:17
    storing wastes underground, **4**:19–20
Radioactivity, defined, **2**:157, **4**:17, and Glossary
Radionuclide, defined, **1**:64, **4**:17, and Glossary
Radionuclides in the ocean, **4:21–26**
    artificial nuclides, **4**:25
    cosmogenic nuclides, **4**:23
    human-made radionuclides, **4**:23
    primordial nuclides, **4**:23
    radioactive decay chains, **4**:23*t*
    radioactive elements, **4**:22*t*
    radioactivity, in general, **4**:21–23
    radioactivity in the ocean, **4**:24–25, 24*f*
    types, **4**:23
Radon
    in drinking water, **4**:19
    in fresh water, **2**:98–99
    groundwater pollution, **3**:218
    and NORM (naturally occurring radioactive material), **4**:18
Rain sensor devices, **1**:208
Rain shadows, **1**:181, 181*f*, **3**:260
Rainbow trout, defined, **1**:32, and Glossary
Rain-fed agriculture, **1**:12–13
Rainwater harvesting, **4:26–29**
    catchment-cistern conveyances, **4**:27, 28
    catchments, **4**:26–27
    cisterns, **4**:27–28
    components of harvesting systems, **4**:26–28
    fog harvesting, **4**:28
    public and domestic water supplies, **4**:148
    quality of harvested rainwater, **4**:28
    withdrawal methods, **4**:28
Rapids. *See* Waterfalls
Rational method (flood flow model), **3**:105
Rays (fish), **2**:*70*
RCRA (Resource Conservation and Recovery Act), **3**:220, 230
RDUs (Rural Development Units), **2**:230
Recharge, defined, **2**:195, **3**:1, **4**:32, and Glossary
Recharge area, defined, **4**:109, and Glossary
Reclamation, defined, **1**:121, **2**:58, **3**:32, and Glossary
Reclamation Act, **3**:29, **4**:31
Reclamation and reuse, **4:29–35**
    agricultural reuse, **4**:*30*, *31*, 32–33
    direct nonpotable reuse, **4**:30
    direct potable reuse, **4**:29
    future of water reuse, **4**:34
    greywater, **4**:34
    improving water-use productivity, **4**:30–31
    indirect potable reuse and nonpotable reuse, **4**:29–30
    industrial reuse, **4**:34
    municipal reuse, **4**:33
    potable or nonpotable reuse, **4**:29–30
    reuse trends, **4**:32–34
    space travel, **4**:105–106
    water reclamation, **4**:31
Recreation, **4:35–38**
    economics of, **4**:35–36
    future trends, **4**:37–38
    Nationwide Survey on Recreation and the Environment, **4**:36–37
    oil spill impact, **3**:178
    water-based opportunities, **4**:36–37
    See also Sports
Recruitment, defined, **1**:218, and Glossary
Red tide, **1**:16–*17*, 24, 25, **3**:82
    defined, **3**:123, and Glossary
    See also Algal blooms, harmful
Redox conditions, defined, **2**:241, and Glossary
Reduction, defined, **2**:98, and Glossary
Reef, defined, **1**:45, and Glossary
Reflection, wave, **4**:226
Refraction
    defined, **3**:53, and Glossary
    wave, **4**:226
Regionalization (flood flow model), **3**:106
Regionalization (water management integration), **2**:222
Reisner, Marc, **4:39–40**
Religions, water in, **4:41–43**
    Hinduism, **4**:*41*, 42
    Judeo-Christian religion, **4**:41, *42*
    Lourdes water, **4**:42
    religion's influence on water management, **4**:42–43
    water, purification, and healing, **4**:42
    water and creation beliefs, **4**:41
    water and its forms, **4**:41

Remediation, defined, **1**:116, **3**:165, and Glossary
Remote sensing, **2**:106–109
   active or passive, **2**:107
   defined, **1**:120, **3**:145, and Glossary
   groundwater supply exploration, **2**:160
   petroleum from the ocean, **3**:182, 185
   radar, **2**:108–109
   satellites, **2**:107–*108*
   sonar, **2**:109
Remotely Operated Vehicles (ROVs), **4**:142
Reptiles, **4**:43–47
   crocodilians, **4**:*46*–47
   lizards, **4**:44–45
   snakes, **4**:*45*–46
   turtles, **4**:*43*–44
Research Experience for Undergraduates (REU) model, **3**:96, 97–98
Reserves, defined, **3**:182, and Glossary
Reservoir, **2**:50–51
   defined, **1**:100, **2**:25, **3**:29, **4**:10, and Glossary
Reservoirs, multipurpose, **4**:47–49
   ecology impacts, **4**:48
   operating challenges, **4**:48
Residence time, **3**:135
   defined, **1**:166, **4**:129, and Glossary
Resource Conservation and Recovery Act (RCRA), **3**:220, 230
Respiration, defined, **2**:92, and Glossary
Restoration of a watershed. *See* Watershed, restoration of a
Retardation factor, **3**:103
Reuse. *See* Reclamation and reuse
Reverse osmosis, **1**:*241*, 242
   defined, **1**:260, **3**:91, and Glossary
Rhine River, **4**:61
Rhyolite, defined, **2**:167, and Glossary
Ricelands Habitat Partnership, **4**:40
Rideau Canal (Canada), **3**:231
Ridge Interdisciplinary Global Experiments (RIDGE), **3**:165
Rights, public water, **4**:49–52
   categories, **4**:49–51
   navigation, **4**:49–*50*
   public interest protections, **4**:51
   public trust doctrine, **4**:50
   reserved water rights, **4**:51
   water law, **3**:27–28
   yesterday and today, **4**:51–52

Rights, riparian, **4:52–54**
   evolution of riparian model, **4**:52–54
   international water law, **3**:19–20
   mill privilege, **4**:53–*54*
   state and local water legislation, **3**:32
   today's framework, **4**:54
   water law, **3**:28
   water markets, **3**:67
Rill, defined, **4**:117, and Glossary
Ring of Fire, **3**:127, **4**:*205*, 206
Rip currents, **1**:73
Riparian, defined, **1**:144, **2**:37, **3**:101, **4**:115, and Glossary
Riparian rights. *See* Rights, riparian
Riparian vegetation, **2**:52, **4**:115, *116*, 123, 218
Riprap, defined, **4**:113, and Glossary
Risk management, defined, **1**:263, and Glossary
River and Harbor Act, **1**:48, **2**:29, **3**:29, 119, 192, 240, **4**:184
River basin planning, **4**:54–58
   basinwide program, **4**:55
   comprehensive regional development, **4**:55
   evolution of U.S. planning models, **4**:56
   multipurpose dam, **4**:55
   Powell, John Wesley, **4**:54–55
   river commissions worldwide, **4**:57–58
Rivers. *See* "Stream" entries; *specific rivers*
Rivers, major world, **4:58–62**
   Africa, **4**:60
   Asia, **4**:59–60
   Europe, **4**:61
   largest rivers by discharge, **4**:59*t*
   North America, **4**:61–62
   South America, **4**:62
   See also specific rivers
Road salt, and water quality, **3**:3
Roads, effects on forest hydrology, **2**:88–89
Rock
   carbonate, **2**:243
   consolidated, **1**:39–40
"Rock cycle," **2**:104
Rocks, weathering of. *See* Weathering of rocks
Roman Empire waterworks, **4**:222–223
Roosevelt, Franklin, **3**:191, 193, **4**:56
Roosevelt, Theodore, **2**:54, **3**:192, **4**:55

Ross Ice Shelf, **2**:*204*, 208, 210
ROVs (Remotely Operated Vehicles), **4**:142
Rowing, **4**:107
Runoff, defined, **4**:115, and Glossary
Runoff, factors affecting, **4:62–66**
   factors influencing infiltration rate, **4**:62–64
   ice and snow, **4**:64
   plants and animals, **4**:*63*, 64
   runoff and flooding, **4**:64–65
   runoff and urban development, **3**:*245*, **4**:65–66
   slope, **4**:64, *65*
   soil characteristics, **4**:63–64
Rural Development Units (RDUs), **2**:230

S

Sacagawea, **3**:40*f*, 41
Safe Drinking Water Act, **3**:30, **4:66–69**
   amendments, **4**:67
   contaminant categories, **4**:68
   drinking-water standards, **4**:67–68
   evolution of law, **4**:66–67
   keeping water safe, **4**:68–69
   source-water assessments, **4**:145–146
   supplies covered, **1**:87–88, **4**:147
Sailing, **4**:*37*, 107
Saint Helens, Mount (WA), **4**:*202*
Sakkia, **4**:12
Salamanders, woodland, **2**:51
Saline, defined, **1**:101, **3**:248, **4**:152, and Glossary
Salinity
   Colorado River Basin, **1**:194
   defined, **2**:39, **3**:139, and Glossary
   extreme water environments, **3**:45
   lake chemical processes, **2**:263, *264*
   life in water, **3**:51
   in sea ice, **4**:76–77
   sea water physics and chemistry, **4**:80–81
Salmon decline and recovery, **4:69–72**
   Columbia River Basin, **1**:195, 196, *197*, **2**:190–191
   historic declines, **4**:70
   modern declines, **4**:70–71
   Oregon Plan for Salmon and Watersheds, **4**:71
   salmon recovery, **4**:71–72

Salmonella, **2:**181*t*, 183
Salt
    desert hydrology, **1:**244–245
    fresh water, **3:**87–88, 137
    ocean, **3:**89
Salt domes, **1:**95–96
Salt fingering, **3:**146
Salt oscillator, **2:**205
Salt-water intrusion, **1:**188
    defined, **1:**51, **4:**30, and Glossary
San Andreas Fault, **3:**202, 205
San Francisco Bay, **1:**68, 69
Sand
    black lava beach sand, **1:**184, 185
    and estuary formation, **2:**38
    mineral resources from the ocean, **3:**90
    mining, **3:**86
Sanitation and sewage. *See* Drinking water and society; Human health and water; Landfills: impact on groundwater; Septic system impacts; Wastewater treatment and management
Sarez, Lake (Pakistan), **2:**249
Satellite altimeters, **4:**75
Satellites
    mapping of the ocean floor, **3:**163
    oceanography from space, **3:**162
    physical oceanography, **3:**167
    remote sensing, **2:**107–108
Saturated thickness, defined, **3:**174, and Glossary
Saturated zone, defined, **3:**12, **4:**62, and Glossary
Scallops, **1:**84–85
Scarcity value, defined, **3:**263, and Glossary
Scavenger, defined, **1:**224, and Glossary
Scientific Ice Expeditions (SCICEX), **4:**142–143
Scientific method, defined, **1:**112, and Glossary
Scree, defined, **1:**1, and Glossary
Scrimshaw, defined, **1:**55, and Glossary
Scuba, **1:**222, **2:**177, **4:**174
    defined, **1:**45, and Glossary
Scute, defined, **4:**44, and Glossary
Sea anemones, **3:**50
The Sea Around Us (Carson), **1:**137
Sea breeze, **1:**182
Sea Grant, **3:**165
Sea ice
    defined, **2:**123, **3:**169, **4:**76, and Glossary
    See also Ice at sea; Sea water, freezing of
Sea lampreys, **2:**58, **3:**213–214
Sea level, **4:72–76**
    beach erosion, **4:**74
    causes of eustatic variations, **4:**73
    and estuary formation, **2:**38
    and glaciers, **2:**119–120, 123
    impacts of rising sea level, **2:**38, **4:**73–74
    mean sea level, **4:**72
    measuring, **4:**75
    post-glacial rebound, **4:**74
    regional processes, **4:**74
    satellite altimeters, **4:**75
    tide gages, **4:**75
    trends for selected U.S. cities, **4:**74*f*
Sea lions, **3:**63
Sea nettles, **3:**49
Sea otters, **3:**63
Sea snakes, **3:**51, **4:**45–46
Sea water, freezing of, **4:76–77**
Sea water, gases in, **4:77–79**
    nitrogen, **4:**79
    noble gases, **4:**78
    oxygen, **4:**78–79*f*
    percentage in sea water *vs.* percentage in air, **4:**77*t*
Sea water, physics and chemistry of, **3:**136*t*, **4:79–83**
    density structure of the ocean, **4:**81–82, 82*f*
    light in the ocean, **4:**82
    vs. river water composition, **3:**134*t*
    salinity, **4:**80–81
    sound in the ocean, **4:**82–83
    temperature and density, **4:**81–82
Seabirds, **1:**81
Seafloor age, **3:**203
Seafloor spreading, **3:**150, 203–204
Seafood. *See* Food from the sea
Seahorse, **3:**50
Seals, **3:**63
    elephant, **3:**62
Seamount, **3:**150
    Axial Seamount, **4:**205
    defined, **1:**128, **3:**161, and Glossary
Seawalls, **1:**75
Seaweed, **2:**80, 81
    kelp forests, **2:**14
    mariculture, **3:**58, 60
Secchi disks, **3:**53
Security and water, **4:84–89**, 199
    countering attacks, **4:**88–89
    military actions, **4:**84–85
    terrorist actions, **4:**85–86
    water as target, **2:**215–216, **4:**86–87
    water as tool, **4:**87–88
    water transportation system, **4:**187–188
    See also Terrorism
Sediment
    bedload, **2:**34
    biogenic, **3:**152, 153–154, 156
    cosmogenic, **3:**152, 154
    defined, **1:**45, **2:**40, **3:**85, **4:**59, and Glossary
    estuaries, **2:**40–41
    hydrogenous, **3:**152, 154
    lake chemical processes, **2:**265
    land use and water quality, **3:**5
    as nutrient source in lakes and streams, **3:**125
    stream channel development, **4:**111–112
    stream health, assessing, **4:**122
    suspended, **2:**34, 35–36
    terrigenous, **3:**152, 153, 154, 156
    unconsolidated, **1:**39–40
    volcanogenic, **3:**152, 154
Sediment burial, **1:**111
Sediment load, **2:**34
    defined, **1:**203, and Glossary
Sediment yield, **2:**34
Sedimentary bedrock, **1:**40
Sedimentation
    defined, **1:**166, **2:**191, **3:**35, **4:**115, and Glossary
    drinking-water treatment, **1:**259
    stream, **2:**34–35
    wastewater treatment, **4:**211
    See also Erosion and sedimentation
Sediments, ocean-floor. *See* Ocean-floor sediments
Seeps, **4:**108, 110
Seiche waves, **4:**229
Seismic sea waves. *See* Tsunamis
Selenium, **2:**99
Semiarid, defined, **1:**12, **2:**229, **3:**174, and Glossary
Senior appropriator, defined, **4:**2, and Glossary
Senses, fresh water and the, **4:89–94**
    carbonation, **4:**92
    color and appearance, **4:**89–91, 93
    hardness and softness, **4:**92–93
    odor and taste of groundwater, **4:**91–92
    sound, **4:**94
    temperature, **4:**94

Septic system impacts, **4:**94–98
    design and maintenance, **4:**96–97
    disposal of household chemicals, **4:**97–98
    drainfield, **4:**95–96, 96f
    improper functioning, **4:**97
    nitrate and groundwater, **4:**96
    as nutrient source in lakes and streams, **3:**124
    preventive measures, **4:**97–98
    septic system components, **4:**94–96, 95f, 96f
    septic tank, **4:**95
    soil, **4:**96, 96f
Sequester, defined, **1:**64, **4:**177, and Glossary
Serotonin, effect on animals, **1:**157–158
Sewage. *See* Pollution of the ocean by sewage, nutrients, and chemicals
Sewage treatment works, **1:**160, **3:**124
Sextant, **3:**121
Shaduf, **2:**232–233, **4:**12
Shale, defined, **4:**215, and Glossary
Shallow well, **2:**83, 229, **4:**130
    defined, **2:**150, **4:**130, and Glossary
Sharks, **2:**70, 71
Shells, on mountaintops, **3:**35–36
Shigella, **2:**181t, 183
Shrimp, **1:**225–226
Silent Spring (Carson), **1:**137–138, **2:**29
The Silent World, **1:**223
Silica, defined, **4:**201, and Glossary
Silicoflagellates, **3:**188
Silicon, **3:**132–133, 133f
Sill, defined, **3:**56, and Glossary
Silt, **3:***162*
    defined, **1:**43, **2:**33, and Glossary
Sink, defined, **2:**131, **4:**178, and Glossary
Sinkhole
    defined, **1:**140, **2:**90, **3:**73, and Glossary
    karst hydrology, **2:**243–244, *245*
    and lake formation, **2:**249
Sirenians, **3:**62–63
Skates, **2:**70
Slope, **3:**16, **4:**64, *65*
Slough, defined, **2:**45, and Glossary
Sludge, defined, **4:**210, and Glossary
Slush, **2:**206
Small Watershed Grants Program, **1:**168
Snail darters, **2:**23–24, 218

Snakehead fish, **2:**258, **3:***210*, 211, 212, 216
Snakes, **4:***45*–46
    anaconda, **4:***45*, 46
    yellow-bellied sea snake, **3:***51*
    See also Reptiles
Snorkeling, 4:173–174
Snow, **1:**181, **2:**210–211
Snow line, defined, **4:**64, and Glossary
Snowpack, defined, **4:**64, and Glossary
Sodium, in groundwater, **4:**91
SOFAR (sound fixing and ranging) channel, **4:**103–104
Softness, of fresh water, **4:**92–93
Soil
    acid rain effects, **1:**8
    buffering capacity, **1:**7, 8
    development, **4:**236
    and runoff, **4:**63–64
    septic system impacts, **4:**96, 96f
Soil science careers. *See* Careers in soil science
Solar humidification, and desalinization, **1:**241
Solar system, water in the, **4:**98–101
    asteroids and comets, **4:**100
    Earth, **4:***99*, 100
    inner terrestrial planetary bodies, **4:**98–100
    Mars, **4:**100
    Mercury, **4:**99
    Moon, **4:**100
    outer solar system, **4:**100–101
    satellites of outer solar system planets, **4:**101, 101f
    types of ice, **4:**99
    Venus, **4:**99–100
Solid Waste Agency of Cook County, Illinois, v. U.S. Army Corps of Engineers, **1:**172
Solubility pump, **1:**109
Solute, defined, **4:**130, and Glossary
Solute transport computer model, **4:**130–131
Solvent, defined, **1:**58, and Glossary
Sonar, **2:**109, **3:**149, **4:**83, *102*
Sorption, of pollutants, **1:**63
Sound transmission in the ocean, **4:**82–83, **101–104**
    SOFAR (sound fixing and ranging) channel, **4:**103–104
    "sound channel," **4:**102–103
Sounding lines, **3:**148–149
Southern Ocean, **1:**77, **3:**129, 141, 167–169, 170–171
Southern Oscillation, **2:**18–19, 122

Sovereign
    defined, **2:**136, **3:**19, **4:**180, and Glossary
Sovereignty
    absolute territorial, **3:**19
    hydrosovereignty, **2:**200
    and water issues, **3:**32
Space, oceanography from. *See* Oceanography from space
Space travel, **4:104–106**
    bioregenerative processes, **4:**105–106
    cooling garments, **4:**105
    humidity, **4:**105
    recycling, **4:**105–106
    spacecraft ecosystem, **4:**104–105
Space zoning, **2:**256
Spas. *See* Mineral waters and spas
Special interest groups, **2:**222, **3:**207
Species
    critical habitats, **2:**22
    defined, **1:**66, **2:**11, **3:**99, and Glossary
    diversity, **1:**77
    euryhaline, **3:**51
    eurythermal, **3:**52
    exotic, **2:**256–258
    indicator, **2:**53
    injurious, **3:**210
    nuisance, **1:**82, **3:**210
    recovery plans, **2:**22
    stenohaline, **3:**51
    stenothermal, **3:**52
    threatened, **2:**22
Specific yield, defined, **3:**175, and Glossary
Spike, defined, **1:**146, and Glossary
Sports, **4:106–107**
    canoeing and kayaking, **4:**106–107
    fishing, **4:**107
    rowing, **4:**107
    sailing, **4:**37, 107
    surfing, **4:**106, *107*
    swimming and diving, **4:**106
    waterskiing, **4:**106, **if:***38*
    See also Recreation
Spreadsheet, defined, **1:**132, and Glossary
Spring, defined, **4:**125, and Glossary
Spring overturn, **2:**268
Springs, **4:107–111**
    age of water, **4:**109–110
    characteristics, **4:**109–110
    coastal, **4:***108*
    defined, **1:**43, **2:**76, and Glossary
    as discharge area for groundwater, **2:**152, 152f

Springs (continued)
	formation, **4:**108–109
	Grand Prismatic Spring (WY), **3:***45*
	gravity, **2:**152, 152*f*
	human interest in, **4:**108
	lakebed, **2:***264*
	large-volume, **4:**110–111
	Metolius Spring (OR), **4:***111*
	mineral, in U.S., **3:**92
	origin of water, **4:**109
	public and domestic water supplies, **4:**148
	temperature, **4:**109–110
Squid, **1:**142–143
St. Lawrence River, **4:**61
St. Lawrence Seaway, **1:**107, **2:**143
St. Petersburg Beach (FL), **2:***238*
Stakeholder, **1:**101–102, **4:**5–6
	defined, **1:**38, **2:**73, **3:**10, **4:**56, and Glossary
Stalactites, **1:**140
Stalagmites, **1:**140
State water legislation. *See* Legislation, state and local water
Static water level, defined, **1:**40, and Glossary
Steady state, **3:**135
	defined, **1:**162, and Glossary
Step tests, **1:**41–42, 42*f*
Steroid, defined, **1:**162, and Glossary
Stock, defined, **1:**196, and Glossary
Stommel, Henry, **3:**142
Storage, chemical, **1:**147
Storms
	and beaches, **1:**73–74
	dust storms, **3:***155*
	human health and the ocean, **2:**178, *179*
	storm surges, **2:**178, **4:***230*
	thunderstorms, **3:**258–259, 258*f*
	waves, **4:**227–229
Stormwater
	defined, **1:**50, **3:**30, **4:**209, and Glossary
Stormwater management. *See* Wastewater treatment and management
Straits, **1:**70–71
	See also Bays, gulfs, and straits
Stratification, thermal, **2:**262, 265, 267–268
Stratosphere, defined, **4:**178, and Glossary
Stream, hyporheic zone of a, **4:**128, **129–132**
	biogeochemical influences, **4:**131
	delineating, **4:**130–131
	ecological and physical influences, **4:**131
	effects on stream temperature, **4:**131
	future research challenges, **4:**132
	hyporheic zone hydrology, **4:**129, 130*f*
Stream channel development, **4:111–114**
	braided channels, **4:***112*, 129
	dendritic drainage network, **4:***112*, 125
	effects of floods on stream channels, **4:**113
	human alteration of stream channels, **4:**113–114
	meandering channels, **4:***112*–113, 128–129
	stream channel geometry, **4:***112*–113
	water and sediment movement, **4:**111–112
Stream ecology: temperature impacts on, **4:114–117**
	assessing stream health, **4:**122
	human influences, **4:**115–116
	hyporheic zone, **4:**131
	mechanisms of temperature change, **4:**114–115, *116*
	thermal pollution reduction efforts, **4:**116–117
Stream erosion and landscape development, **4:117–121**
	consequent, subsequent, and superimposed streams, **4:**120
	erosive and transport capacity, **4:**118
	riverbed erosion, **4:**118–120
	rivers and landscape shaping, **4:**117
Stream health, assessing, **4:121–125**
	aquatic life, **4:**122–123
	bacteria, **4:**122
	natural and human environment, **4:**121
	nutrients, **4:**122
	protecting stream health, **4:**124
	sediment, **4:**122
	temperature and dissolved oxygen, **4:**122
	toxic chemicals, **4:**122
	water-quality parameters, **4:**121–122, *123*
Stream hydrology, **4:125–129**
	flood events, **4:**127
	flow within river, **4:**126–127
	groundwater and baseflow, **4:**126, 126*f*
	hyporheic zone, **4:**128
	measuring streamflow, **4:**126–127
	perennial and intermittent streams, **4:**125
	stream form, **4:***128*–129
	stream interaction with its environment, **4:**127–129
	streamflow, **4:**125–126, 126*f*
	surface runoff, **4:**125–126, 126*f*
Stream piracy, **2:**244–245
Stream pollution. *See* Pollution of lakes and streams; Pollution of streams by garbage and trash
Streamflow, modeling. *See* Modeling streamflow
Streamflow variability, **4:132–136**
	causes of long-term streamflow changes, **4:**136
	downstream water losses, **4:**134–136
	long-term *vs.* short term streamflow changes, **4:**133–134, 134*f*
	short-term fluctuations, **4:**126*f*, 133
Streams
	artificial recharge benefits, **1:**51
	canalization of rivers, **1:**106–107
	consequent, subsequent, and superimposed, **4:**120
	disappearing, **2:**244–245
	fresh water composition, **2:**91–92
	fresh-water ecology, **2:**7–9
	influence of rivers on estuaries, **2:**40
	intermittent, **4:**125
	vs. karst hydrology, **2:**246
	largest rivers by discharge, **4:**59*t*
	National Park Service, **3:**119
	perennial, **4:**125
	river and stream cleanups, **3:**231
	river water *vs.* sea water composition, **3:**134*t*
	rivers and landscape shaping, **4:**117
	rivers as flowing-water habitats, **3:**51
	sandstone streambed, **4:***119*
	stream erosion and landscape development, **4:**118–120
Streams, microbes in. *See* Microbes in lakes and streams
Streams, nutrients in. *See* Nutrients in lakes and streams
Stromatolites, **3:**79, *80*
Structure contour, defined, **2:**193, and Glossary
Structure contour mapping, **2:**194
Study and Interpretation of the Chemical Characteristics of Natural Water (Hem), **2:**162–163

Stumm, Walter, **4:**137
Subduction, defined, **1:**59, **2:**7, **4:**200, and Glossary
Subduction zones, **3:**150–151, 204, 206, **4:**205–206, 205*f*
Submarine volcanoes. *See* Volcanoes, submarine
Submarines and submersibles, **4:137–144**
   Alvin, Argo, Jason, **4:**141–142, *143*
   atomic energy, **4:**140–141
   historical developments, **4:***138*–140
   Hydrolab, **4:**141, *142*
   military and civilian operations, **4:**141–143
   modern submarine, **4:**140–141
   Nautilus, **4:**141
   ROVs (Remotely Operated Vehicles) and AUVs (autonomous underwater vehicles), **4:**142–143
Subpolar regions, **1:**177, 177*f*
Subsidence, defined, **4:**74, and Glossary
Subsidy, defined, **2:**66, **3:**263, and Glossary
Substrate, **2:**53
   defined, **1:**83, **2:**7, **3:**211, and Glossary
Subsurface, defined, **1:**244, and Glossary
Subtropical regions, **1:**177, 177*f*
Sulfate, defined, **1:**9, and Glossary
Sulfide, defined, **1:**2, **2:**170, and Glossary
"Sulfide chimneys," **2:**170, **3:**85
Sulfur, **3:**80, *81*
Sulfur dioxide (SO_2) emission reduction program, **1:**9, 10
Sulfur hexafluoride (SF_6), **2:**157
Supercooling, **2:**206
Supercritical fluid, defined, **2:**102, and Glossary
Superfund, **3:**220
Superior, Lake, **2:**143, 146, 147
Supernatant, defined, **2:**211, and Glossary
Supplies, protecting public drinking-water, **4:144–147**
   developing protection strategies, **4:**146
   focusing on prevention, **4:**145–146
   identifying contamination sources, **4:**146
   identifying source areas, **4:**146
   multiple barrier approach, **4:**145
   pollution sources, **4:**144–145
   source-water assessment program, **4:**145–146
   threats to public drinking-water supplies, **4:**144–145
Supplies, public and domestic water, **4:147–150**
   vs. bottled water, **4:**148
   contaminants in water, **4:**148
   groundwater, **4:**148
   rainwater capture, **4:**148
   sources of water, **4:**148
   springs, **4:**148
   surface water, **4:**148
   then and now, **4:**147
   water protection and conservation, **4:**149–150
   water treatment and distribution, **4:**148–149
Supply development, **4:150–153**
   appropriate technology, **4:**152
   and demand management, **4:**152–153
   future trends, **4:**150–151
   historical perspective, **4:**150–151
   process, **4:**151–152
   water rights, **4:**151
   water-using sectors, **4:**151
Surface spreading technique, for artificial recharge, **1:**50, 50*f*
Surface water
   California, **1:**100
   defined, **1:**253, **2:**94, **3:**11, **4:**193, and Glossary
   drinking-water treatment, **1:**258–259
   hydrologic cycle, **2:**195
   land use and water quality, **3:**4–6
   point sources, **3:**244
   public and domestic water supplies, **4:**148
Surface Water Treatment Rule (SWTR), **3:**75
Surfing, **4:**106, *107,* 227
Survival needs, **4:153–156**
   absolute minimum requirements, **4:**153–154
   future water perspectives, **4:**155
   human right to water, **4:**154–155
   recognition of, **4:**153–154
Suspended, defined, **2:**51, **3:**5, and Glossary
Suspended load, defined, **4:**113, and Glossary
Sustainable, defined, **1:**131, **2:**60, **3:**116, **4:**56, and Glossary
Sustainable development, **4:156–159**
   different perspectives on, **4:**158–159
   ethics and professionalism, **2:**43
   key principles, **4:**156–158
   pros and cons, **4:**159
   water goals for 2025, **4:**159
Sverdrup, **4:**160
   defined, **3:**140, and Glossary
Sverdrup, Harald, **4:160–161**
Sweat, **2:**102
Swimming, **4:**106
SWTR (Surface Water Treatment Rule), **3:**75
Synergistic effect, **1:**148

T

Tablemount, **3:**150
Tahquamenon Falls (MI), **4:***91*
Taking (private property), defined, **4:**52, and Glossary
Talus cave, defined, **1:**140, and Glossary
TAO (Tropical Atmosphere Ocean) array, **2:**20–21
Tapered channel system (TAPCHAN), **2:**25–26
Task force, defined, **2:**220, and Glossary
Taxonomy, **1:**22
TCE (trichlorethylene), **1:**62, 63, **3:**223, 244
TDS (total dissolved solids), **2:**92–93, 100
Tectonic, defined, **2:**38, **3:**92, **4:**59, and Glossary
Tectonic cave, defined, **1:**140, and Glossary
Tectonic plate, defined, **1:**215, **2:**169, **4:**203, and Glossary
Telegraphic Plateau, **3:**149*f*
Temperate regions, **1:**177, 177*f*
Temperature
   fresh water and the senses, **4:**94
   life in extreme water environments, **3:**43–44
   life in water, **3:**52
Temperature, ocean
   global warming and the ocean, **2:**131
   physical oceanography, **3:**166–167
   radiometer data, **3:**161–162
   salmon decline and recovery, **4:**71
   and sea level, **4:**73
   sea water physics and chemistry, **4:**81–82
Temperature impacts on stream ecology. *See* Stream ecology: temperature impacts on

Tennessee Valley Authority, 3:33, **4:161–164**
    dams, 4:162–163
    first two decades, 4:161–162
    hydropower, 4:163
    map, 4:163f
    ongoing programs, 4:163
    river basin planning, 4:55, 56
    TVA v. Hill, 2:23–24
    water resources planning and management history, 3:191–192
Terrace (marine), defined, 3:88, and Glossary
Terrace (river), defined, 3:86, 4:120, and Glossary
Terrorism
    countering attacks, 4:88–89
    environmental, 4:85–86, 207
    threats to water-supply systems, 4:86
    water as target, 2:215–216, 4:68, 86–87
    water as tool, 4:87–88
    See also Security and water
Theis, Charles Vernon, **4:164–165**
Thermal waters. *See* Geothermal energy; Hot springs and geysers; Hot springs on the ocean floor; Mineral waters and spas; Volcanoes, submarine; Volcanoes and water
Thermocline
    defined, 2:19, 3:142, 4:24, and Glossary
    lakes, 2:262, 267
    sea water, 4:76, 82f
Thermoelectric power, as water use category, 4:195
Thermohaline, defined, 2:132, 4:78, and Glossary
Thickness mapping, 2:193–194
Thiomargarita namibiensis, 3:80
Thorium, 4:24–25
Threatened, defined, 2:55, 3:211, 4:44, and Glossary
Thrust faults, 2:174
Thunderstorms, 3:258–259, 258f
Tidal bores, 4:*170*, 171
Tidal pools, 2:14
Tidal power, 2:24–*25*
Tidal waves. *See* Tsunamis
Tide gage, 3:166, 4:75, 171
Tideline, defined, 1:71, and Glossary
Tides, **4:165–171**
    bays and estuaries, 4:170–171
    coastal ocean, 1:185–186
    Coriolis effect, 4:169, 170
    declination, 4:168–169
    dynamic theory, 4:168–171
    equilibrium theory, 4:165–168
    and landforms, 4:169–170
    lunar, 4:166
    neap, 1:186, 4:167–168, 167f
    patterns, 4:168
    prediction and tide tables, 4:168–171
    solar, 4:166–167
    spring, 1:186, 4:167–168, 167f
Tigris River, 2:227, 233
Tilapia, defined, 1:32, and Glossary
Till sheets, 2:116–117
Time zoning, 2:256
Tin, 3:91
Tisza River, 3:*226*
Titanic, 1:45, 2:208, 4:142
Titanium, 3:91
Toad, cane, 3:211, 212–213
Tonle Sap, 2:57, 58, 59
Topography
    altimeter data, 3:161
    defined, 1:71, 2:7, 3:117, 4:109, and Glossary
    global distribution of precipitation, 3:260–261
Total allowable catch, 2:65, 66, 67
    defined, 2:65, and Glossary
Total Coliform Rule, 3:74–75
Total dissolved solids (TDS), 2:92–93, 100
Total Maximum Daily Load, 3:30
    defined, 1:15, and Glossary
Total Water Management, 2:222–223, 3:200
Tourism, **4:172–175**
    canoeing and kayaking, 4:174–175
    cruise ships, 4:172, *174*
    ecotourism, 4:172–173
    iceberg tourism, 2:*209*
    sport fishing, 4:173
    underwater diving, 4:173–174
Tower karst, 2:244
Toxic, defined, 1:6, 2:175, 3:11, 4:108, and Glossary
Toxicant, defined, 1:159, and Glossary
Tracer, defined, 2:246, 4:109, and Glossary
Tracers in fresh water, **4:175–177**
    introduced tracers, 4:176
    oxygen isotopes, 4:176–177
    tritium, 4:177
Tracers of ocean-water masses, **4:177–179**
    chlorofluorocarbons (CFCs), 4:178–179, 179f
    tracer timescales matched to ocean processes, 4:178
Trading permits, 2:136
Transboundary water treaties, **4:180–182**
    benefits and shortcomings, 4:181–182
    international cooperation, 2:198, 199, 225
    International Joint Commission, 4:181
    Law of the Sea, 3:25
    United Nations Convention, 2:201–202, 4:180
Transform boundaries, 3:205
Transform faults, 3:84–85
Transformation product, defined, 1:151, and Glossary
Transmissometers, 3:53
Transparent, defined, 3:207, and Glossary
Transpiration, defined, 2:195, and Glossary
Transportation, **4:182–188**
    early history, 4:182–184
    economics, national security, and environment, 4:187–188
    and foreign trade, 4:184–185
    international maritime fleets and law, 4:186–187
    Marine Transportation System, 4:185–*186*, 188
    twentieth and twenty-first centuries, 4:184–186
    westward expansion, 4:184
    See also Ports and harbors
Trash. *See* Pollution of streams by garbage and trash; Pollution of the ocean by plastic and trash
Treaties. *See* Transboundary water treaties
Trench, defined, 3:127, and Glossary
Tributary, defined, 1:101, 3:99, 4:117, and Glossary
Trichlorethylene (TCE), 1:62, 63, 3:223, 244
Trickling filters, 4:211
Trigonometry, defined, 3:94, and Glossary
Tritium, 4:25, 177
Trophic level, 2:11–12
    defined, 2:10, and Glossary
Tropical Atmosphere Ocean (TAO) array, 2:20–21
Tropical oceans. *See* Oceans, tropical
Tropical regions, 1:176–177, 177f

Troposphere, defined, **4**:178, and Glossary
Trunk stream, defined, **4**:117, and Glossary
Tsunami
    defined, **2**:71, **4**:206, and Glossary
Tsunamis, **4**:**188–191**, 229
    effect on beaches, **1**:74
    landfall, **4**:189–190
    mechanics, **4**:189–190, 189*f*
    mitigation and research, **4**:190
    propagation, **4**:189
    Sea of Japan, **4**:190
Tubeworms, **2**:171, *172*
Tunnels, underwater, **1**:*93*–94
Turbidity
    defined, **1**:258, **2**:254, **3**:56, **4**:66, and Glossary
    impacts of, **2**:35–36
Turbidity current, defined, **3**:154, and Glossary
Turnover, **3**:50
    defined, **2**:265, and Glossary
Turtle (submarine), **4**:*139*
Turtles, **4**:*43*–44
    Eastern box, **4**:*43*, 44
    Kemp's ridley, **4**:44
    See also Reptiles
TVA. *See* Tennessee Valley Authority
TVA v. Hill, **2**:23–24
Twain, Mark, **4**:**191**
Typhoid, **2**:181*t*, 185, **4**:33, 208
Typhoons, **3**:148, **4**:230, 232, 232*f*, 233

U

Ultraviolet radiation
    and algae, **1**:27, 28
    and amphibians, **1**:29–30
UN. *See "United Nations" entries*
Unconfined, defined, **3**:174, and Glossary
Unconsolidated, defined, **2**:5, **4**:112, and Glossary
Underwater archaeology. *See* Archaeology, underwater
Underwater pipelines, **3**:185
Underwater tunnels, **1**:*93*–94
    See also Bridges, causeways, and underwater tunnels
UNESCO (United Nations Educational, Scientific and Cultural Organization), **1**:123
Uniformitarianism, **1**:89, **2**:187
Unit hydrograph method (flood flow model), **3**:105

United Nations, and international agreements, **2**:226
United Nations Conference on the Human Environment, **3**:241
United Nations Conference on the Law of the Sea, **3**:24
United Nations Convention on the Law of the Sea, **3**:25–26
United Nations Convention on the Non-Navigational Uses of International Watercourses, **2**:201–202, 226, **3**:20–22, **4**:180
United Nations Educational, Scientific and Cultural Organization (UNESCO), **1**:123
United States
    aqueduct systems, **1**:100, *101*
    Constitution, and public water rights, **4**:49, 51–52
    drinking-water infrastructure, **1**:253–254
    drought, **1**:262*f*
    factors affecting water-use trends, **4**:196–197
    large caves, **1**:140–141
    major ports, **3**:253–254
    mineral springs, **3**:92
    sea level trends for selected cities, **4**:74*f*
    water pricing in selected cities, **3**:262*f*
    water use trends by category, **4**:195
    water use trends from 1950 to 1995, **4**:195, 196*f*
    water withdrawals in 1995, **4**:193, 194*f*, 194*t*
United States Global Change Research Program (USGCRP), **3**:131
Unsaturated zone, defined, **3**:12, and Glossary
Updrift, defined, **1**:75, and Glossary
Uplift
    defined, **4**:74, and Glossary
    waterfalls, **4**:213–214
Upwelling
    defined, **3**:56, **4**:115, and Glossary
    tropical oceans, **3**:172–173
Uranium, **4**:24–25
Urban areas
    land-use planning, **3**:7, 8
    pollution of lakes and streams, **3**:226
    and runoff, **4**:65–66
Uses of water, **4**:**191–197**
    categories of water use, **4**:193–195
    commercial, **4**:193
    domestic, **4**:193, 195

    factors affecting water-use trends, **4**:196–197
    future water use, **4**:197
    industrial, **4**:193, 195, 196
    irrigation, **4**:193–194, 195, 196
    livestock, **4**:194, 195
    mining, **4**:194
    per capita public water use, **3**:247, 248, 249
    public supply, **4**:194, 195
    thermoelectric power, **4**:195
    trends by category, **4**:195
    trends from 1950 to 1995, **4**:195, 196*f*
    U.S. water withdrawals in 1995, **4**:193, 194*f*, 194*t*
    water-use cycle, **4**:192–193
USGCRP (United States Global Change Research Program), **3**:131
Utility management, **4**:**197–199**
    demand management, **1**:238–239
    infrastructure maintenance, **4**:199
    responsibilities and challenges, **4**:198–199
    security, **4**:199

V

Vadose zone
    defined, **3**:223, and Glossary
    studies, **1**:244
Values, defined, **1**:100, **2**:42, **3**:26, **4**:151, and Glossary
Vaporization, defined, **1**:240, and Glossary
Variability, streamflow. *See* Streamflow variability
Vector, defined, **1**:36, **2**:185, and Glossary
Vendetta, law of the, **3**:22
Vent biocommunities, **2**:171–173
Vent circulation and chemistry, **2**:169–170
Venus, water on, **4**:99–100
Vibrio cholerae, **3**:77, 82
Victoria, Lake (Africa), **2**:56, 58, 59, 60
Victoria Falls (Africa), **4**:213–214, 216
"Virtual water," **2**:85
Virulence, defined, **2**:183, and Glossary
Viruses
    in groundwater, **3**:72, 73
    human health and water, **2**:181, 181*t*
    in marine food web, **3**:188
    microbes in the ocean, **3**:79

VISA (Voluntary Intermodal Sealift Agreement), **4**:188
Volatile, defined, **1**:200, and Glossary
Volatile organic compounds
 defined, **1**:258, **3**:2, **4**:67, and Glossary
 land use and water quality, **3**:3
Volcanism, defined, **1**:200, **2**:112, **3**:163, **4**:203, and Glossary
Volcanoes, submarine, **4**:203–206
 Earth's tallest mountain, **4**:204
 hot spots, **4**:204
 and island formation, **2**:237
 listening in on, **4**:205, 206
 mid-ocean ridges, **4**:204–205
 subduction zones, **4**:205–206, 205f
Volcanoes and water, **4**:200–203
 eruption characteristics, **4**:201–203
 explosive eruption, **4**:*202*–203
 fluid inclusions, **4**:200
 and hydrologic cycle, **2**:196
 lake formation, **2**:249
 magma characteristics, **4**:200–201
 magma viscosity, **4**:201
 non-explosive eruption, **4**:*201*–202
 volcanoes under ice, **4**:203
 water circulation in inactive volcanoes, **4**:203
 water introduction via plate tectonics, **4**:200
 water vapor in magma, **4**:200–201
Voluntary Intermodal Sealift Agreement (VISA), **4**:188
Vostok, Lake (Antarctica), **2**:108, 248, **3**:43–44

W

Waialeale, Mount (HI), **3**:260
War and water, **4**:206–209
 damage to environment, **4**:206–207
 disease, **4**:208
 disputes over water, **2**:198–199
 environmental terrorism, **4**:207
 human impacts, **4**:207–208
 lingering damage to water access, **4**:208
 population displacement, **4**:208
 troop access to water, **4**:207
 water scarcity, **4**:207
 See also Conflict and water
Waste stream, defined, **3**:244, and Glossary

Wastes
 agricultural, **3**:238
 animal, **3**:227–228, 237
 industrial, **3**:239–240
 inorganic, **2**:175–176
 medical, **3**:234
 organic, **2**:175–176, 242
 radioactive, **1**:244, **4**:18–21
 types, **3**:237
Wastewater treatment and management, **4**:209–212
 conventional treatment, **4**:210
 primary and secondary treatment, **4**:*211*
 sludge processing and disposal, **4**:212
 stormwater treatment and management, **4**:212
 tertiary treatment, **4**:211–212
 wastewater discharges, **3**:5–6
 wastewater treatment process, **4**:209–212
Water and agriculture. *See* Agriculture and water
Water and globalization. *See* Globalization and water
Water chemistry careers. *See* Careers in fresh-water chemistry
Water circulation
 Baltic Sea, **3**:56
 Black Sea, **3**:56
 circulation cells, **4**:231
 estuaries, **2**:39–40
 global distribution of precipitation, **3**:259–260
 Gulf of Mexico, **3**:57
 and latitude, **3**:259–260
 Mediterranean Sea, **3**:56–57
 tropical oceans, **3**:172–173
 volcanoes, **4**:203
Water conservation. *See* Conservation, water
Water in the arts. *See* Arts, water in the
Water Infrastructure Network (WIN), **2**:215
Water law. *See* Law, water; "*Law*" entries
Water management. *See* Planning and management, history of water resources; Planning and management, water resources
Water markets. *See* Markets, water
Water masses, role in ocean currents, **3**:143–144, 144f
Water on Mars. *See* Mars, water on
Water permits, **1**:171
Water pricing. *See* Pricing, water

Water Quality Act, **3**:29
Water quality and land use. *See* Land use and water quality
Water quality in a watershed. *See* Watershed, water quality in a
Water resources careers. *See* Careers in international water resources; Careers in water resources engineering; Careers in water resources planning and management
Water Resources Council, **3**:30
Water resources development. *See* Supply development
Water Resources Development Act, **3**:30
Water Resources Planning Act, **2**:15, **3**:30
Water resources planning and management. *See* Planning and management, water resources
Water resources planning and management, history of. *See* Planning and management, history of water resources
Water rights. *See* Prior appropriation; "*Rights*" entries
Water sports. *See* Sports
Water supply development. *See* Supply development
Water table, defined, **1**:40, **2**:157, **3**:173, **4**:74, and Glossary
Water tracing, **2**:246
Water User Associations, **2**:230–231
Water uses. *See* Uses of water
Water vapor, **1**:180, 182, **2**:195, **4**:200–201
Water witching, **2**:160, *161*
Waterborne pathogens, common, **2**:181*t*
Waterfalls, **4**:213–216
 differential valley erosion, **4**:215
 erosion, **4**:213
 formation, **4**:213–215
 ongoing evolution, **4**:216
 river diversion, **4**:215
 uplift, **4**:213–214
Watershed
 defined, **1**:1, **2**:8, **3**:33, **4**:121, and Glossary
 hydrologic addresses, **2**:32
 National Park Service, **3**:119
Watershed, restoration of a, **4**:216–219
 approaches and methods, **4**:217–218
 engineered solutions, **4**:218
 goals and considerations, **4**:216–217

reestablishment, **4**:216
rehabilitation, **4**:216
Watershed, water quality in a, **4:219–221**
   forest hydrology, **2**:87–89
   indicators of water quality, **4**:219–220
   testing methods, **4**:220–221
   water density test, **4**:220
Watershed Atlas, **1**:235
Watershed concept, **2**:196–197
Waterskiing, **4**:*38*, 106
Water-use cycle. *See* Hydrologic cycle
Waterways. *See* Bridges, causeways, and underwater tunnels; Canals; Transportation
Waterworks, ancient, **4:221–224**
   Roman Empire, **4**:222–223
Waterworks, modern. *See* Infrastructure, water-supply; Privatization of water management; Supplies, public and domestic water; Utility management; Wastewater treatment and management
Waves, **4:224–230**
   breaking, **4**:225–226
   characteristics, **4**:224–226, 225*f*
   energy generation, **2**:25–26
   formation at sea, **4**:227
   interference, **4**:228–229
   internal, **4**:229–230
   Kelvin, **4**:229
   refraction, reflection, and diffraction, **4**:226, *227*
   seiche, **4**:229
   shallow-water, **4**:168, 225
   storm surge, **4**:*230*
   storm-generated, **4**:227–229
   surfing, **4**:227
   tsunamis (seismic sea waves), **4**:229
Weather, defined, **3**:116, and Glossary
Weather and the ocean, **4:231–233**
   El Niño and La Niña, **4**:232–233
   heat transfer, **4**:231–232
   hurricanes, typhoons and cyclones, **3**:148, **4**:232, 232*f*, 233
Weathering
   defined, **1**:2, **2**:4, **3**:47, **4**:90, and Glossary
Weathering of rocks, **4:234–236**
   chemical, **3**:153, **4**:234–*235*
   mechanical, **4**:234
   physical, **3**:153
   results and rates, **4**:236
Weir, defined, **4**:115, and Glossary

Wells and well drilling, **4:236–241**
   artesian wells, **2**:152
   casing, screen, and seals, **4**:237–238
   construction, **4**:238–240
   "contributing area" wells, **3**:1–2
   developing, **4**:239
   as discharge area for groundwater, **2**:153
   disinfecting, **4**:240
   existing wells, **2**:159
   flowing artesian wells, **2**:152
   injection wells, **1**:50–51, **4**:176
   logging, **4**:240
   monitoring wells at landfills, **3**:*13*
   parts of well, **4**:237–238, 237*f*
   private wells, **4**:67, 240
   pumps, **4**:238
   repair and abandonment, **4**:240
   safety, **4**:240
   sealing, **4**:239
   siting, **4**:238
   test wells, **2**:158–*159*
   testing, **4**:239–240
   water levels as earthquake detectors, **2**:7
   well drilling, **4**:241
   well interference, **2**:154–156, 154*f*, 155*f*
   well reports, **1**:40, 41, **2**:192–193
   well types, **2**:151*f*, 152
West Nile virus, **2**:184
West Wind Drift, **3**:141, 167, 168
Wet deposition, **1**:6
Wetland, defined, **1**:77, **2**:35, **3**:8, **4**:34, and Glossary
Wetlands, **4:241–247**
   acid bogs and ancient conditions, **4**:243
   classification, **4**:243–245
   coastal, **4**:*244*–245
   conserving and mitigating, **4**:245–247
   fresh-water ecology, **2**:10
   glacial and dune, **4**:245
   legal definition, **4**:244
   mountains, **4**:*242*, 243–244
   National Park Service, **3**:119
   occurrence and characteristics, **4**:241–243
   plant and animal communities, **4**:242–243
   plateaus and high plains, **4**:244
   playas, **4**:244
   preservation, **2**:43–44
   protection program, **1**:172
   river valleys, **4**:244

Whales, **3**:61, 64–65
Whistle-blowing, **2**:44
White, Gilbert, **4**:55, **247–249**
White smoker, defined, **3**:137, and Glossary
Wild and Scenic Rivers Act, **2**:219
Wildlife issues. *See* Fish and wildlife issues
"Willy" (killer whale), **3**:*61*
WIN (Water Infrastructure Network), **2**:215
Wind
   as erosive force, **2**:33
   influence on estuaries, **2**:39–40
   and transport of sediment, **3**:154, *155*
Win–win solution, defined, **1**:205, and Glossary
Women in water sciences, **4:249–253**
   future, **4**:252–253
   then and now, **4**:251
World Bank, **1**:123, **2**:226, **4**:155
World Commission on Environment and Development, **4**:156, 159
World Health Organization, **4**:32, 33
World Summit on Sustainable Development, **4**:156
World Water Council, **4**:157–158

X

Xeriscape, **1**:208

Y

Yangtze River (China), **2**:57, 58, 59, **4**:59–60
Yellowstone Lake (WY), **3**:214
Yellowstone National Park (WY), **2**:167
   Grand Prismatic Spring, **3**:*45*
   Old Faithful geyser, **2**:165, *166*, 167
Yosemite Falls (CA), **4**:215, 216

Z

Zoning, defined, **3**:10, and Glossary
Zoology, defined, **1**:137, and Glossary
Zooplankton, **3**:189–190
   defined, **2**:9, **3**:51, and Glossary
   flotation, **3**:51
   lake, **2**:261, 261*f*
   marine, **2**:13, **3**:49
Zoos, **1**:36–37
Zooxanthellae, **1**:215, 218